Nazi Medicine and the Nuremberg Trials

From Medical War Crimes to Informed Consent

Paul Julian Weindling

First published in hardback 2004

First published in paperback 2006 by
PALGRAVE MACMILLAN
Houndmills, Basingstoke, Hampshire RG21 6XS and
175 Fifth Avenue, New York, N.Y. 10010
Companies and representatives throughout the world

PALGRAVE MACMILLAN is the global academic imprint of the Palgrave
Macmillan division of St. Martin's Press, LLC and of Palgrave Macmillan Ltd.
Macmillan® is a registered trademark in the United States, United Kingdom
and other countries. Palgrave is a registered trademark in the European
Union and other countries.

ISBN-13: 978–1–4039–3911–1   hardback
ISBN-10: 1–4039–3911–X   hardback
ISBN-13: 978–0–230–50700–5   paperback
ISBN-10: 0–230–50700–X   paperback

This book is printed on paper suitable for recycling and made from fully managed
and sustained forest sources.

A catalogue record for this book is available from the British Library.

Library of Congress Cataloging-in-Publication Data

Weindling, Paul.
  Nazi medicine and the Nuremberg Trials : from medical war crimes to informed
consent / by Paul Julian Weindling.
    p. cm.
  Includes bibliographical references and index.
  ISBN 1-4039-3911-X (cloth) 0–230–50700–X (pbk)
    1. Nuremberg Medical Trial, Nuremberg, Germany, 1946–1947. 2. Human
experimentation in medicine–Law and legislation–Germany–History–20th
century. 3. World War, 1939–1945–Atrocities. I. Title.

KZ1179.M43W45 2004
364.1'38–dc22

                                                                    2004049123

10   9   8   7   6   5   4   3   2   1
15   14   13   12   11   10   09   08   07   06

Printed and bound in Great Britain by
Antony Rowe Ltd, Chippenham and Eastbourne

Contents

Figures

1. The Accused

2. The Nuremberg Court

Source: Staatsarchiv Nürnberg, KV-Prozesse Generalia Q-1.

Acknowledgements

Documents on the Medical Trial are widely dispersed. This research would not have been possible without assistance from a small army of patient and exceptionally helpful archivists and librarians. Access to documents required special procedures at the United Nations, the Archives de France, which have the most significant holdings on the Medical Trial, the Colmar archives of the French Occupation of Germany, and the Canadian National Archives. I am grateful to Christopher Laico in giving access at an early stage to the superbly indexed Telford Taylor papers at Columbia University College of Law, and to the biographer of Telford Taylor, Jonathan Bush, for advice. The Stetson College of Law Library made available the Sebring Papers, and the University of Washington helpfully copied Beals materials. Robert Lifton kindly facilitated access to his papers at the New York Public Library. Rockefeller Foundation and Rockefeller University Papers, held at the superb Rockefeller Archive Center in Sleepy Hollow, have proved invaluable.

My research owes a fundamental debt to Angelika Ebbinghaus and Karl Heinz Roth for the superb edition of the proceedings of *The Nuremberg Medical Trial*. This indispensable work covers not only the Medical Trial, but also medicine in the Second World War and the Allied war crimes investigations.

I owe a special debt to the impressive Max-Planck-Gesellschaft project on the history of the Kaiser Wilhelm Gesellschaft, which has provided me with a base in Berlin over the past four years. I would particularly thank Doris Kaufmann, Carola Sachse and Susanne Heim as Directors of the project, and Florian Schmalz for advice on German wartime biochemistry. I have enjoyed considerable support from numerous colleagues in Germany, notably Alfons Labisch and colleagues in Düsseldorf, Wolfgang Eckart, Gabriele Moser and Alexander Neumann (Heidelberg), Sabine Schleiermacher (Berlin), Volker Roelcke (Giessen), Andreas Frewer (Göttingen), and Michael Hubenstorf in Austria. I am also grateful to Michal Šimonek and Petr Svobodný (Prague) for facilitating research there, and to Pietro Corsi, Lion Murard and Patrick Zylberman in Paris.

Canadian research was assisted by a grant for Canadian studies from the Canadian High Commission, and I am grateful to Susan Solomon (Toronto) for help along the way. The bulk of the research was carried out with a grant from the Wellcome Trust, as well as with support for the funding of my post as Wellcome Trust Research Professor at Oxford Brookes University.

Above all, it has been a pleasure and a privilege meeting those with memories of the Trial and the period. I benefited from both much wisdom and personal warmth. I would like to pay tribute to Telford Taylor, Keith Mant and Alice Platen, recognising their formative contributions to the problem of medical war crimes.

Abbreviations and Acronyms

ADIR	Association Nationale des Anciennes Déporteés et Internées de la Résistance
ALSOS	'Grove': Allied Scientific Intelligence Organisation
AMA	American Medical Association
Ashcan	Allied Detention Center at Mondorf, Luxemburg
BAOR	British Army of the Rhine
BIOS	British Scientific Intelligence Organisation
BMA	British Medical Association
BW	Biological Warfare
BWCE	British War Crimes Executive
CIC	Counter Intelligence Corps
CINFO	Combined Intelligence and Detective Agencies of the British and US Armed Forces
CIOS	Combined Intelligence Objectives Sub-committee
CW	Chemical Warfare
DFG	Deutsche Forschungsgemeinschaft. German Research Fund.
DP	Displaced Person
Dustbin	Schloss Kransberg, an Allied Detention centre
G-2	Allied Military Intelligence Division
FIAT	Field Information Agency, Technical
HCN	Hydrocyanic acid
IG Farben	'Interessen-Gemeinschaft' Farben, a German industrial conglomerate
IMT	International Military Tribunal
IRO	International Refugee Organisation
ISC	International Sciencitic Commission
JAG	Judge Advocate General
JAMA	Journal of the American Medical Association
KWG	Kaiser Wilhelm Gesellschaft
KWI	Kaiser Wilhelm Institute
MPG	Max Planck Gesellschalt
MRC	Medical Research Council
NSDAP	Nationalsozialistische Deutsche Arbeiterpartei
OCCWC	Office of Chief of Counsel for War Crimes
OKL	Oberkommando der Luftwaffe
OKW	Oberkommando der Wehrmacht
OMGUS	Office of the Military Government US
OSS	Office of Strategic Services
POW	Prisoner of War

RKI	Robert Koch Institute
RFR	Reichsforschungsrat
SA	Sturmabteilung
SEA	Staff Evidence Analysis
SHAEF	Supreme Headquarters Allied Expeditionary Force
SS	Schutzstaffel der NSDAP
T-4	Tiergartenstrasse 4 [address of the killing organisation]
T Force	Allied military and industrial task force
Tomato	British detention centre at Minden
UNESCO	United Nations Educational Scientific and Cultural Organization
UNO	United Nations Organisation
UNWCC	United Nations War Crimes Commission
USAF	United States Air Force
USFET	United States Forces European Theater
USHMM	United States Holocaust Memorial Museum
USSBS	United States Strategic Bombing Survey
WHO	World Health Organisation
WMA	World Medical Association

Introduction

Science and genocide

On 4 March 1945 liberated Auschwitz prisoner doctors made an international declaration on how prisoners had been treated as experimental animals. They urged the Allies and neutral states to bring to trial those responsible.[1] They hoped that prosecuting perpetrators would prevent coerced human experiments and medical atrocities from recurring in the future. Survivors and witnesses of human experiments called for documentation of Nazi medical abuses, justice and compensation. Their role in alerting the Allies to medical atrocities, and in declaring the need for a humane and ethical medicine oriented to consent of the patient and research subject merits historical recognition.

Survivors created a deep impression on scientific intelligence officers, who set out to collect evidence of human experiments. After inspecting several German university clinics, an Allied scientific intelligence officer – John Thompson (a US citizen in the Royal Canadian Air Force Office seconded to British scientific intelligence) – declared that 90 per cent of the research of leading German medical scientists was criminal. He coined a new concept – that of a 'medical war crime'. He and fellow intelligence officers were aghast at the perversions of science under Nazism. Their mission had been to investigate German scientific achievements in the war. But they all too often found evidence of fragmented and low-grade research, and inhumane practices. Linking medical research to war crimes spurred American, British and French military authorities to document the criminality of medicine under National Socialism.[2]

The wide-ranging investigations of medical atrocities between the end of the Second World War and the onset of the Cold War culminated in the Nuremberg Medical Trial. The Trial, which ran from December 1946 to August 1947, scrutinised German racial research, bacteriology and experimental medicine. Prosecutors and defendants clashed over the place of medical research in the Nazi power structure. The military

1

tribunal set out to unravel complex issues of medical power in the Nazi war machine.

One crucial issue was the extent that the Allies recognised the criminality of eugenics and preventive medicine. The medical war crimes investigators confronted a series of problems about Nazi medicine:

1. What was the relation of medical research to the war aims and racial policies of the Nazi state?
2. Were the German experiments path-breaking science, or depraved 'pseudo-science'?
3. How did the human experiments relate to eugenics and genocide? Could the experiments on poison gas be seen as pilot studies for the gas chambers of the Holocaust?
4. Did the verdict on the 23 defendants apply more broadly to German medicine under National Socialism?

The victims and witnesses of the human experiments, and energetic efforts by intelligence officers, amassed devastating evidence on German medical atrocities. The Allies were faced with a number of alternatives:

1. To view the atrocities as mass murder and launch a series of medical trials for the perpetrators of euthanasia, sterilisation, human experiments, and other medical abuses.
2. To document wholesale German medical crimes and allow scientific experts to evaluate them.
3. To consider the ethical failings that led to the crimes and draw up new ethical guidelines.
4. To regard the perpetrators as mentally abnormal and subject their conduct to psychiatric analysis.
5. To exploit German wartime research for defence-related projects.

The revelations on medical war crimes led to an International Scientific Commission (ISC) to document coercive human experiments. It drew the attention of the Nuremberg prosecutors to the genocidal human experiments and to the ethical problems for medical research. Its documentation, which fed into the Trial, showed how the deaths and maiming of tens of thousands of victims in the concentration camps arose from the unscrupulous callousness of German doctors, whose profession had once led the world in scientific research. Given the significance of German medical research in the modernisation of medicine on an experimental basis, the journalist Werner Süskind commented, the Medical Trial formed 'a dark chapter of modern cultural history'.[3]

Medical knowledge and the legacy of total war

Allied investigators of German military medicine were confronted by the choice of exploiting captured personnel and documents for weapons research, or prosecuting war crimes. The Allies had a high regard for the ability of German aviation medicine to solve problems of high-altitude flight. The Atom bomb required knowledge about the hazards of radiation, and German chemical weapons and nerve gas might be deployed against the Japanese and then against the Soviets. The British and Americans feared that German scientists would opt wholesale to work for the Russians. The Allies faced a conflict between exploiting German medical know-how and prosecuting its criminality.

The American Medical Association (AMA) and British Medical Association (BMA) were concerned that releasing news of the German atrocities would undermine public confidence in medical research. Formulating new ethical standards became a priority to ensure the future viability of research-based clinical medicine. The Nuremberg Code on the conduct of human experiments promulgated at the close of the Trial was a response to such concerns. The consent of the research subject, and the right to know and participate voluntarily in medical research remained central issues in clinical research. At the same time, the Trial revealed much about the structures and values attached to research on both the German and Allied sides. The German defence counter-attacked by challenging the ethical standards and practices in Allied medical research.

The Nuremberg prosecutors saw medicine as a crucial component of the regime's racial and social policies. The chilling mass of documentation on euthanasia and sterilisation was supplemented by the testimonies of victims of racial atrocities.[4] The Trial opened on 9 December 1946, just before the United Nations on 11 December took steps to declare genocide a crime under international law. States should legislate to prevent and punish genocide, and the United Nations was to draw up a Genocide Convention.[5] The two events drew on common impulses: preventing doctors perpetrating mass killing for racial purification was an aim. Raphael Lemkin, who in 1943 first conceived the term 'genocide' to describe the Nazi measures to destroy nations deemed degenerate, was the prime mover for the UN Declaration.[6] When Lemkin advised Mickey Marcus, chief of the war crimes division, he argued for the genocide concept to be applied to medical abuses. Controversy persists over the extent that the Nuremberg Medical Trial was a genocide trial.

While the Medical Trial was the first of twelve US military tribunals at Nuremberg, it drew legitimacy from international law as developed for the four-power International Military Tribunal (IMT). International support came from victim states, notably Czechoslovakia and Poland, as well as from abused ethnic groups. Survivors drew attention to numerous medical miscreants, and singled out Josef Mengele as a prime candidate for eventual trial. The Trial opened up the controversial issue of links between the concentration camps and mainstream academic medicine.

The prosecution accused German medical and scientific research of complicity in aggressive war, crimes against humanity and genocide. Evidence was amassed on how the experiments were to assist the Nazi schemes for conquest and German military endeavours. Suspicion surrounded the Kaiser Wilhelm Gesellschaft (KWG) and the German universities for harnessing science for military expansion, autarky and mass murder of racial and social undesirables. The accusation went to the heart of a German academic enterprise as serving the twin gods of racial research and national glory, and threatened German academic efforts to sustain the continuity of academic personnel and institutions.

The IMT judged the SS guilty as a criminal organisation, but was the Medical Trial a trial of German medicine as a whole? German professorial and clinical leaders denied that they collectively betrayed their ethical calling for the benefit of Nazi racial expansionism. Those on trial were fanatical purveyors of an ideologically debased pseudo-science, while German university professors and physicians acted with dignity and honour under the stresses of Nazism and Hitler's war. As German public opinion became increasingly vocal against 'victors' justice', the German medical profession insisted that it had defended ethical standards with honour and dignity in the face of Nazi barbarism.

Verdicts on the Trial have emphasised a coercive state, rather than how medical researchers and their representative bodies manipulated the state to gain resources for the conduct of ruthless experiments. Since the Trial the paradigm has been one of the state as instigating ethical violations. Some commentators went so far as to see the historical basis for medical crimes in Bismarck's system of state-regulated sickness insurance, introduced in 1883.[7] By way of contrast, the opportunity of resurrecting the idea of an individual citizen's right to health care voiced in 1847 by the Berlin medical reformer Saloman Neumann, who saw health as the 'highest individual right of every person' and an accountable medical profession, was overlooked in favour of a paternalistic professionalism. The World Medical Association took a highly critical view of the socialisation of medical services while defending clinical autonomy. The *Journal of the American Medical Association* in a special issue marking the 50th anniversary of the Medical Trial in November 1996 drew the lesson that 'medicine can be distorted by the state', and that social and political forces distort 'the medical ethos'.[8] Seen at such a generalised level, the conclusion is that the threat to patients and widespread coercion arose through the socialisation of medicine. The danger of such a paradigm is that it could exonerate the medical profession and scientific research under National Socialism as unblemished. Indeed, there has of late been a new stress on the scientific achievements of German medicine under National Socialism, while playing down its criminality and the numbers of victims of human experiments.[9]

Against such retrospective legitimations can be set views of eugenic commitments in German medicine, and the close links between the German medical profession and the Nazi Party.[10] Michael Kater sees the problem in terms of 'medicalising' political ideology.[11] The tendency has been to see the state as the source of abuse, while the profession by definition is benign. A science-based profession held particular dangers, following the critical insights of Ivan Illich and Michel Foucault that medical power and biological knowledge were a potentially lethal combination. Since the 1970s, there has been a critical understanding of ideology in science, and how the rise of German eugenics undermined civil rights. The expansive dynamic of medical science and professional structures posed dangers to patients, the socially vulnerable, and, more broadly, to civil society. This study seeks to critically examine the paradigm of a malevolent and manipulative state exploiting an essentially beneficent medical profession, as researchers actively lobbied the military and SS for research facilities.

On top of professional solidarity against the state came the growing political opposition to Allied war crimes trials. American conservatives objected to the sheer expense of mounting trials, and to the wisdom of prosecuting veteran anti-communists, patriotic generals and enterprising industrialists. The onset of the Cold War meant these conservatives found themselves in league with German nationalists waging an unrelenting campaign against the trials. Nationalists blamed the Allied defeat for depriving Germans of homes, subjecting them to mass starvation and fuel shortages, and detaining and punitively denazifying dedicated public health officials and university academics. Many Nuremberg defendants were rapidly rehabilitated, and further prosecutions lapsed. A coalition of unrepentant ex-Nazis, conservatives, and leading physicians denigrated the trials as 'Victors' Justice'.

The Allied military medical investigators tended to work backwards from the German military and SS structures of control over medicine, which they found in 1945. This revealed the structures of co-ordination welding medicine to a regime geared to aggressive war and racial extermination, and implied that researchers were prepared to exploit the war and concentration camps for advancing medical knowledge in society. Chief Prosecutor Telford Taylor argued that Nazism debased teaching and research in German medical faculties. He cited the views of American, British and French experts that 'practically nothing of value to medicine resulted' from the human experiments in the concentration camps, and medical standards suffered severely under the Nazis.[12] Observing that medical abuses caused a vast loss of life, he placed the prime guilt on the German military–industrial complex, not least because of the rapacious ambitions of IG-Farben seeking to develop new products by means of experiments in concentration camps.

Historians have tended to judge eugenics, euthanasia and sterilisation as marginal to the Medical Trial's prime concern with human experiments. Taylor, however, saw human experiments as pilot studies for genocide. The prosecution provided the chilling details of how Jews were culled from Auschwitz and killed for a gruesome skeleton collection, intended to grace the 'Reich University Strassburg'. The physiological experiments were construed as studies in the physiology of death, and poison gas experiments were pilot studies for the extermination camps. The taking of thousands of brains from euthanasia victims showed how medical research was linked to the psychiatric killings. The experimental X-ray sterilisations and other efforts to make women infertile bridged sterilisation and the Holocaust. The prosecutors claimed – with much justice given the massive scale of Nazi euthanasia – that there were half a million victims of medical abuses under Nazism.

What was at stake was far more than a trial of sixteen Nazi doctors, four non-Party physicians and three SS administrators. Frequently and erroneously described as twenty or indeed 'the twenty-three SS doctors', the defendants were conventionally depicted as servile agents of the autocratic Nazi state. Links were made to the SS, army, air force and state public health institutions. The diagnosis of an excess of state power suited three groups: firstly, those keen to portray the Nazi state as totalitarian – with the implicit message that the Soviet Union and other communist states could repeat such atrocities; secondly, it suited medical campaigners against socialised medical services; and thirdly, it reinforced an interpretation of the Nazi state as ruled by corporate industrial and commercial interests, welded to the war economy. By way of contrast, German medical researchers and practitioners claimed to be innocent victims preyed on by Nazi pseudo-scientists and fanatic racists.

But this diagnosis obscured how the defendants had strong links to the German academic elite, who saw human experiments in terms of qualifications, promotion and the building up of research installations. The key issue was how academics sought to steer and exploit the state's new powers under Nazism. What were the motives for human experiments at the various phases of the war, not least for intensifying medical research during 1944 when Germany was manifestly losing? These issues can be seen as significant for a nation, which prided itself on the exceptional excellence of its research and professional training. The Trial raised issues concerning the role of the university academics in an era of total war, and how the structures of German research fared under National Socialism. In confronting the extent that medicine under National Socialism was a science of mass destruction, the Allies mounted dual policies of exploitation of discoveries and prosecuting perpetrators of inhumane experiments.

Recovering the origins of the Trial

This book is the third volume of an informal trilogy on German medical atrocities. The first of these studies, *Health, Race and German Politics between*

National Unification and Nazism, 1870–1945, assesses the impact of German '"racial hygiene"' on public health, and how widespread chronic diseases like tuberculosis and sexually transmitted diseases were branded racial poisons.[13] Its counterpart, *Epidemics and Genocide in Eastern Europe*, covers the broader field of 'hygiene', involving bacteriology and disease eradication between the triumphs of Robert Koch and the abuse of disinfection in the Holocaust.[14] The three volumes combine to analyse how medicine became caught up in politically messianic schemes for racial rejuvenation, and how these involved extermination of peoples designated as biologically inferior and reservoirs of lethal pathogens.

In 1994 I was researching how German bacteriologists collaborated with the Pasteur Institute in the Second World War. The archivist, Denise Ogilvie, brought to my attention documents relating to the ISC meetings in Paris in 1946. How, I wondered, could a Code requiring consent of the experimental subject have been discussed in the summer of 1946, when the Nuremberg Code dates from August 1947? Rather than generated by court proceedings and promulgated by judges, this earlier Code was formulated by physiologists. I set out to document the links in the chain of evidence stretching from the survivors of the experiments to the eventual Code.

Among the overlooked sources are the papers of Andrew Ivy in Laramie, the voluminous Bayle documentation in Paris, and the records of the British, French, American and Canadian war crimes and scientific intelligence organisations in respective national archives. The papers of scientific intelligence officers and advisors – Leo Alexander and Detlev Bronk – provide a further rich source of information. Christian Pross, Robert J. Lifton and Jürgen Peiffer pioneered studies of Alexander's diaries, as does Michael Shevell's study of Alexander's reports and correspondence. The Alexander papers in Durham, NC, Boston University and those held by his family, but also the rich archives of Pross, Lifton and Peiffer, have proved fundamental resources. Leo Alexander's daughter, Phyllis Grable, the intelligence officer Hugh Iltis, who sorted Himmler's papers, and his colleague Ivan Brown provided vivid memories about this energetic and insightful medical expert. The cornucopia of hitherto overlooked sources allows the Trial to be located within a meaningful context on the inter-Allied investigation of medical war crimes, and shows the intensity, depth and tenacity of the Allied investigations and judicial proceedings.

The history of German medical crimes is still living history. Survivors of medical atrocities are able to confront history and point to the inadequacies of care and compensation. I was privileged to meet survivors of experiments at Sachsenhausen, Ravensbrück and Auschwitz, when in 2002 Hubert Markl as President of the Max-Planck Gesellschaft, reflected on the horror of the scientific atrocities, and apologised for the role of the KWG and the broader community of medical researchers in these.[15] It is salutary for the historian to be reminded that victims are living, perceptive, inquiring individuals.

Sadly, the issue of compensation for the victims of human experiments remains unresolved and an open wound in the post-war process of obtaining justice.

Observers and investigators of the Trial generously responded to my inquiries. Alice Ricciardi-von Platen recollected her 'dark period' of working in Mitscherlich's team of German medical observers at Nuremberg. The Royal Army Medical Corps (RAMC) pathologist and Ravensbrück investigator Keith Mant warmly supported my exhuming the issue of medical war crimes. He drew my attention to the importance of the mercurial John Thompson, who had escaped the attention of historians. Keith's lecture on Nazi medical war crimes constituted the highlight of a symposium, which I arranged in Oxford on 14 March 1997 to mark the 50th anniversary of the Medical Trial.[16] The symposium focused on the Trial as an historical event. Major symposia held in Washington DC, Boston, and by the IPPNW at Nuremberg looked at the ethical legacy of the Trial from the perspective of the abusive state. The Paris symposium raised the survivors' perspective, when the bland apologetics of the ICRC were rebuffed.[17]

I discussed the Medical Trial with Telford Taylor, Chief Prosecutor at Nuremberg, and the medical historian Saul Jarcho (as member of the New York Academy of Medicine's Committee on the Medical Trial Documents) in 1996. The assistant prosecutor, Ben Ferencz, and documents analyst, Hedy Epstein (*née* Wachenheimer) generously responded to my questions. I am grateful to Joseph Meier and to Walter Freud for recollections of their period as war crimes investigators. The families of Leo Alexander and John Thompson (extended to the family and students of his closest friend, the psychiatrist Milton Rosenbaum), and Madame Christiane Lépine, have provided a wealth of biographical recollections. Children of two of the accused shared their memories of the end of the war and the impact of the Trial on their families. The travel and research were made possible by the Wellcome Trust as part of its support for my position as Research Professor at Oxford Brookes, and with a project on Clinical Abuses and Nazi Medicine.

The Nuremberg Trials confronted the problem of whether the state and its agencies were taken over by avaricious professionals, or – to put the matter more indulgently – were well-intentioned physicians manipulated by militant leaders and their regenerative ideology? The medical war crimes investigators' and the Trial's efforts to resolve this problem resonate not only for understanding medical dimensions of Germany in the Second World War, but more broadly in a century in which scientific experts sought power while eschewing accountability. Viewing the human experiments from the perspective of the victim, and survivors' demands for an ethically informed medicine, presents a new framework for analysing medical atrocities. Informed consent permeates modern medicine: an understanding of its meaning and implications in the political ordering of human life requires critical historical analysis of the Nuremberg medical maelstrom.

Part I
Exhuming Nazi Medicine

1
The Rabbits Protest

From 'Operation Anthropoid' to experimental operations

In January 1942 SS *Obergruppenführer* Reinhard Heydrich, the ruthless head of the Reich Security Office, convened the Wannsee Conference to formalise plans for the 'Final Solution of the Jewish Problem'. A few months later, he was the target of a daring assassination mission to show that Nazi leaders could have no immunity from retribution. On 27 May 1942 two parachutists from the Czech Brigade seriously wounded Heydrich, the Deputy Protector of Bohemia and Moravia, on the outskirts of Prague in 'Operation Anthropoid'.[1] At first, Heydrich, who had reached for his gun and chased his assailants, expected to recover. Sudeten German surgeons from the Charles University Prague rapidly operated with apparent success.[2] But the bullet, after hitting the rear axle of his bloated, open black Mercedes, had been diverted vertically through Heydrich's back, carrying cloth, wire and wool from the seat; his wounds became gangrenous.

The bullet had profound medical repercussions, prompting Nazi doctors to argue over how best to treat infected wounds. Heinrich Himmler ordered his escort surgeon, Karl Gebhardt, to Prague to save Heydrich's life.[3] Gebhardt found that the site of the wound and its contamination prevented his operating to remove the damaged spleen. Heydrich's death on 4 June triggered the brutal destruction of the village of Lidice: its men were shot, women sent to concentration camps and children taken for forced adoption. But there was a calculating, medical side to SS brutality, about which the accounts of the Heydrich assassination are generally silent.

Hitler's medical entourage erupted in vicious recrimination. The Führer's doctor, Theo Morell, criticised Gebhardt for not using what he claimed was his more powerful patent form of sulphonamide, Ultraseptyl. The sulphonamides had at first contained the infection, but then rapidly lost their effect.[4] Himmler used the Heydrich incident to gain a hold on military surgery through the SS military branch, the Waffen-SS. He authorised Gebhardt and a team of Waffen-SS surgeons to embark on experiments on the legs of concentration camp prisoners. The surgeons were ambitious to produce a German answer to the British discovery of penicillin (announced in *The Lancet* in August 1940, tests on severe and often fatal infections being publicised early in 1941).[5] Gebhardt claimed to have privileged access to Allied literature on sulphonamide preparations. It was a chance for this arrogant and pedantic surgeon to prove that operative skill had greater value than chemotherapy, while exonerating himself from the charge that he was incompetent in letting Heydrich die.

The Gebhardt and Himmler families came from the Bavarian town of Landshut, and Gebhardt's well-connected Bavarian father was the Himmlers' family doctor. Gebhardt advanced steadily through the academic ranks at Munich, running a sports camp at Hohenaschau for the League of the Child Rich, and assisting in the clinic of the celebrated master surgeon Ferdinand Sauerbruch. In 1932 Gebhardt gained his Habilitation[6] qualification, but the Nazi takeover drew Gebhardt to the northern planes of Berlin-Brandenburg. Reich Physicians' Führer Gerhard Wagner appointed him on 1 November 1933 Director of the rambling Red Cross sanatorium of Hohenlychen. He modernised it as a centre of orthopaedic medicine, emulating the American model of Warm Springs, and founded a medical institute at the Academy of Physical Exercise in Berlin. In 1935 Gebhardt transferred to Berlin to develop sports medicine and orthopaedics.[7] His prestige rose, when he took charge of the medical care for Olympic athletes at the Berlin Games of 1936. Sports were not only a way of legitimating the Nazi regime, but an arena of applied physiology and an opportunity for orthopaedic surgeons to display their mastery of muscular traction. The war gave Gebhardt the opportunity to develop military surgery on an experimental basis.

Sauerbruch rallied to defend Gebhardt. Himmler exploited Gebhardt's insecurities as a means of extending the tentacles of the SS into medicine through military surgery. Himmler and the SS Criminal Police chief Arthur Nebe ordered that wound experiments be carried out in the concentration camp of Ravensbrück.[8] Medical researchers discussed whether it would not be better to treat already infected patients, but Himmler supported those who wanted to replicate ideal laboratory conditions in concentration camps. Whereas the American army – undertaking experiments on treating gonorrhoea, for which there was no animal model – asked for volunteers, the Germans favoured human experiments.[9] The standard method involved comparing results with the experiences of a control group, who were infected but not treated. There was nothing like a randomised control trial.

Gebhardt and Hitler's escort surgeon Karl Brandt had orders to improve surgical services. Military surgeons wanted to settle a debate on drug or surgical treatment of war wounds, which had flared in the faltering Russian campaign. They looked to the experiments to combat a battery of infections threatening frontline troops. These included hepatitis (which German researchers claimed to be an infectious rather than chronic disease), the insect-borne diseases of typhus and malaria (all widespread in North Africa and on the Eastern Front), and wound infections.

Gebhardt joined the SS in 1935, and organised an SS medical station at Hohenlychen. Himmler appointed him his escort surgeon in 1938. The Waffen-SS took Hohenlychen over in 1941 as part of their expanding realm of medical institutions. Gebhardt inspected Waffen-SS field hospitals in Russia,[10] and attained high rank as SS-*Gruppenführer* and *Generalleutnant* in the Waffen-SS.

Hohenlychen was just 12 kilometres from Ravensbrück in the undulating Mecklenburg countryside, and readily accessible from Berlin. The series of experiments went well beyond the immediate aim of testing the efficacy of sulpha drugs in wound treatment. Gebhardt was ambitious to develop Hohenlychen into a major centre of experimental medicine with an institute of cancer research. The Allied bombing of Berlin meant that Gebhardt gained a division of the pathology department from the city's Virchow Hospital.[11]

Relative seclusion from Allied bombs meant Hohenlychen became a haunt of Nazi leaders, and dark conspiracy. Gebhardt treated Hitler's architect and Armaments Minister Albert Speer, the French Minister of Production Jean Bichelonne, the Nazi agriculturalist Walter Darré, and other prominent politicians. That Bichelonne died after his operation and Speer considered that Gebhardt had attempted to kill him while under treatment in January to March 1944 made this a deeply sinister institution.[12] Gebhardt sheltered Himmler's mistress, Häschen Potthast (the 'little hare'), who gave birth to their second child at the sanatorium.[13]

Gebhardt was an overbearing and irascible chief.[14] He ordered Fritz Fischer (who transferred from the Virchow Hospital) to carry out the wounding experiments, and the ambitious Ludwig Stumpfegger undertook bone transplantation experiments. Stumpfegger was appointed first in Gebhardt's place to attend Himmler and then in Karl Brandt's place as Hitler's surgeon in attendance in October 1944. (Stumpfegger remained in the Führer's Berlin bunker and administered poison to the six Goebbels children, and poisoned himself alongside Martin Bormann after a failed escape attempt on 2 May 1945.)[15]

Twenty men were transferred from Sachsenhausen concentration camp on the northern outskirts of Berlin to Ravensbrück for use as experimental subjects.[16] Women prisoners were then favoured subjects, because of the proximity of the camp to Hohenlychen and the mistaken view that women would be more docile. The 74 women were Polish political prisoners, and there were 2 Ukrainians, a Russian, Belgian and 5 German Jehovah's Witnesses.[17]

From 1 August 1942 Gebhardt's underlings selected Polish resistance fighters as experimental targets. Gebhardt took full responsibility and described the operations to Grawitz as if he had done them. When Grawitz came to visit, the victims were covered with sheets with only their legs visible, but saw Gebhardt from under the sheets. The compliant camp doctor, Herta Oberheuser, declared that the Germans had the right to experiment, because the victims were members of the resistance. This was in line with the constantly reiterated German conviction that their experimental subjects were under a death sentence – this assumed that all camp inmates were guilty by virtue of their imprisonment.[18]

At first, Gebhardt's team surgically wounded the prisoners' legs, but Reichsarzt SS Ernst Robert Grawitz ordered that victims' legs be gashed with splinters and glass shards, and infected with tetanus to replicate battlefield conditions.[19] When the Hohenlychen surgeons did not go to this extreme, Himmler was angered that no deaths resulted – the greater the brutality, the more the surgical perpetrators would be bound into the SS. Oberheuser selected camp prisoners with perfectly healthy legs. The Hohenlychen doctors gashed their legs and infected the wounds with wood and glass shards; they broke bones, transplanted them, placed the injured limbs in traction, and destroyed muscles and nerve fibres. Bacteria causing gangrene were put into the wounds of one group, and in another group cultures causing blood poisoning. In the event, 13 of the 74 experimental subjects died from gangrene and tetanus or from loss of blood, and six were executed. Early in 1943 the Rabbits protested in writing to the camp commandant, Fritz Suhren, with one of the victims, Wladyslawa Karolewska, boldly objecting that 'it was not allowed to perform operations on political prisoners without their consent'. Assisted by solidarity among the victims and their fellow prisoners, 55 survived despite severe and crippling wounds.[20]

The sulphonamide and bone transplantation experiments gained notoriety. The victims contemptuously styled themselves 'Rabbits'. In February/March 1943 the Rabbits protested in writing to the camp commandant Fritz Suhren. The coerced subjects signed the protest, and refused to attend the Revier (the camp hospital) for further 'experiments'.[21] They showed extraordinary political acumen in submitting a statement that 'international law does not even permit experimental operations on criminals/political prisoners'. This was an audacious stance, separating their status as internees from their ordeals as experimental victims. While inspired by the Red Cross agreements on the rights of detainees, the International Committee of the Red Cross (ICRC) did nothing to halt the medical abuses, and the camp authorities conceded that the experiments went beyond an appropriate punitive regime.[22] They marched in protest to the camp commandant and demanded that he inform them whether the operations were part of their sentence. It was an epic protest, which the judges cited when delivering judgment on Gebhardt, Fischer and Oberheuser at the Nuremberg Medical Trial.[23]

The Ravensbrück commandant Suhren disapproved of the experiments taking place in the camp, principally because the prisoners' opposition provoked ill-discipline. Suhren declined to hand over further prisoners for experiments. On 15 August 1943 ten prisoners refused to present themselves for more operations. Ten days later Suhren was ordered to report to Richard Glücks (the Inspector of the Concentration Camps), who asked him why he had refused to supply patients. Glücks accompanied Suhren to Gebhardt in Hohenlychen where Suhren was humiliatingly forced to apologise. Himmler ordered Suhren to supply three more human subjects. This was the final group of experiments carried out in the camp, and as Suhren foretold, the girls revolted and were operated on by force in the camp prison.[24]

The Rabbits planned to alert the Allies and the Vatican about their plight. One prisoner, Janina Iwanska, had the idea of writing letters to their families using code words and secret writing in urine. She joined with Krystyna Iwanska, Wanda Wojtasik (a psychiatrist from Cracow) and Krystyna Czyz (a geography teacher from Lublin) in sending messages to their families with details of the 74 Rabbits. [25] They asked that news of these experiments reach the BBC in London, the Red Cross in Geneva, a Swiss Catholic mission in Fribourg and Polish exiles in Lisbon. Labour detachments came into the camp from outside and communicated details of the atrocities.[26] A French prisoner, Germaine Tillion (later a distinguished anthropologist), secretly carried a roll of undeveloped photographic film with pictures of the injured legs from 21 January 1944 until she left the camp.[27]

The Polish underground press published details of the Ravensbrück experiments, and the Lublin command of a resistance group sent the information to London. The Reich Security Office informed Gebhardt that intelligence about the experiments had reached Great Britain and Switzerland, and a delegate of the actively pro-German Swiss Red Cross told him late in 1944 that the Polish government in exile had condemned him to death.[28] In December 1943 a released prisoner who had US citizenship left the camp with a list of the names of the victims and the dates of the operations.[29] Investigating the events early in 1946, Keith Mant, a British army forensic pathologist, proved that the experiments were not instigated by the camp medical staff but arose from Gebhardt's insistence on human experiments.[30]

The Red Cross cover-up

The Rabbits' protest was part of a wider pattern of resistance and sabotage to disrupt German experiments. Prisoner-scientists in Buchenwald risked their lives by supplying the German army and Waffen-SS with ineffective vaccines.[31] The prisoner doctors, who accompanied 20 children transferred from Auschwitz to Neuengamme for injections with tuberculosis bacteria, attenuated the cultures to lessen their effects. Soviet prisoners of war selected for experiments resisted with brute force. Prisoner researchers at Auschwitz and Buchenwald duped their German scientific masters, and relished the opportunity for sabotage and delivering bogus results and useless batches of vaccine.[32]

The problem is compounded by concealment at the time and by the dispersed locations of the experiments. The focus has been on the experiments in concentration camps, but it was equally possible to camouflage medical killing in clinics, POW and slave labour camps. One of the original investigators, Thompson, estimated several thousand deaths from medical experiments, and numbers of victims used for experiments or medical research on body parts amounts to well over 10,000 individuals. Taylor referred to hundreds of thousands of victims of medical atrocities in his opening speech for the prosecution at the Medical Trial. The evidence of survivor testimonies and war crimes trials suggest a substantially larger number of victims than recent minimal estimates of 1,000 deaths, separating the experiments from the Holocaust.[33]

The Rabbits protested that the experiments violated their international rights as prisoners. This drew a stark contrast to the failure of the ICRC in Geneva to halt or condemn experiments on camp prisoners. An ICRC medical commission inspected the SS-Lazarett at Hohenlychen on 21 October 1943, when they were impressed by the results of operations on Waffen-SS soldiers. The inspection failed to detect cases involving prisoners from Ravensbrück.[34] By June 1944 Roland Marti, the ICRC's Berlin delegate, was informed about the Ravensbrück experiments, and the official responsible for civilians and deportees thought the victims would be better off if they could be sent the means to commit suicide rather than food.[35] Late in 1944 the ICRC appointed as a delegate to Germany (all delegates had to be Swiss) the surgeon Hans E. Meyer who had worked as assistant to Gebhardt between 1943 and August 1944. This appointment shows how the ICRC preferred conciliating the Germans to confronting them on their record of atrocities. Meyer dealt directly with Himmler at Hohenlychen to secure release of French prisoners from Ravensbrück.[36] In February 1945, Count Folke Bernadotte of the Swedish Red Cross asked Himmler at Hohenlychen about the fate of the Polish experimental victims. Three Rabbits were smuggled out on a Swedish transport.[37] The victims of the operations were sporadically singled out for execution; but when the German camp staff attempted to kill the remaining Rabbits in February 1945, fellow prisoners hid them.[38] This act of solidarity enabled the survival of experimental victims, who eloquently testified at post-war trials.

While the Rabbits appealed on the basis of the Geneva Convention, which guaranteed the rights of military prisoners of signatory states, the SS exercised a stranglehold on the German Red Cross. One defendant at the Nuremberg Medical Trial, Kurt Blome, was at first groomed to take over direction of the Red Cross, but his refusal to abandon the rowdy SA for the disciplined ranks of the SS meant that he was rapidly dropped in 1934. Another defendant explained that because Himmler's wife, Marga, was a qualified nurse who worked for the Red Cross in Berlin, he kept his distance.[39] The ICRC restricted its monitoring and relief activities to POWs (and even then excluded Red Army prisoners). It tragically downplayed the plight of civilians under German occupation, and of prisoners in the concentration camps.[40]

Gebhardt informed German medical colleagues about the experiments. At the end of May 1943 he and Fischer outlined the results of their experiments to a military medical conference of about 200 doctors.[41] This was one of several occasions when German medical specialists heard about concentration camp experiments. The accusation was later levelled that their lack of protest indicated that they accepted the experiments as medically legitimate. Some countered this allegation made at Nuremberg by indignantly asserting that they had criticised the experiments. The Luftwaffe bacteriologist Gerhard Rose protested about typhus vaccine experiments at Buchenwald, but his muted criticisms were discreetly raised in private. A follow-up conference was held at Hohenlychen in May 1944, a further sign of its importance as a military medical centre.[42]

The plight of the Rabbits gained international notoriety. In October 1944 the International Council of Women in London demanded that the ICRC give 'all possible protection' to the women imprisoned at Ravensbrück. The Council expressed horror at the 'barbarous experiments under the guise of scientific research'. The British and Polish Red Cross forwarded messages of concern.[43] The ICRC replied to the British Red Cross in December 1944 that 'we do not feel the moment is well chosen at present to take up the question once more' until it had resolved general issues of minimum guarantees for internees in concentration camps.[44] In contrast to the ICRC's obsessive secrecy, the BBC broadcast details of the experiments early in 1945.[45]

The ICRC negotiated with SS officers who had ever-diminishing authority, so ensuring that it continued to do nothing for victims of medical atrocities. In January 1945 the ICRC belatedly considered a scheme to send medicines using the Dachau camp as a central distribution point.[46] It focused efforts on providing aid parcels, which the Germans often hoarded or pillaged. Rescue efforts were belated; officially limited to French and Norwegian prisoners, a few of the Rabbits were smuggled to safety.

The ICRC failed to keep any systematic record of medical experiments. Roland Marti, the Berlin delegate of the ICRC, informed the ICRC Secretariat on 12 June 1944 about a special compound where orthopaedic experiments were conducted on the limbs of Polish prisoners. Marti stressed that this information was strictly confidential and was not to be utilised in any way.[47] The ICRC declined to publicise the case. Its inspection in October 1944 noted that experiments were carried out on criminals condemned to death. The Germans widely used this justification not only for the experimental victims, but also for the body parts dissected for anatomical atlases and research. German anatomists traditionally exercised rights to the bodies of executed criminals for research purposes.[48] The Berlin ICRC delegate recorded on 3 January 1945 that the experiments had ceased, although this was at a time when Roma children were being sterilised in Ravensbrück, and when Hohenlychen doctors were experimenting on the 20 children transported from Auschwitz to Neuengamme.[49] The ICRC made no provisions for care for the maimed victims of the human experiments, even though the Germans were known to be shooting persons who were sick or could not walk. An ICRC report on Buchenwald after liberation on 17/18 April 1945 belatedly mentioned extensive human experiments there, whereas its earlier reports had praised the ultra-modern medical provision in the camp.[50] The children experimented on by Heissmeyer were murdered in Hamburg on 20 April. Only on 25 April 1945 did a local ICRC official warn the Ravensbrück commandant not to further imperil the health or lives of camp inmates – the Germans promptly ignored the plea by organising death marches for the surviving camp inmates.[51] An ICRC report of 4 May 1945 on Ravensbrück finally acknowledged the plight of the victims of the surgical experiments, as well as of 60 Roma children sterilised there in January 1945.[52] A few ICRC field officers plucked up courage to intervene in the chaotic situation late in April 1945: one informed the Germans leading what were euphemistically termed 'evacuation marches' of concentration camp prisoners that all SS personnel were known, and that each would be called to account for his crimes.[53]

The failure of the ICRC in Geneva to condemn the Holocaust and rescue victims of racial persecution was replicated by the ICRC's lamentable record on human experiments. If it had documented the full dimensions of the medical atrocities, publicised abuses and admonished the perpetrators, the ICRC could have saved lives during the war and could have provided evidence for an impartial international tribunal on medical misconduct. That the medical condition of prisoners was a traditional concern of the ICRC made it an appropriate humanitarian agency to act against medical torture and killings. Instead, it accepted as legitimate the German view that the victims were in 'protective custody', that they posed a potential threat to German security and that the experiments were conducted on convicted criminals. The ICRC failed to secure guarantees from the Germans to protect child victims of human experiments – Mengele's twins, Roma children targeted for experiments and the 20 Jewish children, whose lives ended just days before the armistice, when they were brutally hanged in a Hamburg cellar of what, with cruel irony, was a school – the targets of the final frenetic period of human experimentation.

The ICRC's Swiss staff were obsequious to the Nazi German Red Cross officials. President Max Huber of the ICRC was also president of aluminium and machine companies, which produced armaments for the German army, while exploiting slave labour in a German subsidiary plant.[54] Reichsarzt SS

Grawitz (warmly addressed by Huber as 'lieber Herr Professor') had been acting German Red Cross President since 1936, making the protest of the Rabbits against their gratuitous maiming particularly poignant. Grawitz killed himself and his family on 22 April 1945.[55] The next day, when Gebhardt visited the Führerbunker in beleaguered Berlin as Himmler's emissary, Hitler apparently (no document survives other than Gebhardt's testimony[56]) appointed him the President of the German Red Cross. On 27 April Gebhardt finally abandoned Hohenlychen and accompanied Himmler – desperate for a pact with the Western Allies – to the northern stronghold of Flensburg on the Danish border.[57] Gebhardt's position as Red Cross President – stated as his occupation on his arrest report – expressed his aspirations to a role in post-war relief work. But Grawitz was a staunch supporter of the concentration camp experiments and Gebhardt a leading human vivisector: Hitler's appointment of them expressed a scathing contempt for the obsequious Geneva committee.[58]

In the final months of the war some national Red Cross organisations urged humanitarian action to rescue concentration camp detainees. The US State Department vainly tried to impress upon the ICRC its role in saving prisoners under German authority by at least inspecting frequently and thoroughly, so that lives should not be lost through starvation, exposure and deliberate neglect.[59] The ICRC remained paralysed by the dimensions of the human tragedy, unwilling to comprehend the plight of civilian and military victims of experiments, and intent on covering up its own lamentable record of inaction. The Swiss Red Cross had supported the German military offensive in the East, and its officers later protested against the prosecution of Handloser at Nuremberg. The ICRC was to inspect scrupulously how the Allies held the accused Germans (a mixture of prisoners of war and civilians) at Nuremberg and in the prison of Landsberg.[60] The ICRC was right to act impartially and monitor the conditions of the German doctors and administrators who were accused of medical crimes; but it did immensely more to assist the German perpetrators both during and after the war than the victims of the human experiments, whose lives were callously sacrificed.

After arrest early in May 1945, Handloser continued to direct public health, research and anti-epidemic measures. On 30 May 1945 (when working under British supervision as a member of the rump Karl Dönitz administration) and again on 27 June (under American supervision in Bavaria), he stressed that the German Red Cross had a vital co-ordinating role for both soldiers and civilians in what ought to remain a centralised system of public health provision. Handloser demanded permission to visit the presidential staff of the German Red Cross and to establish contact with the ICRC.[61] Handloser's demands were typical of the medical internees, who considered that they should remain in commanding positions over German public health and medical organisations.

Other internees thought in similar terms. Rose crossed and re-crossed front lines under the protection of Red Cross insignia. His mission was to explain to the Americans the need to continue malaria fever therapy experiments on psychiatric patients at Pfafferode in Thuringia. But he did not fully explain the dangers of the experiments, or that this was a euthanasia collecting point for eastern slave workers. Karl Brandt had approved supplementary food rations for Rose's 'patients'.[62] After arrest as a member of the German General Staff, he tried to activate ICRC contacts to secure his release. Rose had declared after the First World War that it was necessary to continue to fight the Allies to secure release of all POWs.[63] He resumed this embattled stance, not least to campaign for his own release. Rose's arguments became iconic for the medical detainees, who adroitly assumed the mantle of wronged and innocent victims. The chief Luftwaffe medical officer, Oskar Schröder, protested that the Allies held between 50 and 60 medical officers with the rank of general: it was unjust to hold them as their duties were essentially humane, and their services were needed to support Germany's health.[64] German doctors opportunistically declared themselves victims: their indignation rose at the Allies for violating their academic freedom to conduct research.

2
Allied Experiments

The Allied mobilisation of science for war produced major medical break-throughs, which required clinical tests and human experiments. When the Allied medical investigators reached Germany, they were curious about what their German rivals had achieved. The Anglo-American success with penicillin stands in stark contrast to the German scientific failure to develop penicillin production, as well as to the ethical abuses of wound infection experiments.

Allied military medicine prioritised operational use over theoretical innovations, and the rapid scaling up of pilot projects into efficient mass production. The mobilisation of scientists required a shift from the fragmented, competitive academic world to teamwork, the collaborative pooling of data and technical innovations, and inter-Allied liaison. Large-scale projects involved interdisciplinary research and swift implementation. Applied research involved human experiments in aviation medicine, parasitology, nutrition and toxicology. Human experiments were carried out by researchers on themselves, on soldiers, on volunteer groups of conscientious objectors and in US penitentiaries, but any view that dangerous exposure of experimental subjects was justified by the war was kept in check.

The American Committee on Aviation Medicine was launched in October 1940 by the Committee of Medical Research of the Office of Scientific Research and Development. The Committee co-ordinated research and organised extensive human experiments in pressure chambers and in the air.[1] Scientists spun a web of scientific collaboration across the Atlantic. The Yale physiologist, John Fulton, feared a German invasion of the United States was imminent; he visited England at the height of the Battle of Britain to organise systems of liaison in aviation medicine. Cuthbert Bazett, an Oxford-trained physiologist who held the Chair of Physiology at the University of Pennsylvania, headed north to Toronto University where he reinforced the aviation medicine laboratory, set up by the ferociously patriotic moderniser Frederick Banting, who discovered insulin. The British wanted the Canadians to accelerate their studies of the strategically relevant problem of oxygen deficiency.[2]

Canadian research institutions in medicine and neuro-surgery established themselves as 'a scientific linchpin between the British and the Americans'. Canada provided large-scale resources, expert researchers and a safe haven for the development of aviation medicine. Banting gave energetic support from the Department of Medical Research, Toronto until he was killed in an air crash in February 1941 on a flight to England. Banting was a dedicated self-experimenter: he plastered his right leg with mustard gas so as to observe sores and blisters, and subjected himself to ever higher and more frequent 'trips' in the pressure chamber. The scientist was in the frontline of combat: 'Science plus technical skill is the means by which this war will be won. Unless science is able to counteract German inventions we will be destroyed. ... we must give our pilots better planes, faster planes, more manoeuvrable planes.'[3] His fatal flight to Britain was one more experiment to observe the sensations of flight. Banting's self-destructive endurance stands in stark contrast to the German coercive experiments.

The Allied physiologists pointed out that the Germans had invested in the construction of experimental pressure chambers for testing effects of high-altitude and high-speed flight.[4] The crossing of new thresholds of endurance with speeds of 755 km at a height of 17,000 metres meant that the German aviators were hailed as heroic pioneers, but aroused British concerns given the strategic importance of the war in the air. US neutrality facilitated access to German aviation medicine until late 1941. To reconnoitre the whole area of aviation medicine, he circumvented a German security embargo and obtained early in 1941 Hubertus Strughold's and Siegfried Ruff's *Grundriss der Luftfahrtmedizin*. This overview of German research covered basic problems in aviation medicine. Fulton rated this as a fundamental compendium, because of pioneering experiments and its account of modern research facilities; he consequently arranged for its translation.

In May 1941 Fulton obtained Strughold's Rockefeller Fellowship record. Characterised as 'a nice upstanding German', Strughold had visited US aviation medicine centres in 1928–9. He worked with the physiologists Carl J. Wiggers in Cleveland and Anton Julius ('Ajax') Carlson in Chicago. Strughold attended the Aero Medicine Association in New York in September 1937, and impressed the Air Force doctor and enthusiast for high-altitude studies, Harry Armstrong, who after the war earmarked Strughold for transfer to the US. Fulton reconstructed Strughold's research, and asked about his character and interests. Carlson – defensive about his own anti-war position, as he backed the Chicago-based America First Committee, campaigning for neutrality – expressed the hope that Strughold had remained a 'first-class scientist' rather than having deteriorated into a 'first-class Nazi'.[5]

The Ruff-Strughold 'Compendium on Aviation Medicine' circulated in the United States, Canada and Britain in mid-1942, once the Alien Property Custodian sequestered the German copyright.[6] Committee members ranked the translated handbook as 'splendid', and pored over each issue of the Ruff and Strughold-edited journal *Luftfahrtmedizin*.[7] The US Air Corps wanted 250–300 copies of the translation in June 1942, and there were well over 100 requests for single copies from individual flight surgeons and air force bases.[8] Strughold was to take a major role in US high-altitude aviation physiology after the war, although he became the subject of much controversy concerning his attitude to unethical experiments. By the time Ruff was a defendant at Nuremberg, he was well known in US aviation medicine circles.

The Allied investigators accepted a higher level of risk for self-experiments than the Germans, but generally held to safer levels for experimental subjects. Most American experiments went to 40,000 feet with 47,500 feet as a maximum. The German experiments would generally be from 47,000 feet without supplementary oxygen to 59,000 feet with some subjects reaching 69,500 feet. One crucial difference is that the British and Americans had primates available for the most dangerous experiments. Fulton used monkeys and chimpanzees to test whether pre-breathing of oxygen alleviated the effects of rapid descent and whether air embolism contributed to the animals' death.[9] As food stocks diminished, the Germans became desperately short of animals for experiments; primates were unavailable – Gebhardt explained that the Germans even considered the possibility of kidnapping Gibraltar apes. The Americans continued to experiment on medical students, hospital patients, persons with mental disabilities and prisoners in state penitentiaries. The war added conscientious objectors in Civilian Public Service Camps to the ranks of experimental subjects.[10]

How ethical was such war research? The American Committee for Aviation Medicine noted the training routine in Toronto required for high-altitude experiments: 'Be sure that the subjects understand properly what is expected of them.'[11] This maxim covered the procedures and writing tests in the pressure chamber. Canadian air force researchers recognised the need to inform their subjects.[12] But the question arises as to whether due attention was consistently paid to the quality of consent. Ethical issues were tempered by the requirements of war, and a sense that prisoners and conscientious objectors had a social debt to repay. The physiologist Andrew Conway Ivy even considered the physiological effects of providing a pecuniary incentive.[13] The war effort involved a vast amount of research on military personnel on new therapies like radium irradiation.[14] Jonathan Moreno characterises American standards on consent in military medical research as 'wildly inconsistent'.[15] In Britain the Medical Research Council turned down Kenneth Mellanby's requests to infect human 'guinea pigs' with malaria and typhus after their consent had been obtained. Even so, his scabies experimental subjects risked blood poisoning, not least because their resistance was lowered by the vitamin deprivation experiments of the biochemist Hans Krebs.[16] But compared to the Germans, the level of risk was less severe, and while the Allies organised groups of conscientious objectors and prisoners for experiments, these always expressed a willingness to participate, even though the quality of information about the risks involved had defects.

War crimes investigators of Nazi human experiments had gained direct experience of human experiments and clinical research during the war. Ivy was Secretary to the [United States] Sub-committee for Decompression Sickness from June 1942, and between 1925 and 1946 was professor of physiology and pharmacology at Northwestern University, Chicago, where he claimed his laboratory was 'the largest and best equipped in the world'.[17] He was Director of the Naval Medical Research Institute in 1943, when he supervised tests on making seawater drinkable.[18] *Life Magazine* regularly ran features on experimental breakthroughs in the 1940s and 1950s, doing much to promote the public image of medical experiments. An account of how US shipwrecked sailors desalinated seawater caught Himmler's attention. He demanded a German desalination device, so triggering research, which was to be under contention at the Nuremberg Medical Trial.[19]

Ivy had met Strughold when he was a Rockefeller Fellow. The question was whether Ivy posed a security risk: he had written enthusiastically about the Soviet Union, he was a founder member of the radical American Association of Scientific Workers in late 1938, and supported Carlson's campaign for non-involvement in any European war.[20] Whereas Fulton feared a German invasion via Canada or Mexico, Carlson believed a fifth column of unemployed and native Indians posed the primary risk to American security. Ivy was caught in the crossfire between these two giants of physiology, when in 1942 his security clearance was held up after he was invited to chair the strategically important Committee on Decompression Sickness.[21] Security officers noted Ivy's pacifism, and unorthodox religious views, but eventually cleared him as a capable and patriotic physiologist. Fulton congratulated Ivy on being 'officially decompressed', and invited him to serve on the National Committee on Aviation Medicine of the National Research Council.[22] Ivy declined military rank, as did Strughold for most of the war, so as to assert that a scientist should not be subject to military orders. Ivy threw himself into war-related physiological research, and from 1942–3 was Scientific Director of the Naval Research Institute. He again insisted on retaining civilian status as more befitting a scientist.[23]

Ivy analysed the effects of rapid pressure loss on monkeys. He ran tests on dexedrine tablets as well as how different types of diet, vitamins and caffeine affected human metabolism and could help the frontline soldier overcome fatigue. He examined altitude tolerance, experimenting on the relation between diet and the problems of 'bends' and 'chokes'.[24] He took 93 subjects to 40,000 feet, while 32 subjects were given dextro-amphetamines.[25] He built up a colossal research establishment, applying the physiology of the human metabolism to problems of survival and endurance. He stressed his ethical conduct: 'In order to get the corpsmen interested in serving or "volunteering" as "assistants" or subjects in the earlier experiments at the Institute, I served as a subject in a trial experiment. This was true of the dilute sea water tests, the Goetz water tests and straight sea water test. This, however, has always been my policy in laboratory work.'[26] Ivy abided by a 1943 Navy directive that consent of serving personnel should always be obtained.[27] At the Medical Trial Ivy defended his experiments as safer than those of the Germans. His subjects breathed supplementary oxygen to a height of 47,000 feet. The experiments of the accused Ruff reached a height of 50,000 feet, and then involved a slow – and potentially dangerous – descent, simulating a parachute.[28] (See Table 9.)

Aviation medicine attracted scientists who later became key war crimes investigators. The health, safety and aptitude of aircrew were priorities for all sides. The neuro-physiologist John Thompson had studied in Germany with the celebrated pathologist Ludwig Aschoff, as had a fellow Canadian medical intelligence officer, James Blaisdell. Thompson had the advantage of an international academic pedigree and firsthand knowledge of German medical research. He was until August 1946 a US citizen, born in Mexico in 1906, educated in California, studying biology at Stanford and medicine at Edinburgh. He spent time at the renowned pathology institute at the University of Freiburg in 1932–33 during the Nazi takeover, and with the celebrated neurologist Santiago Ramón de Cajal in Madrid in 1935 at a time when the civil war was imminent. By 1938 he was at Harvard, interested in linking mental illness to respiratory anomalies. He became friends with Leo Alexander and the innovative psychiatrist, Milton Rosenbaum. They deepened his concern with the psychopathology of anti-Semitism, which Thompson regarded as the illness of his age. He became fascinated with Jewish history, theology and culture. With the onset of war, he joined other aviation medical researchers at the University of Toronto, which under the inveterate self-experimenter Banting rapidly became a linchpin in British and American collaboration in research.

Thompson researched on susceptibility to altitude sickness, and, despite his weak lungs, conducted self-experiments. He enlisted in the RCAF, and developed schemes for the pre-breathing of oxygen to prevent decompression sickness. Thompson went on a tour of duty to Washington DC to present Canadian research on altitude sickness to the Fulton-Ivy Committee on Aviation Medicine in July 1943.[29] Ivy and Thompson met after the war to discuss the criminality and ethics of the German human experiments; both understood fully the aims and methods of the German research, although their views on experimental ethics were to diverge. They provided a defining framework for ethical evaluation of the German experiments.

The neurologist Alexander contributed to the psychological evaluation to establish appropriate numbers of missions for airmen. Born in 1905, Alexander grew up in Vienna, where he graduated in medicine in 1929, and was assistant at the Kaiser Wilhelm Institute for Brain Research in 1928, before joining the neurological institute in Frankfurt. He was fortuitously at the Peiping medical school in China when Hitler came to power in 1933, and then settled in the United States. His enthusiasm and intelligence achieved respect among American neurologists, and the Rockefeller Foundation's medical programme officers continued to support this talented neurologist.[30]

Alexander's work in neurology in Boston brought him into contact with Thompson. Thompson observed with disgust violence against Jews in Freiburg. While at the Harvard Fatigue Laboratory he was already preoccupied by the problem of anti-Semitism, diagnosing it as the major illness of civilisation. Remarkably, given their later collaboration, Alexander in 1938 presented Thompson with a *History of the Jews*. Thompson had the idea of establishing a chair at Harvard for the Study of anti-Semitism and tried to interest Harvard President James Conant and the Warburg banking family in the scheme.[31]

During the war Alexander, Ivy and Thompson worked energetically to improve the medical assessment of the fitness of aircrew and their chances of survival. Thompson assisted by developing apparatus to pre-breathe oxygen. Alexander was Associate Professor of Neuro-psychiatry at Duke University School of Medicine between 1941 and 1946, and joined its medical detachment to Europe.[32] His task of monitoring the psychological fitness of American aircrew for bombing missions prepared him for assessing the mental state of captured German aviation researchers. Ivy, Thompson and Alexander were sent on intelligence missions to assess German research, when they encountered the problem of unethical human experiments as medical crimes and acts of genocide. The search for methods and weapons of mass destruction became a quest into the causes, ethics and psychology of genocide.

3
Criminal Research

A debased science

Debate erupted in 1933 about whether Nazism and the maltreatment of Jewish scientists and doctors destroyed German science and medicine. Anti-Nazi critiques like Gumpert's *Heil Hunger!* derided the weakness of German medical research and pointed to increased rates of infections like diphtheria and puerperal sepsis.[1] Robert Brady's pioneering study of German fascism examined how the Nazis reorganised science on the basis of cultural and economic autarky.[2] Nazi propaganda claimed achievements in hereditary research, biochemistry and pharmaceuticals.

Had German medicine under Nazism continued its brilliantly innovative trajectory? German medicine attained world leadership on the basis of applying experimental techniques to understanding human physiology. Patriotic scientists mobilised to work on poison gas, armaments, and disease control in the First World War. The 1920s had seen new national research funding institutions, and the Rockefeller Foundation stepped in to sustain German experimental research. After the war Germans claimed that the Nazi Party and SS interfered to the detriment of research and experimental medicine.

Allied intelligence officers knew German medicine from the inside: the neurologist Alexander (who was expert witness for the prosecution at the Medical Trial) and Thompson (who became the linchpin of the International Scientific Commission set up to investigate medical war crimes) felt acute dismay at Germany's fall from international leadership in medical research, as they found mounting evidence of its deranged atrocities. At first, Allied expectations were of a highly innovative scientific war effort, but the closer their encounter became with the remnants of the wartime research establishment, the more they became convinced of criminality and the sheer craziness of the Nazis. They puzzled over how German research was co-ordinated in the war, the role of the military, the requirements of industry and the demands of the SS and NSDAP for medical support to realise a racial utopia. German scientists and administrators – anxious to maintain their posts, avoid dismissal and prosecution – claimed that they resisted the Nazis. The administrators and institute directors of the prestigious Kaiser Wilhelm Gesellschaft (KWG) alleged that this sprawling conglomerate of laboratories and research stations was neither nazified nor involved in racial and armaments research, nor in servicing the needs of war production.

A handful of dissident insiders and critical observers accused German research of being geared to mass destruction. This view was taken by the Soviet-appointed KWG president and chemist, Robert Havemann; by the scientific intelligence officer Samuel Goudsmit for the physical sciences, by the neurologist Alexander, who characterised the human experiments as 'thanatology' (an 'idolatrous delight in death'), by Thompson who argued that most clinical and laboratory research fell into the category of 'medical war crimes'; and by critically minded dismissed scientists such as Otto Meyerhof, formerly director of the KWI for Medical Research.

After a hiatus between the 1950s and 1980s when German research remained largely under the leadership of wartime elites and their student disciples, the debate resumed as to how mainstream was the murderous medicine of the Nazis, and on the authoritarian potential of German science. A new wave of social historians of medicine and gender established that eugenics and racial medicine were core concerns in the 1920s, and realigned in 1933 to support Nazi racial policy.[3] Robert Lifton examined the psychology of the scientific perpetrators in the 1970s and early 1980s, observing how German physiology, bacteriology and biology had become corrupted by ideas of inferior life as a rationale for human experiments.[4]

Benno Müller-Hill analysed the German geneticists as purveyors of 'Murderous Science', based on fallacious theories like those of the biochemist Emil Abderhalden's specific protein reactions. He confronted the indignant outrage of Adolf Butenandt, who as a veteran KWI director and president of the Max-Planck-Gesellschaft (MPG) in the 1970s, found efforts to scrutinise scientists' conduct under Nazism distasteful. In 1997 Ernst Klee denounced the Deutsche Forschungsgemeinschaft for providing funds for murderous research aimed to boost German war potential.[5] Karl-Heinz Roth developed the idea that scientists offered a 'final solution to the social problem'; he and Angelika Ebbinghaus have characterised the research of two leading biochemists, Butenandt and Richard Kuhn, prevented by Hitler from accepting Nobel Prizes, as 'Vernichtungsforschung' – exterminatory research.[6] Such accusations have led to painstaking research to accurately establish what motivated these scientists, and the extent to which they worked on military assignments and conducted coercive human experiments, and how racial medicine reflected the faction-ridden Nazi leadership and power blocs. Questions arise about the extent to which the research institutes and clinics employed slave labour, and initiated or condoned coerced experiments. This has prompted renewed scrutiny of the vast research conglomerate of the KWG, and of the extent to which it mobilised for military and racial research.[7]

Allied investigators became convinced that the historical reality of Nazi science was of a fragmented, poorly co-ordinated set of research initiatives. In the physical sciences there were pockets of innovation – the development of torpedoes or rocket research. The classic enigma has been whether the Germans actually had an atomic bomb project. How medicine reinforced the German war effort posed intriguing problems in terms of unravelling the interactions of civilian, military and SS organisations and projects. The Allies focused on determining the scientific basis of new therapies – for example, to manage war wounds and cure infectious diseases – and to support military technologies like high-altitude flight. The Allied investigators examined whether German research improved the survival rate of soldiers, faced with extremes of cold or high altitude, or lack of food and water, or when gravely wounded. The grim low-pressure and freezing research of Sigmund Rascher was condemned as 'pseudo-science' and sadistic gratification, but the collaborating physiologists raised the possibility that such research went to the heart of mainstream German physiology. The issues crystallised into a general evaluation of German medicine, and German professional leaders felt that their profession as a whole was on trial.

Allied scientific teams monitored German medical innovations, as well as captured medical supplies. Intelligence units screened medical publications to find out about the incidence, therapy and prevention of infectious diseases as well as about scientific innovations. They analysed the interrogation records of prisoners of war (POWs) for information on medical equipment and procedures, and assessed captured pharmaceuticals. The British decrypted police reports about epidemics in concentration camps. The Allies found out about the medical conditions advancing troops might encounter, in case there were epidemics of typhus or influenza as there had been after the First World War. Botched German attempts to culture penicillin and operationalise the Swiss discovery of DDT suggested a flawed research organisation. But the Allies expected to uncover fundamental innovations in German pharmaceuticals, clinical medicine, and chemical and biological weapons.[8]

The interrogators and intelligence gatherers built up a picture of German wartime research. The Germans complained that in 1939 researchers were drafted into military posts. The conviction grew during the war that the innovative capacity of German research was being disrupted. Reich Marshal Göring took over as president of the Reich Research Council in July 1942 to redress the fragmentation of German research and to energise it by setting strategic targets. The SS physicist Rudolf Mentzel aligned research with rearmament as president of the reorganised German National Research Council. The SS administrator Wolfram Sievers (a defendant at the Medical Trial) was his deputy.[9] Research on cancer provided a cover for bacteriological warfare.

The prioritising of research went so far as to allow drafting in prisoners from the concentration camps. Himmler ordered on 25 May 1944 that trained scientists in the camps be formed into research groups in mathematics, chemistry, physics and (a month later) in medicine. A mathematics group was convened in the Sachsenhausen camp.[10] The Waffen-SS bacteriologists put together a prisoner research team in Buchenwald for typhus vaccine research and production. In 1944 expectations that the research revival was about to yield momentous discoveries meant intensification of human experiments. The ruling elite hoped for production of a miracle weapon, and scientists looked forward to salvaging their careers. The Medical Trial defendant, surgeon and Berlin Faculty Dean, Paul Rostock, explained how in autumn 1943 his former student Karl Brandt recruited him to advise on how to revive basic research.[11] The implication was that the German defendants were valiant guardians of scientific traditions rather than complicit in genocidal crimes.

Refugees from Nazism and governments in exile lobbied the British and American governments to prosecute those responsible for racially motivated crimes. By the time of liberation, the lawyer Raphael Lemkin was pressing for the charge of genocide, and the Allies took account of 'persecutions on racial and religious grounds' at the London conference in 1945 and in the preparations for the IMT.[12] It was at this point that liberated doctors and prisoner medical auxiliaries from concentration camps, survivors and a nucleus of committed and conscience-stricken Allied medical investigators called for documentation and prosecution of human experiments. Few German scientists had any ethical scruples in taking advantage of the Nazi appetite for results in physiology and bacteriology. The unedifying spectacle of the German academic scramble for power and influence under Nazism suggested that the time had come for a fundamental reappraisal of the ethical basis of medical research.

What did the Allies know?

Wartime Allied medical intelligence monitored and predicted the diseases advancing troops might encounter. More was known about outbreaks of disease in operational theatres than about the deportations and murder installations of the Holocaust. The raging typhus epidemic in Belsen showed how unscrupulous the Germans were in allowing disease to decimate camp populations. The risk was that rats, refugees and the destruction of water and sewage systems would cause an epidemic conflagration. As the likelihood of a typhus epidemic receded, attention shifted to restructuring German health care provision. This in turn raised the problem of what to do about the nazified medical profession. As Allied medical surveillance assessed German research, there was a series of revelations on inhumane experiments.

German medical researchers clung to traditionally fragmented academic structures. The Inspector of the Medical Services of the Luftwaffe lamented that research was hindered by 'vanity on the part of individual scientists, every one of whom personally wants to bring out new research results', and that this was true also for the SS scientists undertaking human experiments.[13] Karl Brandt and his surgical colleague Rostock failed to co-ordinate the efforts of the 15 or so departments competing to produce a viable form of penicillin.[14] German communications on penicillin were intercepted, and transmitted to British scientists.[15] Gruesome experiments were undertaken to find an equivalent to penicillin by intravenous and intramuscular injections of pus, followed by chemotherapy – at Dachau 800 prisoners were injected with 'Phlegmone' (infectious pus meant to be analogous to penicillin) in 1942–43.[16] The combination of academic recklessness and the disregard for 'worthless lives' meant that clinical researchers did not flinch from lethal human experiments.

In December 1943 an editorial in *The Lancet* deduced that vaccine trials, which were described in a leading German journal of bacteriology by one Erwin Schuler, involved deliberate infection of human subjects.[17] The inference drawn – that the experiments were on POWs – showed that the Allies' primary concern was for their servicemen and women. *The Lancet* envisaged that the experiments were in a POW camp and that British soldiers were the victims (indeed, British prisoners on Crete were used for hepatitis experiments). What was not known was that hundreds of Russian POWs were used in the often fatal typhus vaccine experiments. Schuler was a Waffen-SS medical officer in charge of a compound for viral and rickettsial research and serum production in the concentration camp of Buchenwald, which was conveniently close to the University of Jena, itself a centre of Nazi and SS medicine.[18] Schuler was identified as 'Dr Ding' arrested on 16 April 1945.[19]

Some of the earliest evidence on medical atrocities came from the Soviet Scientists Antifascist Committee under the National Commission on Nazi War Crimes. Stalin established this in November 1942. The Soviets did not join the United Nations War Crimes Commission (UNWCC), which was founded in October 1943 as an inter-Allied legal authority to co-ordinate war crimes documentation. Although neither the US nor British diplomats much liked the UNWCC with its predominance of victim countries, the Soviets were kept at arm's length as they insisted on representation for seven constituent republics, including the annexed Baltic states and Finno-Karelia. The Soviet Commission was a large-scale and public investigative enterprise with a central committee, regional and local committees, and special commissions. They collected evidence on German atrocities, taking statements from witnesses, and conducting forensic and scientific investigations, so that between 1943 and 1947 the commission prepared a staggering 250,000 reports on the basis of forensic investigations and interrogations of German prisoners – although the Soviet Commission had a strongly ideological stance. The Soviets mounted the first war crimes prosecution of the Second World War – which was significantly for a German massacre of psychiatric patients at Kharkov in the Ukraine. Other allegations of medical atrocities linked to mass extermination included executions at the Institute of Hygiene in Kiev.[20]

Soviet Academicians involved in public health and medical research accused German physicians of performing experimental operations on healthy prisoners of war and of using civilians as guinea pigs in testing new medicines and poisonous substances, and for experimental operations.[21] In May 1944 the Commission called on 'doctors of all united Nations to place [a] brand of shame on German army doctors who have converted medical practice into weapons for destruction of life'. The Soviets cabled a report on the crimes of German doctors to the Royal Society for Tropical Medicine in London, where the inter-allied health charter movement had organised a 'communications centre' to collect data on clinical abuses.[22] This centre linked British and exiled doctors who publicised data on the German scientific abuses in specialist publications in 1944. What was missing was any co-ordinated and systematic effort to keep a central record of medical crimes, so that these could be investigated as the Allies advanced.

By the time of the German capitulation, Churchill, Roosevelt and Stalin had not agreed whether there should be summary executions, trials or some other means of justice and retribution. In August 1945 the four-power basis for the IMT Nuremberg was agreed. But there was increasing distrust between the British and Americans on the one side and the Soviets on the other, and competition to seize German scientists. Until the disarray at the top political level was resolved, there was no agreement on medical atrocities. This was compounded by Western suspicion that the Soviets were recruiting German scientists to develop biological and gas warfare. The Russians were interested in Hans Reiter of the Reich Health Office in May 1945 for information on biological warfare. Numerous medical crimes were committed on Soviet territory, and Soviet expansion and military occupation increased the responsibility to investigate medical atrocities. But the Soviet judicial authorities fabricated accusations against German medical scientists: one was of conspiracies to kill by using poisoned vaccines; another – seen in the Russian trial of the Berlin bacteriologist Heinz Zeiss – was of a vast conspiracy by Germans and Russian traitors to undermine Soviet rule by subversive germ warfare.[23] Allied competition for intelligence on chemical and biological weapons of mass destruction diverted attention from criminal experiments. The incipient Cold War rivalry with the Soviets to seize and exploit German scientific know-how softened Allied attitudes to the criminality of German aviation physiology and bacteriology. These sciences were regarded as strategically relevant in building up the West's defences.

Criminal revelations

When the war in Europe ended on 8 May 1945, the Allies lacked any policy for dealing with perpetrators of forced experiments and medical atrocities. There was no intention of holding a trial, no systematic collecting and dissemination of evidence to ensure that those responsible would be on wanted lists, and no concept of a 'medical war crime'. The Allies failed to compile a consolidated war crimes list during the war. Medical atrocities were absent from war crimes investigations, which were generally limited to crimes against Allied servicemen and women, and the murder of civilian hostages. By way of contrast, the Allied military authorities were keen to identify – and exploit – war-related medical innovations and scientific discoveries made by the Germans. Allied scientific intelligence targeted medical research installations and sought out research scientists and organisers of the German scientific war effort. The Allies had high expectations of German pharmaceuticals, aviation medicine, nerve gases and defences against biological weapons.

The manifold abuses of medicine under Nazism appeared as an ill-defined component of violence, executions and mass killings. During the war, the Allies had reliable intelligence on abusive experiments, sterilisation, euthanasia and the genocidal gas chambers. But the liberating forces were scarcely prepared for their encounters with victims of sterilisation, forced abortion, starvation and infectious disease. The troops that liberated Buchenwald and Dachau were surprised to find specially sealed off compounds for human experiments. The skeletal survivors urged the Allies to investigate German racial policies, genocide and unethical medical practices.

The Allied heads of state had declared in Moscow in October 1943 that perpetrators of war crimes would be returned to the scenes of their crimes and tried there.[24] The discussion ran in general terms of war crimes against Allied soldiers and civilians, rather than anything specifically medical, and ignored the concentration camps and crimes against civilians. The UNWCC did not recognise medical atrocities as a distinct category of war crime.

On 8 August 1945 – three months after the armistice – it was finally agreed in London to establish the IMT for war crimes and atrocities against civilians. 'Crimes against humanity' – a catchall legal construct – enabled the military authorities to investigate and prosecute atrocities against civilians under German occupation, and covered murder, extermination of ethnic groups, torture and other types of inhumane, racist acts. The rapidly convened tribunal at Nuremberg publicised the Final Solution and a range of crimes planned and executed by the Nazi military and political leadership; lesser crimes were to be tried in the countries where they had been committed.[25] There was no declared intention to hold trials of offending medical personnel.

The Germans tried to keep human experiments a closely guarded secret. The experimental installations were securely guarded enclaves within camps. Even if the subjects survived an experiment, promises of release were ignored and the surviving human guinea pigs were liable to be killed – after all, they might warn fellow prisoners of the experiments – or died from infections and their weakened condition. Incriminating medical records vanished. The Kaiser Wilhelm Institute (KWI) for Anthropology destroyed papers relating to Auschwitz research.[26] The Buchenwald doctor running the experimental station changed his name from Ding to Schuler, and began to burn documents. His prisoner secretary, Eugen Kogon, who was to emerge as a key witness at Nuremberg, persuaded this Sturmbannführer to make concessions to the political resistance in the camp, and to preserve documents.[27] The Dachau malaria researcher Karl Claus Schilling claimed that in mid-March 1945 an order came from Berlin to destroy the records of his experiments, although an assistant saved his personal notes.[28] He was outraged when documents were presented against him in court.[29] His colleague, Rose, wanted to write up his notes while in American custody.

The Nazi leadership became anxious to liquidate underlings with incriminating knowledge. Viktor Brack, administrator of the T-4 euthanasia programme, was arrested with his superior, Philipp Bouhler, in April 1945 and condemned to death. Waldemar Hoven, the Buchenwald camp doctor, had been in custody since 1943. Sigmund Rascher was arrested with his manipulative wife Karoline ('Nini'), who had been Himmler's mistress in the 1920s. The SS ordered the arrest of the Raschers in April 1944 for the abduction of one child. They were released but rearrested when it transpired that all their four children had been abducted, and some subsequently exchanged. They were held in prison in Munich, but at the beginning of 1945 Rascher was sent to Buchenwald and his wife to Ravensbrück. He was last seen in Dachau on 26 April 1945. At first doubts surrounded whether he had been killed or escaped, but the consensus grew that he had been summarily shot on Himmler's orders.[30] When beleaguered leaders could no longer exert effective political control, they relished purging their sycophantic followers.

Hitler ordered the arrest of the leading medical administrator, Karl Brandt, on 16 April 1945. Brandt had been Hitler's devoted escort surgeon from 1934, overseer of euthanasia from September 1939, and co-ordinator of civilian and military medical services from 1943.[31] But when he objected to Hitler's doctor, Morrell, prescribing strychnine in anti-gas pills to treat stomach spasms in October 1944, Hitler was furious.[32] Speer intervened to have Brandt released – they had collaborated on building prefabricated emergency hospitals, and Speer shielded him during the final collapse.[33] Brandt lost a final round in the power struggle among Hitler's medical entourage, after earlier confrontations with the Party and SS figures,

Leonardo Conti, Bouhler and Himmler. The pragmatic Brandt provided medical and surgical services in the civilian and military spheres, as Germany's military position deteriorated.[34]

The criminality of German wartime medicine remained uncertain. Was a war crime restricted to violations against an Allied soldier; or did acts against civilian victims also count? Could sterilisation, forced abortion, maiming and murderous experiments and euthanasia committed by the Nazi medical establishment against Germans be ranked as war crimes or as genocide? Did it make a difference whether crimes were motivated by research, or did all medical abuses – sterilisation or draining litres of blood – merit prosecution as criminal? Uncertainty arose over how to deal with racial crimes, and whether a victim's nationality, particularly whether they were an 'Allied national', made a difference. There was no systematic effort to determine the scale and momentous nature of the medical crimes.

Concern with Nazi racial atrocities resulted in meetings and manifestos against Nazi racial ideology as pseudo-science on the Allied side.[35] The anthropologist and political activist, Gene Weltfish, saw the need to inculcate a sense of humane responsibility among scientists. She reviewed the issue of how scientists devised techniques of mass destruction in a speech to the American Association for the Advancement of Science in 1944. She suggested that there be a 'dishonor roll' for scientists – like the immunologist Alexis Carrel – who collaborated with Nazism. She also recommended 'a Scientist's Oath' – analogous to the Hippocratic Oath – to pledge that knowledge should not contribute to the techniques of mass extermination.[36] These were far-sighted remedies, and although she was a noted campaigner for race equality and advised the OSS, her radicalism debarred her from effective influence. Her call for a Scientist's Oath was taken up by Aldous Huxley, the author of *Brave New World*. He argued that in the wake of the horrors of the concentration camps, it was necessary to inculcate a sense of responsibility among scientists.[37] The Allies lacked any sort of systematic blacklist of scientific criminals, and at first saw no need to tackle the ethics of the science of mass destruction.

Captured German medical scientists faced a lottery: some gambled on surrender to one or other of the Allies in the hope that they would be treated as informants about wartime medical research and allocated new responsibilities, rather than being held as prisoners, or worse still as war criminals. Some were drafted into clinical and public health services. The senior medical officer Handloser at first retained executive powers over the remnants of a capitulated German military medicine. But in autumn 1945, he was abruptly treated as a war crimes suspect.[38]

A British analysis of the Nazi system of medicine appeared in the *Basic Handbooks* series for briefing the occupation authorities. Compiled in late 1944, the report presented a well-informed analysis of Nazi administrative and medical structures. It highlighted the conflict over public health administration from 1942 between the Reich Health Führer Conti and Karl Brandt as the Plenipotentiary of the Führer for Medical and Sanitary Affairs. The report noted at least 100,000 compulsory abortions and sterilisations (in fact there were over 350,000 sterilisations, and we have no reliable estimate on forced abortions), and that 'hundreds of thousands, if not millions, of "racial and genetic enemies" have perished in the prisons and concentration camps.'[39] It was becoming clear that sterilisation involved coercion and trauma, and Nazi surgeons conducted a dangerous operation when 5 per cent died.

The *Basic Handbook* mentioned criminal-biological research on children of concentration camp prisoners. Nazi medical science was depicted as 'brutal and ruthless'. The report failed to mention euthanasia in the section on psychiatric hospitals and the clearing of hospital beds for war casualties. There was just a passing reference to a department for 'experimental research on the standard operations and the use of X-rays for sterilisation and abortion'. The glaring omissions indicated that what one arm of Allied intelligence knew was not divulged to another. But the abuse of science, and particularly genetics, was stressed: genetics was linked to killings at the extermination camp of Maidanek and to slave labour.[40] The message was that advancing troops should be alert to Nazi medical crimes.

The US War Department's Medical Intelligence Division included from 1943 two budding historians of medicine. Captain Saul Jarcho held qualifications in tropical medicine. An avid classicist, he compiled reports on medical hazards in the Eastern Mediterranean. This erudite bibliophile spotted a forthcoming publication on Japanese biological warfare research, which could threaten national security.[41] This incident showed how monitoring epidemics spilled over into intelligence on coerced experiments.

Captain George Rosen applied expertise in social science and history. Rosen had the benefit of having studied in Berlin in the 1930s, so that he had an insider's understanding of the Nazi medical system. Jarcho and Rosen deployed linguistic, medical and historical expertise, and Alexander shared a medical historical perspective on Nazi medicine.[42] The medical investigators found the more that they pieced together the fragmentary evidence on health conditions amidst the devastation of war, the more they encountered evidence of criminal experiments. The sense of history provided a yardstick to measure medical misconduct.

Rosen analysed German medicine under National Socialism, contributing to a Civil Affairs Guide of 26 May 1945 on *Denazification of the Health Services and the Medical Profession in Germany*.[43] In March 1945 he interviewed public health experts as part of the colossal military medical monitoring operation.[44] He evaluated the KWI for Medical Research in Heidelberg, which was suspected of complicity in Nazi military projects.[45] He laid the foundations for follow-up investigations by

Alexander.[46] A Basic Handbook issued in April 1945 referred to the 'Use of Prisoners for Scientific Experiments':

> This has taken various forms. Thus, at Mauthausen, certain prisoners were compelled in 1942 to undergo tests in connection with new war gases, their effects, and the efficacy against them of existing protective devices.... It has been reported, also, from a number of camps, that their medical officials have used prisoners as 'guinea-pigs' in the course of researches, including vivisection, experimental injuries, inoculations, and so on.

In June 1945 Rosen interrogated the Luftwaffe bacteriologist Rose about tropical medicine and public health. The loquacious Rose made copious observations about disease control in frontline Russian locations of the Crimea and Kuban. He outlined what diseases German frontline troops faced, and what therapies were to hand. His admissions of human experiments were noted in the report without any special comment, suggesting that experiments were considered standard military medical practice.

Rose reported on malaria therapy that, 'Sontochin and brachysan have not been used in the field, only in experiments with mentally ill patients at Mühlhausen'. When called up for service in the Luftwaffe, he converted his department at the Robert Koch Institute (RKI) Berlin into a department for military hygiene, which was relocated to Thuringia. He used patients evacuated from the state asylum at Arnsdorf in Saxony. He tested 200 patients, dosed with different drugs, exposed for eight weeks to malaria-carrying mosquitoes, and then kept under observation for 400 days. He recorded that when malaria broke through the 'protective screen' of the drug Atebrin, the cases were mild. This was a remarkable admission of deliberate experimental infection. How the malariologist Schilling injected patients with sporozoites was discussed. Rose's criticisms were scientific, and he made no comment that Schilling's subjects were prisoners at Dachau and that he often visited Schilling there.[47]

Rose used various types of typhus vaccines, which he tested in controlled experiments. Again, the implication was of human experiments. Reference was made to Ding's experiments on '150 prisoners who had been condemned to death'; 33 out of the 35 unvaccinated 'controls' died. Rose mentioned a certain Hauer in Berlin who tested new drugs on cases of amoebic dysentery. Finally, he offered to prepare a paper on infectious hepatitis for the Americans. He was to use this as evidence in his defence when tried at Nuremberg.[48]

The hepatitis problem involved the search for a viral cause of jaundice, as the disease was widespread on the Eastern Front. Infectious jaundice was disabling but generally not fatal. Rose mentioned experimental infection of volunteers, and that Dohmen of the RKI and Eugen Haagen at Strassburg isolated viruses. The usually loquacious Rose covered up his silence on human experiments by saying that his statement was necessarily incomplete as made without notes. The prosecution case at Nuremberg established that Dohmen and Haagen carried out experiments at the concentration camp of Sachsenhausen, and that Friedrich Meythaler experimented on British prisoners of war in Crete.[49]

Rosen was more interested in public health systems and social medicine than problems of experimental medicine. This raised the issue of the criminality of German eugenics and racial research. The Czech émigré Hugo Iltis alerted the UNWCC that scientific representatives of German race theory should be dealt with as war criminals. Franz Weidenreich listed 44 racial scientists suspected as war criminals.[50] Racial victims of sterilisation – the Rhineland half-caste children (the children fathered by black African French troops and German women in the Rhineland occupation after the First World War) sterilised clandestinely in 1937, and Jews who had been sterilised for racial reasons – evoked the concern of the Allies for victims of medical crimes. The bulk of the so-called 'legal sterilisations' were condoned as committed in Germany and under German law (the situation in Austria was altogether less clear).[51] Sterilisation victims began to lobby the war crimes authorities with evidence of their ill-treatment.

Rosen faced the problem of how German public health and medical science prioritised race over health. In May 1945 there was an inter-Allied conference on the dangers of racial theory.[52] In June 1945 the US War Department Guide on German medicine[53] showed how the Allied monitoring edged towards the view that German medical practice was inhumane. It dealt with euthanasia, the abuse of the sterilisation and castration laws for political purposes, and the battery of laws on racial purity and population policy. The Guide followed the British Basic Handbooks by elaborating on the bitter conflicts between Conti and Karl Brandt.

The Guide recommended that records on race and genealogy be seized. Allied search teams located enemy documents, records and archives. The Americans established document centres in Heidelberg, Munich and Berlin, and in Austria. The laborious process of indexing and cataloguing began, as a preliminary for documenting Party and SS membership, and reconstructing Nazi power structures, and weapons technology. Teams of analysts searched for evidence for each of the Nuremberg Trials. Paper was held to be more reliable than personal testimony, and documentation provided an acid test of the reliability of captives' testimonies. Once the medical investigators had their leads, they could extract evidence from the ever-growing pile of documents, and reconstruct the context and extent of the experimental programmes. In turn, this evidence meant that interrogations could be conducted with an informed authority.[54]

The US report on medical denazification prescribed assessing 'the degree of culpability of the individual physician or member of the auxiliary medical profession'.[55] The Report estimated that about half of German physicians were 'proven Nazis' (i.e. about 24,000 in all – a figure contrasting to the profession's later view of only 350 criminal doctors).[56] They should at first be kept in concentration camps; but most could be redeemed: 'It is not impossible that considerable numbers may be able to work their way back into acceptance as decent members of society and respectable members of their professions.'[57] The policy adopted was of selective trials for the worst culprits, and wholesale denazification to comprehensively screen the German medical profession.

Denazifying doctors was undermined by the need to maintain medical services, but even so, doctors were dismissed and fined. The case of the Rhineland-Palatinate is instructive. Here the French kept records on every physician, noting political activities under Nazism and the verdicts of the denazification tribunals. Activism in Nazi organisations generally meant dismissal from any state post. The case of Walter Christ in November 1946 exemplifies the tough measures. His instruments and practice were confiscated, and he was denied the right to practise medicine for five years and to participate in politics; he was authorised to open a private practice only in September 1949.[58] The Rhineland-Palatinate suggests that just as the Medical Trial began in Nuremberg, there was a mass dismissal of doctors.

There was a double standard, favouring scientists. The French were keen to retain the biochemist Butenandt, despite his Nazi sympathies suggesting that scientists were dealt with leniently.[59] The British and Americans were spurred on by a growing sense of rivalry with the Soviets in acquiring German medical technologies and personnel. Political calculations meant that at first some of the worst perpetrators of human experiments went undetected.

The German doctors' response varied from defiant suicide to denial and evasion. Leading Nazi doctors like Grawitz committed suicide in the death throes of the Third Reich in despair that their idyll had been shattered. Karl Astel, racial researcher and rector of the University of Jena, which had supported medical research in Buchenwald, committed suicide on 5 April 1945.[60] Herbert Linden, the euthanasia administrator and responsible for hospitals, ended his life on 27 April 1945, one of a spate of suicides by euthanasia doctors.[61] Suicide meant defiance of the Allied liberators. Maximinian de Crinis – the psychiatrist, euthanasia expert and ministerial adviser on medical appointments – shot himself in May 1945. The anatomist August Hirt, who organised the Jewish skeleton collection at the Reich University Strassburg, went into hiding in the Black Forest where he shot himself in June 1945.[62] The Auschwitz doctor, Hans Delmotte, shot himself when about to be captured.[63] Enno Lolling, the physician in charge of concentration camp medical services, committed suicide at Flensburg on 27 May 1945, and the Auschwitz camp doctor Eduard Wirths hanged himself after capture in September 1945. Lifton's in-depth analysis shows

Wirths was tormented by guilt and feared the disgrace of a trial.[64] Suicide expressed contempt for Allied promises of justice and a fair trial, and given Himmler's belief in reincarnation, a sense that it was time to move on to a new existence. While Lifton gives instances of suicides motivated by a wish to avoid the shame of being found guilty, it was not until the summer of 1945 that a trials policy was in place.

On 16 April 1945 US forces in Weimar arrested the SS bacteriologist Erwin Schuler, who had published the infamous paper commented on by *The Lancet* in 1943. The British landed a large haul of medical experts who fled to Holstein to escape the Soviets. Conti and Karl Brandt were prime targets for arrest, and it was only a matter of weeks before they were located in the Flensburg area, where the Waffen-SS held out. The British arrested Conti on 19 May 1945. Four days later they captured his arch-rival Karl Brandt as part of the operation to take Speer, who had befriended Brandt following his release from Hitler's custody arrest from 16 April to 3 May 1945. The rump Dönitz government was tolerated under Allied surveillance at Schloss Glücksberg until 23 May.[65] The British also captured the Waffen-SS medical commanders Gebhardt, Genzken and Joachim Mrugowsky, and Generalarzt Handloser in the Flensburg area. The circumstances of capture followed the factionalist fault-lines in the Nazi hierarchy: Brandt as Hitler's escort surgeon protested his distance from the SS, whereas Gebhardt, captured on 17 May 1945, was escort surgeon to his school friend 'Heini' Himmler, and was involved with the Reichsführer SS's final abortive flight.[66]

His hopes to exonerate himself were dashed by the arrival of the former Buchenwald camp doctor Hoven at the Freising camp. Wirths committed suicide in British custody,[67] and Conti in US custody at Nuremberg on 5 October 1945. Greater determination to prevent prison suicides would have meant that by the summer of 1945 the Allies would have been in an optimal position to mount a medical trial for human experiments, euthanasia and medical involvement in genocide.

What the Allies lacked was a master-list of war crimes suspects to apprehend leading medical miscreants. How Josef Mengele outwitted the Allies pointed to weaknesses in the war crimes machinery, notably underresourcing for operations to search and prosecute, gullibility in accepting the protestations of Mengele's family that he was dead, and the prioritising of strategic intelligence over justice. The Americans efficiently secured Nazi administrative records, but after the early successes of ALSOS were less effective in locating and arresting suspects. By contrast, the British had success in search-and-capture operations, but their legal procedures were too brisk to permit in-depth analysis of documents. Although the French were punitive against Nazi camp staff, they established cordial relations with German scientists so that their zone achieved the reputation of an 'Eldorado for National Socialists'.[68] Allied judicial efforts were hampered by the drive to secure German scientific expertise and intelligence on weapons of mass destruction.

It took months for intelligence to be pooled and acted on. The US Office of Strategic Services (OSS) drew up a 'List of German Doctors Buchenwald', which on 12 May 1945 was passed to the US War Crimes Office in Washington. A key informant, code-named 'Bruno', was the psychiatrist Theo Lang.[69] He defected to Switzerland in ca. 1941, and this founder of the League of National Socialist Physicians had been a convinced advocate of sterilisation of homosexuals. The Czechoslovak representative of the UNWCC received his report through the Czech Minister in Berne. Lang suggested that discussions on an unobtrusive method of mass sterilisation of racial undesirables by means of a drug or X-rays went back to 1937. The Czechs proposed approaching military authorities in Germany 'with a view to interrogating the persons indicated by Dr Lang and generally investigating the whole question'.[70] French diplomatic representatives in Switzerland raised the issue of tracing doctors involved in the criminal sterilisation of Jews, French nationals, Ukrainians and Poles.[71] Fortuitously, the lurch from sterilisation to euthanasia had occurred around the time that war began: Hitler backdated his euthanasia decree to 1 September 1939, entrusting the programme to Karl Brandt and to his Chancellery official, Bouhler. In 1941 the American journalist William Shirer speculated that the killings could be to test new poison gases and death rays, although he thought a radicalisation of eugenics the more likely reason.[72] Lang submitted a detailed statement on the murder of ill and aged in Germany, mentioning the role of the psychiatrists Hermann Pfannmüller and Werner Heyde, and of Conti; he passed this to the British secret service in Switzerland.[73]

Bruno denounced his former chief, the psychiatrist Ernst Rüdin, and the medical administrator and NSDAP activist, Kurt Blome.[74] An American report of 8 September 1945 ran: 'A Professor Rudin, a specialist in Nazi Race Hygiene, has been located in Munich. He is 73 and confined to home with heart trouble. Capt. Brinckerhoff advises that Rudin does not appear willing to talk but probably can be made to do so under sufficient pressure. If you feel that this subject is of sufficient interest to the case, I will request him to go ahead with a full interrogation.' The order came that 'Subject should be interrogated as suggested, but should not be transferred to Nuremberg until some estimate of his value to the case is determined.'[75] Rüdin was arrested on 20 December 1945, and indignantly pointed out that sterilisation was widespread in the United States.[76] He was released on 1 November 1946 when it was clear that the sterilisation laws were not to be part of the criminal prosecution at the Medical Trial.[77] When it came to the human geneticist and sponsor of Mengele, Otmar von Verschuer, he was under house arrest in the critical months of July and August 1946, when the Americans decided to mount the Trial.[78] The eugenicists received lighter treatment, as its criminality appeared less certain and of marginal value to the prosecution of the Nazi leadership at Nuremberg.

By August 1945 the drive to investigate racial and medical crimes had taken shape. The OSS analysed 'Principal Nazi Organisations Involved in the Commission of War Crimes, Nazi Racial and Health Policy'. The records and files of every *Gesundheitsamt* were to be seized, as the health offices were the focus of racial policy. Other targets included the files of eugenics organisations, notably the German Society for Race Hygiene, and the Institute of Hygiene Berlin, which had a key role in medicine in the occupied East.[79] The report prescribed rounding up leading scientists: 'Such individuals as Ernst Rüdin and Professor Rostock [arrested on 5 May 1945] will undoubtedly claim that they were interested in such programs merely as "scientists". There is no evidence that the scholars did not know of Nazi aims of eliminating racial or political undesirables.' A strategy of tackling not only the perpetrators but the bureaucratic go-betweens for Himmler and the scientists, the special experimental facilities and ideological background organisations began to unfold. Targets included the files of the Reich Chamber of Physicians and of the medical publisher J.F. Lehmann, who specialised in 'pan-germanist literature in scientific and pseudo-scientific literature on race'.[80] Operatives were to seize files on euthanasia, and 'investigate all orphan asylums, insane asylums, mental institutions, old age homes'.[81] The strategy involved collecting evidence from the victims: 'ex-inmates of concentration camps would provide abundant evidence concerning the medical practices of these institutions. Records of the medical Department of the concentration camps would also be helpful ... . Foreign workers would testify to the lack of any medical attention rendered them.'[82] Chilling details of human experiments began to emerge from Buchenwald and Dachau survivors in April 1945.[83] Another source were the interned academics rounded up from the Reich University Strassburg and Heidelberg (liberated on 30 March 1945). The information was crucial in shaping US policies on the handling of Nazi science and medicine.

Initially, the Allied investigators were more interested in scientific results than the experimental victims. ALSOS caught up with Schilling on 5 May, and concluded that here was a prime example of the caricature German scientist 'who pursues his legitimate scientific ends without thought of the ways he is pursuing them'.[84] CIOS reviewed Schilling's malaria inoculation research in Dachau on 7 June 1945. He insisted that 'his subjects were willing and cooperative', and loyal helpers countered an order from Berlin to destroy evidence. He attributed the deaths among his subjects to the appalling general conditions in the camp, and that 'in no instance was the disease a primary cause'. He overcame scruples over human experiments and set about developing an anti-malaria vaccine, after meeting Conti in 1941. Schilling viewed experimental chemotherapy as a battlefield with the doctor seeking to exterminate parasitic microbes. Chemotherapy was a military arsenal, and the physician and pharmaceutical manufacturers combined to develop therapeutic weapons. Aggressive experimentalism was linked to eugenics and racial notions of disease – the Nazi medical researchers viewed typhus as a 'Jewish disease', oblivious to how they caused it by savage measures of delousing and segregation. Schilling – who held no Nazi Party membership – typified how few doctors involved in human experiments and genocide had any sense of guilt.

The CIOS officers recommended Schilling receive assistance from a translator and that an expert be delegated to liaise with him as regards dissemination of his findings.[85] This positive view was echoed in July 1945 when the British Foreign Office ranked Schilling's research as 'technical intelligence': 'With the possibility of an Allied victory the German authorities ordered that the results of these investigations were to be destroyed. Fortunately, however, several papers were saved and further interrogation of Professor Schilling and closer study of his work, will probably disclose other interesting results.'[86] After interrogation at Nuremberg about his experiments, Schilling was held in solitary confinement from mid-August 1945 when he continued to write up his research. In September 1945 his papers were confiscated, and he was tried at Dachau. The scientific informant became a war criminal.[87]

German scientists presented themselves as victims of Nazism. They refused to acknowledge complicity in Nazi racial policies, and considered that every colleague interned by the Allies or placed on trial for scientific crimes had worse treatment than under the Nazis. The Allies continued a dual policy on German science – tempering suspicions of criminality with strategic priorities of military technology. The German scientists were potential assets against the Soviets. When the Americans withdrew from Saxony and Thuringia in June 1945, they evacuated hundreds of scientists to keep them from assisting the Russians.[88] The British and Americans launched programmes of wholesale transfer of German personnel to counter Soviet offers. The British began measures to revive German science. Such developments were set to disrupt ethical evaluation of German medical and scientific crimes.

4
Exploitation

Forming field teams

As Allied screening of German medical research intensified, what came to light was not the expected cornucopia of medical discoveries but an unfathomable mire of criminal atrocities. Scientific intelligence officers were shocked by how Germany 'succeeded in focusing every aspect of scientific activity, within the framework of a planned organisation, to waging war. She was perhaps the only nation who carried out the prostitution of science to this extremity, and these facts must be taken into account in any consideration of how best to deal with German research.'[1] The Allied Control authorities debated whether German science and medicine should be subjected to tight control or allowed a free rein to develop, providing that weapons research was not involved.[2] The House of Lords in May 1945 feared the prospect of German scientists waging a war of revenge with new secret weapons.[3] The problem was whether German medicine was so compromised by war crimes that stronger corrective measures were necessary.

Despite the realisation that Nazism was a corrupt ideology that had engendered immoral scientific practices, the Allies authorised the utilisation of German research findings. Control Council Law No. 25 specified either an institute would be earmarked for liquidation; or its work was categorised as being of strategic utility to the Allied war effort against Japan. The physiologist Ivy explained that: 'As our troops advanced into Germany, a number of trained scientists accompanied the advanced patrols to obtain information that might be of assistance in completing the war against the Japanese.'[4] German physicists were of special interest, because of the suspected atomic bomb research. The Allies also screened German biological and medical research for military applications.

The acronyms of the Allied investigative organisations, ALSOS (code-named after the ancient Greek for 'grove', as L.R. Groves was the commanding officer of the Manhattan atomic bomb project), CIOS (the Combined Intelligence

Operations Services) and its British counterpart BIOS, USSBS and the inter-allied FIAT, and the rhetoric of 'targets' and 'operations' convey the ethos of secrecy, military security and operational priorities. The United States Strategic Bombing Survey (USSBS) reconstructed the wartime role of some of the (later) Nuremberg defendants, as it assessed the effect of intensive bombing.[5] The objective was to exploit whatever could be scavenged from the caches of documents in caves and mines, as well as from academic installations for armaments, economic reparations, or for scientific and technological value. The field teams rounded up German scientists, inspected research institutes, accumulated documents and data, and inquired about research work. They hoped to uncover devastating secret weapons as well as fundamental medical discoveries.

As the Allied Control officers puzzled over the remains of a gargantuan German scientific establishment, they were dismayed at its decay and shocked by the criminality of Nazi medical research. In the course of these intelligence operations, some of the German medical scientists who at first were treated as informants became suspects as medical criminals; a wider circle of suspect scientists emerged as having information about criminal experiments. The initial expectations of scientific achievements under Nazism gave way to a view that the German medical elite had been involved in widespread criminality. The interrogators combined the hunt for useful weapons with a concern to locate evidence of unethical human experiments. The issue was central to interrogations, and the constant probings were met by denials. The captives sensed that any admission might amount to signing their death warrant.

The integrated Anglo-American command structure of SHAEF established a G-2 section, characterised by General Eisenhower as 'a small scientific intelligence section'; CIOS was in fact staffed by hundreds of scientific officers. It aimed to prevent German scientists engaging in weapons research and to acquire strategic intelligence by means of interrogations and reconnaissance missions. In May 1945 SHAEF weighed the issue of what should be done about the German laboratories and research institutes once the counter-intelligence programme ceased. It wished to know whether all German scientists and technologists should be detained in custody or placed under restriction, or whether only certain categories should be so treated and if so, what categories? The solution was to assess the capabilities of individual German scientists, as well as to procure high-quality industrial and scientific intelligence from Germany.[6] As G-2 sifted evidence on aviation medicine, and biological and chemical warfare, the future defendants at the Nuremberg Medical Trial came to the investigators' attention.

The medical branch of CIOS ran from the summer of 1944 until the summer of 1945. Its reports on 3,377 'targets' included medical installations, research and personnel. The parallel British organisation, BIOS, had

3,000 men who investigated 10,000 targets.[7] The US sent 12,000 investigators to comb through the debris of German science and technology installations. Some reports were perfunctory and naïve, and the German scientists evasive. But resourceful officers began to probe their politics, ideology and criminal activities.[8] The British and Americans found that they had the lion's share of the scientific spoils, because the Germans evacuated institutions westwards from Berlin and preferred to avoid surrender to the Soviets. The Military Medical Institute's toxicologists landed in Giessen, and the KWI for Biochemistry in the French zone at Tübingen. The Allies were faced with the problem of disentangling strategic from normal research, and animal experiments from human research.

The medical section of the Combined Intelligence Priorities Committee set the agenda for field units in July and August 1944, and the [British] Medical Research Council made recommendations.[9] Field teams, sent to Paris in August 1944, aimed to unravel the extent that French medical research had collaborated with the Germans on biological warfare, vaccine research and penicillin.[10] While the French had embarked on a complex set of pharmaceutical collaborations with the Germans, they did not wish to be passive channels of information for the Allies and started their own monitoring operation for German medicine.[11]

The British and Americans opened an internment camp, code-named 'Dustbin', in 1944 at Chesnay near Versailles for captured German scientists. The CIOS programme was taken over by FIAT (Field Information Agency Technical) during 1945, and 'Dustbin' moved to Schloss Kransberg (a former headquarters of Göring, refurbished by Speer). Kransberg was near the FIAT base in the chemical and pharmaceutical factories of Hoechst, a part of the IG-Farben conglomerate at Wiesbaden.[12] 'Dustbin' was a well-equipped interrogation centre. FIAT transferred Reich Research Council documents there on capture in April 1945. FIAT officers detained 'parties' of scientists specialising in atomic research and poison gas for interrogation, and set as 'homework' the task of reporting on their wartime activities.[13] The British intelligence staff who ran 'Dustbin' until August 1946 scrutinised Nazi scientists, and set out to reconstruct the Nazi system of medicine, and the extent that it was geared to military needs.[14]

'Dustbin' was part of a web of Allied interrogation centres. The German political leaders were first held at 'Ashcan' (actually the Palace Hotel of Mondorf in Luxembourg), where Karl Brandt was transferred along with Göring.[15] German medical experts were interrogated at Beltane school near Wimbledon from August to October 1945.[16] For medicine, there was nothing as focused as the Farm Hall interrogation centre for atomic scientists, where the British secretly recorded inmates' stunned reactions to the news of the dropping of the bomb on Hiroshima on 6 August 1945.[17] But surveillance was intensive, as Allied scientific intelligence puzzled over why the Germans had failed to deploy poison and nerve gas, or attempted germ warfare, and whether this was due to scientific incompetence, strategic decisions, the understanding of operational problems, or moral scruples. They were astonished when they stumbled on the devastating nerve agents Tabun, Sarin and Soman. The Germans pretended these were mere analgesics.[18] The Americans secretly recorded conversations of the German malariologist, Ernst Rodenwaldt, when he referred to experiments on malaria control, and the fondly held dream of German settlement of the colonies.[19] Interrogators set out to determine the extent that human experiments were part of programmes to produce biological and chemical weapons of mass destruction.

The German scientists became objects of intense suspicion and curiosity. Hundreds were interned and interrogated. An Allied directive of 13 April 1945 commanded the arrest of all professors in military service appointed since 1941. Overall, the Allied investigations had large manpower, but lacked co-ordination and a concerted policy on what to do about German medical abuses. An inmate of 'Dustbin', the physiologist Johannes Ranke, was set 'homework' of writing up his research at the Physiological Institute of the Military Medical Academy. He was alert to how the Allied medical interrogators were looking for evidence of criminal experiments, and his report was a careful exercise in self-exoneration. He affirmed that when making human experiments he secured agreement of the individual subject, and made animal tests only when there were dangers (as for carbon monoxide poisoning). 'I did not initiate or permit the initiation of any class of experiment that might have the slightest detrimental effect on individual well-being until I had satisfied myself by personal tests of the harmlessness of the reactions and very likely it was on this account that no serious incidents occurred.'[20] What this reveals is that by late 1945 a set of ethical protocols had emerged, although whether these really governed wartime science was highly dubious.

The military investigators had an arsenal of powers. Alexander had 'full powers to investigate everything of interest, and to remove documents, equipment or personnel if deemed necessary'.[21] But all this was for evaluation of German medical achievements rather than for war crimes purposes. Military priorities meant that civilian atrocities ranked as secondary to crimes against service personnel. In effect, this continued the wartime priority of concern for Allied POWs over civilian detainees and Holocaust victims. Criteria to identify war crimes – or at least a liaison mechanism with war crimes authorities – were lacking and much depended on the expertise and acumen of the field officers.

In comparison to the scientific intelligence operations, war crimes units were far fewer. They faced problems of under-resourcing, the legal quandary of defining war crimes, and of comprehending the magnitude of the atrocities and identifying perpetrators while operating in the unfamiliar territory of a shattered nation. The Americans preferred to search for, retrieve and analyse documents. After some successful search and arrest operations of Nazi criminals, the British favoured using military police; even though lacking linguistic competence, they worked effectively. Part of the skill was to select reliable informants.

The former Ravensbrück prisoner, Carmen Mory, became a British informant. She was a Swiss doctor's daughter with a taste for intrigue, dalliance and the role of *femme fatale*. The Germans and French had suspected her of being a double agent since the outbreak of war.[22] Mory assisted the British Field Security Unit 1001 with the arrest of Herta Oberheuser on 20 July 1945.[23] She provided background information on the surgeon Gebhardt, unearthed his assistant Fritz Fischer, who was arrested on 3 August, and resolved the confusion over the arrest of his namesake, the historian.[24] Mory emerged as a compromised informant, because she had been befriended by the camp doctor and human vivisector, Percival Treite. The British arrested her in November 1945, accusing her of killing mentally ill prisoners. She was dramatically convicted on the basis of her defence evidence at the Ravensbrück Trial, when a co-prisoner thanked her for the gift of a fur coat, which had – as it emerged – belonged to another prisoner, whom Mory kicked to death.[25] Mory and Treite cheated the gallows by suicide on the night of 8–9 April 1947.[26]

The Mory case shows that the informants might be compromised. Further problems were that investigation teams were constantly changing; civilian life beckoned, and post-war financial retrenchment meant that war crimes investigations were selective and time-constrained. Above all, the priorities of the emerging Cold War and political stabilisation undermined the hunt for culprits, as the Allies began to recruit German scientists for strategic research. Yet, despite all these bad omens, recognition of medical war crimes crystallised in the autumn of 1945.

ALSOS: the Haagen affair

The quest for strategic science led to revelations of German human experiments. In the final months of the war, the Allies feared that the Germans were preparing a last, desperate counter-attack using an atom bomb or germ warfare. The small, energetic and highly expert American ALSOS organisation confirmed that the physicist Werner Heisenberg had not completed work on a nuclear reactor, so that the suspected bomb project remained fragmentary.[27] ALSOS tracked down researchers suspected of involvement in biological warfare (BW), not least because of the fear that the Germans might deploy these. The scientific targets included Butenandt and Kuhn, as well as the hygiene expert Rodenwaldt. The ALSOS team reported that: 'prior to the extension of field activities of the Mission to Germany a target list was drawn up which included a wide variety of sites of potential BW research. It was assumed that such potential sites might be located in any of the Bacteriologic and Hygiene Institutes of the German Universities, in the biologic institutes of the Kaiser Wilhelm Gesellschaft, in the pertinent laboratories of commercial firms producing biologicals, etc.' Virological research was especially suspect.[28]

At first, information was fragmented and unreliable. The British gained information from Gauchard (a manufacturer of sprays and atomisers) in France that Heinrich Kliewe, professor of hygiene at Giessen, was a key figure in German BW. The ALSOS target list included university institutes for hygiene and bacteriology, biological institutes of the KWG, and veterinary and commercial biological laboratories. ALSOS inspected four concentration camps to assess the biological research. It located Kliewe at Tutzing in Bavaria, and interrogated him on 13 May 1945. He stressed the defensive nature of German BW since 1941 following Hitler's orders. ALSOS took his files to the US Army Documents Center at Heidelberg for evaluation. Other documents seized by ALSOS revealed medical crimes. These included the *Reichsforschungsrat* 'Correspondence from SS-Anthropological Institute', and 'with Dr Rascher concerning obtaining equipment' for the freezing experiments.

ALSOS discovered letters from Haagen to Rose concerning vaccine tests in 1943.[29] Rose (now quiet about his malaria ward at Pfafferode) fell into US hands on 25 May 1945. His medical research for the Luftwaffe was ostensibly concerned with aerial spraying and vaccines. He typifies the trajectory of being interrogated initially for his knowledge of germ warfare and for military medical research. Only later – and much to his surprise and indignation – did his casual mention of deliberate infection for the purpose of experimental study of disease transmission and therapy attract the interest of war crimes investigators. Rose was articulate, scientifically well-informed, critical of various aspects of German wartime organisation and massively opinionated about his colleagues (he was in dispute with the director of the Robert Koch Institute), and the failings of German epidemic control. He was enraged at the iniquity of his detention by the Allies. He was a member of the secret BW Committee, code-named *Blitzableiter* (i.e. lightning conductor), but said that he was critical of BW on grounds of its impracticality. Rose's change in status from informant to suspect exemplifies the transition from intelligence gathering to war crimes prosecutions.

The problem was to fit haphazard documents into a bigger picture of wartime research. The historian encounters the experiments at different points: firstly, through surviving evidence from the time of the execution of the experiments; secondly, through the eyes of the Allied investigators; thirdly, through the courtroom discussions and survivors' testimonies; and fourthly, through the presentation of the experiments by other historians. Some of the worst atrocities were the Dachau freezing and pressure experiments, conducted by Rascher, a young Luftwaffe and SS medical officer. Rascher secured the support of Himmler for experiments in Dachau. Himmler wanted to prevent Waffen-SS troops in Russia from freezing. He ordered Rascher to experiment further on dry cold and to draw up recommendations for field action of re-warming.[30] Rascher resented working with academic physiologists, and to evade their controls wished to conduct dry cold experiments in Auschwitz or Lublin, as well as loathing his Luftwaffe scientific superiors.[31] Himmler ordered the SS Ahnenerbe Research Organisation to support Rascher, and Sievers was helpful. Hitherto historians have viewed these deadly experiments as 'pseudo-science' (as they were characterised by Wolfgang Benz and Michael Kater). Himmler demanded homeopathic tests on digitalis and other plant extracts.[32] The procedures were as rigorously experimental as for other types of deadly physiological tests. Evaluating them as inhumane research supported by a scientific rationale, allows one to establish the extent to which the experiments found academic support.

Rascher approached Himmler to request professional criminals or imbeciles for potentially lethal high-altitude experiments. Himmler (swayed by Rascher's scheming wife Nini) approved the Dachau experiments on low pressure on 15 May 1941, but they took place nearly a year later, between March and April 1942.[33] Although Rascher pointed out the pressing needs of air combat, it took several months to start the experiments. The delay suggests that they met not with overt resistance but non-co-operation. The dean of the Munich medical faculty endorsed the view of the University's aviation medicine specialist, Weltz, that they could not provide the elaborate equipment to conduct the experiments. Weltz (born in 1889, and no SS member) attempted to exclude the inexperienced Rascher (born in 1909) from the experiments, but the SS insisted on his participation. The SS found it opportune for specialist German aviation research groups to supply equipment and expertise for the pressure experiments.

The follow-up series of wet cold experiments (from August to October 1942 in Dachau) and dry cold experiments (from February to April 1943 in Dachau) were initiated by physiologists. Himmler insisted on Rascher's participation. The experiments caused similar problems of resourcing, the procuring of equipment for measuring body temperature by inserting thermopiles in the stomach and rectum, respiration and heart action, and analysis of blood, urine and spinal fluid.[34] The Deutsche Forschungsgemeinschaft (DFG) declined to supply apparatus on 9 October 1942.[35] The SS therefore requisitioned elaborate apparatus for the wet cold experiments from the Dutch universities of Leyden and Utrecht.[36] The dry cold experiments were in collaboration with the military Mountain Academy at St Johann in the Tyrol.

The planning, support and reception of the experiments aroused a complex set of responses. Rascher boasted how the experiments involved autopsies to study internal organs while they were still functioning. Himmler exulted in a gruesome Agfa-produced film of the aviation experiments.[37] Air

Field Marshal Milch absented himself, although 30–40 experts attended.[38] The Chief Luftwaffe medical officer, Erich Hippke emerges as ambivalent, privately denouncing the experiments as amoral, but dutifully thanking the Reichsführer for their significance. (This was ironic, given the later prosecution of Milch and non-prosecution of Hippke.)

Rascher had to accept collaboration with other researchers, because of the complex physiological issues, equipment and measuring techniques. But he proudly informed Himmler that only he could obtain the 'extreme results' required, and this had to be done when the conventional physiologists, Holzlöhner and Romberg, were absent. He condemned Romberg for taking too indulgent a view of the experimental subjects.[39] Rascher's reports to Himmler and Sievers derided senior physiologists and Luftwaffe officers for their Christian morality and their 'Jewish' attitudes. But the security service reports showed Holzlöhner to be an active supporter of nazifying Kiel university.[40] When Rascher denounced a Danish expert in Eskimo research as an uncooperative Jew, the SS administration under Werner Best proved that the accusation was groundless.[41] That Romberg was awarded a war service medal by the SS in October 1942 suggests that his collaboration with 'comrade Rascher' was closer than either Rascher or his Nuremberg defence conceded.[42]

Rascher's reports to Himmler, Rudolf Brandt and Milch aroused the jealousy of SS researchers and physicians. The surgeon Karl Gebhardt attacked Rascher for contravening the medical hierarchy within the SS, as did the Reichsarzt SS Grawitz and his deputy Poppendick.[43] Gebhardt intimidated Rascher, while arguing that he ought to have a university assistant's post. Rascher appealed to Sievers for protection against the jealous ranks of academics. ALSOS soon alighted on Rascher, documenting his dual academic and SS career, and how the Raschers found resolute support from Sievers and Himmler's secretary, Rudolf Brandt. Sievers enjoyed the hospitality of the Raschers on repeated visits to Dachau from 1942, and reciprocated with gifts for the Raschers' growing brood of children. The SS supported Rascher's habilitation research and links to Ruff at the aviation research laboratories at Berlin-Adlershof.[44] The Ahnenerbe established an Institute for Military Science, which sponsored researches by Rascher on wet and dry cold, May on insect control, Haagen on typhus vaccines, and Hirt on mustard gas and Jewish skeletons. Rascher reported how, in the first half of 1943, he carried out 309 experiments, and made 2,855 measurements on the effects of freezing water, cold air, and for re-warming, stomach, insulin and hot water experiments.[45] Rascher's quest to gain a Habilitation revealed that he was unacceptable even to academics, who were SS sympathisers. Hippke, the chief medical officer of the Luftwaffe, was reluctant to lose Rascher, writing effusively to Himmler on Rascher's positive qualities.

The matter came to a head when Rascher wished to report on the experiments at the Nuremberg meeting of 26–27 October 1942 on survival at sea and in extreme cold for military medical officers. The meeting was designed to operationalise 'the new experiences' on wet and dry cold. Holzlöhner delivered the paper on the freezing experiments, and referred to the collaboration of Rascher and Finke in investigating humans. Rascher immediately waded in, making it crystal clear that experiments with multiple types of measurements and controls were involved.[46] Rascher complained to Himmler that Holzlöhner and Finke wished to marginalise him and downplay the significance of the results. Rascher characterised them as scared that the human experiments would ruin their scientific reputation. Although they recognised the gains of the experiments, they shied away from taking responsibility for having carried them out.[47] Heinz von Diringshofen of the Office for the Examination of Fliers expressed admiration and wished to visit Dachau.[48] Roth shows that the physiologists repeatedly tried to involve Rascher and establish a dialogue with him, only to be rebuffed.[49] Ruff developed an ejector seat, based on the discovery made in Dachau of a few vital seconds of consciousness before rapid descent from heights above 15 km.[50]

Once the first revelations on the human experiments and skeleton collection at Strasbourg reached the Allied press, the German scientists were alarmed that they could be arrested not as POWs but as war criminals. Hirt – who had fled to Tübingen and then into hiding in the Black Forest – drafted a robust defence refuting the allegation that he killed to order in order to build up the skeleton collection. He defiantly asserted that he obtained the corpses as part of routine transfers of executed prisoners to his institute for dissection.[51] This defence – that the Allies were making false allegations of mass killings when only normal research was involved – was to be used at Nuremberg. The German Foreign Office issued commands to trace, select and destroy documents on bacteriological research, chemical warfare and aviation medicine. Hörlein, director of pharmaceutical research at IG Farben, gave orders to destroy all documents ranked as secret. Major collections concerning key medical institutes at Berlin, Munich and Posen could not be traced, so impeding the war crimes investigations. To compensate, the Allies found not only NSDAP and SS records, but also Himmler's intact correspondence.

Allied medical intelligencers set out to establish what therapies, medicines and vaccines the Germans used, and how and where they were produced. The ALSOS sweep of leaders of research organisations revealed much about Nazi medical research and the role of the *Reichsforschungsrat* and SS-Ahnenerbe, the Ancestral Heritage research organisation. The Ahnenerbe 'Reich Business Führer', Sievers was questioned at 'Dustbin' between 1 and 17 August 1945. The UNWCC circulated the results of Sievers' interrogation, as this revealed how the SS Ancestral Research organisation supported a vast range of medical projects. Sievers' diary, along with Himmler's papers, turned out to be major resources for unravelling the command structure for medical crimes. This prompted ALSOS to investigate Ahnenerbe medical and biological research at the Dachau concentration camp.[52] ALSOS produced a series of memoirs on 'Physiological Experiments on Human Subjects' covering incidents at Strasbourg and Theresienstadt.[53]

The investigation of the bacteriologist Haagen and associates at Strasbourg illustrates the problems of inter-Allied and inter-agency co-ordination when dealing with a vain and unrepentant scientist. The ALSOS mission benefited from the strategic skills of Boris Pash, who got the investigating scientists to the right place at the right time.[54] The mission included British officers, who gathered scientific intelligence on German biological warfare from November 1944 to June 1945.[55] Its investigators combed through the scientific installations in liberated Paris. The nazified Reich University Strassburg was a target for ALSOS investigations as a centre of research in theoretical physics and medicine. Goudsmit, the scientific director of ALSOS, commented, 'Only the politically absolutely safe, or trusted Nazis were selected for this important propaganda outpost'.[56] After entering Strasbourg on 24 November 1944, ALSOS BW experts seized Haagen's stocks of germ cultures, and setting up headquarters in Haagen's apartment, secured his papers. His experimental facilities and nearby military installations were disguised as an 'Electro-technical institute'. The ALSOS mission first interrogated Haagen on 21 April 1945 and arrested him at a hospital in Saalfeld in Thuringia in June 1945 as part of its sweep of German BW experts. He was promptly interrogated concerning vaccines and pharmacological research, notably on penicillin. But the interrogators were unaware of his typhus vaccine experiments on special transports of Roma prisoners, sent from Auschwitz to Natzweiler.[57]

Haagen was well known to American bacteriologists, as he had worked in New York from 1928 until 1934. The Reich Health Office had seconded him for training in cell culture and virus research at the Rockefeller Institute for Medical Research. Initially, the Institute declined to offer bench space, but the persistent Haagen stressed his request's official backing. Thomas Rivers the virologist (whose name later came up as a potential adviser to the court at Nuremberg) overcame his dislike of Haagen, and conceded that he was hardworking, capable – and ambitious. They joined forces to tackle the problem of immunity to yellow fever.[58] From January to December 1933 Haagen was a special member of the International Health Division of the Rockefeller Foundation. Here, he was the first to culture the yellow fever virus, and experimented on vaccination and virus attenuation. Haagen's work for the Rockefeller Foundation was brought to the Tribunal's attention on 17 April 1947.[59]

Haagen was a convinced racist and Nazi. He used blood samples from 'colored patients' in various American hospitals.[60] He became involved in the pro-Nazi Bund while in America.[61] In August 1933 he told his American hosts how Hitler's Berlin was 'in a much better condition', and he marvelled at the achievements of the Nazis.[62] By 1934 he was back at the nazified Reich Health Office, praising Hitler's cures for 'the new Germany', and planning to continue yellow fever studies, which found their culmination in human experiments at Natzweiler.[63] Haagen tried to use his pre-war contacts to resume his research and exonerate him from charges of war crimes.

It turned out that Haagen (who from 1941 was professor of hygiene at the Reich University Strassburg) conducted reckless human experiments on hepatitis and typhus vaccines using live cultures.[64] His connections to the SS arose through the professor of anatomy Hirt, who was responsible for the killing of 115 victims brought from Auschwitz for the Jewish skeleton collection. Captured documents indicated how he had the support of Wolfram Sievers, who since 1935 had been administrative Secretary of the SS-Ahnenerbe research organisation. The Americans concluded that Haagen was deeply involved in the atrocities in Strasbourg and in the nearby concentration camp of Natzweiler/Struthof.[65]

Himmler approved of Haagen's research requiring hundreds of experimental subjects to be infected, and then vaccinated with experimental live typhus vaccine. He ordered that Oswald Pohl (the head of the economic office of the SS) and Enno Lolling (medical inspector of the concentration camps in Pohl's office) supply the experimental subjects. Rather than take French prisoners, who were regarded as valuable slave labourers, they arranged for 100 Roma from Auschwitz to be sent to Natzweiler. Haagen expressed dissatisfaction at their physical condition: they were starving so they could not match soldiers physically. Having been condemned as useless for research, they were killed. Haagen ordered new batches of 100 and then 200 experimental subjects. But to ALSOS he admitted to experiments on only 80 subjects and denied that there were any deaths.[66] ALSOS could not find evidence of Haagen's involvement in biological warfare, although they remained convinced that he was a leading force in this type of work. ALSOS discovered evidence concerning links between Haagen and Gerhard Rose concerning the testing of a Danish typhus vaccine.[67] These documents were sent to the Nuremberg prosecutors.[68]

French intelligence found that ALSOS officers in their haste had overlooked documents. Grombacher explained that ALSOS had seized the accounts and correspondence of the Institute:

> Nevertheless I undertook a new investigation which met with success. I found some documents, some in a drawer in the secretary's desk and some in the refrigerator in the basement. These documents appeared to me sufficiently conclusive as to constitute a record of war crimes.[69]

Although the documents were sent to Washington, the US authorities ingenuously released Haagen on 15 June 1946, permitting his return to Saalfeld in Thuringia, which was now in the Soviet zone. His release – just before medical suspects were evaluated for the Medical Trial at Nuremberg – showed the lack of co-ordination between agencies. His entrapment and rearrest by the British was spurred on by concern that Haagen should not remain in Russian hands.[70]

The Nuremberg prosecutors were dissatisfied with what had been discovered about the wartime activities of Haagen. They interrogated Sievers in August 1946 about Haagen's research, but Sievers claimed that the vaccine research was funded by the military and the Reich Research Council rather than by the SS. The contacts arose through the chemical weapons toxicologist Wolfgang Wirth (rather than the murderous Hirt). Sievers claimed that all the experimental subjects had volunteered in line with Himmler's stipulations that prisoners would receive a pardon, and no deaths resulted from Haagen's experimental typhus research in Natzweiler.[71] Sievers had been incarcerated with Ding/Schuler, the Buchenwald typhus researcher, who had informed him of the immense extent of experimentation on the disease.[72] Sievers here covered up his involvement with murderous research.

When ALSOS seized the papers of the former head of the Strasbourg Policlinic, Otto Bickenbach, on 9 June 1945, he was ranked as a convinced Nazi activist and anti-Semite, as well as a highly capable research scientist.[73] ALSOS failed to link him to human experiments at the Natzweiler concentration camp. Bickenbach petitioned that he had worked to serve the health of his fellow human beings, and arrangements were made for his release on 10 August 1946 just as preparations for the Medical Trial began.[74] Bickenbach was not yet identified as part of the chemical warfare researchers at Strasbourg, and the ALSOS report on Natzweiler overlooked his experiments with Haagen.[75] His case exemplifies both the limitations of ALSOS, and lack of co-ordination among the Allies.

The interrogators of the British FIAT, Scientific and Technical Branch, Majors Gill and Kingscote, noted on 21 August 1945 that Karl Brandt 'directed the extremely suspicious researches of a Prof. Otto Bickenbach at a prisoners' camp at Strasburg'.[76] This evidence was at first applied to Karl Brandt rather than to Bickenbach. Only on 21 November 1946 did the American prosecutors visit Strasbourg, and find that Bickenbach was part of a circle conducting human experiments at Natzweiler on the effects of poison gas.

British intelligence lured Haagen into West Berlin on the pretence that his sister was ill, and arrested him on 16 November 1946.[77] Although this was just a few weeks too late for his inclusion in the Medical Trial as defendant, he was kept in isolation, then handed over first to the French and then to the Americans as a potential defendant in a second medical trial.[78] Haagen was transferred back to the French on 14 August 1947 when the Medical Trial proceedings were at an end.[79]

On 1 October 1946 Haagen went to work at an institute for virus and tumour research under the Soviets at the former KWI at Berlin-Buch, which (as he pointed out when again in US custody) had been built with funds from the Rockefeller Foundation. His defence lawyer unsuccessfully approached the Rockefeller Foundation for a testimonial as to his scientific importance. But the Foundation officers were disgusted at what was emerging about wartime German science.[80]

Haagen's work was supported by the German Academy of Sciences in the Russian zone.[81] He later claimed that his re-arrest was prompted by his collaboration with the Soviet military, as his new institute was suspected as being a germ warfare centre. Somerhough, the energetic head of the British war crimes office, had decided to trap Haagen and offer him to the surprised French on the basis that the Strasbourg police issued an arrest order in 1945.[82] Haagen protested that he had been illegally 'kidnapped', and that the British threatened to kill him when transferring him through the Soviet zone.[83] The German defence lawyers accused the Allies of deliberately concealing Haagen from them in French custody, so that he could not be called as a defence witness for his fellow Luftwaffe bacteriologist Rose.

Haagen pulled legal strings in the United States after his transfer, as American lawyers pressed the Nuremberg authorities to make documents available to Haagen's defence. They believed that these showed that there were no fatalities from Haagen's experiments.[84] Haagen insisted on his status as a scientist, when interrogated by Alexander and Natzweiler survivors in July 1947: '"What I have done, I did for the sake of humanity," he said, but Mr Hulst interrupted him and said that experiments against the will of the patient are plain vivesection. Dr. Haagen did not agree.'[85] Haagen – and Bickenbach – were tried by a French military tribunal in Metz in December 1952.[86] The French were reluctant to impose a long sentence on this recalcitrant and unrepentant scientist, who considered that the victims should be grateful for the immunity conferred by his experiments, and that he should have received a Nobel Prize for yellow fever vaccine research rather than a custodial sentence.[87] Haagen soon returned to virus research in West Germany. He escaped lightly given the Natzweiler deaths. The Haagen case exemplifies the pitfalls and successes of Allied war crimes investigations.

ALSOS captured a major installation at Geraberg in Thuringia in April 1945, where equipment from Nesselstedt near Posen had been evacuated in mid-January. This had been planned as a 'dual-use' institute combining cancer and biological warfare research.[88] In July 1945 the Deputy Reich Physicians' Führer, Blome, declared that he disapproved of the way advances in medical science were used for atrocities like mass sterilisation and gassing Jews.[89] Along with his wife Bettina, he provided information about the intended BW researches on plague by an Austrian SS officer, Karl Josef Gross. Joachim Mrugowsky, a leading SS bacteriologist – also interviewed by ALSOS – seconded Gross from the mammoth Hygiene Institute of the Waffen-SS. Blome (as an SA member) distrusted Gross: he believed that the SS used Gross (who later mysteriously disappeared) as an informer.[90] Blome was Director of the Central Institute for Cancer Research, which was massively funded by the German Research Fund section for medical research.[91] The authorisation came from the Berlin surgeon Sauerbruch, who had from 1923 to 1928 run the Munich surgical clinic where Gebhardt and Karl Brandt had trained.

The ostensibly defensive German germ warfare programme emerged as fragmented and disorganised.[92] Telschow, chief administrator of the KWG, was arrested in Göttingen on 10 April 1945. His interrogation led to the capture of the scientific administrator and the head of the planning bureau of the *Reichsforschungsrat*, Werner Osenberg, in June 1945.[93] His bulky files, brimming with details of wartime medical and scientific research, were at first kept intact at the original location, while he was extensively interrogated in Dustbin.[94] The interrogator ranked him as a 'pseudo-scientist' and his group as consisting of 'common mischief-makers or even gangsters'.[95] Disruption followed when Supreme Headquarters insisted on moving this crucial collection on Nazi research to Paris. It emerged that in July 1943 Göring, as President of the Reich Research Council, had given Osenberg the task of co-ordinating wartime scientific and medical research. Osenberg had the confidence of Himmler, and occupied a strategically key position between basic research, industry, the military and the SS.[96]

Osenberg surrendered files on 10,000 scientists in 1,400 institutes to the Americans, who found documents on extensive human experiments funded by the Reich Research Council.[97] A document of 4 October 1943 listed three projects for research by Rascher, all on the topic of re-warming after exposure to severe cold. The other project of the same date was for the Strasbourg anatomist Hirt to undertake research on the physiological effects of chemical weapons.[98] Allied interrogators reconstructed the German networks of military medical research involving human experiments. Osenberg's interrogators probed the activities of the organic chemist Richard Kuhn, who claims that the Gestapo had forced him to decline a Nobel Prize in 1939, and of the biochemist and fellow Nobel laureate Butenandt. Neither scientist joined the NSDAP.[99] British BW experts attached to ALSOS found Kuhn as suspiciously nervous with a 'nasty rat-like expression', concluding he was a 'good chemist and a good Nazi'. It emerged that Kuhn had developed a deadly nerve gas, Soman.[100] The German scientific elite saw interrogations as an opportunity to protest their innocence, and blamed incompetent functionaries of the NSDAP and SS. A FIAT report of late July 1945 saw a fundamental divide between valid pure research and the SS researchers who engaged on pseudo-scientific experiments at Dachau.[101]

Osenberg identified Hitler's surgeon in attendance Karl Brandt, the surgeon Paul Rostock and Kurt Blome as leading figures in BW. ALSOS swooped on Blome's protective wife in Geraberg, and were the first to recognise Blome's significance.[102] ALSOS interrogated Mrugowsky, and Wolfram Sievers,[103] who were to join the Medical Trial defendants, along with Handloser. The Giessen disinfection expert Heinrich Kliewe was a key informant on the German BW programme and never prosecuted.[104] Karl Brandt was first interrogated by ALSOS at Ashcan for information on Hitler's doctors,[105] and then held in 'Dustbin' from 8 July, where he outlined his efforts in co-ordinating German public health, and that he resisted the extremes of Nazi ideology.[106] The FIAT interrogators in August 1945 were sceptical of Brandt's allegation of Himmler's animosity and of his protestations of ignorance of SS medical research.[107] BIOS interrogators noted how, when asked about his concentration camp activities, he answered 'tersely and negatively'. His interrogators felt that they needed to locate 'definite evidence of complicity' in human experiments to destroy his denials.[108]

Once the Allies had convinced themselves that the Germans were not about to unleash a deadly weapon, the priority was to exploit German discoveries for the war in the Far East. Colonel Childs of the British Ministry of Supply investigated German biological and chemical warfare, particularly the deadly nerve gases Sarin and Tabun, originally developed as insecticides, which affected the nervous system. These were developed by IG Farben at Leverkusen and at a special plant in Silesia. British chemical experts were promptly on the scene when Kurt Gerstein (a delousing specialist from Mrugowsky's Waffen-SS Hygiene Institute) began to speak about his role in the use of poison gas to kill the Jews. What interested the British was whether he had information about chemical weapons experiments. Gerstein ended his life in a French prison cell, traumatised by his arrest and by his bearing witness to the gassing of millions.[109] The Allies continued to be more interested in German scientific weapons technology than the Holocaust.

In September 1945 a BIOS team investigating chemical warfare scrutinised German chemists and bacteriologists. The German scientists mutated from being colleagues and potential allies in any future conflict against Bolshevism to perpetrators of 'war crimes experiments' on human subjects. The British attempted to prove the criminality of the toxicologist Wirth, the bacteriologist Gross and the IG Farben research director Hörlein and the truculent pharmacologist Fritz Hildebrandt, who conducted digitalis experiments for the SS, who falsely denied research on nerve gas and demanded release on the basis of being indispensable for the nation's health care.[110] Five (Karl Brandt, Blome, Rostock, Mrugowsky and Handloser) of the 20 doctors eventually prosecuted at the Medical Trial were assembled in 'Dustbin', because of their suspected involvement in chemical and germ warfare.[111] This shows the importance of the scientific intelligence operations, which exposed as false the denials of research on chemical and biological weapons. But it leaves the problem of why others who conducted human experiments on prisoners went unprosecuted. Wirth knew of concentration camp experiments, but there was no proof that he was involved in them.[112] Moreover, there could be extenuating circumstances. Under the Nazis the dissident chemist Robert Havemann had been shielded from execution by the pharmacologists Heubner and Wirth, although the execution of another radical, Georg Groscurth, could not be prevented by his work in a military installation for chemical experiments.[113]

The United States Strategic Bombing Survey set out to assess the impact of the bombing on the civilian and military health services.[114] In July 1945 it visited 'Dustbin' to interview Karl Brandt, Handloser and Schröder to see how German medicine had responded to the severe disruption of the bombing from 1943. The three German doctors told of an organisational battle to overcome the obstructive Conti, and how the SS and Waffen-SS under Genzken refused to accept any provisioning arrangement. Handloser and Schröder stressed the positive achievements of the *Aktion Brandt* (an emergency hospital programme, which controversially has been interpreted as a euthanasia operation), because of the introduction of a new generation of functionally designed hospitals.[115] The priorities of therapy, the constructive interaction with confessional hospitals by leading surgeons like Brandt, Gebhardt and Rostock, and efforts to conserve and distribute clinical supplies were to be a cornerstone of the defence at Nuremberg.

The Allied scientists searched for evidence on the military use of poison gas. Chemical warfare experts scrutinised the gas chambers of the Holocaust for information on the toxicity of the different gases. British chemical warfare experts from Porton Down puzzled over a putative Dachau gas chamber in March 1946. While Zyklon – a well-known disinfectant, in common use for insect eradication from the mid-1920s – did not arouse concern among the British during the war, after the war there was interest in its neurological effects. D.C. Evans, the British officer who had located the SS medical officer, engineer and informant on the Nazi gas chambers Gerstein, had a dual mission of procuring data for the Nuremberg prosecutions and obtaining details of the German experiments for the British.[116] He reported in January 1948: 'Considerable time was spent in trying to discover reliable data on which to calculate what concentrations of HCN were used for the extermination of victims. The hunt was on for evidence of experiments with nerve gases so that such chemical weapons could be developed.'[117] By the time of the Medical Trial, the suspicion grew that the poison gas experiments were pilot studies for the gas chambers of the extermination camps. While it was conclusively shown that the experiments were part of systematic racial killings, the chain of evidence linking the poison gas experiments by the German army with the gas chambers of Auschwitz was never conclusively established.[118]

The interrogators were more interested in application than misapplication of medical expertise – in the medical underpinnings of biological warfare than in human experiments and genocide. The head of the Hygiene

Institute of the Waffen-SS, Joachim Mrugowsky, found that his network of institutes (particularly those sited at Buchenwald under Ding/Schuler and Auschwitz) were objects of intense suspicion, and the G-2 section of SHAEF earmarked him for interrogation in Dustbin.[119] The interrogators reconstructed the labyrinthine military and SS research initiatives. The Nazis were masters of camouflaging atrocities under apparently benign medical organisations. Blome outlined relations between the Reich Chamber of Physicians, medical experiments and research. He opportunely shifted the blame for ruthless experiments onto Conti as a key figure in assigning research tasks. Blome became plenipotentiary for cancer research in 1943, due to an initiative of Göring, in part as a cover for BW research.[120] The siting of research installations in the concentration camps drew the Allied scientific intelligence officers into a criminal world of murderous human experiments and genocide.

Survivors' testimonies

Liberated prisoners contributed sober, objective accounts of medical abuses in the concentration camps to medical journals, and meetings. Survivors presented their evidence out of a sense that their experiences should contribute to medical science. Primo Levi co-authored his first memoir on Auschwitz in *Minerva Medica*.[121] The bacteriologist Lucie Adelsberger published on conditions in Auschwitz in *The Lancet* in March 1946,[122] and the clinician Alfred Wolff-Eisner analysed starvation and disease in Theresienstadt.[123] Medical contributions appeared on the self-organisation of ghetto health services, clinical research and resistance to the attempts to kill by starvation and the withholding of medical resources, and on the human experiments.[124] These papers presented survival in scientific terms, and demonstrated that the struggle against Nazi human experiments was part of a broader pattern of resistance, subterfuge and protest to maintain human life, dignity and rights in the camps.

Journalists realised that the experiments represented a distinctive form of atrocity, which merited publicity. When the camp of Struthof-Natzweiler was liberated, *Time* magazine reported on 15 January 1945 that rumours of Nazi scientists using prisoners as human guinea pigs were confirmed by a French investigating commission. Hundreds of men and women were tortured and killed 'in order to supply data for Nazi science'.[125] The international press revealed that Buchenwald (liberated on 1 April 1945) and Dachau (liberated on 29 April) were major experimental centres.[126] Whether medical war crimes should remain a military secret, confidential to the military medical teams, or be given the widest possible publicity led to a series of policy changes.

The prisoners' testimonies transformed the Allied scientific monitoring operation into the hunt for medical war criminals. Survivors of Auschwitz demanded justice to prevent the recurrence of medical atrocities.[127] A liberated prisoners' newssheet provided details of atrocities by the camp doctors Mengele and Fritz Klein.[128] Former Auschwitz detainees provided details of the experiments on 600 women at Block 10, and on serological experiments. The report urged the Allies to arrest the culprits.[129]

In May 1945 liberated prisoners from Poland, Czechoslovakia, Yugoslavia, the Netherlands, Italy and Hungary organised an 'International Investigation-Office for Medical SS-Crimes in the German Concentration Camps, Dachau'. They appointed an International Secretary, and issued an appeal on 11 June 1945. They explained that two Polish priests, Minkner and Wegener, had 'contrived to form a regular information-office and to secure a really considerable amount of acts. In these we possess a collection of surprising dates and details'; and that other national groups of prisoners intended to add to this collection of evidence about the experiments. They produced a form 'concerning the physical and psychical damages done' for victims and witnesses to complete.[130]

The prisoners' aims included that:

'every victim of any nationality can be interrogated and examined by a medical authority';
'every case of death caused directly or indirectly by an SS-experiment can be properly established, so that widows and orphans can be put into the rights of legal heirs'; and that
'every case of total or partial invalidity can be treated in a proper way.'

'The task of the International Investigation-Office will be a gigantic one. It will be necessary to take up connection with the representatives of all other camps and their national groups.' They appealed to the US Military Government for resources.[131] The Dachau scheme was part of a broader movement for compensation for victims of slave labour. The victims of experiments lobbied to obtain recognition as economically exploited for the commercial interests of companies like IG Farben. Medical victims took a lead.[132] The Dachau scheme was a precursor of the International Scientific Committee for the Investigation of War Crimes.

There were similar organisations for Buchenwald and Neuengamme victims of human experiments. The concentration camp of Neuengamme near Hamburg was a prime location of human experiments. A British war crimes investigating team on 29 June 1945 reported on its search for evidence 'as to medical experiments carried out at Neuengamme, viz. artificial infection of tuberculosis, removal of glands from Jewish children, artificial insemination of female prisoners, sterilisation of homosexual prisoners'.[133] Survivors sent in testimonies to the Allied legal authorities. The Nuremberg prosecutors received testimonies from victims of sterilisation, forced abortion and human experiments.[134]

Time and time again, the liberated prisoners brought the issue of medical atrocities and human experiments to the attention of the Allied medical personnel. The survivors' testimonies and their physical condition made a strong impression on Allied investigating officers. The survivors realised the importance of allowing their injuries to be assessed in medically objective terms as injured cases, but they took an active role in impressing on Allied medical investigators the rectitude of their demands for justice, compensation and a re-evaluation of medical ethics. These victims shifted Allied policies from strategic evaluation of wartime German medical science to prosecuting the perpetrators of criminal experiments, and providing ethical safeguards.

Medical investigators extended investigations to medical conditions and human experiments in the camps. The Allied officers responded sympathetically, and going beyond their formal orders provided assistance and supporting documentation, and in a determined effort to cut through military red tape and blimpish command structures, impressed on their superiors the need to conduct trials of the German physicians and their henchmen. The encounter with survivors left an indelible and deep impression on the investigators. The medical atrocities galvanised the war crimes authorities into instigating prosecutions of the perpetrators of the human experiments at Nuremberg and in related trials concerning Dachau, Ravensbrück and Neuengamme. The trials gave surviving experimental subjects the opportunity to testify that the experiments were coercive, and that the experiments met with resistance, protest and sabotage.[135] The UNWCC catalogue of experiments compiled by Auschwitz prisoners was marked as secret, as the Allies still preferred to treat human experiments as strategic intelligence and war crimes, rather than as atrocities meriting wide publicity.[136] But overall, Allied policies turned from strategic evaluation to the criminality and ethics of the experiments.

Survivors and witnesses of human experiments called for documentation of Nazi medical atrocities, justice and compensation. The released prisoners organised committees and issued newsletters about the experiments in the camps. Former prisoner doctors and nurses, and other witnesses, provided extensive testimony of the human experiments and widespread abuses of medicine in the camps. Their ranks included the two politically astute *'Arztschreiber'* (concentration camp slang for medical secretaries) Eugen Kogon from Buchenwald and Hermann Langbein from Auschwitz. Kogon's book *Der SS-Staat* was based on a briefing report for Allied intelligence, and was published in 1946.[137] Langbein's *Die Stärkeren* drew attention to constant resistance and sabotage against the experiments. We see the former victims taking an active role in doing their best to prevent and disrupt the human experiments as these were being carried out, and after the war to document and call for justice. These initiatives show a remarkable capacity for self-organisation and altruism.

Women political prisoners from France publicised the experiments, which they had witnessed. Some liberated Polish prisoners remained in France. Others were at Lund in Sweden, to where they had been evacuated by the Swedish Red Cross. Survivors who had witnessed the experiments – such as Denise Hautval and Germaine Tillion, and the Norwegian Sylvia Salvesen – and the Rabbits stressed the necessity of bringing the perpetrators of human experiments to justice.

The British army pathologist Keith Mant investigated the executions of women SOE officers in Ravensbrück, as Officer in Charge of the Pathological Section of the War Crimes Investigation Team. The evidence of the survivors of medical atrocities prompted him to extend his investigations to the human experiments and murderous medical conditions at Ravensbrück. He located survivors in France, Belgium, Norway and Sweden, recording their testimonies and compiling case histories of their injuries. His evidence was used in the successful prosecution of the sulphonamide experiments of Karl Gebhardt, Fritz Fischer and Herta Oberheuser at the Nuremberg Medical Trial.[138]

Ravensbrück was in the Soviet zone, and Mant was denied permission to visit (although FIAT units did work in the Russian zone from November 1945 to February 1946[139]). Mant proved the culpability of Gebhardt, and his evidence persuaded the British to mount the major Ravensbrück trial in Hamburg, involving other human experimenters. Mant wrote an influential report on medical conditions at Ravensbrück, and he became secretary of the International Scientific Commission for Medical War Crimes, which aimed to collect and evaluate details of all human experiments. Survivors' evidence was reinforced by war crimes investigators, who corroborated testimony by means of documents and affidavits, so that witness statements should not be dismissed in court as inadequate hearsay or exaggeration.[140]

For the war crimes investigators the scenes in the concentration camps were a turning point. Troops had some briefing about war crimes, but not about the specifics of human experiments. A Canadian airforce unit (which Thompson briefly commanded) was briefed about 'legal action to be taken against Luftwaffe personnel – grades of summary trials, courts martial, etc., German mentality, lessons learned by the Control Commission after the last war, treatment planned for war criminals'.[141] This suggests that troops going into Germany were routinely informed about war crimes, albeit in narrow military terms.

Alexander's sensitivities were heightened, when in May 1945 he saw a film about the camps. 'Although we knew what was going on for a long time, even since before the war, being confronted with the actual pictorial evidence is a grim and necessary experience, in order to understand the full importance of dictatorship, and in particular what German dictatorship stands for. It is the greatest crime against humanity ever perpetrated. We must see to it that it will never happen again, anywhere.' Similar film had stunning effect at the IMT Nuremberg.[142]

The encounter with victims of coercive experiments shocked Allied medical investigators. Alexander experienced a psycho-physiological reaction, as the blood drained from his hands when he smelled blood by the Dachau gas chamber.[143] Thompson felt that his first help for Belsen victims marked a turning point in his life. War crimes teams were alerted to the criminality of escaping SS doctors, who sought refuge in residual enclaves in Holstein and Bavaria. German doctors involved in camps further east like Auschwitz and Ravensbrück were washed up in the British zone, and came to the attention of British medical investigators. Mant was stationed at a former Luftwaffe hospital at Brunswick, which received cases of (mainly) British personnel infected with typhus from Belsen.[144] The Czech Women's Camp for DPs thanked Thompson for his saintly assistance:

> We take a wellcome [sic] opportunity to thank you [for] the kind deeds for our hospital and for our camp-people, who are mightily estimating your soft and consoling words to them ... we feel in your presence what a great man of history called: 'Ecce homo!'[145]

Thompson interrogated the Belsen doctor Fritz Klein, who conducted experiments with mescalin and the drug Rutenol in Auschwitz.[146] These encounters with the Holocaust galvanised Thompson to draw attention to medical war crimes. He concluded that science was no longer possible after the depraved world of the concentration camps, and that it was necessary to heed the survivors' calls for justice and ethics. The chain of events culminated in the Nuremberg Medical Trial.

5
Aviation Atrocities

A forensic odyssey

German physiologists and biochemists gravitated towards aviation medicine as it offered unrivalled opportunities to test the human metabolism's ability to withstand extreme conditions. Allied investigators were confronted with the conundrum of whether the German researchers were patriots who had placed their skills at the service of the war effort, unscrupulous seekers after scientific truth or zealots serving racial ends. German scientists claimed that war service with the Wehrmacht or Luftwaffe was a way of evading the Party and SS. Countering these excuses was the evidence of a depraved and debased science.

The overall structures were highly politicised. Aviation medicine came under Reich Marshal Göring, who headed the German Academy of Aviation Research. Its ranks included leading scientific figures, notably the biochemist Butenandt, the chemist Kuhn, the physiologist Rein and the high-altitude specialist Strughold. The KWG was deeply involved in armaments and military medical research. The Americans led the campaign for disbanding the KWG. Persistent suspicion clung to the KWG for being involved in some way with the Dachau aviation medicine experiments. But despite Allied scrutiny, arrests, interrogations and damning evidence of euthanasia, brain and concentration camp research, no KWG scientist was ever prosecuted for war crimes. Strughold has remained controversial, as knowledge of Nazi concentration camp research tarnished his credentials as a pioneer of space medicine.

At first the Germans appeared to be open about their wartime research, but on closer acquaintance the Allied officers found them resentful, deceitful and engaged in petty rivalries. One solution was to establish an Aero Medical Center under US military supervision. This would allow the Germans to complete research on potentially useful projects like high-altitude ejection seats. But six aviation physiologists were arrested as potential defendants for the forthcoming Medical Trial.

The take-off of German aviation medicine was linked to the strategic importance of the Luftwaffe in *Blitzkrieg* strategies. Göring became a scientific dictator through the Reich Research Council and by creating a network of research institutes and contracts.[1] The Allies set about decoupling research from military strategy. The British and the Americans regarded Göttingen (renowned as the cradle of aviation research and of German nuclear physics) as the key location for reviving German science. On 16 and 24 April 1945, CIOS filed highly positive reports about the outstandingly well-equipped Physiological Institute at Göttingen, where the physiologists Rein and Strughold were located. CIOS recognised the Institute as 'very valuable as a research centre', having a pressure chamber with cooling facilities.[2] The Americans obtained a copy of the 1944 edition of the prized Ruff-Strughold compendium on aviation medicine, identified the location of Strughold's library and resources, and evaluated experiments on explosive decompression and oxygen deficiency.[3]

The Allies were faced with a dilemma. They wanted militarist German research to cease, but also wanted to exploit its findings and to know of potentially useful projects, which might still be completed. Even if the blustering Göring negligently failed to inspect academic centres, the self-mobilisation of scientists for war work meant that aviation medical research had been shaped by military priorities.

Research into aviation technologies like fuel and aerodynamics opened the problematic issue of adapting these to human use. Aviation medicine involved experimental projects for the Luftwaffe. Milch increased the autonomy and facilities available to researchers. The Air Chief Medical Officer Hippke allocated resources for research at the Reich Universities of Posen and Strassburg.[4] The situation facilitated human experiments.

Doubts lingered around the eminent physiologists, Rein and Strughold. Both attended the Nuremberg conference of 26–27 October 1942 on Winter Hardship and Distress at Sea, when Rascher reported on his deadly freezing experiments. While Strughold remained a prisoner of war until November 1945, Rein became Rector of Göttingen University in 1946. He was celebrated for research in aviation medicine and as an author of physiological textbooks. He insisted that he had found Rascher's experiments scientifically flawed and unnecessary. But he had served in a flight squadron with Göring in the First World War, and had been a supporting member of the SS.[5] A cloud of suspicion hung over Rein, as it was thought he was involved in criminal human experiments. At the same time his eminence protected him, as the British opted in June 1945 for a liberal, hands-off approach to German research, predicated on the view that control of science would work best when the Germans were encouraged to develop new research initiatives.[6] He staunchly defended German medicine claiming it had resisted the demands of the Nazi order.

Strughold has long been suspected of initiating or utilising results from criminal human experiments.[7] The issues revolve around whether he had greater knowledge of the unethical human experiments than he divulged to his interrogators, and whether he failed to condemn the murderous methods of obtaining the results. Strughold, Rein and the pathologist Hans Büchner maintained that they condemned Rascher and those physiologists who collaborated with him. Their excuse was that the results of experiments on 'prisoners condemned to death' were a top secret '*Geheime Reichssache*'. They considered that they could not take their criticisms any further.[8]

Alexander was convinced that Strughold had positively refused to become involved in coercive experiments. The accused researchers at the Medical Trial, Becker-Freyseng and Schäfer (involved in seawater experiments) both worked under Strughold, who, they said, received reports on all experimental work, as he was in Berlin until the winter of 1943–44.[9] Strughold reported regularly to Hippke's successor, air chief medical officer Oskar Schröder, about research at his Aviation Medical Institute, and in May 1944 demonstrated the seawater purification research of Schäfer. But crucially there is no evidence that Strughold discussed coerced human experiments.[10] It escaped Allied notice that Strughold facilitated the supply of a pressure chamber for experiments on children and epilepsy to the KWG geneticist Nachtsheim.[11]

For most of the war Strughold declined the rank of officer in the Luftwaffe to protect his scientific independence. Although a member of the National Socialist Flying Corps from 1937 to 1943, and the Nazi Welfare League, he never joined the Nazi Party, and had declared for the Catholic Centre Party in 1932–33.[12] He guarded his status as a scientist; but the crucial issue as to his knowledge of the criminal experiments remained open.[13] The Americans transplanted Strughold to Heidelberg, where he directed the German contingent at the American Aero Medical Center.[14] Arrangements were made with BIOS and FIAT to interview the German medical personnel assembled at Heidelberg. British Medical Research Council representatives conducted the interviews, and the results were circulated to the Americans and Canadians.[15]

Alexander emerged as one of the most thorough and effective investigators of human experimentation. His tour of aviation medicine installations developed from identifying clinical and scientific innovations into the reconstruction of criminal experiments. He arrived in Munich on 23 May 1945, and began to investigate German centres in neurology from 28 May 1945 – one month after the liberation of Dachau.[16] He had a unique opportunity to examine the destructive psychology of a totalitarian state, not only by interviews of German medical scientists but also 'on the streets, railroad trains and farms of Germany'.[17]

Alexander's role as medical adviser to the prosecution at the Medical Trial drew on his neurological mission, and on his wartime medical assignments. He joined the Duke University medical detachment in 1942, and served at the US VIII Air Force at Diss in East Anglia, where he assessed the psychological fitness of aircrew.[18] He became interested in the survival of rescued personnel from the sea. He had a canny sense of where to probe the German interviewees, and made informed comparisons with American and British research. He spot-checked experimental data from immersion experiments in Dachau with his own and colleagues' experiences with rescued patients treated by the 65th General Hospital to prevent post-rescue cardiac collapse.[19]

Alexander's interests ranged from psychotherapy to physical treatments like deep sleep and insulin narcosis to treat extreme anxiety states.[20] He kept extensive clinical notes, carried out Rorschach tests (the psychological interpretation of inkblots) to determine prognosis and choice of treatment, studied servicemen as normal controls, and reflected on the public health implications of untreated cases of the traumatised. He hoped for a year to write up his war research and to produce 'a tremendous book' on Psychoneurotic and Psychopathic Reactions in Air Combat. But the psychology of Nazi criminality diverted him.[21]

Neurology gained enhanced military support, because of the problems of war injuries.[22] Alexander conducted a survey of the neuro-psychiatry and neuro-surgery of the German army, airforce and civilian institutes.[23] CIOS Trip No. 268 was to evaluate therapeutic procedures, notably head injury treatments.[24] His energy, acumen and keen sense as to what was legitimate medical research meant that he cast a critical eye over German medical achievements. Alexander's pre-war evaluation of sterilisation meant that he understood much about Nazi eugenics. But he was horrified to discover the role of German psychiatrists in compulsory euthanasia and human experiments. What began as a reconnaissance mission to document neuro-surgical innovations became an odyssey into the destructive mentality surrounding human experiments and the medical exploitation of brains of euthanasia victims. He encountered survivors, interrogated German researchers and explored Nazi documents to find out about the criminal experiments and when and why they were carried out.

Alexander came from an assimilated Jewish family in Vienna, and entered American medicine and society as a neurologist at Boston and Harvard in 1934–41. His grasp of English, spoken and written, was excellent, and he and his American wife settled on Unitarianism as a suitably respectable and indefinite creed. In 1945 he re-established contact with one sister who had survived the ordeal of the occupied Netherlands, though the Holocaust had reaped its toll amongst those in the family who had remained in Vienna.[25] He was deeply moved by the survivors' testimonies, and elated at their triumph over death. When he arrived in Munich he reported that 'you can see small groups of freed concentration camp inmates wandering about proudly wearing parts of their prison garb, and I have talked to many of them. A group of 15 year old boys from Hungary was particularly touching.'[26] He sifted the debris of wartime research records and became fascinated by its mentality. While the ordeal of the survivors set the tone for his mission, Alexander set out to understand the psychology of the perpetrators and to document what they had done.

Alexander's task became that of connecting elite German scientific institutes with the sinister labyrinth of the concentration camps. His starting point was the 'big institute in Munich', the German Institute for Psychiatry, which had developed links to aviation medicine. The scientists presented themselves as victims, complaining of the oppressive burdens of the Nazi state and the denunciations of academic rivals, while demanding resources for experimental research. Alexander approached them on equal terms as a colleague in neurology: at this level he expected a frank exchange of information. His military mission to secure technical documentation meant that he could demand a full account of past research. He could feign a degree of unknowing in the hope that this would release indiscretions and confidences, while the interviewee hoped for favours. The brain anatomist Julius Hallervorden complained that Alexander appeared to be a scientist, but was in reality a policeman.[27] Once Alexander had evidence of criminal experiments, his encounters with the perpetrators became more problematic.

The German response to the investigators was rarely open and truthful.[28] German scientists claimed to have been victims of Nazi oppression. The neurologist Georg Schaltenbrand complained about his misfortunes, although he conducted some of the earliest coerced experiments in 1939 and had proudly communicated news of these abroad. Another stance was of having been dedicated to patient care. The neurosurgeon Tönnis adopted this position in explaining how he organised treatment of head wounds. Tönnis was a highly respected neuro-surgeon in Berlin, and he also directed a department for tumour research and experimental pathology at the Kaiser Wilhelm Institute for Brain Research.[29] His clinical and organisational achievements earned praise from Karl Brandt in 1944 as unique and exemplary.[30] Tönnis had been a candidate for membership of the NSDAP, and his Luftwaffe connections meant some indirect links to the perpetrators of the human experiments. He was on good terms with the aviation medicine researchers, and he attended the military medical meeting at Hohenlychen in May 1944.[31] The British neurologist Hugh Cairns recollected how Tönnis had expressed 'great hatred' of Britain, and that he visited the offices of the Rockefeller Foundation as a German officer in occupied France.[32]

An evasive strategy was to blame distant or deceased researchers (notably the spectral Rascher) as sadistic monsters and incompetent pseudo-scientists.

Competitors were denounced for defective science and collusion with Nazism – Rein blamed the physiologist Weltz (tried at Nuremberg) as incompetent. Others blamed Rein for collaboration with Nazi aviation research. The alert interrogator could reconstruct the criminality of German doctors from the discrepancies in their narratives, using protestations of how responsibility for any misdeeds lay elsewhere. The German self-image of the scientist as victim began to crumble.

Alexander correlated evidence from German scientists, survivors and captured documents. He researched the pathology of Nazi medicine by constructing a topographical anatomy of the German scientific crimes and by localising the functionaries. He placed scientific institutions within the broader structure of Nazi racial policy and examined each component. He probed different sections of the KWI for Brain Research, where he had been a student in 1928. The section for brain anatomy and general pathology under Hugo Spatz, who worked for the Luftwaffe, was dispersed from its lavish modern buildings in Buch (where Haagen resurfaced) on the outskirts of Berlin to three locations. The vast dispersal of specialist institutes from Berlin in 1943 meant that new links were forged with university academics and researchers in such locations as Giessen and Tübingen. The German Institute for Psychiatry, Munich was the main recipient of the relocated laboratories. The section for brain pathology came under Hallervorden, who conducted brain autopsies for the army, while continuing his scientific and clinical roles. He secured the evacuation of an immense collection of brain specimens to Dillenburg in Hessen-Nassau in May 1944. The section for neuro-physiology went to the massive Physiological Institute of Rein in Göttingen.[33]

Hallervorden was a punctilious East Prussian with a vast appetite for dissecting brains. He had been appointed professor for neuro-pathology at the KWG in 1938, while retaining his position as Prosector of the Brandenburg psychiatric institutions, giving him access to an immense supply of 'fine' and 'unique' specimens. He evacuated his stockpile of brains, gleaned from collections including those of his dismissed Jewish predecessor, Max Bielschowsky. His conversation with Alexander came to focus on child euthanasia victims. While deprecating euthanasia as damaging German psychiatry, he regretted that full autopsies were not carried out on each victim as this would have increased the scientific value of the specimens. He brazenly justified how he had harvested vast numbers of victims' brains. Most of Hallervorden's 697 brains came from the psychiatric hospital of Brandenburg-Görden. This was under the euthanasia enthusiast Hans Heinze, who took a grim interest in children with cerebral palsy. He sent children to be killed at a prototype gas chamber set up in January 1940 in the Brandenburg prison. The official explanation was that their prognosis was poor, but Hallervorden explained that the children were rapidly selected to clear wards, and often those whose behaviour was deemed difficult were chosen. Hallervorden and Heinze correlated clinical observations with anomalies in brain tissues.[34] Hallervorden selected 'beautiful cases' of feeblemindedness, malformations and early childhood diseases, but rejected brains of schizophrenics as of less scientific interest.[35] Buch and the Brandenburg state prison were in the Soviet zone, so that Alexander could not readily inspect them. Heinze was in Soviet custody from October 1945 and sentenced to seven years' hard labour in March 1946.[36]

The German Institute for Psychiatry had since 1933 been an object of controversy for its role in sterilisation, and Alexander searched for evidence of its Director, Rüdin's involvement in euthanasia.[37] The Institute had since 1938 also played a role in aviation medicine, especially through the physiologist, Fritz Roeder, who was also a member of Weltz's aviation medical institute.[38] Neurologists explained that they were critical of the Nazi sterilisation laws as excessive, and that this explained why the laws allegedly lapsed in Germany just as they had done in America.[39] While the arrest of Rüdin on 20 December 1945 resulted from Alexander's report, his release in August 1946 arose from the sense that prosecution for sterilisation would be too difficult other than for the clearly illegal X-ray sterilisations. Rüdin covered up his contacts to the murderous T-4 organisation and his exploitation of euthanasia for research. He exonerated his support for sterilisation by stressing the legality of sterilisation in the US and his international collaboration.[40] Remarkably, the eminent physicist Max Planck certified that Rüdin had never acted from political motives.[41] The whitewashing of politically suspect colleagues was part of a broader defence of the autonomy of the scientist from political interference. German conservatives equated the Nazis with the Allies, undermining Allied denazification and justice. Defending the freedom of science was to become the new Cold War orthodoxy, conveniently exonerating past crimes.

Alexander condemned medical justifications for euthanasia of the mentally ill and disabled, given the brutality, coercion and massive scale of the killings. He viewed their administration (erroneously) as under SS control, not realising the role of the clandestine T-4 organisation with its panel of psychiatrists, who cursorily meted out death sentences on 90,000 victims. He noted that the professor of psychiatry in Heidelberg, Carl Schneider, was a war criminal, and the locations of the killing centres.[42]

Anton Edler von Braunmühl, the Director of the State Hospital of Eglfing-Haar near Munich, passed on the secret files of the former Director and euthanasia supporter, Hermann Pfannmüller, to the Americans.[43] Alexander was struck by the subterfuge and callous bureaucratic procedures of the involved senior psychiatrists. He drew a contrast between concerned relatives – for example, a Jewish relative distressed at the disappearance of a patient's funeral shroud – and the high-handed, duplicitous psychiatrists, whose anti-Semitism came on top of a malign view of the patients.[44] One

Nazi psychiatrist declined a job because he recognised that he would become emotionally too attached to the child victims.[45] Alexander concluded by listing 26 persons implicated in the killings for the information of war crimes prosecutors.[46]

Alexander evaluated Spatz's neuro-anatomical research on over 3,000 brains of German airforce personnel. At the time Spatz, a convivial Bavarian who retained his rank as Oberfeldarzt, was in American internment at the German Institute for Psychiatry in Munich. He then transferred to the Aero Medical Center Heidelberg. He had been in charge of the Aussenstelle für Gehirnforschung des Luftfahrtmedizinischen Forschungsinstituts under the overall command of Strughold, and the Heidelberg internment represented continuity.[47] Spatz showed Alexander a film on the motor control centres of sexuality.[48] He claimed that Alexander regarded his research as in the interests of the Allies and requested his release so that he could resume his research work for the KWG.[49]

At Bad Ischl near Salzburg, Alexander evaluated head injury treatments in a vast military hospital. Tönnis, who was Chief Consulting Neurosurgeon to the entire German armed forces since 1943, made a positive impression on Alexander as an intelligent and effective organiser of therapy.[50] But Alexander and two British head injury specialists found the work of Tönnis and Spatz dated in their obsession with brain infection, and far behind neurosurgical practice in Britain or the US with penicillin, superior equipment and more refined surgical techniques. When interrogating Gebhardt, Alexander could refer magisterially to the US surgeon Elliot Cutler, who had found that sulphonamides were no substitute for good surgery.[51] In January 1946, when there was renewed concern with medical war crimes, Tönnis was briefly arrested by the British authorities when he was working at the Krankenhaus Bergmannsheil in Bochum – ironically where Karl Brandt had trained as a surgeon. Tönnis was prevented from returning to his university clinic in the Soviet sector of Berlin, while the British supported him for a combined university post and Max Planck Institute for brain tumour research at Cologne.

Alexander's probings of German neuro-psychiatry took him to Hans Luxenburger, whose antipathy to coercive sterilisation forced his move from research on the genetics of mental diseases to air force psychiatry.[52] Alexander focused on how neuro-anatomists obtained brains for dissection.[53] He found it disturbing that the German military did not discharge patients for neuro-psychiatric reasons, and the organic approach to disease was motivated by the prevailing bias against neurosis.[54] Alexander diagnosed how the SS brutalised 'normal, stable and solid people of the conforming, authority respecting type'.[55]

Col. Prentiss of CIOS extended Alexander's mission to include 'experiments of explosive decompression and very high altitudes conducted by Stabsarzt Wolfgang Lutz who also invented [a] pressurised suit, and the work of Weltz on the effects of immersion'. The order described the aviation medical research as conducted in a Kaiser Wilhelm Institute. Given that Spatz ran a brain research institute for aviation medicine, located at the Deutsche Forschungsanstalt für Psychiatrie, it was understandable that other related institutes for aviation medical research should be misconstrued as coming under the KWG umbrella.[56] The Spatz institute had satellite institutes at Dillenburg, at the physiological institute in Göttingen and Bad Ischl, which were all inspected by Alexander. Although Lutz and Weltz were not KWG affiliates, it emerged that some of their associates were.[57]

Alexander's quest for evidence on resuscitation drew him further into the grotesque realm of criminal experiments. When an advanced CIOS field team reported on the target Institute for Aviation Medicine at Freising, near Munich, they outlined Georg Weltz's discovery of rapid re-warming by immersion in hot water. The report spoke at times of 'the patient or animal', but the experiments were described as solely on animals.[58] Alexander suspected lethal experiments. Weltz explained his test for aviators' reactions to oxygen deficiency. The loss of aircrew in the Channel during the Battle of Britain prompted him to research the physiological basis of resuscitation from cold. Weltz and his Austrian co-worker (and former SS officer) Wolfgang Lutz gave details of animal experiments to solve the problem of the nature of death from cold.[59] Weltz found that speedy re-warming was effective in small animals. 'Dr Lutz was then questioned as to the application of his findings to man ... and whether experimental work was carried out on human beings.'[60] Alexander noted that the equipment for experimenting on large animals like adult pigs was curiously absent, as were statistics on the use of his novel method. 'He was then asked whether he had, or whether he knew if anyone had performed any experimental work along these lines on human beings. The question was again repeated during a subsequent private interview without witnesses, and denied on both occasions.'[61] Alexander was careful to allay Weltz's suspicions so that evidence should not be destroyed.

Despite the denials by Weltz and Lutz, Alexander was convinced that experiments on human beings were being concealed.[62] Alexander heard about the Dachau experiments, when an army chaplain mentioned a radio broadcast on immersion of prisoners in tubs of freezing water.[63] Alexander gained a sharper view of contentious aspects of the work of Weltz, when physiologists at Göttingen explained that the Munich research was unorthodox from a neuro-physiological point of view. Weltz offered collaboration with the US, hoping for funding from the Rockefeller Foundation with which Alexander had good contacts.[64]

Alexander's understanding of the chilling world of human experiments was deepened when he took evidence from survivors, and saw the death, disease and mutilations among victims. When he visited the concentration camp of Ebensee (a satellite camp of Mauthausen, packed with survivors of forced marches from Auschwitz and Neuengamme, and liberated on 1 May) the scars of Nazi atrocities were fresh.

> 31 May 45 Ebensee Concentration Camp: A real inoculation against the German deference and charm which they are turning on. Grim. 18,000 people (a large portion of Hungarian Jews taken from Hungary a year ago), under the whips of SS men and of 2000 convicts (mostly crimes of violence such as murder and robbery) who acted as foremen. The mortality was 300 per day, rising to 1,000 a day for one week. Now, after 24 days of American care, the mortality is down to 70 per day. We saw the bodies of 3 people who died today in a large hall at the crematorium.[65]

He talked to prisoners about their loss of parents or children, and of medical selections for the gas chamber: 'The doctor whom we saw later (Dr. Tot) told us that he saw 3,000 women and children killed by gas in one night, in groups of 200; that he and the others had to carry out the bodies to pits; he cried when he recalled it. On rainy days when trains arrived and the officer did not want to stand in the rain, all arrivals went into the gas chamber.'[66]

Alexander now had two lines of war crimes investigations and took on the role of a scientific detective. First, he had to resolve the issue of whether there were human experiments on cold immersion treatment, and second concerning the brains harvested from euthanasia killings. He decided to leave out the 'anoxia brains' transferred to Büchner at Freiburg. Given Büchner's senior position as Luftwaffe pathologist, and Rascher's status as an airforce officer, he might have transferred brains to Freiburg, rather than to Hallervorden and Spatz in Berlin or retained them for dissection in the Dachau pathological laboratory. Moreover, Büchner's condemnation of euthanasia indicated that he was the sort of Catholic moralist whom Rascher held in deep contempt.[67] It later emerged that Rascher's collaborator, Hans Romberg, was assistant in pathology to Büchner, before he went to Freiburg.[68]

Alexander's mission remained at one level an investigation of criminality of the Kaiser Wilhelm Institute in the spheres of brain anatomy and neurology. After meeting neurologists at Heidelberg on 10–12 June Alexander interrogated the neurologist Karl Kleist in Frankfurt, when poignant memories of his years as an assistant flooded back. Kleist told of euthanasia by neglect and provided a set of military psychiatric reports.[69]

Alexander travelled via the 'Annihilation Institute' of Hadamar where psychiatric patients were killed, to Hallervorden at Dillenburg.[70] Although the Director of the Kaiser Wilhelm Institute for Brain Research, Spatz, was still in Allied internment, Hallervorden received Alexander as a colleague and former affiliate of the Institute. He confided that he had collected 200 brains from Jewish typhus victims from Warsaw, and 500 brains came from the killing centres for the insane.[71] Hallervorden enthused that: 'There was wonderful material among those brains, beautiful mental defectives, malformations and early infantile diseases. Where they came from and how they came to me was really none of my business'.[72]

Hallervorden's admission reflected the German view that the medical researcher could legitimately research on the body parts of executed Nazi victims. Researchers who had harvested the German concentration camps for specimens adopted this justificatory stance to exonerate themselves and their colleagues. This is how Nachtsheim judged Mengele's collaboration with the human geneticist Verschuer. Sometimes the researchers examined the victims prior to death, then correlated findings about disabilities with autopsy evidence. It brought home to Alexander how medical research fed off Nazi mass murder.

The criminality of medicine was becoming clear at a number of levels: murderous experiments reported as scientific findings at conferences, the scientific use of body parts of persons who were criminally killed, orders to undertake human experiments and receiving reports on them, and direct involvement in human experiments. On 16 June 1945 Alexander arrived at Göttingen in the British zone. The university had a distinguished reputation in mathematics and medicine, and as the most prestigious and least war-damaged university in the British zone it was pivotal in the British strategy of resuscitating German science. At the renowned physiological institute, Strughold told Alexander that Rascher and the Kiel physiologist Ernst Holzlöhner reported their Dachau experiments at the meeting on survival at sea in Nuremberg in October 1942. Strughold was quick to condemn the experiments on the basis of morals and medical ethics: 'Any experiments on humans that we have carried out were performed only on our own staff and on students interested in our subject on a strictly volunteer basis.'[73] Strughold's unequivocal condemnation set him apart from the evidence of criminality on the part of colleagues in aviation medicine.

The eminent physiologist Rein expressed scepticism about the researches of Weltz, Lutz and their co-workers in Munich. He condemned scientific flaws, as experience from the Russian Front suggested that there were major differences in tissues, which were rapidly frozen rather than subject to prolonged chilling. Rein characterised Rascher as a 'nasty fellow', and the emerging picture of corrupt ambition, brutality, greed, sexual perversion, sadism and deception gained depth from the Himmler documents.[74] Rein differentiated between the legitimate experiments of Strughold and the crude butchery of Rascher, who was contemptuous of academic physiology.

The information that Rascher was an SS officer prompted Alexander to delve into the holdings of the 7th Army Documentation Center at Heidelberg where Himmler's recently discovered papers were being sorted for the Nuremberg tribunal.[75] The correspondence between Rascher and Himmler became the basis of an extensive collection on human experiments and euthanasia with the generic name 'Case No. 707-Medical Experiments', which became known as the 'Heidelberg Documents'.[76] These grouped documents on human experiments in Dachau, Buchenwald and elsewhere with documents on euthanasia and sterilisation, so forming a reference collection for prosecuting criminal abuses of medicine.

Alexander became convinced that Rein and Strughold were concealing Rascher's collaborative links with academic physiologists.[77] He learned of research by the Military Medical Academy on the effects of war wounds on the central nervous system. Strughold had responsibility for satellite institutes in Silesia and Bavaria, and at his large Berlin institute Konrad Schäfer conducted experiments on desalination.[78] The question arose whether Strughold's collaborator Ruff was involved in the Dachau pressure chamber experiments.

The human experiments under Rascher at Dachau were astonishing acts of cruelty. Alexander retraced his steps, primarily searching for documents. He visited the US Document Center in the University Library, Heidelberg, where truckloads of Himmler documents were coming in, and the 1280th Intelligence Assault Force was assigned to examine thousands of files to document Nazi atrocities. Hugh Hellmut Iltis (the son of the Czech anti-Nazi geneticist who had alerted the UNWCC to German racial criminals) was already on the Rascher case, and found 'incredible documents' for Alexander.[79] Iltis and Alexander together broke the original seals on Rascher's paper on the Dachau experiments.[80]

Iltis was documenting Nazi atrocities for the imminent war crimes trials. When he found photos of the liquidation of the Warsaw ghetto uprising, he admired the defiant stance of the women victims being marched to their death. He discovered the reports on the deadly experiments in the mass of Himmler's papers. These pointed to military and medical involvement in Rascher's human experiments and revealed the concern among the SS that the Luftwaffe authorities would condemn the experiments as 'amoral'. Their start was delayed until March 1942 when Ruff and Romberg (associates of Strughold at the aviation research institute in Berlin-Adlershof) supplied a low pressure chamber.[81] The chamber was illustrated in the 1944 edition of the Ruff-Strughold textbook on aviation medicine.

Alexander found that Rascher's experiments were authorised by the German airforce, and he discovered how noted physiologists were involved. He reconstructed the administrative hierarchy reaching up to Air Marshal Milch. The chief Luftwaffe medical officer Hippke ordered the freezing experiments to be carried out with the help of the Kiel physiologist Holzlöhner (who was an airforce medical officer), the Innsbruck pharmacologist Jarisch and the Munich pathologist Singer.[82] Hippke was soon to be relieved of his post as not sufficiently in line with Nazism. He insisted that he had doubts about experiments on prisoners. He was critical of Rascher, warning him that if he left the air force for the Waffen-SS he could lose scientific credibility. But Hippke still thanked the SS for its support.[83] Alexander noted how the SS feared that Christian moral scruples would undermine the experiments, and how this resulted in tensions between the SS on the one side, and the Luftwaffe and involved physiologists on the other.[84] Holzlöhner reported the experiments to Hippke's successor, Schröder, but appeared to have been psychologically broken by his complicity in scientific murder.[85]

Alexander learned of tensions between Reichsarzt SS Grawitz and the Ahnenerbe SS secretary, Sievers – so providing a glimpse into the tension-riven SS state. Rascher proposed experiments on concentration camp prisoners in igloos in the Austrian Alps to assess exposure to extreme cold.[86] In the event, the experiments on leaving naked prisoners outdoors were carried out in Dachau, although the callous Rascher preferred the more easterly situated Auschwitz camp.[87] He exploited sexual physiology by using the 'animal warmth' of naked female prisoners, transferred from Ravensbrück to Dachau; they were forced to have sexual intercourse as an experimental re-warming technique.[88] Rascher's cruelties were driven by his self-heroising as a relentless experimenter.

On 20 June Alexander visited the bombed Physiological Institute at Munich, where he found the charred remains of a low-pressure chamber. Returning to the aviation medicine institute, Weltz's collaborator, Lutz, 'came clean' about the human experiments.[89] He explained that he had declined to participate in the experiments; although an SS officer, he was 'too soft'. The young assistant Romberg attempted to moderate the risks in the experiments.[90] Holzlöhner rationalised his role as holding in check Rascher's cruel excesses.[91] The claims to have restrained Rascher were developed at the Medical Trial.

On 21 June Alexander returned to Dachau to find witnesses and survivors of the cold and pressure experiments and recorded survivors' testimonies on the conduct and organisation of Nazi experiments.[92] He gathered details of Rascher's experiments on Jews, Sinti and Roma, Catholic priests, criminals and political prisoners. Alexander promptly issued a report for CIOS and the US Army Document Center, providing a list of names and addresses for the War Crimes Commission. The list of suspect criminals included Ruff, Romberg and Weltz as collaborators of Rascher, as well as the orthopaedic surgeon Gebhardt and Walter Neff (Rascher's prisoner assistant), Lutz and the pathologist Singer, who were later witnesses at Nuremberg. Again, he urged further investigation by the War Crimes Commission.[93]

Weltz was arrested on 11 September, and Romberg on 9 October 1945 for conducting experiments at Dachau. Other suspects joined the Aero Medical Center from October 1945. On 6 October 1945 the *Süddeutsche Zeitung* published on the use of Dachau prisoners for human experiments. Weltz protested that his experiments were on animals, and that he had been at loggerheads with Rascher.[94] His defence was already robust.

On 22 June 1945 Alexander returned to the Heidelberg Document Center, where he and Iltis found more material on the freezing and pressure experiments. Himmler's papers revealed much about Rascher and his associates. Alexander located Himmler's copy of the final secret publication on cold experiments, received on 21 October 1942 (a week before the Nuremberg military medical conference): 'Bericht über Abkühlungsversuche am Menschen', by Holzlöhner, Rascher and E. Finke. Alexander found that the triumvirate of physiologists convicted themselves of dozens of murders in charts on blood CO_2.[95]

Alexander's ethical awareness can be contrasted to Bryan Matthews, the British aviation physiologist, who located Holzlöhner at Kiel on behalf of FIAT. Matthews did not recognise the criminality of the experiments. Holzlöhner regarded the final report on freezing experiments with Rascher and Finke (also from Kiel) as their death sentence in the event of losing the war. Shortly after Matthews' visit, Holzlöhner committed suicide with his child and attempted to kill his wife: he felt unable to live in a society not ruled by the National Socialists.[96] Finke was listed as missing at the end of the war.[97]

Alexander's verdict was that the pressure experiments inflicted much unnecessary pain, suffering and death, but failed to add anything significant in terms of knowledge to the original animal experiments.[98] He strongly endorsed the value of the medical evidence for rapid re-warming (although by 1949 he condemned the research as unreliable).[99] He roundly condemned the experimental practices, but supported the scientific validity of the conclusions drawn by Lutz and Weltz.[100] He evaluated data on exposure to cold, resuscitation, altitude tolerance and explosive decompression from the point of view of operationalising German research work.

Alexander's mission to document medical achievements uncovered atrocities, despite a barrage of denial. He did not take the initial excuses at face value, but scrutinised testimony in the light of hard documentation. He was then in a position to re-interview the suspects and take evidence from victims and witnesses of the experiments. He telegraphed his shocking findings about Rascher to CIOS.[101] Alexander's verdict on the grim spectacle of German medicine was damning: 'German medical science ... remained essentially static and became comparatively incompetent, and second, because it was drawn into the maelstrom of depravity of which this country reeks – the smell of the concentration camps, the smell of violent death, torture and the suffering of victims will not out of one's nostrils.' He was astonished at the depraved scientific curiosity of German medical scientists. What he had found far exceeded his worst expectations.[102]

Reviving German aeromedical research

Alexander's mission had a counterpart when, in July 1945, the American physiologist Detlev Bronk, Chief of the Division of Aviation Medicine, headed a team to evaluate German aviation medicine.[103] Bronk was an academic high-flyer and linchpin of US aviation medical research. He had the idea of a German-American research centre prior to his reconnaissance. Some of those interrogated had been investigated by Alexander between 28 May and 20 June 1945. No sooner had Alexander returned to London in early July 1945 than Bronk's team set to work between 7 July and 7 August 1945. They interrogated Ruff, Rein and Strughold, and Lutz at Freising on 22 July. There was a delicate balance between criminality and strategic exploitation of new research. Strughold retained the confidence of the American Air Force Aero Medical Center at Heidelberg, where he was in charge of the German staff; Rein remained University Rector in Göttingen. These leaders of aviation physiology kept their positions despite evidence of their knowledge of widespread medical atrocities.

Bronk perceived the dark shadow of criminality and political fanaticism behind the façade of normal medical research. Only one of Alexander's reports – the crucial 'The Treatment of Shock from Prolonged Exposure to Cold, Especially in Water' – was written by 8 July. It took some time for the report to reach Bronk and his colleague Howard Burchell, who were working for the US Air Force rather than CIOS. The Bronk-Burchell interrogations set out to obtain scientific results from the Germans. The record stated: 'No effort was made to assess their political and ethical viewpoints, or their responsibility for war crimes.'[104] The disclaimer suggests suspicion, but also a sense that if the war crimes element would be allowed to predominate then this could close off inquiry into lines of scientific value. The interrogators prudently liaised with American counter-intelligence for information on scientists' careers and commitment to Nazism. But they were unable to confirm Alexander's revelations about the Dachau pressure chamber or the involvement of Ruff with human experiments: Burchell was curious 'to know what the war crimes commission will do about Alexander's findings'.[105] In October 1945 a Naval Technical Mission assessing the Dachau experiments referred to violations of 'the Oath of Hippocrates and the flouting of humanitarian principles'.[106]

Despite scientific priorities and moral disclaimers, Bronk and Burchell were alert to political attitudes. They noted Rein's extensive support from the Nazi state and his overarching influence on personnel, research facilities and research projects, and Strughold's cunning as an organiser of aviation medical research: 'Strughold was not always quite honest in presenting the true significance of the work which he supported'.[107] Rein raised the issue of Holzlöhner's research in concentration camps: the ostracising of Holzlöhner implied the ethical probity of mainstream aero-medical research. The report on Ruff highlighted his efforts in practical applications; later, he served as medical adviser to Lufthansa, and developed catapult ejection seats.

The physiologist Theodor Benzinger had liaised extensively with Ruff on topics like explosive decompression. Benzinger explained that he had gone up to 20,000 metres in a pressure suit. He showed evidence of experiments on the fatal effects of injections of air in lung veins.[108] An associate of Strughold, Dr Desaga, admitted that he had researched on histological change in the testicle of an individual who 'had been provided by the Nazi Party'. Evidence on human freezing experiments came to light regarding information from Schröder, Becker-Freyseng and Rose. Becker-Freyseng in his defence shifted the initiative back to the mercurial Strughold, as well as to Weltz and his assistant Lutz in having undertaken cold experiments.[109]

In May 1945 Major General Harry Armstrong organised the first conference on 'Biological Aspects of Space Flight' at Randolph Field. Ivy attended this.[110] The Aero-Medical Research Section of the US Air Forces in Europe set out to locate and interview German research workers. They drew up a list of 115 researchers, and by mid-1945 53 had been interrogated. The list included Becker-Freyseng, Rose, Ruff and Schröder.[111] A team under Armstrong reported on 16 July 1945 on the interrogations of Schröder, his assistant Leinung, Rose and the 'extremely well-informed' Becker-Freyseng – three were future candidates for trial. Rose reported on the experimental transmission of infectious jaundice by Dohmen, Haagen and by German naval physicians at Bucharest.

Rose gave an account of his experimental infections of inmates of the psychiatric hospital at Pfafferode in Thuringia by subjecting the subject to the bites of 20 infected mosquitoes in one day: 'PW says that insane people were selected as subjects because they would be eligible for fever therapy in the treatment of the disease anyhow.'[112] Rose and Oskar Schröder outlined German 'defensive' measures in germ warfare. Alexander interrogated Eyer and Rodenwaldt for the VIIth Army as part of a survey of bacteriologists.[113] At the Medical Trial Rose claimed that the Americans commended his experimental research at Pfafferode.[114]

Burchell felt that interrogations did not yield the full capacity of aviation medical research.[115] He proposed that an Institute could be established under American control in the British zone at the Göttingen Physiological institute. Here they could benefit from Rein's ingenious battery of devices to measure respiration and cardiovascular function. Burchell's second choice was to locate the Aero Medical Center in Heidelberg. The aim was to secure a compliant corps of German aviation researchers. But by September the visiting American officers, Burchell and Otis B. Schreuder, were in conflict over the setting up of the new Center. It became clear to Burchell that science and politics could not be entirely divorced.[116] Strughold as Harry Armstrong's protégé was made the scientific director of the Center. While work on the new Strughold compilation and analysis of documents continued, Schreuder felt constrained as to whether other forms of collaborative work were possible at the Center.[117]

The Center was approved in late August 1945; in mid-September facilities were requisitioned at the well-equipped Kaiser Wilhelm Institute for Experimental Medicine in Heidelberg; and it opened on 19 October 1945.[118] Burchell was concerned at the upheaval it meant for the incumbent Bothe (who had on balance an anti-Nazi record), whereas the politically adept Kuhn, the director of the KW Institute, was doing everything he could to accelerate the dismantling of Bothe's department. American interrogators ranked Kuhn as a borderline case because of his willingness to serve Nazi ends, and his war-related scientific research.[119] Burchell was shocked at the German scientists' selfishness and the lack of concern for victims of the war, whether survivors of air raids or concentration camps. German scientists complained about minimal inconveniences to their work, while they forgot about the German treatment of universities under German occupation. Bronk secured a mobile high-altitude chamber as part of the battery of equipment, and the Center collected a wide range of German wartime scientific documents. He thought that transplanting a few Germans to the School of Aviation Medicine at Wright Field would be advantageous, providing the US Congress would not object.[120]

The KWI for Medical Research was involved in war research on chemical weapons. The biochemist Kuhn continued as institute director, despite his record of anti-Semitic outbursts, Nazi sympathies and involvement in poison gas research.[121] The aims of the Aero Medical Center were avowedly strategic under the Direction of Col. Robert J. Benford of the Westpoint Military Academy and the Mayo Hospital.[122] This knowingly flouted the restriction that Germans were not to engage in strategic research. This was a sort of medical Farm Hall (the detention centre of Heisenberg and other German atomic researchers); at Heidelberg the Americans could establish not only actual knowledge of technical expertise on strategic problems but also gain an idea of individual psychology and attitudes. The American observers did not always like what they saw when it came to the attitude of German researchers. The US political authorities at first raised no objection, although they began to dig into the political past of the involved aviation medical researchers.[123] They included the Berlin aviation physiologists Ruff and Schröder, who were prosecuted at Nuremberg. Burchell doubted Ruff's involvement with human experiments at Dachau. He ranked him as a 'serious man' with expertise on catapult ejection seats and competent in the whole field of aviation medicine.[124] Burchell alerted Bronk to the publication of Alexander's CIOS report on Dachau. Yet he took the view that while figures like Benzinger, Ruff, Weltz and Strughold were problematic in having reached an accommodation with Nazism, they could still be valuable scientific collaborators. By September 1945 the more critical Bronk and the conscientious Burchell opposed the full-blooded support by Schreuder and Armstrong for American exploitation of German aviation medicine.[125] Later, Benford supported Becker-Freyseng with an affidavit at the Medical Trial.

American intelligence objected to the employment of Ruff and Benzinger as Nazi Party members, and also to Strughold because of his key role in Nazi aviation. It was dawning on the American scientific organiser Burchell that Weltz (who had joined the NSDAP in 1937) was a rabid nationalist. Burchell's delving into German aviation medicine publications exposed the vehement Nazism of the scientists. The US air force physiologists could use 'exploitation' as grounds for short-term employment, but their intention was long-term research collaboration. By this time (September 1945) Alexander's CIOS report on research at Dachau was available, and Burchell drew Bronk's attention to it. Burchell quipped he was ready to murder Benzinger and Strughold.[126]

Before the AAF Center was fully established, we can see that its creators were becoming disillusioned, as their close acquaintance with the German researchers revealed an ingrained nationalism and academic arrogance. The convivial Kransberg interrogators and intelligence officers, Tilley and Thompson, agreed with this view. Tilley became concerned with growing German insolence and the dangers of exonerating the Germans because they were potential allies against communism. The interrogators Tilley, Kingscote and Gill were deeply sceptical of the German scientists' denials about human experiments on offensive nerve gases. Tilley battled hard with the pharmacist Hildebrandt concerning phosgene research to disentangle narcotics research from nerve gas experiments, confronting his interrogator with strenuous denials that he had even heard the term 'nerve gas'.[127] Tilley exposed a cunning attempt by the Nuremberg defence lawyer Gawlik to infiltrate Kransberg and strengthen the internees' resolve that they were dedicated patriots, innocent of crimes.[128] The danger was that the Allied authorities would take an indulgent view of atrocities committed by the Germans.

The Alexander reports

By 8 July 1945 Alexander was back at the London base of the Office of the Surgeon General. He presented his highly critical findings on German science in a series of CIOS reports, written between late June and August 1945. His first compendious report on 'The Treatment of Shock from Prolonged Exposure to Cold, Especially in Water', submitted on 10 July, was the incisive account of medical crimes at Dachau. It outlined the tiers of German scientific organisation, as well as citing many involved medical experts and witnesses. Alexander included the prisoners' testimonies as persuasive evidence. Although he endorsed the German finding that rapid rewarming by means of immersion in hot water was the best method of recovery, his report was a penetrating exposé of German scientific crimes.

His reports on 'Neuropathology and Neurophysiology, including Electroencephalography, in Wartime Germany', completed on 20 July,

'German Military Psychiatry and Neurosurgery', completed 2 August, and the later 'Miscellaneous Aviation Medical Matters' dealt with the potential validity of some German wartime research. 'The Medical School Curriculum in War Time Germany' and 'Methods of Influencing International Scientific Meetings as Laid Down by German Scientific Organizations' dealt with how far Nazism prevailed among medical students and researchers. The bulky 'Public Mental Health Practices in Germany: Sterilization and Execution of Patients Suffering from Nervous or Mental Disease' addressed the criminality of Nazi medicine. Alexander found himself torn between his academic interests and fascination for the innovative, and horror at the ghastly crimes committed.[129]

By 24 August 1945 Alexander could write in triumph, 'In the past seven weeks I have written seven reports, totalling 836 printed pages (more than 1,500 typewritten pages). It was a hard grind, but it is done.'[130] OSS in Paris decided that there was no further need of his services.[131] The reports were circulated as restricted publications to the Medical Intelligence Division of the US Surgeon General, and – along with Rascher's report to Himmler – to the war crimes executive in Paris.[132]

Alexander's reports reached the medical programme officers of the Rockefeller Foundation by November 1945. He briefed the Foundation regarding the criminality of the German scientists, notably about the Leipzig physiologist Thomas, whose institute the Foundation supported before the war.[133] He hoped to arrange through the Surgeon General's office to have his reports released for publication and to bring them out in book form.[134] In September 1945 the UNWCC gave the Dachau report wider circulation, praising it as 'a narrative of skilful detective work'.[135] The reports showed investigative flair, scientific acumen and ethical awareness, and excelled over many more perfunctory CIOS reports.

The Dachau report began to achieve sensational status. On 9 November 1945 the Pentagon reversed its objections to publicising scientific crimes, and released 'The Treatment of Shock ...' for publication.[136] By coincidence it was on 8 November that the Commander of the Aero Medical Center Heidelberg sent the Air Surgeon Washington 19 German reports on experiments in chilling human and animal subjects. That these were considered relevant to 'the effect of operations in arctic climates on man' shows the continuous 'doublethink' on strategic exploitation and criminality.[137]

The priority of technical intelligence was overridden by the need to uncover Nazi criminality for the IMT at Nuremberg. The Technical Intelligence Branch of the Joint Intelligence Objective Agency passed the report to the *Washington Post*, which made the most of its scoop, running the story as 'Lurid Nazi "Science" of Freezing Men'.[138] It was only when Alexander returned to Washington that he managed to overturn the embargo on publicising his findings.[139] The magazines *Coronet*, *Harper's* and *Newsweek* featured Alexander's devastating findings on the massive sacrifice of life for research.[140] The Munich-based *Süddeutsche Zeitung* ran a story on 'Human Beasts' on 6 October 1945 which accused Weltz (arrested in September 1945) and Rascher of criminal experiments.[141] The climax came when the IMT used Alexander's report as evidence.[142]

Alexander's investigations were part of a wider effort to evaluate German military medicine. Not all clinical experiments were criminal, but they intrigued the US medical experts. On 13 October 1945 a list of experiments was passed to the US public health officer Major John Monigan, who placed the aviation medical research of Lutz, Ruff and Rein and Strughold among 24 other sets of experiments on altitude and extreme temperatures. Karl Wetzlar of the Frankfurt Physiological Institute researched the reaction of the human organism to extreme temperature. Ruff was noted for experiments on acceleration and pressure chamber experiments to determine altitude illnesses and recuperation during parachute jumps from high altitudes without an oxygen mask. Ruff, Benzinger and Strughold all researched on altitude adaptation. Büchner curiously researched the 'Pathology of the lack of sperm in causing flight accidents due to the influence of cold'.[143]

With the mounting evidence of medical crimes, Alexander hoped to be retained for psychiatric study of the perpetrators. But in November 1945 the US War and State Departments dispensed with his services. He was disconcerted by the sudden publicity for his investigations, and the lack of support for his scheme of 'neuropsychiatric-sociologic' analysis of war criminals.[144] He began research and private practice in Boston from January 1946.[145] Alexander thought that the military medical chapter of his life was over.

Alexander discussed the state of post-war German science with the Rockefeller Foundation. Its officers, Alan Gregg and Warren Weaver, had done much to fund German research institutes before the war; they were shocked at the complicity of medical researchers in Nazism and opposed renewal of funding for German neurology and psychiatry. They placed a brake on John D. Rockefeller III's willingness to support medical initiatives in post-war Germany. When the Foundation was sounded out in November 1945 about the liquidation of the KWG, it made no attempt to press for its retention. The Foundation commissioned a report on German science and medicine, which included a review of the former KWG: the Foundation officers decided in April 1947 not to support what had formerly been a favoured recipient of funds because of concerns over ingrained German militarism.[146] There was revulsion against the misdeeds of 'Hitler's Professors'.[147]

FIAT: From investigations to arrest and trial

Knowledge of German criminal experiments was spreading as other investigative agencies discovered evidence. On 3 August 1945 the US Information Section Counter-Intelligence Branch circulated a report of 12 September

1944 on 'The Use of Human-Beings as Guinea-Pigs by the SS' in its 'Spotlight' series. This identified Ding and Mrugowsky as SS doctors acting in conjunction with the police and SS chemist, Albert Widmann of the Criminal Technical Institute, in conducting experimental shootings with bullets poisoned with acontine.[148] Counter-Intelligence claimed this was the first documented proof of human experiments by the SS, although by this time ALSOS completed investigations of Strassburg and the Natzweiler camp. The report marked a renewed interest in human experiments. Widmann was involved in developing the T-4 carbon monoxide gas chambers to kill psychiatric patients, and thereafter in setting up the gas chambers of the *Aktion Reinhardt* extermination camps in eastern Poland.[149]

FIAT was the successor organisation of CIOS. By August 1945 its military mission was completed, and so scientific priorities came to the fore. Its programme covered human anatomy, physiology, bacteriology, and medical and veterinary science.[150] One aim was to secure 'intellectual reparations' by procuring German wartime technical and scientific publications.[151] Its massive reconstruction of German science and medicine ran into the problem of Nazi human experiments. Although in June 1945 the British and US elements of FIAT were divided, Anglo-American liaison continued, and the French FIAT was established on 5 September 1945.[152] 'Dustbin' went on producing interrogation reports, not least due to the efforts of seconded Canadian personnel. By May 1946 nearly 30,000 reports had been completed. The next stop of the germ warfare internees Osenberg, Blome and Karl Brandt was Nuremberg where they were held as witnesses for the IMT.[153]

Human experiments were ever more suspect as medical crimes. On 4 September 1945 Major A.A. Kingscote of the British Scientific and Technological Branch of FIAT evaluated SS medical research on the basis of the ALSOS reports, the Osenberg documents and interrogations of captured SS doctors. He set out to determine what was criminal and what was harmless. The report highlighted experiments on 'human guinea pigs' in concentration camps. While the SS Hygiene Institutes were naively ranked as primarily for monitoring health conditions, Sievers, the administrator of the SS Ahnenerbe research organisation, was characterised as a 'cunning liar' and in control of lethal research complexes.[154]

The Kingscote Report highlighted atrocities in Auschwitz, based on information supplied by the camp doctor Fritz Klein, an ethnic German from Romania. Klein had experimented with psychotropic drugs to prevent malingering. He was arrested and interrogated at Gifhorn military hospital on 14 June 1945.[155] Klein had been evacuated to Belsen along with the Auschwitz bacteriologists Bruno Weber (who also conducted experiments with hallucinogenic agents), the controversial Hans Münch (acquitted after extradition to Poland) and Hans Delmotte (who committed suicide in 1945). The evidence was supplied by Alexander's former Boston colleague,

Squadron Leader John Thompson, who was a Royal Canadian Airforce intelligence officer, attached to 84 Group RAF with a brief to dismantle Luftwaffe installations. It was a turning point for Thompson, whose incisive grasp of scientific issues brought him promotion as Chief of the Scientific Branch of the British FIAT on 3 October 1945.[156]

Thompson collaborated with James Blaisdell, an aviation physiologist, who joined FIAT in August 1945 as part of a team of Canadian scientific intelligence officers. Both Blaisdell and Thompson had studied pathology with the renowned Ludwig Aschoff in Freiburg.[157] These two Canadian physiologists at FIAT gained praise for the 'utmost value' of their scientific and technological mission.[158]

Thompson's interrogation of Klein became a key source on the role of doctors in mass gassing in Auschwitz. He found out about the human experiments conducted by 'Mangerlay (or Mengle)' as well as by Clauberg, Entress, Koenig, Münch, Weber and Wirths at Auschwitz. The acute and sensitive Thompson rapidly elevated the problem of human experiments as a pivotal issue in the rehabilitation of Germany. It required collaboration between the technical and war crimes departments on an inter-Allied basis. The revelations of German crimes made Thompson lose faith in experimental science, but prompted him to confront leading intellectuals with the problematic legacy of the Nazi experiments.

Thompson was an ideal liaison officer who had wide scientific, linguistic and cultural interests. His task as Chief of the Scientific and Technological Branch of the British FIAT from November 1945 was to make available the results of German scientific and medical research to the Allies. He cooperated with the Medical Research Council (MRC) to assess applied physiology, the Agricultural Research Council, and Department of Science and Industrial Research.[159] Thompson sporadically consigned German scientists to the 'Dustbin' interrogation centre. He secured reports from the pharmacologist Wirth and from Otto Ranke, a physiologist at the Military Medical Academy. Thompson developed the 'Homework' system by initiating the landmark series of FIAT reports.[160] These were part of a comprehensive screening programme for German science under the FIAT umbrella. On 10 April 1946 he gained the support from the German Scientific Advisory Council in Göttingen, when he outlined a scheme for self-reporting by German scientists on their war work. The reports allowed the Germans to present a positive view of their work, in keeping with the British policy of reviving German research.

The FIAT reports – 60 were initially planned – were instructive for the Allies and therapeutic for the Germans in prompting reflection and restoring self-esteem. The ambivalence of British policy can be illustrated by attitudes towards the physiologist Rein, who was on the Scientific Advisory Committee as a distinguished academic, although Thompson suspected Rein of complicity in criminal human experiments.[161] Thompson's strategy was multi-layered, covering justice for victims, therapy for the perpetrator and advancing scientific understanding.

Scientific reports flooded into BIOS in London, and to Ottawa and Washington. There was a colossal microfilming operation on 'Investigation of German Medical Science'. Between November 1945 and March 1946 Blaisdell sent 593 documents to Ottawa on German Medical Intelligence. Over 13,000 pages outlined structures of German research, provided details of medical schools and research institutes, and covered strategic areas like war gases, or less obviously strategic topics as the sexual lure of the silkworm moth.[162] Blaisdell, who like Thompson had postgraduate training with Aschoff in Freiburg, compiled over 1,000 medical reports: 'Through his efforts FIAT (British) possesses one of the most complete files on German medical intelligence in existence.' The list of reports assembled by Blaisdell and Thompson included interrogations with persons implicated in war crimes (eg no. 157 was on the euthanasia psychiatrist Pfannmüller and no. 158 on Claus Schilling, and the protocols of the *Arbeitstagung Ost*). The batch included the reports of Alexander, whose findings came to the attention of the British Control Commission.[163] By December 1946 he covered medical targets for BIOS and the Ministry of Reconstruction, Ottawa – over 10,000 pages by December 1946 were ready for microfilming. Blaisdell's German Medical Documents were extensive, including items 1–593 in the Frankfurt series and 1–437 in the Kaiser Wilhelm Institute series.[164]

The German research reports provided disturbing evidence of human experiments, involving wounding or sacrifice of the experimental subject. In December 1945 Major-General Lethbridge noted that 'FIAT are on the track of a great deal of important information regarding the guinea pig experiments on human beings'. The FIAT report by Schaltenbrand on the KWI for Brain Research prompted the psychiatrist Werner Leibbrand to level accusations of criminality.[165] There was a dual strategy: to establish whether 'the purely scientific results of these experiments, now they have taken place, can be of benefit to mankind'. Secondly, to gather information on the doctors, surgeons and scientists who carried out experiments with a view to their arrest for war crimes.[166] Thompson promised to secure outstanding German scientists for Canada, and the British drew up their own list of desirable personnel.[167]

Thompson clamoured that the authorities failed to provide any directive as to what to do when faced by unethical experiments. He astutely placed his superiors under an obligation to formulate a policy on criminal experiments. He contacted the US chief of the Scientific and Technological branch of FIAT to discuss the matter. Then, on 29 November 1945, Thompson first deployed the concept of a 'medical war crime' in a memo to the US military legal authorities. He declared that 'the sacrifice of humans as experimental subjects' was widespread, and had been well-nigh universal in Nazi Germany. He demanded arrest, interrogation and prosecution of the culprits. To investigate but merely to let the Germans realise that the Allies knew about the ethical violations would be to condone criminality. There was a grave danger that unethical practices might continue in Germany and spread to other countries.[168] The future of clinical research was jeopardised by leaving Nazi experiments unprosecuted.

The Legal Division of the British Element of the Control Commission for Germany was keen to take rapid action and favoured a trial of Karl Brandt because of his seniority. But the Foreign Office and the Judge Advocate General's office in London rejected Thompson's estimate of 90 per cent of German doctors as criminally involved as a 'gross exaggeration'.[169] The Foreign Office justified evading the issue, in that too extensive investigation of German medical war crimes would take out much needed medical manpower in the midst of winter.[170] The civil servants considered that further probing of criminal atrocities in hospitals and universities was undesirable, because of the manpower involved, the difficulty of securing convictions and, 'if undertaken on a large scale it might result in necessary removal from German medicine of large number of highly qualified men at a time when their services are most needed'.[171]

Thompson's memo made an impact: the new policy on medical atrocities was 'to bring one or two conspicuous cases before Mil[itary] Courts', leaving the others to be dealt with by German courts with a British Observer.[172] Although the new policy on medical war crimes was not comprehensive, selective trials marked an improvement on the hitherto total neglect of this distinctive branch of criminal abuse.

Twelve future Medical Trial defendants came to the attention of American and British investigators during 1945: the physicians Blome, Karl Brandt, Handloser (who abruptly lost his position overseeing German public health) and Mrugowsky – all for biological and chemical warfare. To these can be added Rose, who was interrogated by George Rosen in June 1945. The SS academic administrator Wolfram Sievers was identified as having a key role in organising criminal medical research. The aero-medical investigations of Alexander placed Ruff, Oskar Schröder and Weltz under a cloud of suspicion, which soon enveloped Romberg. The British round-up of Ravensbrück perpetrators meant the sulphonamide wound treatment experiments of Gebhardt, Fischer and Oberheuser came under investigation. While these internees were at first informants for technical and strategic information, their association with concentration camps, human experiments and euthanasia prompted the formulation of the concept of medical war crimes by December 1945.

Alexander, Mant and Thompson provided crucial impetus to the new wave of investigation, arrest and trial. Thompson's concept of a medical war crime linked revelations of medical atrocities to war crimes investigations, and provided a basis for prosecutions. The investigators tied the human experiments to German war aims: they reconstructed how prisoners were wounded, subjected to pressure and freezing, and to drugs for experimental purposes. On the one hand, the military relevance of this work meant strategic interest in exploiting the results. On the other, the involvement of high-ranking Germans meant the cases became of interest to war crimes prosecutors. The chains of military command and academic responsibility were to be the basis of the prosecution case at Nuremberg. The Chief Nuremberg Prosecutor Telford Taylor stressed the importance of documentation drawn from Himmler's files and how the IMT proved involvement of key military and SS leaders in ordering experiments.[173] By the end of 1945 there was ample evidence from victims of the experiments, and the Allied forces mustered a cohort of suspect doctors and administrators, dug up a body of incriminating documents and set in motion further investigations – these became the life and blood for the Medical Trial.

Part II
Medicine on Trial

6
From the International Tribunal to Zonal Trials

Medical atrocities featured prominently at the International Military Tribunal (IMT) at Nuremberg. The prosecution of 24 leading politicians, officials and generals publicised the criminality and ethical violations of the Nazis. But it was slow-motion drama with sporadic release of revelations on Nazi misrule. The IMT lasted from 20 November 1945 until 1 October 1946. Its start coincided with the first discussions of the problem of 'medical war crimes', and its close with preparations for the Medical Trial. There was a transfusion of expertise, evidence and the culpable, as witnesses became defendants.

The IMT had a crucial role in shifting Allied priorities from strategic exploitation of German medicine to its evaluation for criminal and ethical violations. The prosecutors revealed a grim catalogue of 'crimes against peace', covering the charge of conspiracy to wage aggressive war. The related charge of 'crimes against humanity' – a concept introduced in 1943 by Hersch Lauterpacht – covered atrocities against civilians, and especially Jews.[1] A staggering quantity of documents and witness testimony revealed how the Nazi leadership waged an aggressive world war with devastating and unprecedented atrocities. Medicine was demonstrably a component of Nazi genocide with medical involvement in gas chambers, chemical warfare, sterilisation, euthanasia, human experiments and plans to eradicate 'racial degenerates'.

The IMT was 'experimental' in prosecuting political leaders for violating world peace.[2] Ideas of human rights, rooted in natural law, were pitted against the view that states were inviolable. By extension, the objective and technical sphere of science, medicine and biology could be exposed to legal scrutiny. The prosecution placed much effort into reconstructing the workings of the Nazi state. Massive charts were hung in court to depict command structures over medical organisations used to execute racial policy. It was within such a structure that the Nazi leaders Himmler and Göring, in his capacity as commander in chief of the Luftwaffe, supported lethal human medical experiments in pursuit of war aims. The experiments were precisely recorded, photographed and filmed, and presented at scientific meetings to obtain medical approval and advance clinical procedures. The prosecution used the experiments to develop their cases of aggressive war and crimes against civilians, as well as to show the malign effects of Nazi power structures. Medicine was central to the criminality of the Nazi state.

The IMT claimed status as an international trial, as victim states and neutral countries endorsed its statutes. But lacking precedents in international administration and law, it was in practice a highly testing exercise in Four-Power collaboration. Taking place in the American zone at Nuremberg, the US authorities had extra prominence and the trial was mainly financed with American resources. The Chief of Counsel at Nuremberg, Robert Jackson, had acted on President Truman's decision that a trial was the best means of dealing with captured Nazis. Jackson led the Four-Power negotiations to establish the tribunal and made much of the initial running in terms of procedure. He took a crucial role in reaching the decision to try not only individuals but also organisations – notably the SS and military high command, and in adopting 'crimes against humanity' as a major charge.[3]

The charge of war crimes against civilians included experimental surgical operations in concentration camps. The operations, shootings, cold and pressure experiments were part of a larger picture of beatings, tortures and killings.[4] Jackson's opening address described how 'To clumsy cruelty, scientific skill was added', and how 'to cruel experiments, the Nazi added obscene ones'. He cited how Air Marshal Milch authorised the re-warming of frozen bodies by 'four female gypsies'. Medical atrocities showed how 'Germany became one vast torture chamber'.[5]

The prosecution linked activities of SS doctors to racial philosophy. The Americans presented evidence on links between slave labour policies and euthanasia of chronically sick prisoners, and on the experiments of Rascher in Dachau. The evidence came from Alexander's CIOS investigations, ALSOS and the Dachau trials.[6] The IMT accelerated the hunt for evidence of criminality. ALSOS had investigated the Reich Research Council as the key agency linking civilian, military and racial medical research. Its files contained evidence on Rascher's human experiments at Dachau and Strasbourg, and on links between Haagen and Rose. Late in 1945 these files were re-evaluated as regards human experiments, and crucial documents went to Robert Jackson's Nuremberg prosecution team. These included Haagen's correspondence with Rose on typhus experiments using different vaccine strains. The letters were kept on file by the war crimes officers who used them for the case against Rose in the Medical Trial.[7] A group of medical culprits emerged: the aviation medicine researchers Romberg, Weltz and Holzlöhner; Karl Brandt and the Ahnenerbe manager Sievers were key organisers; Mrugowsky and Ding were cited as involved in experiments on poisoned bullets.[8]

The War Crimes authorities at Dachau scrutinised Weltz in March 1946, when he insisted that he experimented on animals and that he and Rascher were at loggerheads.[9] His defence was essentially that which earned him acquittal in the following year. The first Dachau trial ran from 15 November 1946, and among the 40 defendants were the concentration camp doctors Fridolin Puhr, Wilhelm Witteler and Hans Eisele. Although Eisele conducted experimental operations and gave lethal injections in other camps, he was given only a nominal sentence. The prosecution of the malariologist Schilling was the first of the post-war trials to confront the issue of medical experiments as murder. Although neither a Nazi Party nor SS member, Schilling's research was supported by Conti and Himmler, and involved infection of between 1,000 and 1,200 inmates, and – because of falsified death certificates and secondary causes of death – an unknown number of fatalities.

Schilling's death sentence provided a rallying point for medical colleagues. Nauck of the Hamburg Tropical Institute, the Berlin pharmacologist Heubner and Nachtsheim, the Director of the KWI for Biology, were indignant at the prosecutions. Scientists protested that Schilling was a humane researcher and a conscientious physician. He was executed on 28 May 1946. The Schilling case was in many ways a forerunner of the Medical Trial, and acted as a focal point for a self-righteous German scientific community, who accused the Americans of human experiments.[10]

The main US witness for the human experiments at Dachau was a Czechoslovak physician Frantiseck Blaha, who testified on 11 January 1946 about his experiences as a prisoner from April 1941. He provided a staggering catalogue of experiments resulting in deaths of thousands of prisoners. His evidence that Dachau was 'an extermination camp' appeared grim and precise. Leading Nazi officials, military and medical men visited the camp as if 'an exhibition or a zoo', and the experimental stations of the aviation medical researcher Rascher and Schilling formed a grotesque tour itinerary. The visitors included the defendants Kaltenbrunner, Frick and Rosenberg, indicating how devastating experiments were linked to the highest levels of the Nazi state.[11] Blaha was challenged in court on whether Frick and Rosenberg really visited the camp, but otherwise his evidence was allowed to stand.

Blaha overstated the numbers involved in some experiments and exaggerated Dachau's role as an extermination camp. He alleged that the camp had a functioning gas chamber (in fact, although installed in 1943, the chamber was used for experimental gassings with an unknown number of victims). Neither the prosecution nor defence knew enough about the concentration camp experiments to challenge Blaha's confident assertions. He came to disagree with Alexander over the scientific rationales of the experiments, and Alexander and Mant expressed no confidence in this 'obvious faker'.[12] Blaha was a committed communist (at the time of interrogation he was a member of the Czechoslovak parliament), who criticised the US lawyers for their ostentatious display of their legal prowess under the sensationalising glare of international publicity. He clashed with Alexander for ascribing too much scientific value to the German 'research'.[13]

The French prosecution dealt with medical experiments as part of their case against criminal groups in the Nazi state. The French presented Alexander's CIOS report on 'Neuropathology and Neurophysiology' to the court on 7 February 1946, and cited how the Director of the Kaiser Wilhelm Institute for Brain Research, Hallervorden, harvested the brains of victims who were experimented on at his institute. The prosecutor deployed this as evidence against Göring in that the brains were used to assess effects of air accidents, and established a link between Göring and Hallervorden.[14] Hallervorden wanted to be interrogated at Nuremberg in order to refute the view that he experimented on victims, who were then killed at his request.[15] Whether the killings were at Hallervorden's express wish remained controversial.[16] However, Hallervorden never produced any documentation to refute his first admission to Alexander. Neither Alexander nor the Americans pressed for Hallervorden's prosecution, whereas the French considered him culpable.

The French contributed documents on human experiments and the Jewish skeleton collection at the former Reich University of Strassburg, and at Natzweiler and associated camps in Alsace. Marie Claude Vaillant Couturier described the X-ray sterilisation experiments at Auschwitz, and the operations on the Polish 'Rabbits' at Ravensbrück. On 29 January 1946 Alfred Balachowsky, a parasitologist and resistance activist from the Pasteur Institute, testified on the human experiments under Ding/Schuler. The Buchenwald experiments to compare the effectiveness of different types of typhus vaccines resulted in the death of 600 prisoners who were used as a culture medium for the infectious pathogen, as well as other non-vaccinated 'controls'.[17]

The Soviet counsel, allocated the prosecution of crimes committed in the East, provided massive documentation and eloquent witnesses on atrocities on a scale so vast that even the hardened (and initially sceptical) US and British judges and prosecutors were stirred. They included evidence on euthanasia, deliberate infections with typhus and malaria, efforts to induce cancer, and germ warfare experiments, which put 'to a shameful and evil use the great discoveries of Robert Koch'.[18]

The Soviet prosecutor called the bacteriologist Major-General Walter Schreiber as a surprise witness. To considerable effect, and at a late stage in the trial – on 26 August 1946 – Schreiber testified that German military medical research contravened 'the unchangeable laws of medical ethics' with biological warfare research and experiments on human beings. As a German, his testimony on Nazi biological warfare could not be dismissed out of hand as communist propaganda. He undermined the defence evidence that Hitler opposed gas and biological warfare. Schreiber was a suspected war criminal, and this made him a pliable tool for the prosecution.[19]

Schreiber described an institute undertaking germ warfare experiments, sited at the Reich University of Posen in occupied Poland under the Deputy Reich Physicians' Führer Blome, who had been appointed by Göring. Blome attempted to conceal the installations for human experiments as the Germans retreated. Schreiber testified that as late as March 1945 Blome contacted him so as to be able to continue research on plague cultures. He cited lethal head operations by Gebhardt at Hohenlychen, and freezing experiments by the physiologists Holzlöhner and Kurt Kramer.[20] Schreiber's statement incriminated Blome, although partially exonerated the medical commander-in-chief Handloser. Schreiber's concern was that any artificially induced epidemic would backfire as it would infect German soldiers and civilians, particularly with the westward movement of refugees.[21]

These issues gained prominence in the widely disseminated IMT reports. The veracity of the accusations was left to the Medical Trial to resolve, when Blome and Karl Brandt were prosecuted for chemical warfare research. Whether Schreiber was pressurised by the Soviets into making false allegations has remained controversial. He had been taken to Moscow for indoctrination and then returned to a senior position in the Soviet zone.[22] Ironically, Schreiber was to defect to the Americans, who whisked him to the USA; after accusations of his participation in human experiments at Ravensbrück, he was dispatched to Argentina.[23]

At first the commitment of the British government and military authorities to war crimes trials was tenuous. But once they engaged with the evidence, British lawyers contributed much to securing convictions at the IMT. The British took on crimes against peace, and presented evidence of how military needs resulted in the calculated cruelty of the human experiments. The British cross-examined Air Marshal Erhard Milch, who was called as Göring's witness. The prosecutor pinpointed Milch's scarcely credible explanation of how he corresponded with Himmler and the SS about human experiments without – apparently – remembering anything about the matter or reading the letters he signed.[24] When concluding the British case, Elwyn-Jones accused the SS and German armed forces of human experiments as part of a broader pattern of aggressive warfare.

The IMT assembled Nazi medical criminal suspects from 'Dustbin' and various other internment camps for interrogation at Nuremberg. The IMT proceedings cited activities of 14 physicians, who would be defendants at the Medical Trial (Beiglböck, Blome, Karl Brandt, Fritz Fischer, Karl Gebhardt, Karl Genzken, Siegfried Handloser, Joachim Mrugowsky, Adolf Pokorny, Helmut 'Poppendiek', Hans-Wolfgang Romberg, Konrad Schäfer, Oskar Schröder and Georg-August Weltz). Herta Oberheuser and the SS administrators Viktor Brack, Rudolf Brandt and Sievers were also in the court prison. The virologist Haagen was held as an IMT witness for two months in the winter of 1945 until 5 January 1946. The trial drew attention to links between IG-Farben and human experiments through the Frankfurt professor, Carl Lautenschläger.[25] The IMT generated a criminal nucleus of 17 medical suspects for subsequent prosecution, as well as a battery of charges. The war crimes authorities were reluctant to release the medical suspects until the issue of their criminal activities had been resolved in court. The physicians Hoven, Rose (who expected his release[26]) and Rostock were in custody and interrogated by separate agencies, and Becker-Freyseng and Ruff were collaborating with the American Air Force. Their eventual inclusion in the Medical Trial arose from issues raised at the IMT, rather than from any new lines of inquiry. Overall, the IMT was formative for the prosecution of medical crimes. But at the time most prisoners were viewed as only temporarily at Nuremberg. The Hohenlychen group of Oberheuser, Fischer, Gebhardt and Treite were transferred to British custody in January 1946.[27]

Sievers was the only one of these witnesses to be cross-examined in court. He appeared on 27 June and 8 August 1946 as part of the cases against the SS, and against Göring. Sievers claimed that he penetrated the sinister SS in order to disrupt its murderous workings. The prosecution depicted him as an unscrupulous supporter of mass murder based on pseudo-science. A charitable explanation, developed by Michael Kater, is that he was a romantic elitist, who saw the SS as a means of realising a racial order based on national culture, history, honour and duty.[28] Sievers presented evidence to the IMT on how the SS collected skulls of 'Jewish-Bolshevik' commissars as a degenerate species. His evidence outlined a chain of responsibility for this macabre medical atrocity linked via Rudolf Brandt to Himmler. Elwyn-Jones used Sievers' diary to interrogate him on his role in a series of human experiments, notably the Jewish skeleton collection at Strassburg and Rascher's experiments. Elwyn-Jones accused Sievers of engineering 'the perversion of science' to assist the SS in enforcing German occupation.[29] Horst Pelckmann, the counsel for the SS, argued that although Sievers was 'chargeable with guilt', Göring was ignorant about the gruesome experiments.[30]

Sievers turned out to be an extraordinary witness. He denied any role in instigating experiments. His denial of individual responsibility was typical of the defendants, and he refused to estimate the numbers of victims of human experiments. He insisted that the real culprits were the deceased Hirt and Rascher.[31] This fitted the defence view that the SS was a fragmented organisation.[32] He testified to a sceptical court that he was part of an organised anti-Nazi conspiracy, in effect turning the conspiracy charge on its head. He declared that in his conscience, he 'personally rejected the experiments'.[33] The number of issues requiring resolution in a follow-up tribunal was mounting.

The accusations of the criminality of human experiments spurred the German lawyers to formulate defence strategies. Servatius (who emerged as one of the toughest defence lawyers at the Medical Trial) was defence counsel for the leadership corps of the Nazi Party. He argued that public health doctors and military medical advisers should not be condemned without establishing individual complicity in crimes. He pleaded that the IMT had to verify the connections between the actions of political leaders on trial and the NSDAP and the complicity of the professionals. Moreover (and by this time the Medical Trial was decided on), the task of verification should not be left for successor trials.[34]

But, despite the mounting evidence of medical atrocities, proving responsibility for what had taken place was by no means easy. There were so many grim medical abuses – lethal injections, blood transfusions and other forms of medical killing – that among the copious evidence certain incidents may have been wrongly construed. For whenever the defence lawyers had no interest in refuting allegations, then the witnesses' statement was allowed to stand. The exaggerations made by Blaha illustrated this, as neither the defendants nor their legal representatives had the knowledge to mount a challenge. The SS judge Georg Konrad Morgen contested Blaha's evidence that Dachau was an extermination camp.[35] A prosecutor might calculate that even though a piece of evidence was dubious, it could be strategically advantageous to have it entered in the court record.

The allegations regarding Pokorny and herbal sterilisation exemplify the problem. Fortuitously. Karl Tauböck, an IG Farben botanist, who was resident in Nuremberg, had made a statement to the CIC and then to Jackson's prosecution staff, alerting them to the sterilisation research.[36] Pokorny's memo to Himmler suggesting sterilisations of Jews by means of the plant *Caladium seguinum* was read in full to the court on 9 August 1946. The circumstances of his writing the memo, and the repercussions, were not dealt with at the IMT, so that the denial in court that the herb extracts were used in the mass sterilisation experiments was not properly scrutinised.[37] A similar problem arose with the Tribunal's finding that medical experiments were performed on Soviet POWs as part of bacteriological warfare tests by Blome at Nesselstedt/Posen. The American prosecutors then had the task of substantiating allegations that took place in a location to which the Soviets denied access.

At a more general level the evidence showed how – in the words of the French chief prosecutor – the killing became 'all the more brutal because more scientific'. The link was tightly drawn between science and genocide, and medicine as a component of the totalitarian state.[38] The racial classifications of Hans F.K. Günther and the perverted Nazi 'ethics of life' led to 'sterilization, physiological observations made in the camps, and 13,000,000 dead'.[39] The Soviet prosecutor Marshal Roman Rudenko took a similar line: that scientific methods of extermination were used to implement genocide.[40] He cited evidence from Blaha concerning the torture and killings of Soviet soldiers, and accused Speer of responsibility for production of poison gas and for backing of chemical warfare.[41]

The defence counsel for the SS exploited Rascher's Dachau atrocities to establish that the SS adhered to strict legal principles. The SS magistrate Friedrich Karl von Eberstein testified that he arrested Rascher so that his 'criminal deeds should be punished by a court trial'. In effect, it was the SS which first conceived of a special medical trial for this key human vivisector. To refute the claim of legality, witnesses testified that Himmler knew of the murderous cruelty of the experiments, and that SS Ahnenerbe records showed that experiments continued on inmates after the arrest of Rascher and his wife. They were accused of deceitfully removing babies, rather than of maltreating prisoners.[42]

Dr Paul Hussarek (a Dachau internee) informed Alexander that Rascher had been executed by the SS. Other victims passed evidence of atrocities to the prosecution at Nuremberg. The IMT elicited vocal statements from witnesses of coerced experiments. The statements were sent to various Allied authorities or were issued by ad hoc medical committees. In December 1945 Karel Sperber, a Czechoslovak physician (at the time in New Zealand) sent a deposition to the British war crimes authorities on how he had been captured in the Far East and sent to Auschwitz, where he witnessed medical atrocities. In February 1946 the British Judge Advocate General received Sperber's report on human experiments at Auschwitz.[43] Gisela Perl and fellow medical survivors of Auschwitz were keen to testify against Mengele. A Committee of Czechoslovak doctors reported on German Medical Science as Practised in Concentration Camps and in the so-called Protectorate.[44] As many victims and witnesses were doctors, they provided the possibility of expert testimony, which was underused at the Nuremberg tribunals.

Major extermination centres were in the US zone (notably Hadamar, Grafeneck, and, until US withdrawal in favour of the Soviets, Bernburg; Hartheim was also in the US zone of occupation in Austria). The first Hadamar Trial took place in Wiesbaden from 8 to 15 October 1945, amidst a barrage of publicity, and with an invitation for 'all thinking Germans' to attend.[45] Because of the uncertainty whether the Allies could legitimately prosecute the killing of Germans as a war crime, the trial dealt with the killing of Polish and Russian forced labourers, who had been murdered on the pretext that they were suffering from 'incurable' TB – a pretext contradicted by the forensic pathology of exhumed corpses. The killing of the supposedly tuberculous at Hadamar became an issue at the Medical Trial, when the prosecution argued that it was an extension of the euthanasia programme.[46]

The euthanasia killings of some 70,000 German victims under the T-4 programme were left to the reconstituted German judicial authorities.[47] In February 1946 the Staatsanwaltschaft Frankfurt am Main began to investigate the staff of Hadamar and certain T-4 personnel. The SS officer and Eichberg asylum director, Friedrich Mennecke, was tried for serving the T-4 organisation by selecting victims in hospitals and concentration camps. The Eichberg trial, which concluded at Frankfurt on 21 December 1946, implicated Viktor Brack, for having recruited physicians to serve as adjudicators for euthanasia killings.[48] Mennecke provided early and decisive testimony at the Medical Trial.[49] The trial of asylum staff at Hadamar for the killing of 10,000 patients between 1940 and 1945 opened on 24 February 1947, and ran until 21 March 1947 in the shadow of the Nuremberg Medical Trial.[50] The Hadamar Trial focused on the physicians, nurses, administrative staff and technical staff within the Institution. One consequence was that the Wiesbaden district official, Fritz Bernotat, responsible for psychiatric institutions in Hessen, was never in fact prosecuted. These prosecutions highlighted Karl Brandt and Brack as involved in the T-4 euthanasia organisation.

In June 1946 the revelations of the cruel TB experiments on 20 children at Neuengamme, and their gratuitous killing along with two French doctors (Professeur Florence from Lyons and Dr Quenouille of Villeneuve St Georges) who protected them provided a shocking instance of the cruelty of medical experiments. The British placed two of the camp doctors on trial, one of whom – Alfred Trzebinski – was involved in the murders. The British did not have the key doctor involved in the atrocity, Heissmeyer, whom they characterised as a doctor from the University at Berlin.[51]

The IMT commanded widespread attention in terms of publicising Nazi atrocities. The prosecution spectacularly accused Wilhelm Frick, the Reich Minister of the Interior, who had inspected the killing installation of Hartheim, of facilitating 275,000 deaths by euthanasia. But the accusation was made in vague and imprecise terms, without outlining the different phases of Nazi euthanasia. The British considered Conti, the Reich Health Führer, a prime target for a future trial, but his suicide in October 1945 prevented this. This meant the loss of a key witness. His interrogations thrust the issue of human experiments to the forefront of the Allies' attention. Conti accused Blome, the Deputy Reich Medical Führer, of carrying out human experiments. But his final note left the issue tantalisingly open: 'Blome told me of his intentions to experiment on humans ... I had to do anyhow, what he did. I have never learnt whether he actually made experiments.'[52] When the decision to hold the Medical Trial was reached, the IMT's voluminous documentation provided a foundation for the prosecution case.

The IMT tentatively explored links between the racial ideology of the NSDAP and medical abuses of human experiments and euthanasia. Mengele was mentioned on four occasions, but as a camp doctor rather than for experimental butchery.[53] The IMT criminalised only those aspects of sterilisation which did not come within the Nazi sterilisation law. Rüdin, the psychiatric geneticist and architect of the 1933 sterilisation law, did not appear in the IMT protocols, whereas Pokorny (who proposed sterilisation by dosing with a herbal extract) was associated with the Auschwitz X-ray experiments. When it came to euthanasia, the IMT followed a lead supplied by Alexander and linked brain research to 'the Kaiser Wilhelm Institute'.[54] This was the one occasion at the IMT that the KWG was considered in the broader context of the Nazi military and political structures. The IMT set a pattern for the Allied doublethink on German medical research: by leaving the mainstream scientific structures intact, this highlighted concentration camp atrocities and the role of the SS.

Locating the medical experiments in the Nazi state, and establishing the link to racial ideology and interests raised the issues of totalitarianism and genocide. Whereas the defence insisted on fragmentation – for example, that the Waffen SS was distinct from the general SS – the American and British prosecution case was based on the interconnectedness of all aspects of the Nazi state as part of a system of repression and coercive power. The Western Allies drew on the totalitarian paradigm, which although first applied to Italian fascism, gained widespread currency in the incipient Cold War. Jackson broadened the charge of conspiracy to an analysis of the Nazi social system. A US prosecutor analysed the *Führerprinzip* as totalitarian, stretching into all spheres of public and private life.[55] Major atrocities were ultimately attributed to Nazi leaders and their organisations, and were part of a conspiracy to wage ruthless war. The core Nazi leaders Hitler, Himmler and Göring ordered the experiments, and administrators like Sievers and Rudolf Brandt operationalised the orders. Organisations, notably the NSDAP and SS, proven guilty of criminality by the IMT, carried forward the relentless procedures of medical destruction. Against such a streamlined view of concerted conspiracy, the defence lawyers outlined a fragmented system, in which they denied that those accused could be convicted of criminal acts, and (with some success) refuted the notion of conspiracy.

Genocide was a divisive issue. Lemkin hovered on the fringes of the trial, as adviser to the US War Department. He scored a minor victory as 'genocide' was included in the indictment.[56] On 20 November 1945 the opening charge of the IMT included 'deliberate and systematic genocide, viz. Extermination of racial and national groups ... particularly Jews, Poles and Gypsies, and others'.[57] This pioneering use of genocide was massively publicised.[58] The prosecution sporadically deployed the term. Lemkin was unhappy that the concept of crimes against humanity was too weak, because it failed to draw attention to the rationales for mass killings as part of a planned effort to eradicate undesirable races and cultures from Europe. The prosecutors referred to the victims in terms of national origins rather than as Jews or Roma. The more that the IMT was limited in its depiction of the Holocaust, the more critical Lemkin became of the prosecution.[59]

One consequence of the IMT's status as a military trial was that eugenics made only a limited appearance. The interest of the SS in eugenics and issues to do with eugenic social policy was recognised – indeed, eugenics occupied a box in a gigantic courtroom chart detailing the SS power structure. But the trial skirted round the issue of sterilisation carried out in accordance with the law of July 1933. The IMT focused on euthanasia in the concentration camps rather than in German psychiatric hospitals.

The broader issue was whether the medical experiments were part of a system of planned genocide. The prosecution strategy was oriented to establishing links to the Nazi state and to its organisations. The link to Himmler and the SS opened up the issue of the experiments as racial violence, but Lemkin considered that the role of medicine in bringing about the Holocaust was not fully articulated. More attention was given to the strategically oriented experiments (involving military and air force commanders) than to the role of doctors in mass killings in Auschwitz and other extermination camps. The IMT set the direction of prosecuting human experiments to demonstrate the criminality of Nazi leaders and their organisers.

'The shape of things to come'

The IMT proceedings had an overwhelming importance in shaping issues and procedures at subsequent war crimes trials, and denazification tribunals. Whether there should be a second Four-Power trial remained unresolved until the summer of 1946. Gustav Krupp von Bohlen und Halbach (a major patron of brain research) was too sick to appear at the IMT, while his son Alfried Krupp – despite responsibility for armaments production and slave labour – remained unprosecuted. This lapse was an incentive to the French and Soviets to press for a second international trial of German financiers and industrialists. It was only at a late stage that a medical trial came onto the agenda.

Linked to the second IMT issue was what to do with the burgeoning numbers of German prisoners. Assistant Secretary of War, Howard C. Petersen, remarked in June 1946 that there were 90,000 'potential war criminals' in US custody.[60] The British had large numbers of criminal suspects. Just as the Medical Trial started in December 1946, 'Operation Fleacomb' reduced numbers in detention. By July 1947 the British had released 2,297 of 4,261 persons screened, including numerous concentration camp medical personnel and nurses.[61] War crimes trials were intended only for major criminals. The Allies adopted a highly selective policy of bringing only the most conspicuous perpetrators to trial, leaving the bulk of presumed petty criminals to de-nazification tribunals.[62]

While Allied politicians prevaricated over the next step after the International Military Tribunal, a small but dedicated group of military investigators defined and gathered evidence on 'medical war crimes'. The discussions of a follow-up to the IMT show that a Medical Trial was not on any agenda. Instead, Jackson favoured mounting an 'economic case'. On

1 December 1945 it was agreed that the Office of the United States Chief of Counsel 'should continue in existence beyond the present trial, and take control and general responsibility for all further war crimes proceedings against the leaders of the Axis powers'. The Allied Control Council Law No. 10 of 20 December 1945 laid the basis for war crimes trials in each of the four zones of occupation. Telford Taylor, who had conducted a spirited case against the German General Staff, was appointed as Jackson's successor, and began making preparations for future trials.[63]

On 30 January 1946 Taylor prophesied that 'the shape of things to come' would include 'One more international trial, at which the list of defendants will include a heavy concentration of industrialists and financiers', as well as 'a series of trials of other major criminals to be tried in American courts in the American zone'.[64] He did not envisage a medical trial. A formal agreement had been made to try Alfried Krupp. The French wanted to prosecute the armaments manufacturer Hermann Röchling as the Reich plenipotentiary for iron and steel in Lorraine between 1940 and 1942, and for employing prisoner and slave labour. The question of a second IMT was complicated by dwindling financial resources on the part of the Allies for mounting prosecutions. The launching of de-nazification tribunals to tackle mass membership of Nazi organisations provided an alternative to trials on the basis of individual responsibility.[65]

During the first half of 1946 preparations began for a second international military tribunal. The chief British prosecutor at Nuremberg and Attorney General, Sir Hartley Shawcross, hoped that the second IMT could start in October 1946.[66] He initially considered that Conti, the former Reich Health Führer, should be included among a group of leading Nazis for trial. Conti (nominally a British prisoner) was under interrogation in Nuremberg, but the planned trial was prevented by his suicide in October 1945. In December 1945 the British requested Karl Brandt's return from Nuremberg for trial as a war criminal. In April 1946 Shawcross commissioned Elwyn-Jones to prepare the British prosecution of a group of ten industrialists, thereby complementing US preparations under Taylor.[67] At a meeting of the Chief Prosecutors of the Four Powers on 5 April 1946, Rudenko argued for a second IMT.[68] The French lawyers (whose government included some communists) were also 'taking a very determined attitude in favour of another trial'.[69] Testimonies against the atrocious exploitation by German industrialists raised medical issues of conscripted workers as victims of human experiments, and of death through lack of food and medical care.[70]

In February 1946 Jackson was sceptical about a second international trial owing to doubts about American political support, and because 'the Russians were almost certain to insist that any second trial be held in their territory and presided over by a Russian judge'.[71] Jackson was in any case disenchanted with the trial procedures after Göring's robust defence. He observed that, 'a four power trial was more expensive and prolonged than a single power trial'.[72] Separate zonal trial programmes were already in prospect. In July 1946 the Soviets hardened their position on the necessity for an international trial of the industrialists Alfried Krupp, Hermann Schmitz and Georg von Schnitzler of IG Farben, Röchling, and the banker Kurt von Schröder.[73] The suggested location was Berlin, which was the agreed permanent seat of the IMT.[74] The spectre of a trial in the Soviet zone made the British and Americans uneasy that this would be a show trial. The choice of financiers and industrialists was ideologically problematic with the onset of the Cold War.

British and US diplomats and pragmatically-minded US lawyers opposed a second IMT.[75] The liberal-minded Shawcross prudently cautioned that if any Allied power renounced the London Charter, this could jeopardise zonal trials.[76] But he was out of step with the British Foreign Office civil servants, who worked behind the scenes to persuade the Americans to adopt a unilateral policy of a series of trials rather than a second Four-Power trial. This tactic got the British government off the hook from publicly reneging on the London Charter.

Taylor was in Washington between February and March 1946, rekindling political support for a new series of trials and supervising recruitment of a replacement US legal team.[77] He informed the Secretary of War on 29 July 1946 that the British were 'not enthusiastic' as regards a second trial, but (mindful of Shawcross) they did not want the opprobrium of terminating the London Charter of 1945 as the legal basis of Allied war crimes trials.[78] Reinvigorated by Taylor, the US war crimes staff favoured prosecution of IG Farben.[79] Taylor opposed both the Four-Power and any other international scheme. He disliked 'continental and Soviet law principles unfamiliar to the American public'.[80] By March 1946 Taylor was recruiting 'top notch lawyers' for successor trials. He mentioned as German target groups lawyers, officials, financiers and industrialists (he was himself the son of a General Electric engineer) – but not doctors. The scheme for a series of special trials under the jurisdiction of the US Commander-in-Chief arose five months before the decision to hold a Medical Trial.[81]

The legal basis remained that of the London Charter of 8 August 1945, but Taylor believed American judges and prosecutors should follow procedures 'consistent with the laws of military government'.[82] The Four-Power structure of the IMT meant much repetition of evidence in court and was administratively cumbersome. The difficulty was that industrialists and financiers were too prominent in the wrangles over whether to hold a second international trial. Taylor outlined 'a balanced programme covering representatives of all segments of the Third Reich'.[83] His scheme analysed how Nazi power structures involved key socio-economic and professional groups. Nuremberg was to be kept as 'a going concern', but what had been referred to as an international firm was converted into an all American enterprise.[84]

The shift to specialist trials was ratified in June 1946, when the US adopted the 'zonal courts' policy.[85] President Truman ordered preparations for the US trials programme, while he delayed a decision on a second IMT to avoid international complications.[86] Taylor was informed on 16 August 1946 that the US President was personally to decide on subsequent trials. His orders were to prepare all trials other than a second international trial.[87] Nazi industrial leaders were transferred to the zonal programme. Taylor planned to have six courts operating simultaneously for trials of the military, SS, industrial leaders and (significantly) 'medical experiments'.[88] Taylor's powers as Chief of Counsel for War Crimes were consolidated by a US military decree of 24 October 1946, when he was appointed Chief Prosecutor, and brought into the structure of the US Military Government for Germany.[89]

Once the decision to launch the Medical Trial was reached in mid-August 1946, the prosecutors were faced by the short time to prepare their case, and whom to prosecute. Prominent Nazi medical personalities were missing: a US press release gave these as Conti, Grawitz, Hippke, Rascher, Ding and Bouhler.[90] Scientists who had died at the end of the war included Flury the poison gas expert (suspect for having supported carbon monoxide gassings[91]) and Gildemeister of the RKI. Others, like Conti and the Auschwitz camp doctor Eduard Wirths, committed suicide in British custody in 1945.[92]

Walter Rapp, Director of the Evidence Division, began in late August 1946 to prepare the prosecution of Handloser and Mrugowsky for human experiments.[93] By September 200 personnel were employed in preparation of the Medical Case.[94] In contrast to the protracted IMT, the special trials were meant to take just two to three months. To secure the necessary resources from Congress, the lawyers had to promise efficiency and results. In all there were 100 prosecution lawyers, and (in what represented a large financial commitment) an overall staff of 1,776. Taylor managed a system of streamlined mass production of justice.

Why was the Medical Trial such a latecomer, and yet the first of the special trials? By August 1946 the logical requirement was a US military trial in Nuremberg to prosecute a group other than the politically problematic case against financiers and industrialists. The US war crimes department postponed the Flick/Krupp trial as politically too sensitive, and looked for a trial which could demonstrate rapidly and conclusively Nazi guilt for atrocities. Taylor's team planned 18 trials at Nuremberg, but for reasons of cost 12 prosecutions tackled major sectors of the Nazi regime. These covered the judiciary (with 16 defendants), IG Farben (24 defendants), the Ministries (21 defendants) and High Command (14 defendants). When the legal analyst Ben Ferencz found devastating evidence against the Einsatzgruppen mass killings in Russia, Taylor authorised this additional trial of 24 defendants.

Expediency, and the politics of prosecuting industrialists and financiers meant these cases were postponed, but the US trials programme had to be rapidly launched.[95] During August 1946 Taylor decided that the first of the US-sponsored trials in Nuremberg was to target the doctors. There was little time to settle the target group to be prosecuted. The Foreign Office made clear that Britain would be only too happy to hand over any industrialists to her (Western) Allies to answer any case, and extended this policy to medicine.[96] The Foreign Office's analysis of the onset of the Cold War shaped the policy of British assistance for the Medical Trial. This policy was at the cost of alienating the British prosecution team at the IMT, and Hartley Shawcross, the Attorney General, withdrew by the end of September 1946. Tensions between the restrictive Whitehall civil servants and its highly motivated legal team in Germany were to continue.

French war crimes investigations gained momentum. On 26 February 1946 General Koenig, the French Commander-in-Chief, ordered that Control Council law 10 should apply in the French zone. The French adopted the policy of zonal trials, and established a court at Rastatt, which delivered 22 judgments in 1946 and 59 in 1947.[97] While some nurses were tried, and medical abuses appeared as part of the regime of brutality and starvation in the camps, human experiments were only fleetingly an issue at the Rastatt trial of Ravensbrück atrocities.

British military prosecutors from the Judge Advocate General's department had been preparing a medical trial centred on the sulphonamide experiments at Ravensbrück. The French and the Poles demanded extradition of the Ravensbrück doctors and nurses, and the French demanded transfer of the perpetrators of the Buchenwald experiments. The French insisted on participation in all non-French war crimes trials, and made sure that French witnesses could give evidence.[98] Relations became tense with the French in September 1946 at the time of the handover by the British to the Americans of Fischer, Gebhardt and Oberheuser. Somerhough agreed with his French counterpart, Charles Furby, that the French should provide a judge and an assistant prosecutor at the trial in Hamburg as key witnesses were from France. Suddenly, in early September, the French suggested a reverse arrangement, proposing an international court with a British judge and prosecutor as part of a French prosecution team in a trial held in the French zone. The scheme, conceived to rekindle British support for a second IMT, became untenable, as the French held no Ravensbrück camp staff.[99] The issue raised was whether a zonal court could be constituted as an international court. In the event, the British conducted the brisk Ravensbrück Trial in Hamburg between 5 December 1946 and 3 February 1947 with a French and a Polish judge, and French observers. This was part of a series of British trials, involving medical atrocities – starting with the Bergen-Belsen trial from September to November 1945. The efficient trial of Bruno Tesch for supply of Zyklon B ran from 1 to 8 March 1946.

The Ravensbrück trial complemented the Nuremberg Medical Trial, and secured convictions of all defendants (11 out of 15 were to be executed), even though the prosecution case was at times in doubt. The British handling of the Ravensbrück case elicited considerable criticism from the French observers. Geneviève de Gaulle (niece of the General) had been incarcerated in Ravensbrück from February 1944 for resistance activities. She led the French observers who expressed outrage at the British conduct of the Trial, because of their lack of concern with the killings of detainees.[100] Aline Chalufour condemned the main British concern of proving ill-treatment and death in the camp rather than tackling the issue of the Nazi plan to exterminate the resistance elites.[101] Instead of producing a comprehensive overview of the atrocities at the camp, the British prosecution presented evidence to obtain the maximum sentence against each of the accused. It meant a strong series of cases, but was not the comprehensive indictment of the camp within a Nazi system of extermination and genocide. French internees like Germaine Tillion demanded further trials.[102] The difference of opinion exposed the sense of dissatisfaction of victims with judicial procedures.

'Medical war crimes': origins and policy

In 1945 the British invading forces were in general unprepared as to how to respond to German atrocities and what to do when the concentration camps were liberated. The liberation of Bergen-Belsen in April 1945 forced a decision on what authority was to investigate and conduct war crimes trials. The Judge Advocate's Office, the army's legal department, was made responsible for war crimes trials. Colonel Gerald Draper, who worked on the Bergen-Belsen case, reflected: 'We were not geared, or trained or qualified or had enough resources to do the job. It was a makeshift, hurried and ad hoc decision and we had to do the best we could.'[103] Shawcross, the Foreign Office civil servant Patrick Dean and Labour Foreign Minister Ernest Bevin favoured trials, whereas the War Office officials and army establishment were impassively indifferent.

The war crimes investigation teams were placed under a central command, and Group Captain Tony Somerhough commanded a small British office to gather evidence, identify and arrest culprits, and mount trials. This was located at the British headquarters at Bad Oeynhausen in Westphalia, characterised by the poet Stephen Spender as 'a large sprawling nineteenth-century health resort, full of ugly villas ... like middle aged over-dressed women', but set in a German fairy-tale landscape. The atmosphere in Oeynhausen was 'somewhat like that of an English public school or university', combining convivial team spirit with chronic shortages. Somerhough had a wide brief in that he was to investigate crimes against foreign nationals as well as against British service personnel, but had only sparse resources.[104] At Christmas 1945 Somerhough sent a plaintive message to the War Office, 'We have nothing, not even a typewriter.'[105]

The British and Americans drew up lists of suspect criminal doctors, and the British began to prepare a special Medical Trial. The British (in an operation masterminded by Somerhough's energetic deputy Draper with assistance from one of Sigmund Freud's grandsons, Walter) prosecuted the suppliers of Zyklon gas in Hamburg in March 1946, proving that Tesch supplied the gas in the knowledge that it was used for mass murder.[106] But the main interest in medical matters was to evaluate the utility of German wartime research, and – as the war in the Far East gave way to the Cold War – to exploit the findings for strategic purposes.

The Foreign Office had already thwarted the moves by Shawcross and Dean for a more energetic policy on war crimes. More than ever during 1946 it prioritised the need to check Soviet belligerence. The War Office was concerned primarily with crimes committed against British servicemen and women, and the legal prosecution and investigation teams took ethical concerns on board. The war crimes investigators were highly committed, and medical experts gave crucial backup. They widened their brief from atrocities against Allied personnel to medical war crimes, inflicted on concentration camp prisoners and civilians. The War Office favoured selective prosecution of the worst cases: 'Our policy on trial of atrocities is to bring one or two conspicuous cases before Military Courts leaving others to be dealt with before German Courts with (initially) British Observer and review.' British war crimes investigators singled out a group of five doctors who had worked under Gebhardt at Hohenlychen and were currently in Nuremberg under interrogation. The Foreign Office agreed that 'It would be desirable to give the proceedings very wide publicity'; but the civil servants felt that further investigations of hospitals and universities were undesirable, because of the manpower involved, and the difficulty of securing convictions. A Foreign Office civil servant objected, 'Also if undertaken on a large scale it might result in necessary removal from German medicine of large number of highly qualified men at a time when their services are most needed'.[107]

Thompson's energy and acumen overcame official inertia and bureaucratic red tape by bringing together the medical investigators with the war crimes authorities. He was impressed by Alexander's CIOS report on Treatment of Shock from Cold, and evidence from IG Farben concerning poison gases. He lobbied for further investigations 'by competent authorities in law and in medicine', and approached the British, US and French war crimes staff.[108] He did not let the matter rest with the complacent and cost-conscious legal and diplomatic civil servants in London, who regarded investigations of medical experiments as 'undesirable and unproductive', but tenaciously pursued his aim of a legal medical conference.[109] Thompson won the war crimes lawyers over to his side, and worked tirelessly to build up files on the German human experiments.

Thompson's first step in realising his visionary plan to prevent atrocities was to call for an inter-Allied meeting between lawyers and medical scientists to provide evidence for prosecutions. He hoped for high-level scientific support. He contacted the Director of the National Institute for Medical Research in London, and Edward Mellanby, the head of the Medical Research Council. The MRC was more interested in scientific intelligence than ethics. It requested from Thompson copies of all medical intelligence reports, but declined to press for trials for breaches of medical ethics.[110] Its director, Mellanby, recorded in January 1946: 'I shall be glad to see Thompson when he comes here, though I feel that this is a matter for Cabinet decision, rather than for people like ourselves.'[111] In the event, the MRC declined an invitation to the legal-medical conference.[112]

Thompson approached the British War Crimes Executive at Nuremberg on 12 December 1945. He claimed it 'probable that the majority of top-ranking German research workers have been involved in unethical experiments on living human subjects'. He pressed for collecting evidence 'against such scientists who may number several hundred'. Thompson created a stir among War Office civil servants, when he alleged that 'something like 90 per cent of the members of the medical profession at the highest level were involved in one way or another in work of this nature'.[113] Thompson's original estimate stands against the tally of only 350 criminal physicians, which the German medical representatives suggested to minimise connections between medical research and medical atrocities.[114]

The BWCE considered whether there should be trials in a military court, or whether such cases might be left to German courts with British observers. The dilemma was whether to organise a trial of the five Ravensbrück doctors and of other medical criminals in British hands, giving the proceedings wide publicity. But the BWCE doubted whether other than 'notorious cases' should be pursued, as it would consume immense resources and could result in the imprisonment of most of the German medical elite.[115] The Foreign Office rejected Thompson's estimate of 90 per cent of German doctors as criminally involved as a 'gross exaggeration'.[116] The civil servants found Thompson's analysis challenged the practicalities of maintaining order.

Undeterred, Thompson secured the attention of war crimes lawyers at a zonal level. He nudged the British into action by bluffing that the Americans were proposing to hold a medical conference in Germany.[117] An advantage of the FIAT organisation was that it covered three zones and provided a structure for inter-Allied liaison. What was necessary was that war crimes expertise be injected into the investigations of medical atrocities.[118] Thompson contacted war crimes officers for evidence on medical personnel in detention or wanted in Austria, France and Poland.[119] French military medical authorities provided details of 324 medical atrocities, confirming that here was an area of abuse demanding expert scrutiny.[120]

Permission to hold the legal medical conference required authorisation from British Counter-Intelligence, something that took until April 1946 to come through.[121] While the issue of the conference was gestating, the British reviewed the evidence against Karl Brandt in February 1946, pointing out his centrality in chemical warfare. Major Edmund Tilley of FIAT provided details of his links to chemical warfare experiments at Spandau prison and Sachsenhausen concentration camp. Tilley warned that if the Russians became involved they would merely cart the accused off to 'the Great Beyond' rather than making thorough scientific investigations.[122] The French and Czechoslovaks were also angling for Brandt, who was a prize catch.[123]

Thompson's aim was an inter-Allied meeting involving medical and legal experts to discuss policy on unethical experiments. This posed the challenge of engineering an inter-disciplinary, inter-departmental, and inter-Allied conference. Thompson contacted Charles Fahy, the legal adviser to the US Deputy Military Governor, as well as British and French legal officers.[124] Somerhough, the dynamic head of the JAG's office at the British HQ in Bad Oeynhausen, backed Thompson's initiative.[125] Somerhough was a liberal-minded lawyer who appreciated the need to tackle the human experiments for humanitarian reasons. Although lamentably understaffed and overwhelmed by the enormity of the crimes, which cried out for investigation and prosecution, he committed resources to the task. The pathologist Mant gave tenacious support at the British War Crimes Investigation Unit.[126] Thompson enlisted Sydney Smith, professor of forensic medicine at Edinburgh, to review his card index on medical war crimes.[127] Smith had particular expertise on bullet wounds, and advised Mant on technical issues of evidence of the coercive experiments. Thompson's tenacity over the early months of 1946 accelerated investigations of medical war crimes.

Thompson's inter-Allied meeting on medical war crimes finally took place on 15 May 1946 at the FIAT offices, located (appropriately given IG Farben involvement in medical experiments) in the Hoechst chemical works in Wiesbaden.[128] Four American, two French and nine British FIAT officers discussed the problems of 'scientific information gathered during investigation of war crimes'. The Nuremberg prosecutors were invited but declined to attend, indicating that medical war crimes still appeared to be a marginal issue.[129] They had no inkling before mid-May 1946 that medical crimes were to become the focus of their attention as a result of the momentous FIAT conference.

The head of the British FIAT, Maunsell, explained that the technical brief of FIAT precluded it from becoming a war crimes agency, but advice was necessary as to what to do about this distinct type of war crime. He asked whether a new type of quadripartite agency was required and what its composition should be. Thompson established that material existed on unethical experiments; but often the evidence was fragmentary. German scientists like the chemists Flury and Wirth constantly denied having used human subjects, and evidence was often destroyed. However, persistent questioning of German detainees resulted in admissions that human experiments had taken place. He suggested that the leading German physiologists Strughold, Ruff and Rein (all at this point in academic positions) would have been among the many Germans who knew about the human experiments on freezing and poison gas.[130]

The pathologist delegated by Somerhough to attend the meeting was the capable and energetic Major Mant, attached to the BAOR War Crimes Investigation Unit. A[rthur] Keith Mant began his career in pathology by examining exhumed bodies of Allied military personnel in Germany in 1945. As pathologist to the War Crimes Group BAOR, his task was to establish whether airmen had died from injuries on crashing or whether the Germans executed the airmen after capture. Over three years he examined over 150 corpses, often from unmarked graves, and found 49 were executed by a *Genickschuss* – shooting through the neck or the back of the head.[131] He investigated medical war crimes at Ravensbrück as part of a brief from the British Special Operations Executive. Vera Atkins, a formidable organiser of SOE operations (the Special Operations Executive had supported resistance activities and dropped combat personnel behind enemy lines), initiated investigations of 118 missing agents, and managed to trace the fate of 117 of them when she went to Germany in 1945. She was concerned about the deaths of 13 of 'her girls', who had been killed at Natzweiler, Neuengamme and Ravensbrück concentration camps. Mant was asked to determine the circumstances of the women killed at Ravensbrück.[132]

Mant broadened the scope of investigations from the killings of the SOE operatives to reconstructing the full range of medical abuses at Ravensbrück. He extended the investigation to Auschwitz, taking on board sterilisation and infertility experiments by Carl Clauberg and twin experiments by Mengele, and analysing the German motives. Drawing on the testimonies of concentration camp survivors, Mant listed no fewer than 12 categories of experiments, which he attributed to war priorities (as typhus experiments, war gases, war surgery), racial theories or to the whims of Himmler.[133] Mant's findings on Ravensbrück prompted him to argue that there were in effect two cases against the doctors: one for ill-treatment of prisoners and a 'second powerful case against them in that they were the instruments of Professor Gebhardt of Hohenlychen, and as such carried out medical and surgical experiments in the camp on his directions'.[134] His two substantial reports dealt with the separate categories of medical abuse:

1) Report by Major Arthur Keith Mant, RAMC on the Medical Services, Human Experimentation and Other Medical Atrocities committed in Ravensbruck Concentration Camp.[135]

2) Experiments in Ravensbruck Concentration Camp Carried out under the Direction of Professor Karl Gebhardt.[136]

Tilley, the FIAT officer responsible for poison gas investigations, pointed out that if a suspect was charged, it was easier to extract scientific information. The meeting agreed that scientific data should be gathered in the course of criminal investigations.[137] Thompson cited evidence against a number of persons, who were to be Medical Trial defendants: not only Karl Brandt and Gebhardt, but also the aviation physiologists Ruff, Romberg, and Weltz. The scheme arose for co-operation between the British, French and US investigating authorities.

The British agreed that war crimes lawyers should keep in touch with FIAT, which should provide scientific assistance and documentation.[138] Colonel Clio E. Straight, head of the Legal Branch of the US War Crimes Wiesbaden, admitted that the US authorities did not have evidence 'as extensive as that in British hands', and that trials at Dachau were 'more on the grounds of mass murder than for scientific crime'. It was agreed to conduct separate trials in each zone and that scientific personnel rather than laymen should compose the investigating teams.[139] The forensic pathologist, Smith, who chaired the meeting, favoured a special commission on a Four-Power basis, and involving medical and legal specialists. Much impressed by Thompson's card index on the experiments, he recommended 'critical examination by scientists' and that Major Mant 'should be given the main responsibility with adequate staff and authority'.[140] Smith's proposal for an international authority for investigating medical war crimes led to the founding of an International Scientific Commission.

Thompson concluded that further meetings should be held every two months when the British, French and US medical teams could meet at FIAT. Here other national representatives could also attend. Mant was to centrally co-ordinate all data, and began to collaborate with Lépine in interviewing French victims.[141] Thompson showed Mant FIAT reports from which he selected over 100 names of possible witnesses and accused.[142] With this began the fruitful collaboration of FIAT and the War Crimes Executive, bringing together the mercurial Thompson and the incisive and methodical Mant. The International Scientific Commission for War Crimes was launched at the follow-up meeting on 31 July 1946.

The French supported the scheme for an inter-Allied scientific commission. The immediate post-war years saw a strong current of outrage against Nazi atrocities, particularly from the very large numbers of former medical internees in concentration camps.[143] Bacteriologists from the Pasteur Institute endorsed scientific investigation of medical crimes. Pierre Lépine and René Legroux (who ran the Institute's microbiological service[144]) were both from Lyons, and had rallied to the resistance from a nationalist stance. Lépine was a close associate of René Du Roc, who led the Croix de

Feu movement, which was anti-German and ultra-nationalist. Lépine had expertise in tropical medicine and parasitology, and from 1941 developed the Institute's virological researches. He came from a well-connected medical dynasty, his father at Lyons having held major support from the Rockefeller Foundation, and in the 1930s supported Jewish medical refugees. In May 1945 Pierre Lépine was already in London on a scientific mission, and in June 1947 he was at the Rockefeller Foundation in New York.[145]

At the Hoechst meeting Lépine expressed his shock at Nazi atrocities; he considered that scientists should gather evidence on specific categories of abuse (for example, the typhus vaccine, or cold experiments) and then pass it on to legal authorities for prosecutions. He believed that the French had gathered sufficient evidence for one such prosecution, thanks to the Pastorian parasitologist Balachowsky, who gave evidence on Buchenwald atrocities, and other interned French doctors and medical scientists could provide firsthand evidence. Indeed, Balachowsky achieved the status of a semi-official spokesman on Nazi crimes, as he lectured in America.[146] Lépine also wanted a moral condemnation by scientists in the name of the Four Powers.[147]

The Americans favoured zonal trials. Straight considered that rather than formal, Four-Power Nuremberg proceedings, each country might take up one case, follow it up and arrange for a trial in its zone. For example, the US authorities might be allotted the case of freezing experiments. Evidence would be collected and witnesses brought from other zones. Here in embryo was the scheme for a US-sponsored trial of Nazi human experiments in preference to the Four-Power arrangements of the IMT. But the meeting favoured an inter-Allied co-ordinating structure to supervise the zonal programme, and that FIAT investigators should provide guidance on cases.[148]

The point at which the US authorities began to prioritise medical atrocities can be precisely located to the day after the FIAT conference. On 16 May 1946 Colonel David 'Mickey' Marcus, the Chief of the War Crimes Branch, supported American participation in the FIAT scheme to condemn the 'vicious medical experimentation by the Germans'. Marcus was committed 'heart and soul' to war crimes trials, and he ordered energetic action on the medical experiments cases.[149] He drew on the advice of the genocide theorist Lemkin, and maintained a vigorous dynamism until resigning in 1947 to join the fight for an independent Israel. General Lucius Clay (Military Governor of OMGUS, the American administration of Germany) and Marcus supported the idea that German human experiments be condemned 'by the United Nations or by leading national medical associations of the several Allied countries'. Marcus requested 'the designation of a representative of some appropriate American Fed. Agency, such as the American Medical Association or the US Public Health Service to attend conference in Hoechst, Germany'.[150]

Official US support for a follow-up meeting signalled that Thompson's FIAT initiative was at last bearing fruit. On 11 May 1946 OMGUS (the Office of the Military Governor US) contacted the US War Department about a meeting 'to examine evidence collected by FIAT and discuss possible international action re scientific and medical experiments on live human being[s]. Although War Crimes Prosecution of German Experimenters and current German medical ethics are involved, chief object is public condemnation by UNO or leading National Medical Associations of vicious experimentation'.[151] On 17 May 1946 the American Medical Association nominated the physiologist, Ivy, as official consultant to the Nuremberg prosecutors.[152] The Surgeon General endorsed the recommendation, and Marcus appointed Ivy as Special Consultant to the Secretary of War.[153] Until July 1946, the US war crimes officials envisaged an international tribunal of medical experts evaluating the evidence of human experiments, rather than a trial of medical miscreants.

The nomination of Ivy to evaluate the records of medical research under Nazism and to attend the scientific commission increased US government interest in 'war crimes of a medical nature'. The next stage was for the Nuremberg prosecutors to become involved. Taylor's group held a special meeting, coinciding with the decision not to have a Four-Power trial. Ivy reported in August 1946 that, 'a plan of responsibility, procedure, and strategy for the Medical trials was discussed. It was tentatively suggested that General Taylor's group would try the medical cases.' Ivy indicated that the plans were still not settled: 'I was told by General Taylor and McHaney in General Taylor's group that a meeting would be held in about three weeks to determine who would be responsible for trying the cases, where they would be tried, and the general strategy of the trials ...'[154]

Taylor, who was recently converted to the idea of a medical trial, struck a deal with the British authorities.[155] The Foreign Office was pleased to unload the task of a medical trial onto the Americans. The British forensic pathologists would turn over their evidence to the US agencies for prosecution, while the British medical scientists would supervise the writing of a scientific and ethical report.[156] The US war crimes authorities agreed that the British and French could share any scientific data discovered in the course of investigation. Ivy even obtained a Russian report for transmission to the US war crimes authorities. Innovative research was to be reported to the Medical Research Council, and to French and US equivalents. Panels of medical researchers were to assess data scientifically and ethically.[157]

The planned British trial of medical atrocities at Ravensbrück was the first of the zonal trials to emerge from the flurry of interest in medical war crimes. Mant's careful investigations were crucial. As Ravensbrück was in the Soviet zone, Mant worked by interrogating captured camp staff, and by tracing survivors, collecting their testimonies and by medical appraisal of their wounds. He also studied documents, some held by Polish organisations and others captured from the Germans. Between June and October 1946 he collected evidence on human experiments in France, Belgium, Denmark, Sweden and Norway. He interrogated the self-styled 'Rabbits', documenting their wounds and combining humane sympathy with clinical precision. On 24–28 June 1946 Mant and Lépine took affidavits from Janina Iwanska and Helena Piasecka.[158] These affidavits were reproduced as part of the series of American war crimes documents, and were used at the Nuremberg Medical Trial on 2 January 1947.[159] Mant then went to Hirson in northern France to collect a witness statement from Denise Fresnel, who was the doctor in Block 8.[160]

A number of Ravensbrück survivors had found refuge in Sweden, and Norwegian prisoners supplied significant testimony. Mant took a deposition from Sofia Maczka on 15–16 April 1946 with the assistance of the British consul, Stockholm – this document was also to be used at the Nuremberg Medical Trial. On 11 July 1946 Mant's convoy reached Lund where he interviewed Irena Stanislawa Suchon.[161] On 15–19 July he was in Stockholm.[162] After attending the International Scientific War Commission in Paris from 31 July to 2 August 1946, on 9 August he went to Brussels to collect further evidence.[163] On 20 August 1946 Mant visited the formidable Sylvia Salvesen in Oslo; Gerald Draper (the hawkish chief British prosecuting lawyer) had already interrogated her about the Ravensbrück Revier.[164] Salvesen was to be a powerful and articulate witness.[165]

During June 1946 British investigations of the 'Hohenlychen Group' and the medical experiments at Ravensbrück coincided with the agreed increase of US-British-French liaison on war crimes. The scheme was to pool general investigations concerning the organisation of human experiments, while particular experiments fell into a similar category as incriminated persons at concentration camp trials. Other occupied or neutral countries sent observers or were asked to co-operate.[166]

Although the Allies failed to comprehend the full range of medical research, torture and genocide in Auschwitz, certain clusters were investigated. The British focused on the Hygiene Institute of the Waffen-SS, and on the bacteriologists Bruno Weber and Hans Münch, and the gynaecologist Carl Clauberg. The British were primarily concerned with Clauberg's Auschwitz sterilisation experiments, because he had links with the SS surgeon Gebhardt at Hohenlychen, and because the Soviets appeared to be shielding Clauberg. As the British were keen to undermine Franco-Soviet collaboration (manifested in their joint call for a second IMT), and to spur the US to continue at Nuremberg, the Clauberg case demonstrated that 'The Russian zone may now be considered as sanctuary for those German criminals that enter it.'[167]

The Americans evaluated medical experiments at Dachau to see who might be placed on the Wanted List for illegal experiments on human beings.[168] The Anglo-American agreement of 3 June 1946 established a significant new priority: to bring to trial the persons responsible for policy and direction of the experimental programmes, notably Karl Brandt and the SS-Ahnenerbe official, Sievers, who had responsibility for medical atrocities at Dachau and Natzweiler, and was scheduled to appear in the full glare of IMT publicity on 27 June.[169] At the same time the British prioritised prosecution of Ravensbrück experimenters, broadening policy to non-British victims. At this stage it looked as though the British would hold a medical trial. The British investigations by now covered much of Northern Europe, and Mant covered thousands of kilometres with his forensic convoy. The Soviets intimated that they had no objections to the British mounting a trial for Ravensbrück personnel, given that Hamburg had what was in effect a permanent war crimes court. But when it came to permission to conduct autopsies, they rescinded Mant's access to their zone.[170]

The War Crimes Investigation Unit, BAOR, distinguished between the human experiments initiated by Gebhardt from Hohenlychen/Ravensbrück, and the general brutalities at the Ravensbrück camp. Mant expected that a panel of Allied medical experts would assess the human experiments.[171] The British had in May and June 1946 rejected the alternative of a handover of the Hohenlychen group to the Polish judicial authorities, although most victims were Polish. The question arose of how the British would respond to the new US initiative for a trial of the human experimenters at Ravensbrück/Hohenlychen.[172] By late August the revised plan of action was to hand the medical experiments group to the US for a special trial.

The British retained the Ravensbrück camp doctor, Percy Treite, characterised as 'by far the most intelligent of all the accused'. Although Treite, an SS officer and Privatdozent at Berlin, denied taking part in experimental operations and declared he was an anti-Nazi, the investigators regarded the evidence on his role in experiments as 'overwhelming'. The case was sensitive because the SOE agent Mary Lindell (de Moncy) testified that Treite ran a clean and effective camp hospital.[173] Salvesen had used her contacts with Treite as a cover for subversive activities and saving prisoners from execution.[174] Treite was found guilty, but escaped execution by committing suicide on 9 April 1947 on the same night as the errant Mory.

The French strengthened the initiative on medical war crimes. The next meeting of the panel convened by Thompson was at the Pasteur Institute in Paris.[175] While all concerned relished French cheese and wine, and ballet and opera, and Smith reconnoitred the Moulin Rouge, there were sound reasons for meeting at this prestigious research institute. On 19 June 1946 a French government decree appointed four doctors, a biologist and the Director of the war crimes investigations to form a Commission on Scientific War Crimes, headed by the bacteriologist Legroux.[176]

The new consensus was that medical crimes required special expertise, and war crimes services were underperforming. In October 1944 the French had established a war crimes investigations service under the Ministry of Justice.

It relied on regional chiefs in France as well as investigative teams in Austria and Germany. The judicial administration contrasted to the Anglo-American preference for military investigation. The French were not as effective in bringing cases to trial, and 'the long and delicate research procedure' under-produced in terms of results. There was reluctance to probe too deeply in a France having to resolve the split between collaboration and resistance, and to prosecute too intensively in the zones of occupation. Another reason was the sheer inexperience of the responsible magistrates.[177]

The idea of a special medical service for war crimes posed problems to the French justice administration, just as it had to the US and British military legal establishments. The celebrated neurologist, Pierre Behague at the time at the climatic health resort of Pau (in the Basse Pyrenées) suggested on 20 August 1945 establishing 'une Commission Scientifique de Recherches des Criminels de Guerre' to take charge of medical and scientific investigations.[178] He saw this as a panel of distinguished scientists who would study the evidence from the viewpoint of physiology, and evaluate mistreated patients. He gave as an example that medical expertise was necessary to interrogate a German doctor about a new procedure for sterilisation operations. But on 27 August 1945 the Director of war crimes objected to a special medical and scientific commission, as this would fragment the war crimes investigations machinery. He recommended instead that investigating magistrates call in medical experts whenever needed.[179]

French medical concern with German scientific crimes intensified. On 21 June 1946 the *Union des médecins français* organised a meeting on German scientific atrocities.[180] Charles Richet, a survivor of Buchenwald, took a leading role with articles in *La Presse Médicale*. The French took an increasing interest in medical atrocities, investigating Plötner's mescalin and blood styptic experiments in Dachau, when to their embarrassment he escaped from Rastatt.[181] Such debacles prompted the view that the French war crimes service required an overhaul on more centralised lines, which, it was hoped, would allow France to collaborate on an international basis, and increase its general effectiveness. Although the French Communist Party had a massive popular vote, the French insistence on Four-Power collaboration in an international scientific commission alienated the British and US governments. The French intended that the scientific commission should form the basis of a Four-Power International Scientific Commission for the Investigation of Medical War Crimes, and for joint prosecutions. The aim of the Paris Commission was to conduct a process of peer review, in which the Nazi research would be judged by the procedures of science.[182] The French expectation was that Britain and the US would convene separate national commissions. As two of the French members of the commission were bacteriologists from the Pasteur Institute, it explains why the Institute hosted the momentous conference on 31 July and 1 August 1946.[183]

The US Secretary for War sent the physiologist Ivy (representing the AMA) on a mission to Germany and France from 18 July to 12 August 1946 to study the problem, liaise with the British at FIAT, and attend the follow-up meeting at the Pasteur Institute. Ivy first went to Hoechst to meet his former colleague in aviation medical research, Thompson. He commended Thompson's 'excellent records and a large file on war crimes of a medical nature. The files were in the process of being sent to Major Mant at the HQ of BAOR. I found that Dr. Thompson held views very similar to those I had formulated relative to the problem of war crimes of a medical nature.'[184] The Ivy-Thompson conference was crucial for US involvement in evaluating Nazi medical war crimes, and shaped the idea of a new code on human experiments.

Ivy combined scientific, strategic and ethical concerns. He visited the Kaiser Wilhelm Institute for Medical Research at Heidelberg, where since October 1945 the AAF Aero Medical Center was located with 'the outstanding German scientists on aviation medicine' under Strughold's direction. These included some suspect scientists and the neuro-anatomist Spatz, who had received brains of the Dachau pressure experiment victims and was currently studying the brains of air crash victims. In August 1946 there were 27 German scientists engaged on a mix of experimental projects and documentary analysis. Among these were Becker-Freyseng and Ruff, who compared the American and German wartime research on decompression sickness.[185] Ivy commented revealingly (given the illegality of military research by Germans): 'This group is performing experiments of great value to our Air Force at a time when aviation medical research is relatively dormant because of rapid demobilisation.'[186] The following day he attended the conference on medical war crimes in Paris. Ivy well shows how strategic evaluation and ethics characterised Allied policies towards German medicine.

Ivy's visit coincided with a major American initiative on aviation medicine. The Canadian associate of Thompson, Lt Col. Blaisdell reported on a 'Tentative Monograph on Aviation Medicine' in July 1946 with contributions from leading German scientists at Aero Medical Centre. Strughold, whose aviation medicine compendium with Ruff had been re-issued in 1944, was to be editor-in-chief. The work included a section by Becker-Freyseng on Selection, Training and Maintenance in which he contributed on medical selection of flying personnel, and on air accidents. Ruff was to contribute on acceleration and vibration, O. Schröder on air evacuation, and Rein on 'New Methods of Gas Analysis'.[187] Three defendants at the Medical Trial were thus involved in a project, which confirmed the academic ties to the physiologists Rein and Strughold. The plans were sent to Ivy on 21 August 1946.[188] The book was eventually published in a modified form in 1950.[189]

While the Strughold project was a priority for the American Air Force, its employment of German researchers in Heidelberg lacked a legal basis.

Public Law No. 25 prohibited military research, and this was precisely what was going on in the Aero Medical Center. OMGUS and the US military command (USFET) insisted that the law applied to research initiated by an Allied military agency. The Center was in the potentially embarrassing position of contravening the law at a time when three of its members were placed on a high-profile public trial. Its commander, Col. Benson, accepted that all laboratory work should cease by 31 October 1946 (well in time for the start of the Medical Trial). But the decision required that the Strughold monograph be completed and that a band of German scientists be transplanted to the United States.[190]

Here one sees the origins of the 'Operation Paperclip' transfers to strategic installations in the United States, as an exercise designed to circumvent the prosecution of aviation medical research. Six of the Heidelbergers were arrested and taken to Nuremberg.[191] The announcement of a parallel trial of Air Marshal Milch increased the vulnerability of aviation medicine. Benson established that 'none of the men remaining will be summoned to court at Nuremberg'. At the same time, Benson did not regard the German aviation researchers as particularly trustworthy, and they were demonstrably lacking in intellectual honesty. He was also sensitive to the unsuitability of those who had once approved of Nazi aims. On the other hand, the Center offered a better basis for learning about German aviation medicine than interrogation. As Strughold held the chair in physiology at Heidelberg, he needed special inducements to collaborate with the Americans. By November 1946 Strughold emerged as a prime candidate for transfer along with his former assistant Ulrich Cameron Luft, who had already been interrogated in the US.[192] In February 1947 (when Becker-Freyseng, Ruff and Schröder were protesting their innocence at Nuremberg), Strughold visited the USA to oversee the book's production – and so began his transfer to assist US high altitude flight and space research.[193]

For the military lawyers Taylor and Somerhough, there was none of the doublethink on strategic and unethical research. The dynamic head of the JAG War Crimes Investigation Unit, Somerhough pointed out that the British had assembled enough evidence to prosecute the 'Hohenlychen Group', along with Karl Brandt. The British were determined to try this case as involving 'gross breaches of medical ethics'.[194] This was the closest that the British came to mounting a specifically medical trial (although medical atrocities figured at the Belsen, Neuengamme, Ravensbrück and Rühen 'Baby-Farm' trials), as Somerhough's team was overtaken by their success. The British civil servants saw their opportunity to prevent the second International Military Tribunal by passing the medical prosecution materials over to the Americans, so that their Nuremberg program could be rapidly launched. Taylor promptly secured Elwyn-Jones's agreement for a medical trial at

Nuremberg. Taylor's deputy, James McHaney, was apprehensive that the British war crimes officers might be reluctant to transfer the Hohenlychen Group, and he approached Somerhough to confirm the handover.[195] Thus the scheme for the Medical Trial finally took shape in early August 1946, just after the medical experts met at the Pasteur Institute.

Thompson was commissioned to write a survey of the evidence, because of his ability to combine scientific with ethical perspectives. Thompson had an impressive grasp of essential issues, and could communicate these to pivotal decision-makers. He was regarded as vital for the success of the medical prosecutions, and was transferred from a Canadian special mission to the British payroll.[196] Somerhough suggested that the commission should have a 'field member' to evaluate evidence, advise investigators on medical experts and brief the authority conducting the trials, and that this 'working member ... could only be Major Mant'. Additionally, Somerhough insisted Thompson was 'irreplaceable'.[197] The French warmly supported Mant's co-ordinating role in war crimes investigations, and it was hoped that the Americans and Russians would appoint medical representatives to work with him. On 8 November 1946 Médecin Principal François Bayle (the French military delegate to the Nuremberg Medical Trial and to the ISC) joined the field agency to work with Mant, ensuring a steady supply of documents (and cigars) from Nuremberg.[198] Bayle was a neuro-psychiatrist, and had long experience as psychiatric expert of the French naval courts, and was expert in the analysis of handwriting.[199]

Thompson was an ideal secondment to the International Commission, because of his American and British medical links, and competence in romance languages.[200] Mant's forensic talents as a field investigator were important for the Americans. As Taylor explained to Somerhough, the US prosecution team only had resources for analysis of documents rather than for field investigations.[201] But Mant still favoured a British administered medical trial. McHaney's impression was that 'Major Mant is quite reluctant to turn the medical war crimes cases over to us. Optimistic that Somerhough will intercede. I should think it desirable that Major Mant should continue his work and assist us, both because of his familiarity with the field and his liaison with the French committee.'[202] Mant became liaison officer between the British and the Americans, and scientific officer for the ISC.

Somerhough reported to the JAG in London on the Pasteur Institute meeting, suggesting that there be a 'field member' of the scientific and medical commission, 'to advise the authority conducting the trials of the medical aspects'. Somerhough conceded that the trial work 'would be carried out by Brigadier Telford Taylor's Team at Nuremberg and the accused be tried under [Control Council] Law No. 10'.[203] It took the next few weeks to bring the British war crimes investigators into line, and to secure their willingness to support the handover of defendants and evidence for the US-sponsored

Medical Trial. Only on 27 August 1946 did the (British) Judge Advocate General's office confirm its support for the American trial, while informing Somerhough: 'If the responsibility for the trial of this type passes to the Americans, there would, in my opinion, still be some work to be done by your investigation team, since, obviously, we should have to give the Americans all the help that we can.'[204]

On 24 October the Subsequent Proceedings Division became the Office, Chief of Counsel for the Prosecution of Axis Criminality (OCCWC). This was a sign that an autonomous US series of prosecutions was under way.

When the Nuremberg prosecutor Hardy attended the ISC meeting in Paris on 16–17 October, the American interest was in its 'producing evidence for our cause'. Somerhough explained his aim of capturing any German medical criminals who strayed across zonal frontiers in Berlin (hence the arrest of Haagen), and the French offered access to evidence on medical experiments at Strassburg and Natzweiler in exchange for allowing French judicial and medical representatives access to the Nuremberg evidence. The British drew the attention of the Americans to the Neuengamme TB experiments. The ISC took a dynamic role by pooling information on medical crimes.[205] (See Table 10 for ISC meetings.)

The timing of the US decision to launch 'The Medical Experiment Case' can be traced from the activities of the Nuremberg prosecutors. Although the British had back in December 1945 expressed the wish to prepare a case against Karl Brandt, the Chief of Counsel at Nuremberg never handed him back. Karl Brandt was a natural point of departure for US investigations of criminal experiments, and from June 1946 Hardy and McHaney were investigating his role. The documents analyst, Manfred Wolfson, who was based at the Berlin Document Center, linked Brandt, Oswald Pohl (the SS treasurer) and Himmler to Nazi sterilisation experiments using X-rays and herbs. He submitted several 'SEAS' (i.e. staff evidence analyses) on these topics between late June and early August 1946. On 31 July details of Alexander's reports and an overview on SS medical research were communicated to the Nuremberg prosecutors.[206]

These investigative activities reinforced the IMT. On 8 August 1946 Sievers gave evidence to the Tribunal on SS involvement in human experiments. Taylor might have been eyeing up Karl Brandt as the linchpin of a medical trial, or Brandt might have been included at a second IMT, just as the British had intended with Conti. The preparations for the Medical Trial began when documentation had to be urgently assembled from 13 August, suggesting a firm decision to hold a trial just a few days earlier. Documents from the War Crimes Group File on Medical Experiments were sent to the US prosecuting lawyers McHaney and Hardy. The IMT prosecuting lawyer Alderman issued a note on 19 August concerning the whereabouts of Weltz 'as one of the medical personnel they expect to try'. The requisitioning of the Conti file was ordered on 20 August, and there came renewed interrogations of Sievers from 19 August.[207] These developments show that the Medical Trial was a firm objective. The Foreign Office was grateful to the Americans for relieving it of an onerous and costly task, while it could force the narrowing of British efforts to 'other cases of war crimes against British nationals'.[208] The Foreign Office undermined the broadening of the war crimes programme, while intensifying British–US liaison with the onset of the Cold War.

An alternative was to have the UN administer an international trial as a successor to the IMT. This had its counterpart in discussions as to whether the UN should run an international scientific commission. The UN War Crimes Commission in London had a brief limited to co-ordination and monitoring. The Danish General Medical Association asked the Ministry for Foreign Affairs, Copenhagen, for support for a survey of the breaches of the medical 'Code of Ethics' committed by the German medical profession. In July 1946 the Danish delegate noted 'a general feeling among scientists in Denmark that it would serve a very useful purpose to have a survey made of the crimes perpetrated by The German Medical Profession'. Instead of confidential reports restricted to Allied authorities, there should be a widely circulated definitive survey of 'the Crimes committed by the German Medical Profession during or just prior to the War, in the Concentration Camps or elsewhere, especially in regard to medical experiments carried out on human beings'. The Danes demanded a shift from strategic and criminal investigations to history and ethics. The Danish Medical Association 'stated that on historical and other grounds it may be of great importance for scientists throughout the world if an authoritative account of the said crimes could be compiled – an account demonstrating the destructive influence of the power of the authoritarian state on something so essential for the whole of mankind as the medical "code of ethics"'. The UNWCC agreed to compile evidence from trials and from scientific documents.[209]

The UNWCC research officer Lt. Col. H.H. Wade, who was in touch with Lemkin, believed that the material was so technical that the task could only be undertaken by a medical expert. He suggested that Leo Alexander (at the time back in civilian life) would be eminently suitable. During the summer of 1946 the UNWCC Chairman, the Australian Lord Wright, consulted with Taylor in Nuremberg who explained that the issue of a Four-Power or US military prosecution remained unresolved. Taylor was anxious to obtain UNWCC evidence. Wright suggested that the UNWCC appoint an observer or indeed a small commission to monitor the preparations for the Medical Trial. However, Sir Robert Craigie, the UK representative, pointed out that Thompson's investigating commission was the more relevant, and Wade was instructed to liaise accordingly. The call for a medical expert meant that the proposed UN survey of medical crimes was deferred to the emerging ISC.[210]

The ISC was driven by medical and scientific expertise, whereas the UNWCC was staffed by lawyers. In the summer of 1946, Lemkin was consulted by Lt. Col. Wade, the Research Officer of the UNWCC, about euthanasia and 'scientific killing'. Lemkin recommended Professor Sydney Smith, the forensic pathologist at Edinburgh, whom Thompson had recruited. It was at this juncture that the UNWCC learned of Thompson's initiative on medical crimes, and requested a report on medical crimes committed in Germany. Wade was astonished that the UNWCC (based at Berkeley Square, London) knew neither of the Frankfurt nor of the Paris ISC meetings on medical war crimes.[211] The difference was that the ISC did not examine all aspects of medical crimes, but focused primarily on the Nazi sacrifice of life and limb for research. But the UNWCC recognised the ISC's head start and began the long wait for the report.

Smith encouraged the UNWCC and Thompson to enter into contact.[212] Smith would have made an outstanding ISC chairman. But what was meant as a constructive initiative to give international authority to the ISC became its downfall. The UNWCC began to collude with the Judge Advocate's Office in London, which was less enthusiastic about the medical war crimes issue than Somerhough's German team. At the suggestion of the JAG's deputy, Henry Shapcott (who had a track record of obstruction and sabotage when it came to war crimes), Lord Wright wrote to the Prime Minister and to the Foreign Office recommending the appointment of Lord Charles Moran, who had some celebrity as Churchill's doctor and author of the celebrated *Anatomy of Courage*. Despite his eminence, his nickname of 'Corkscrew Charlie' suggests a reputation for deviousness.[213] Shapcott had consistently opposed Somerhough's demands to bring cases to trial. Now, an initiative associated with Somerhough's investigators could be effectively torpedoed. On the one hand, this meant formally constituting a British Committee on Medical War Crimes; on the other, it placed the whole issue in the hands of Moran, who was blundering, vain, an inveterate intriguer, and ready to connive with establishment demands to suppress war crimes evidence.[214] Moran alienated the American trials staff by criticising Alexander for speaking to the journalists of *Harper's Magazine*, *Newsweek* and *Time* on human experiments.[215] His organisation became moribund, thereby falling in with the machinations of those London-based civil servants preferring inaction on war crimes. The UNWCC continued to monitor the progress of medical war crimes investigations. But it took the view that the ISC ought to take the lead; as the ISC was disabled by Moran, efforts to investigate medical war crimes were thwarted.

UNESCO – the new United Nations Organisation for Science – turned out to be disappointing. Thompson contacted Julian Huxley, who was appointed Director General of UNESCO in December 1946, to suggest that there be a UNESCO observer at the Medical Trial. Huxley had a creditable record in drawing attention to Nazi 'pseudo-science' and to German atrocities against scientists.[216] He declined to send a UNESCO observer, but hoped that Thompson would report on the trials. He asked the British and French government to forward reports from technically qualified observers.[217] In December 1946 Huxley conferred with the senior British scientists, A.V. Hill and Dale, who wanted to liberalise British controls on German science.[218] Huxley prioritised scientific progress over Nazi crimes. He turned a blind eye to involvement in racial policy. He energetically supported the animal behaviour expert, Konrad Lorenz, whose Nazi past was used to block his appointment at Graz.[219] Huxley's broader aim was to formulate a biologically based ethics, which stood in contrast to the ethical, legal and historical endeavours at Nuremberg.

The disinterest among Allied politicians and Foreign Office officials in war crimes contrasted to field investigators' sustained concerns with medical war crimes. The initiative to investigate medical war crimes was pressed by a handful of energetic medical investigators. Without Thompson's initiative on 'medical war crimes' and the FIAT conference of May 1946 there would have been no Medical Trial. The demand for a Medical Trial came from below, while the high-level IMT created a responsive climate for some sort of follow-up trial in late 1946. Taylor took a hurried but momentous decision to hold the Medical Trial as Case No.1 in August 1946. It appeared as if the Medical Trial 'happened to be ready to go as the first case'. Taylor viewed the human experiments as uncomplicated – when compared to the potential complexities of the IG Farben case – and as a good way of starting the whole series of trials.[220] For political reasons the medical experiment case was an opportune way of extricating the Western Allies from Four-Power prosecutions. Thanks to the efforts of Thompson, Mant and Somerhough, the medical atrocities were well documented, and the imminent trial looked set to be clear-cut and swift. Politics drove the decision to locate the Medical Trial at Nuremberg, and defined its scope and participants. The first chill of the Cold War diminished the ardour to hold a second International Military Tribunal. But it provided incentives to investigate human experiments, and set in motion events leading to far-sighted ethical discussions.

7
Pseudo-science and Psychopaths

Internationalism and interrogations

Faced by scientific crimes, Allied intelligence attempted to sift the scientific wheat from the 'pseudo-scientific' chaff. These efforts ran parallel to Nuremberg trials. The prestigious KWG possessed a degree of insulation from the Nazi state, but far less than its wily administrator, Ernst Telschow, suggested. The KWG was involved in war-related research in the field of armaments, racial policy and *Ostforschung*. It supported research involving human experiments, and brains and body parts from euthanasia victims, selected by scientists. Luftwaffe-sponsored research involved the KWG brain anatomists Hallervorden and Spatz and the neuro-surgeon Tönnis. The KWG had longstanding interests in aviation research at Göttingen, and Air Marshal Milch was on the governing senate of the KWG. Telschow had links to Blome, and both were involved in the establishing of a Reich Institute for Cancer Research. When Otto Warburg was dismissed in 1941, Brack claimed to have restored Warburg's post.[1]

Poppendick's training in heredity and human genetics at the KWI for Anthropology linked the Medical Trial to the KWG. Chemical weapons research was also suspect. Himmler authorised research on phosgene and mustard gases by Hirt at the Natzweiler concentration camp. Richard Kuhn, of the KWI for Medical Research, was in touch with Hirt in his capacity as section chief for organic chemistry of the DFG.[2] That Karl Brandt was given a co-ordinating role in chemical weapons research in 1944, raised the issue of human experiments in Natzweiler. Further collaboration came to light at the trial of IG Farben, as KWG researchers undertook weapons research at the Institutes for Leather Research and for the Physiology of Work. The KWG developed dubious links in its support of war-related projects. One was with Eppinger, the internist at Vienna, who wanted to give hospitality to a KWI for nutritional research.[3] KWG botanists supported the plant research department at Raisko near Auschwitz.[4] Human experiments, slave labour and support of the Nazi war machine characterised the KWG in the

Second World War. But by 1946 German scientists argued that they were victims rather than agents of Nazi persecution.

The KWG president and industrialist, Albert Vögler committed suicide on 14 April 1945, when arrested by the Americans.[5] But the Allies did not close a single KWI before the general dissolution of the Society in July 1946. The KWIs for Psychiatry and Anthropology remained open, despite their support for sterilisation as part of Nazi race policy. Allied policy to German science pursued contradictory aims. The physiologist and president of the Royal Society, Sir Henry Dale, took a key role at the crucial meetings with the interned German atomic scientists at the Royal Institution on 2 October 1945 and 2 January 1946. The physicists Otto Hahn, Max von Laue and Werner Heisenberg met Dale, Hill, Sir Charles Darwin and George Thomson to discuss the reconstruction of German science. These meetings cleared the way for the Foreign Office Committee on German Science.[6] Dale convinced the British to abandon ideas of restrictive control of German science. His strategy was elitist in that he considered that if the figureheads of German science were supported – by this he meant notables such as Heubner, Rein, Max Planck and Hahn – then the rest of German science would be in a healthy state. The net effect was to insulate leading scientists from prosecution, while investigating minor figures.

The British took the lead in resuscitating the defunct and dismembered KWG as the Max Planck Gesellschaft (MPG) in the British zone. The Allied death sentence on the KWG posed severe problems for the German scientific elite. The US authorities were keen to dismantle the KWG, and the law for its termination was only held up by US-Soviet legal wrangles.[7] While the British established the MPG on 11 September 1946 as a measure to revive German research, they took the view that although many leading German medical scientists were compromised, one could not realistically control every aspect of scientific activity. Channelling scientists' energies into producing reports could be illuminating as to the nature of their war work. The consolidation of new research structures coincided with preparations for the Medical Trial. Condemnation of the worst of the Nazi scientists could clear the way for re-establishing research in the Western zones.

The UK and USA shared a fear of German scientists continuing clandestine military research – so it was necessary both to find out what they had done and to prevent them continuing it. Soviet recruiting of German and Austrian scientists resulted in policies to build up science in the Western zones. The Americans were nominally restrictive – but flouted their own rules with the Aero Medical Research establishment at Heidelberg. The British decided against a policy of control and were more in favour of positive inducements. The Research Branch developed Göttingen as a major scientific centre as a hub of the reincarnated MPG.

The issue of human experiments was potentially disruptive to the relaunch of German research. It fuelled the scepticism of the US critics of

German science, just when the British wanted to bring them on board. German scientists sensed the potential danger, posed by uncovering human experiments. Hahn lobbied the scientific control officer Bertie Blount and Dale for the resuscitation of the KWG in July 1946.[8] Dale became a target for unscrupulous machinations among German scientists seeking to rid themselves of guilt by denial and self-righteous indignation. The general chronology indicates how certain key British scientists preferred to take a positive view of German research, at a time when the evidence was being amassed for the Medical Trial.

In July 1946 when the Royal Society honoured Planck by inviting him to its 300th anniversary celebrations, Dale and the physiologist A.V. Hill received a distress signal from fellow German Nobel laureates concerning the demise of the KWG.[9] Dale conceived the new name, the Max Planck Gesellschaft, for the reconstituted institutes in the British zone. Planck became an icon for a new scientific organisation: the name change symbolised a shift away from imperialist and latterly Nazi militarism to the idea of a free association of scientists, running their own affairs. The first MPG meeting in the British zone was on 11 September 1946, and the new body was entered into the *Vereinsregister* on 23 November 1946.[10] Support for such an interpretation comes from evidence linking the medical war crimes trials to the nascent MPG. This concerns the pivotal role taken by Dale, who as chair of the Committee on German Science swung round to the idea of 'restarting the motors of German science' as a passionate advocate of experimental medicine. On 2 March 1946 he lectured to students at Cambridge on the history of the experiment in medicine: 'The future is bright with promise, indeed, if mankind can be brought to forsake the folly of using the gifts of science for its own destruction.'[11] His guarded reference to wartime abuses of science rejected Aldous Huxley's demand for a Hippocratic Oath for biologists.

Medical madness

Scientists favoured alternatives to cumbersome legal procedures. For if Nazi medical research was pseudo-science, then what were the causes of this collective mental aberration? Psychiatrists were concerned with Nazism as a deviant psychology since the 1930s, and after the war much attention was given to medical reports on the mental state of captured leaders. The IMT formed a significant link in a chain of events linking wartime psychological assessment of the enemy with medical observations on the defendants at the Medical Trial. The Nuremberg Trials provided an ideal opportunity for analysing the psychology not only of Nazism but also, more widely, German national psychology .

Psychologists advised the Allies on the German mentality, and isolated the traits of aggression and servile obedience. Wartime work on Nazi psychology focused above all on Hitler's psychopathology. One aim was psychological warfare, and it was hoped that the dictator's mesmeric hold on the German people could be broken if its basis could be discovered. Whoever had encountered the Führer – especially those acquainted with the young Hitler – was interviewed about Hitler's personality. Walter Langer prepared an analysis – 'The Mind of Adolf Hitler' – for the OSS in 1943, and predicted Hitler's suicide.[12]

The flight of Rudolf Hess to Britain raised the question of the sanity of Nazi leaders. This time psychiatrists and psychologists had a case they could directly observe, and his enigmatic behaviour preoccupied a stream of Allied experts. The aim shifted from determining the motives for Hess's mission to whether he was sane enough to stand trial. Lord Moran and Thompson were among those who interviewed Hess for the British.[13] It was necessary to assess whether Hess was mentally ill or suffered from mental defects (he was examined at Nuremberg by seven doctors from five nations) and whether Julius Streicher (examined by three doctors) was sane. The prison psychiatrist, Douglas Kelley, found Hess to have 'a normal mental state' accompanied by a curable amnesia.[14]

The psychology of Nazism became a widespread preoccupation during the war. Alexander mulled over the sanity of the Nazis. He noted how, in 1925, Georg Soldan anticipated that the coming war would be ruthless in its use of lies in the battle for survival. He analysed this as the perversion of superego values.[15] He consulted Karin Horney's analysis of *The Neurotic Personality of Our Time*.[16] He kept abreast of the discussions of the Committee of Mental Hygiene on German national psychology, as a number of his and Thompson's Boston associates were involved.[17]

By 1944 representatives from national associations of medical psychology and social science formed the Committee on Postwar Germany, which was sponsored by the Joint Committee on Post-war Planning. A range of American organisations for mental hygiene, mental defect and neurology were involved, and the leading figures were all medically qualified: Alvan Barach, Carl Binger, Richard Brickner, Frank Fremont-Smith, Putnam and Adolf Meyer (both Thompson's mentors), John A.P. Millet, and George Stevenson. The Committee organised conferences on the structure of the German character. A key issue was the recurrent appearance of aggressive leaders in Germany, and the social pathology of mass obedience. The idea was to find a scientific basis for policies of the reorganisation and re-education of Germany. Some general works appeared, such as Brickner's *Is Germany Incurable?*, which pointed to the flawed German super-ego. Justice Jackson and Alexander both studied Brickner's analysis.

The liberation of Germany was an opportunity for collective therapy and diagnosis. The Norwegian author Sigrid Undset suggested that psychiatrists should accompany the Red Cross to offer psychological counselling to victims, as she was convinced that the crimes of the Germans were an indication of insanity: 'The trial of war criminals should everywhere be conducted with the aid of psychiatrists and specialists from several branches of medicine. And the forces of occupation should be accompanied not only by regular Red Cross units but by a body of alienists and neurologists.'[18]

The IMT offered a chance for psychologists to deliver their verdict on Nazi pathology. Justice Jackson's attention was drawn to the analysis by Bruno Bettelheim on 'Individual and Mass Behavior in Extreme Situations' with the recommendation that Bettelheim would be an appropriate witness. This prompted Jackson to reflect that 'possibly a group study should begin at once'.[19]

The New York psychiatrist John Millet and Lawrence Frank, chairman of the Committee on Postwar Germany, asked Jackson to authorise psychiatric examination of war criminals. The Committee was keen to dissect the brains of the executed. Millet requested that defendants be executed by being shot in the chest, rather than the head, so as not to damage brain tissue; Jackson replied that hanging was a more likely means of execution as shooting inferred death with honour.[20] Jackson insisted that psychological examinations of the defendants could take place only after conclusion of the trial.[21] Millet submitted a plan for in-depth socio-psychological studies of the Nazi leadership. This was to include books written by defendants, their diaries, medical records, court records, CIC and OSS papers and interviews with the prisoners and their relatives and close associates.[22] After conferring with Allied representatives, Jackson responded that although sympathetic to the general aim, it would be best if no psychological interrogations were carried out before the trial ended. The defence might regard these as court records and exploit any academic disagreements, as well as prejudice any courtroom discussions of the issue of insanity.[23]

Burton Andrus, the American military prison commandant, took the view that psychological observation was part of the monitoring of prisoners' health, particularly when suicide had to be prevented and order maintained. The IMT prosecution valued liaison with the prison psychologists Gustave Gilbert and Leon Goldensohn, as they found this useful in planning prosecution tactics. Andrus extended Gilbert's remit to that of psychologist, and he visited the prisoners daily and observed their individual reactions to the prosecution's accusations and development of the defence positions, as well as the interactions among the prisoners. Gilbert interpreted his role as 'participant-observer' in a 'well-structured social situation' providing opportunities for depth analysis. While the prisoners were in solitary confinement prior to the opening of the case (and oblivious to the cautious position of Jackson), he tested their memory and intelligence, and asked them to interpret the Rorschach inkblots.[24]

In December 1945 Jackson authorised the psychiatrist Col. Paul Schroeder to undertake confidential psychiatric assessments of the defendants in the period after conclusion of the case and the delivery of judgment.[25] Schroeder made personality studies of the defendants at the Dachau Trial, including the malariologist Schilling.[26] By June 1946 five military psychiatrists (Kelley, Goldensohn, Schroeder, Nolan D.C. Lewis and D. Ewen Cameron) – joined by the psychologist Gilbert from 20 October 1945 – had collected observations on the IMT defendants.[27] When Millet sought reassurance from Jackson that psychiatric studies were being made between the verdict and the day of sentence, the brief had been fulfilled, not least by Gilbert's sustained studies.[28] Gilbert terminated his observations on 16 October 1946, and Jackson granted him permission to publish his *Nuremberg Diary*, which appeared in 1947, on the basis that it contributed to science and history.[29]

The Freudian Mitscherlich found Gilbert's use of IQ and Rorschach tests mechanistic and superficial. He felt that a deeper method of analysis was required that could reach the inner person, so as to understand the psychology of perpetrators and bystanders.[30] His article on psychoanalysis and history in the *Schweizer Annalen* led to permission to interview the defendants.[31] Gilbert's work was a model for Alexander's related series of observations. Both were able – to a degree – to gain the confidence of the defendants. Gilbert approached the aggressive dictatorship and authoritarian mindset of the Nazi leaders from a primarily social psychological standpoint. Alexander blended psychology with psychiatry and in-depth analysis, coming closer to the expectations of Millet and Mitscherlich.

When Alexander joined the prosecution team, he took full account of the publications by Kelley and Gilbert. He pointed out that his predecessors dealt with individuals rather than the organisations judged as criminal by the IMT. He sent his CIOS reports to psychiatric colleagues Myerson, Merrill Moore, Lyman and Putnam.[32] He approached the SS through the political concept of totalitarianism. He re-echoed the views of the Millet committee on the relevance of psychological studies for re-education.

Kelley's study of *22 Cells in Nuremberg* was overshadowed by the psychologist, Captain Gustave M. Gilbert, who took a highly publicised role at the IMT.[33] The amiable Gilbert provided prisoners with intellectual diversion, and a sympathetic opportunity for reflection. As a German émigré in US military service he had a useful bridging function between defendants and their prosecutors. The prosecution valued his privileged contacts, his constant monitoring of the psychological effects of the trial proceedings on the group dynamics of the defendants, especially in attempts to isolate the robustly unrepentant Göring. Gilbert merits comparison with Leo Alexander in terms of analytical methods. They were interested in the authoritarian personality as a type susceptible to Nazism and in providing a political explanation of medical atrocities.

A few German medical scientists, who had been outcasts from Nazism, viewed those on trial as pathological specimens. The brain anatomist, Oskar Vogt, who had been commissioned by the Soviets to dissect Lenin's brain and had then been harried by the SA, declared a wish to dissect the brains of executed war criminals. He believed that a specific brain form was responsible for Nazi crimes, and that the heredity and Nazi stock could be identified on a biological basis. His dissections would save humanity from future crimes.[34] The Frankfurt University brain researcher Paul Klingelhöfer contacted the prosecution lawyer Kempner in April 1946 on Vogt's behalf, requesting that the brains of the executed be sent to Vogt's institute for study.[35] The incident marked a reversal of values: whereas medicine had been downgraded under Nazism into a tool of the racial state, medical researchers asserted that they could discover the causes of Nazi dementia. Science claimed to be in the best position to judge Nazi medical crimes.

8
The Nuremberg Vortex

The prosecution team

Nuremberg was a war-scarred wasteland. Allied bombing devastated its medieval heart and industrial peripheries.[1] What had once been a Nazi rallying point was now an uninviting location for the series of trials. Alexander was disparaging: 'The whole thing is gray and dirty with the air of an abandoned slaughter house.'[2] Accommodation was scarce and miserable. The German medical observer, Alice von Platen, who stayed in the dilapidated inn 'Zum Schlachthof' (literally, by the slaughterhouse), recollects the gloom, the thick dust clouds, discomfort and the hostility of the locals to the Trials.[3] Obtaining food, fuel and hot water posed problems in the bitterly cold winter of 1946–47. Venturing out was considered dangerous, because predatory vagrants lurked among the ruins. The court and its staff were beleaguered and heavily guarded, and the Americans with their strange fashions, exotic jazz and dancing styles, and inexhaustible stocks of cigarettes were viewed with a mixture of hostility and envy. A grenade was thrown in the restaurant of the Grand Hotel in protest against the food situation, as the court appeared less as a symbol of justice and more one of opulence and conquest.[4] The marathon court sessions and ponderous routines represented a brave attempt to make sense of the historical debris of human destruction, rather than simply to consign atrocities to oblivion. While the prosecution tried to account for the hundreds of thousands of 'nameless dead', Mitscherlich reflected how the oppressive cold of Nuremberg evoked the human coldness in modern society and medicine.[5]

The clinging ethos of Nazism found expression in attempts to discredit the Nuremberg Trials as 'victors' justice'. Right-wing nationalists have relentlessly attacked the Trials as a left-wing and Jewish conspiracy, aiming to discredit every aspect of the trials from the initial interrogations to the eventual sentences. The nationalist onslaught continued, against Taylor and his staff. Taylor was a pragmatist in terms of court procedure, combining the form of a military tribunal with the ethos of an international court, and developing a socio-legal analysis of Nazi power structures. He understood the need to balance legal efficiency by using familiar US court procedures, and involving representatives of Allied and victim nations.[6]

To understand German administration, law, economics and medicine, technical advisers were necessary. In the eyes of unrepentant nationalists, the émigré advisers were vengeful and vindictive Jews. Their language and cultural skills were vital for the non-German-speaking prosecutors. The émigré advisers contributed to the broad endeavour of humanitarian justice. In contrast, Milch foamed that the whole prosecution team was Jewish. The conservative right similarly denigrated Taylor's team as communists.[7] Milch's bitter fury infuses David Irving's attack on the Nuremberg Trials and his efforts to vindicate the prosecuted.[8]

Taylor realised that zonal trials meant efficiency and securing justice for greater numbers within a limited timescale. He was one of the few seasoned staff members to remain from the IMT. He rapidly recruited a new prosecution team. He appointed James M. ('Jim') McHaney, described by Alexander as 'a smooth Southern lawyer from Arkansas', and the 'very smart' Alexander G. ('Sandy') Hardy from Boston as his deputies.[9] Once Taylor settled the main issues of the case, he left the preparations and conduct of the Medical Trial to McHaney and Hardy. He reflected, 'I could not be a wizard on all fronts. The first day after I presented the opening case for the doctors, I had to turn my attention to the on-coming eleven [trials].'[10] The prosecution team was faced with the issue of how far the Trial was one of experimental killings for genocide, or how far to treat the experiments as self-contained acts of murder.

Former refugees and escapees from Nazism were at a linguistic advantage and highly committed to their work; they supplied the prosecution with devastating evidence, inaccessible to a non-German speaker. (Neither Taylor, Hardy nor McHaney knew German.)[11] The defence lawyers accused the former refugees (they came to Germany not as returnees but as assimilated US citizens) of being bent on vengeance. Certainly, the émigrés had lost relatives in the Holocaust, and sifting through the Nazi documents was spurred by the wish to understand what had happened. But the accusation of vengeance was fuelled by anti-Semitism. In contrast, Holocaust historians have criticised the Nuremberg trial team of downplaying the evidence for genocide.[12] This raises the issue of how far former refugees shaped the prosecution agenda of the Medical Trial.

The neurologist and CIOS investigator, Leo Alexander, was by 1946 thoroughly socialised in American culture through professional experience in Boston and at Duke University, and in military service. His Austrian and German professional training in neurology meant that he could critically appraise German research and size up personalities. German defendants and informants appreciated Alexander's sympathetic understanding as an interrogator, while detached scientific aims spurred him to analyse the psychological dynamics of the perpetrator. Alexander's identity was that of an American officer with the expert acumen of a psychiatrist.[13]

Anti-Semites attacked the prosecution as being in the hands of Jewish communists. The Austrian defence lawyer Steinbauer contemptuously spoke of Alexander as 'the Viennese emigrant'.[14] It would be a distorting caricature to depict him as 'The Austrian Jew', or as a fat, slovenly and philandering spy – by the time of the Trial he was neither Austrian (the resurrected country denied citizenship to its émigrés); nor was Alexander in formal terms Jewish, although he freely acknowledged his cultural heritage. He was always open about his role as an intelligence officer. He prioritised an ethical analysis of medical atrocities. Alexander was historically minded, and his professionalism meant that while personal knowledge could open windows of understanding – for example, in the Viennese Nazi connections of Beiglböck – the sense of personal loss of family killed in the Holocaust had to be set aside in reaching general conclusions as to action and motivation. He grappled with the ethical dilemmas surrounding clinical research, hoping that the Trial could prevent abuses in the future. As he probed the perpetrators' psychology, he felt ever more estranged from the arrogance of the Nazi medical elite.

Ivy suggested on 6 August 1946 that a physician-scientist reinforce the prosecutors. Alexander was ideally suited for the posting. The UNWCC drew attention to the unique significance of his investigations.[15] Alexander was appointed Chief Medical Expert for the Prosecution in early November 1946. He showed tenacity and skill in compiling the CIOS reports, which had a stunning impact at the IMT. He became Associate Director of Research with the neurologist Abraham Myerson at the Boston State Hospital from 1 January 1946.[16] Although he believed that his military service was finally over, he remained 'convinced that more investigative work of a neuropsychiatric-sociologic nature should be carried out in Germany', and felt it should include personality studies of the Nazi medical criminals. He kept in contact with the medical sciences programme of the Rockefeller Foundation: his advice confirmed the tarnished state of German medical science.[17]

When recalled in October 1946, Alexander was aware that he had a unique opportunity in forensic psychology. His interest and skills made him ideally suited as adviser to the prosecution, given his background in Austrian and German medicine. His expertise in psychiatry meant that he had 'knowledge of personality patterns' of SS scientists and leaders, and he could analyse normal and deranged personalities.[18] Family tragedy had made him familiar with the courtroom: a paranoid patient had murdered his father, Gustav, a distinguished Viennese specialist in otology in April 1932. Alexander was unflinching in his sympathy for anyone showing signs of psychological disturbance, and explained to his son, Gustave, that he was determined to understand the rationales and motives of the defendants.[19]

Other émigrés on the US prosecution team had legal expertise or assisted with evaluating evidence. Alexander was pleased to find on the trials staff a former school acquaintance from Vienna, the lawyer John H. Fried, an expert on slave labour.[20] Fried emigrated to the United States in 1938, and received his doctorate at Columbia University in 1942. He served as a special consultant to the US War Crimes Tribunal at Nuremberg from 1947 to 1949, and afterwards as a professor of political science at Lehman College of the City University of New York and with the human rights and technical assistance divisions of the United Nations.[21]

The prosecuting lawyer, Arnost Horlik-Hochwald, was from Moravia. He was originally called Ernest Hochwald, and then – to emphasise the chasm between Sudeten Germans and the German-speaking, Jewish Czechs – took a Czechoslovak name. He represented the Czech government at the IMT. But the communist rise prompted his move to Taylor's staff. He prepared a highly effective euthanasia case, linking the killings of psychiatric patients with selections in the camps and X-ray sterilisation.[22]

The interrogators, Walter H. Rapp (Director of the Evidence Division), Herbert Meyer, Joseph Maier, Fred Rodell and Walther Kauffmann were seasoned by their IMT experience, and focused on the command hierarchies and atrocities of the SS. The documents analysts Manfred Wolfson and Hedwig Wachenheimer[23] ferreted out material in the Berlin Document Center. They worked under the energetic Ben Ferencz, who discovered the records of the *Einsatzgruppen* massacres, and convinced Taylor that he should take this on as an additional case.[24] The Austrian émigré Charles Ippen transferred from the Canadian army to the document analyst team at the prompting of Thompson. The refugees had the advantage of being bilingual, and could judge academic and social structures; they could form relations of confidence with the accused, but they were also tough. In Ferencz's words, 'We were young, brash, eager, and under time pressures.'[25]

The émigrés' linguistic skills were vital. Wolfe H. Frank – a British officer of German origins – was the acknowledged ace of the simultaneous translators, who had translated Göring's belligerent offensive, as well as the IMT sentences.[26] Producing an agreed text involved the interaction of the prosecution, defendants, lawyers and the translators of the Language Division. The Trial used simultaneous translation, and the court reporters set out to produce a dual German and English text within 24 hours.[27] Often phrases were hard fought over, as much depended on the lethality of the diseases, and whether experiments or just routine use of a drug were involved. The authentication of documents, and their translation and duplication, were labour-intensive requirements, as the defence complained that documents were available too late.

The émigrés investigated, interrogated and translated, and in Hochwald's case, prosecuted. Rodell had experience in working in wartime anti-German propaganda, and was confident in interrogating suspects at Nuremberg. His view was that, 'The medical cases were different, because you had the top medical people of the world, experts in their field throughout the world – Rose lectured at Harvard, like Prof Karl Gebhardt a bone expert also came to the US.' In his view they were 'Not just simple country doctors'; 'They had conducted these experiments on people in the camps. It was like they were children who had been playing with toy trains and then were offered real trains. They had been testing mice and rats all this time and then were offered all these people in concentration camps [to] go ahead.'[28]

The refugee Joseph Maier complained that his role was to analyse transcripts of interrogations rather than confront perpetrators. He had a strong moral commitment to the task, commenting that the words of some under interrogation made him feel 'physical disgust', so strong as to make him vomit.[29] The issues of race and genocide were firmly on the trial agenda.

The new series of Nuremberg trials were designed to demonstrate American fairness and justice. From early September 1946 Taylor demanded the rapid procurement of high-quality judges, although judges from the Supreme Court were unable to serve at Nuremberg.[30] The Medical Trial judges were appointed on 25 October 1946. Walter Burgers Beals of Seattle was the presiding judge; a venerable 70 years in age, he had the edge over the senior military doctor, Handloser, aged 62 at the start of the Trial.[31] Justice Beals was a military veteran, a Roman Catholic and a freemason. He maintained strict impartiality and authority in court. His alert and engaged questions indicated a sense of duty to establish an impartial historical record. To acquaint himself with the background to the Trial, he requested Lemkin's *Axis Rule in Occupied Europe*.[32] The Medical Trial judges showed none of the blatant bias of judges at later trials, who disregarded prosecution evidence on genocide and Nazi atrocities.[33]

Harold Leon 'Tom' Sebring (of the Florida Supreme Court) and Johnson Talmadge Crawford (of the Oklahoma District Court in Ada) joined Beals on the bench. Victor Clarence Swearingen transferred from the US war crimes staff to take office as alternate judge, and was familiar with procedural issues. These three judges were aged between 47 and 48 at the time of the trial.[34] When the proceedings were under way, Taylor observed to Jackson: 'It is quite a sight to see sixteen State Court Judges from all over the United States suddenly flung together in Nuremberg to grapple with German history and other novel problems that the cases here present.' The expectation was of a trial in 'the fair American Way, without bias and prejudice'.[35]

Judge Sebring had experience of Europe as a veteran of the First World War. He had been severely burned by mustard gas – experiments on mustard gas were to figure prominently in the Trial. He upheld judicial ideals of honesty, integrity and public service, and was unfailing in his courtesy to lawyers, witnesses and defendants alike. The indelible impression of the Nuremberg revelations meant he later attached importance to making students aware of the horror of Nazi atrocities.[36]

The judges could not speak German, and in many ways felt adrift. They pleaded that the émigré lawyer Robert ('Bob') Kempner be seconded to the trial secretariat. He had scored several successes at the IMT as assistant prosecutor under Jackson in the case against the Interior Minister Frick, involving euthanasia. Kempner's knowledge of German politics and legal procedures was valuable for the prosecution team. But he was reluctant to return to Germany with its shortages and bitter past – he was twice detained in German concentration camps before arriving in the United States in September 1939. He was the eldest son of two bacteriologists, Lydia Rabinowitsch-Kempner, who was one of the first women medical scientists in Berlin, and Walter Kempner, assistant to the professor of hygiene, Carl Franken, at Halle.[37] He eventually returned as Chief Prosecutor for the trial of German Foreign Office diplomatic staff.

The judges found a confused state of organisation. They established rules of procedure by 2 November 1946, although further points were thrashed out as the trial proceeded. The indictment of the accused followed rapidly on 5 November. The German defence had one month to prepare their case, and the defendants entered pleas of not guilty on 21 November.[38] A press release announced that 54 German civilians, including mayors and city counsellors, attended the arraignment of the '23 doctors', to show that the defendants were to get a fair trial.[39]

Selecting defendants

During September 1946 the US prosecutors selected the defendants from the ranks of interned medical scientists and Nazi administrators. The military lawyers wanted clear-cut and winnable cases. They took the view that human experiments were an extreme form of assault and murder, carried out by a co-ordinated group of SS and military medical officers and Nazi-affiliated medical researchers. The aggressive militarist and genocidal aims of the Nazi state provided the overall context. Medical research amounted to unscrupulous murder for political and military ends, and the prosecution hunted for links to SS leaders and ultimately for evidence involving Hitler. The totalitarian paradigm reinforced the view of medicine as an arm of Nazi racial expansion. Legal and political considerations ranked above medical ethics, which had only secondary relevance in drawing attention to issues of coercion and the lack of care in the design of the experiments.

In September 1946 the prosecutors mulled over lists of proposed defendants, and set about locating evidence and determining their strategy. The use of the IMT courtroom at the Nuremberg Palace of Justice dictated a maximum number of 24 defendants for trial. In the event only 23 were indicted for the Medical Trial. The court provided legitimacy conferred by the IMT's status as an international trial, which was useful given the changed political status of the successor trials. The idea was that the Medical

Trial should follow as soon as possible after the IMT, once death sentences were carried out on 16 October 1946.

The Nuremberg Palace of Justice was well suited to the mass production of war crimes cases. It had escaped the bombing of Nuremberg's medieval centre. Built as a bastion of law and order in Imperial Germany, it had the capacity for simultaneous trials. Its cavernous prison with its four radiating wings could hold 1,200 prisoners, and it was conveniently linked to the court.[40] Prisoners referred sarcastically to their Spartan accommodation as 'The Grand Hotel'.[41] A psychiatrist at Nuremberg characterised it as 'a tough gaol' with 24-hour guards, cell searches, restrictions on sleeping posture, and the confiscation of shoelaces and braces.[42] Rostock realised that conditions on the outside with shortages of food and fuel could be worse. Prisoners were given adequate food – in marked contrast to the starvation rations allocated in the concentration camps – an exercise yard and a gym.[43] Prisoners had reading matter ranging from Kogon's *SS-State* to Jakob Burckhardt on art history and American periodicals like *Life*, *Reader's Digest* and *Time*.[44]

The psychological problem of isolation, the general strain of the Trial with an uncertain outcome, and a freezing winter all took their toll, and some prisoners fell ill. Genzken (who was brain-damaged in a political brawl in 1931) suffered from hypertension, and Oberheuser was often absent from court, requiring hospitalisation and an operation.[45] Weltz had a bad attack of asthma in prison. Rudolf Brandt complained that he suffered a general debility with the strain of the trial.[46] Rose ended up in a state of exhaustion. Alexander interpreted these illnesses as a result of emerging from the sublimation of the self in the group psychosis of Nazism. The prisoners felt isolated by the ban on family visits until mid-July 1947 (when two visits of two hours were allowed each month) and the rationing of outgoing post. But otherwise the conditions were correct.[47]

The IMT refurbished the court with a modern décor (since then, the Bavarian state has exorcised the war crimes ethos by reverting to ornate decoration in the original Imperial style). The Americans added a gallery for the press, Allied and German observers, a platform for film cameras, and a screen for films and large charts delineating the German administrative hierarchies. After six months of the routines and spatial strictures of the courtroom, all sides were on a familiar footing. The defendants made solicitous suggestions concerning an attack of neuritis suffered by Beals, and when Hardy pierced himself with a pencil, Rostock asked if a doctor was needed, pointing out (in reference to the sulphonamide experiments) that several present had experience in treating wounds.[48]

Col. Clio Straight of the American JAG sent a list of the names of 140 doctors and scientists who ought to be held as 'perpetrators in medical experiment cases'. Mengele's name was prominent among these. Hardy selected 27 individuals including the psychiatrist Rüdin and surgeon

Sauerbruch, and Karl Brandt, Genzken, Handloser and Rose. Depositions from victims and witnesses of the human experiments at Auschwitz had identified Mengele and the fertility researcher Clauberg as prime culprits.[49]

Taylor was optimistic that the Medical Trial was to be swift and decisive, and looked for clear-cut, winnable cases. On 2 September 1946 he explained, 'we plan to join [Karl] Brandt with Sievers, Mrugowsky, Haagen, Bouhler, Brack, the "Hohenlychen Group", and perhaps Rudolf Brandt and Field Marshal Milch.'[50] Taylor's strategy linked medical experiments with Nazi criminal organisations. Karl Brandt was pivotal with his personal links to Hitler. Journalists invariably described him from the pre-trial phase to his execution as Hitler's doctor rather than as 'escort surgeon or physician' with the limited role of providing emergency treatment in the event of an accident.[51] Hitler found this energetic and devoted young surgeon congenial and useful. First ordered to inspect a disabled baby, and then assigned joint responsibility for euthanasia, on 28 July 1942 Hitler vested Karl Brandt with the powers of 'General Commissar' to carry out special tasks and to settle problems of resourcing the military and civilian health sectors. On 5 September 1943 Hitler extended the remit to 'Medical Science and Research'. On 25 August 1944 Brandt became 'Reich Commissar' for the duration of the war, authorised to act on medical matters in the state, party and armed forces. The prosecution saw him as carrying full responsibility for medical genocide. Brandt argued that the administrative structure remained fragmented, and he never had full executive powers. He explained that his primary concerns were measures to treat military casualties in emergency hospitals, and civilian medical provision as a result of Allied bombing of Germany.

Hitler's decree of July 1942 established a new post of Chief of Military Medicine with power over the Waffen-SS medical services. The prosecutors selected the chief military medical officers, Handloser, and Schröder as chief of the Luftwaffe medical services, but did not prosecute his predecessor Hippke, who was in office at the time of Rascher's most vicious experiments. Taylor selected their counterparts in the SS, Genzken and Mrugowsky, who appeared dogged and dour, but inclined to unconventional philosophical views.

Three non-medical defendants were chosen, all key administrators: Brack of the Führer's Chancellery, Rudolf Brandt of Himmler's personal staff, and Sievers of the SS-Ahnenerbe. Brack had initially gained release under the name of Hermann Ober (his wife's family name). He kept a low profile as an agricultural worker, but in June 1946 Allied efforts to prosecute euthanasia prompted his re-arrest.[52] (See Table 5.)

Pokorny was arrested from his post as dermatologist at the Munich Health Office on 5 September 1946. He was the sole defendant hauled in from a civilian medical post, and prosecuted at Hardy's insistence.[53] By early September 1946 the prosecution tackled the aviation experiments. McHaney contacted Colonel Straight's war crimes office to obtain the Rascher documents, and witnesses were interrogated about Rascher's conduct, connections to Himmler and Rascher's ultimate fate.[54] The shadow of Rascher as a demonic, manipulative and hugely ambitious scientist loomed large in the proceedings.

Taylor informed Assistant Secretary of War, Howard Petersen, on 30 September, that there would be between 20 and 24 defendants. Because the evidence was plentiful and sensational, he expected that of all the planned trials, 'It should be a rather easy one to try and to decide, and therefore I think a good one to start with'.[55] His optimism arose from his view that the human experiments amounted to direct murder, and the experiments were so clearly a product of the Nazi totalitarian war machine. Hitler's architect and armaments minister, Speer, had concluded his defence at the IMT on 31 August 1946 by pointing to the dangers of modern science in the hands of dictators.[56] Taylor did not anticipate how complex the issue of experiments and ethics was to be.

Ivy's meeting with the prosecutors in early August 1946 was crucial in the decision to hold the Trial, but the selection of defendants was made without the advice of investigating scientists. McHaney updated Ivy on 30 September, on the 'tentative list of the doctors whom we plan to try in the first case, to begin about 15 November 1946. This time schedule will necessitate filing of an indictment on or before 15 October, so you will understand that time is exceedingly short.' McHaney explained that an all-out effort had to be made in field investigations.[57]

Air Marshal Erhard Milch was the subject of a simultaneous but shorter trial. This focused on the air pressure and cold experiments at Dachau. The prosecutors contacted the Berlin Document Center early in October 1946 to dig up evidence that Milch received reports on the Dachau experiments.[58] The interrogators worked hard to reconstruct the chain of command over aviation medical services reaching up to Milch.[59] Taylor was convinced that the Dachau cold and pressure medical experiments would be a major basis for conviction. But there were weighty issues concerning Milch's role in air armament, which were out of place in any medical trial. Milch had the distinction of the only single-person Nuremberg trial. This took place between 2 January and 17 April 1947. Milch had deliberately offered autonomy to researchers as an incentive to obtain their support, but this made prosecution all the harder.[60] His aloofness meant that it was difficult to convict him as responsible for aviation medical experiments.

The preparing of five more courtrooms in the Palace of Justice for simultaneous trials gave the prosecutors opportunities to try other categories of defendants for human experiments, and genocide. Nineteen of those prosecuted were held in Nuremberg as 'unfriendly witnesses'. Other witnesses could provide damning evidence. Adolf Murthum, departmental head of the Reich Commissar for Eastern Territories, and Werner Christiansen of the Department of the Reich Ministry of the Interior were involved in suppression of epidemics in the East, and could assist Mrugowsky.[61] The trial of Milch, the SS Economic Office Trial revolving around Oswald Pohl and the IG Farben trial dealt in part with criminal experiments, and Trial 8 on the Race and Settlement Office raised eugenic issues. Defendants in other trials supplied useful evidence. Pohl settled old scores as he had resented the experiments as a diversion away from slave labour and as undermining SS administrative hierarchies. In mid-1946 there was every reason to be optimistic about a large-scale trials programme to deal with key groups of decision-makers and perpetrators of Nazi war crimes.

McHaney's list of prospective Nuremberg defendants differed from Taylor's by including the anatomist August Hirt (*in absentia*) as the perpetrator of the Jewish skeleton collection at the Reich University Strassburg, and the Ravensbrück camp doctors Rolf Rosenthal, Treite and Hertha Oberheuser.[62] Twelve were in British custody; Rosenthal and Treite were eventually tried at the parallel Ravensbrück trial in Hamburg, along with three camp doctors and the prisoner-nurse Mory. This ran over 44 days from 5 December 1946 to 3 February 1947.[63] The 17 accused were charged with killing and ill-treatment of Allied nationals. It was a far brisker affair than at Nuremberg without elaborate translation facilities. Witnesses revealed a chilling picture of the lethal medical conditions in the camp hospital: the experiments were part of a regime of filth, freezing, overcrowding, starvation, beatings, sterilisation, lethal injections, gassings and infectious disease. The experiments exposed Oberheuser's brutality; other camp doctors resented her keenness to take part in the experiments.[64]

The Eppinger affair and medical side-effects

The informative Sievers provided details of Beiglböck's role in the seawater experiments. This prompted fresh inquiries in Graz and Vienna.[65] Since March 1946 the Austrian police held the internist Beiglböck in southern Austria. He was a physician at the military hospital in Lienz on the Drau (in the Osttirol) in the British zone. A former Dachau prisoner, Albert Gerl, recognised Beiglböck, and alerted the Austrian police to his role in the saltwater experiments on Roma, claiming that at least two deaths resulted.

These allegations resulted in a probe of Austrian medicine. Tension was rife between political forces wanting to purge the universities of all Nazi remnants, and medical elites seeking to consolidate their position after a purely nominal removal of a few Nazi doctors. It transpired that Beiglböck was a longstanding member of a Nazi organisation at the Vienna General Hospital since October 1932, of the illegal Austrian Nazi Party since June 1933, and of the SA since September 1934. He had run an illegal Nazi cell from 1935 to 1937 at the medical clinic of the celebrated Hans Eppinger. Nazi officials praised his activism and dedication to Nazi causes.[66] The

British authorities decided in July 1946 to hand Beiglböck over to the Austrian authorities for trial in Graz, but then transferred him to the Americans.[67] The timing was unfortunate for Beiglböck – any earlier or later he might have avoided trial in Nuremberg. The Austrian police and British interrogations formed the basis of the seawater case.[68] The cantankerous Rose claimed that he became a defendant at short notice as a result of the vacancy left by Eppinger.[69]

The transfer of Beiglböck for the impending trial precipitated two eminent medical casualties. Sensing that he was also a prime candidate for any medical trial, the renowned Vienna professor for medical pathology and therapy, Eppinger, committed suicide by hydrocyanic acid poisoning during the night of 25–26 September 1946. He was already in disgrace, when about to be arrested for interrogation in Nuremberg, just a few days before the climactic IMT judgments. Despite attempts of medical circles to insist that dismissing Eppinger would discredit Vienna medicine in the eyes of the world, he lost the legal battle to quash the verdict removing him in June 1945 from his chair in the Vienna faculty. Beiglböck was also deprived of his post as professor of internal medicine.[70] The reasons were that both were members of the (illegal) Nazi Party in Austria before the *Anschluss*. Eppinger had been a clandestine member of the nationalist Deutschen Klub in Vienna, and – although he denied this – had paid dues as a member of the illegal Austrian branch of the Nazi Party since November 1937. He joined the right-wing diehard anatomist Eduard Pernkopf in enforcing a ruthless Nazi ethos in Vienna's medicine.[71] Despite his efforts to strengthen Viennese medicine by building up research facilities, Eppinger encountered opposition from Nazi circles, complaining that he was brutal towards patients, using them primarily as test objects, that he was an inveterate intriguer, that he looked Jewish and had formerly employed excessive numbers of Jews, that he had been three times to the Soviet Union where he had treated Stalin as a patient and advised him on medical research – and that he was a reckless driver.[72]

This eminent medical casualty indicated that the Medical Trial cast aspersions on German (and Austrian) medicine. Eppinger's research on the pathology of the liver, the circulation and respiration was acclaimed as one of the great achievements of twentieth-century scientific medicine. He had declined to allow his name to go forward for the position as director of the Rockefeller Institute of Medical Research in New York, and had directed medical clinics in Freiburg and – until forced out by the Nazis in 1933 – Cologne. It meant that he had strong links to German and American medical research. His Vienna clinic delivered a modern, first-class specialist training for a generation of leading clinicians.[73] His dismissal in August 1945 meant that he lost an institute and clinic, where he was an absolute autocrat, and his exclusion from a world in which science was sovereign troubled Eppinger more than the accusations of criminality.

Eppinger was highly placed in German medical networks. He was a prime mover in negotiating with the KWG in having an Institute for Nutritional Research established alongside his clinic. He encouraged the Luftwaffe engineer, Edvard Berke, based at the Technical University in Vienna, to develop a desalination process based on simply adding sugar and tomato extract to saltwater, and had offered his clinic for tests on military volunteers. But the Luftwaffe research administrator Becker-Freyseng insisted on subjecting Dachau prisoners to the extremely painful experiments, which caused weakness, cramps, blackouts and – although this remained a matter of argument – death.[74] Just as Rascher fell from favour, the air force initiated a new wave of experiments.

Eppinger vehemently denied in September 1945 that he had ordered Beiglböck to undertake seawater experiments, and pointed out that Beiglböck left his clinic in 1941 to work for the army and Luftwaffe. Eppinger's story became increasingly untenable. The Vienna police asked the police in Carinthia to arrest Beiglböck, fearing that he would flee the British camp in Lienz, where as camp doctor his status was that of surrendered personnel. Josef Tschofenig, a witness from Dachau, said that the suffering of the victims was so great that only a dedicated Nazi could undertake such a violation of medical ethics.[75]

Austrian police investigated Eppinger in May 1946 and seized Beiglböck's records. The communist official, Eduard Rabofsky, and the Secretary of State for the Interior, Honner, shielded Eppinger. The protection of a distinguished professor was linked to the interests of the Russian zonal administration. Eppinger was at the time medical adviser to the Soviets in Vienna, and was treating the Soviet commander and his staff. The Austrians struck Beiglböck's statements on Eppinger from the interrogation reports, which were forwarded to the Allies.[76] When Beiglböck again mentioned Eppinger's crucial role, the Americans summoned Eppinger for interrogation. Eppinger swore an affidavit on behalf of the devoted Beiglböck that there had been no fatalities, contradicting allegations of deaths among the Roma experimental subjects.[77]

The Americans interceded that the case was relevant to the IMT and to the planned Medical Trial, and Beiglböck was handed over by the Graz police to the CIC on 24 October 1946.[78] Beiglböck was the only one of the 23 defendants whose case was first prepared by civilian police and judicial authorities. Austrian police in Styria, Carinthia and Vienna in February to May 1946 collected evidence against Eppinger.[79] By way of contrast, the German police failed wholesale to identify a single perpetrator of human experiments. Victims who turned up at police stations with evidence of human experiments met the icy silence of a system unwilling to respond to medical criminality.[80] The Medical Trial prosecutors found the Austrian police investigations helpful; the Russians failed to respond to Alexander's investigations in Vienna in May 1947, whereas the British assisted in Klagenfurt.[81]

Two days after Taylor's opening address, the professor of psychiatry at Heidelberg, Carl Schneider, committed suicide on 11 December 1946. He exploited euthanasia for psychiatric and eugenic research, facilitating the use of victims' brains for pathological research. The defeat of Germany plunged him into a mental crisis: after a brief internment by the Americans, he was released and entered the psychiatric hospital at Erlangen as a patient, where the Americans arrested him with a view to his testifying at the Eichberg trial of the euthanasia doctor Friedrich Mennecke. He committed suicide by hanging while in US custody on 11 December 1946, thereby depriving both the Eichberg and Medical Trials of a major source on the link between eugenics and euthanasia.[82] The *Badische Zeitung* interpreted the suicides of Eppinger and Schneider as linked, and as expressions of remorse.[83] This view may be too charitable, as the suicides could be seen as acts of defiance by physicians who considered themselves unjustifiably ousted from their clinics, where they were sovereign rulers over their medical staff and over the life and death of their patients. The post-war world was profoundly unappealing to such dedicated experimentalists once they were deprived of their status and powers.

Alexander had visited Schneider's clinic on 9 June 1945, and although Schneider mysteriously disappeared at the time, Alexander was well aware of his strong support for sterilisation, of links between his clinic and euthanasia killings of adult and child patients, and of his views that Nazism solved the problem of war neurosis.[84] His suicide constituted an admission of complicity, but also defiant rejection of the Allied jurisdiction. The suicide of the physiologist Holzlöhner added to the sense that the German academic community was in crisis. The suicides of prominent medical professors showed how the Trial placed the record of the whole German medical profession under National Socialism under scrutiny.

Other German academics resisted being called as witnesses. The aviation pathologist Büchner from Freiburg slipped away from an American military transport. Some scientists intent on their post-war careers refused to admit any association with concentration camp and SS doctors. Albert Demnitz, Director of the Behring Works department of IG Farben had to be reminded by Mrugowsky's defence lawyer that he could be taken forcibly to Nuremberg, if he continued to feign ignorance of Mrugowsky's interactions with IG Farben regarding the testing of vaccines and pharmaceuticals.[85] The prosecution lost patience with the evasiveness of the Behring Works bacteriologist Richard Bieling, when he denied he knew Ding. The interrogator was armed with documents to the contrary.[86] Bieling was under consideration for an appointment as professor in Vienna, and the Trial jeopardised his bid to join the ranks of the Vienna Medical Faculty.[87] Overall, the prosecution strategy was to demonstrate links of the human vivisectors in the camps to the SS administration and the euthanasia bureaucracy.

9
Internationalism and Interrogations

The Western Allies

The IMT sentences were pronounced on 1 October 1946, and within two weeks 11 hangings had been carried out. The Medical Trial was scheduled to start mid-November 1946, placing the prosecution team under pressure to gather evidence. Taylor and his staff were keen to work on an inter-Allied basis to gather evidence as rapidly as possible, and to give the Trial international status.

The British readily assisted with evidence, investigations and defendants. The Trial was an exercise in the politics of Bizonia, the Anglo-American political conglomerate. The financially hard-pressed British found it expedient for the Americans to shoulder the burden of a major trial. The British transferred 11 prisoners for the Medical Trial, and agreed to the Milch Trial, though Milch was a British captive. Karl and Rudolf Brandt (nominally British prisoners) were already in Nuremberg. Beiglböck was held by the British in Austria. The chief military medical officer, Handloser, and the Waffen-SS medical officers Genzken and Mrugowsky were British captives. Mant documented the criminality of the 'Hohenlychen Group' of Gebhardt, Fischer, and Oberheuser. The aviation physiologist Romberg felt aggrieved that Ruff and others in American captivity had enjoyed research opportunities at Heidelberg while he languished in a British camp.[1] The SS racial expert Poppendick was held at Neumünster camp. The transferred British captives felt bitter that their efforts to escape arrest by the Soviets ended with the indignity of their prosecution. Gebhardt argued that the Trial should have a papal nuncio as observer and be located in neutral Switzerland.[2] He claimed that his heroic struggle against communist collectivism and its supra-individualism merited his pardon.[3]

Somerhough recognised the special ethical nature of the medical atrocities, and energetically supported the American plan for the Medical Trial. By early September a handover of prisoners was agreed in outline. At first, the group consisted of Karl Brandt, Gebhardt, Romberg, Fischer and Poppendick.[4] On 6 September 1946 Somerhough urged Taylor to 'put a liaison officer from his team at this HQ'.[5] Taylor and McHaney visited Bad Oeynhausen, and made a further selection for a 'Tentative List of Defendants for Medical Atrocities Case'. For a while the Americans considered prosecuting Becker, Brunner,[6] Treite, Auguste Hingst and Rolf Rosenthal from the 'Hohenlychen Group'.[7]

The assistant US prosecutor McHaney handled the negotiations with the British forensic pathologist, Mant. 'In view of the time limitations, I am extremely anxious that we receive the benefit of your files and extensive knowledge in this field.' McHaney looked to Thompson's files for evidence against Karl Brandt, Handloser and Rostock.[8] Mant obligingly sent McHaney the testimonies of five victims and witnesses, as well as his report on medical atrocities at Ravensbrück.[9] He assisted the Americans in preparing their case, while collecting material for the British prosecution of the Ravensbrück case. The US judges permitted him to take voluntary statements with the consent of the defendants.[10] In November an OCCWC officer visited London to gather evidence.[11] Mant fixed up an office in the Court House for Wing Commander Thompson, whom he briefed on his return from Ottawa.[12] Mant and Thompson had the status of British trial observers, and Thompson proposed that the stalwart Mant should assist the Americans at Dachau in preparing evidence for the Buchenwald trial, while collecting materials for the ISC.[13] Taylor invited the British military lawyers Somerhough and Lord Russell of Liverpool to attend the opening of the Trial, and provided facilities for the British war crimes investigators.[14]

The French had to be brought on board, because the second IMT issue remained contentious. The French Ministry of Justice proposed in October 1946 to delegate two attorneys to join the American prosecutors. McHaney explained that 'there was no provision in our ordinance to allow for the participation of the prosecutors of another nation', and suggested that they be mollified by inviting them to act as advisers. Two French ISC members were given hospitality at Nuremberg, in the hope that they could locate evidence concerning the atrocities at the Natzweiler camp near Strasbourg.[15] The French failed when they claimed Mrugowsky and Hoven had been involved in experiments at Buchenwald where there were many French prisoners.[16] In November 1946 Hochwald gathered evidence at Strasbourg relevant to the typhus vaccine charges against Karl Brandt, Mrugowsky, Rose and Rostock.[17] After a spell at Nuremberg, the virologist Haagen was returned to French custody.[18]

Among the accused was one (former) Austrian and one (former) Czechoslovak, and two defendants were born in Alsace. Taylor was resolute in his conviction that this was an international trial with a basis in inter-Allied agreements. He encouraged foreign observers at the trials. The Soviets, French, Czechs, Poles, Yugoslavs and Dutch established offices in the Palace of Justice by November 1946. But most were interested in the impending industrialists' trials rather than medical crimes.[19] On 14 January 1947 Taylor ceremoniously introduced 'the observers of the

Allied nations' to the court from Czechoslovakia, France, the Netherlands and Poland, and Lord Wright of the UNWCC visited on 9 May.[20] There were medical observers: Thompson for Britain, Bayle for France, and General-Major Charles Sillevaerts, the Surgeon-General for Belgium.[21] Sillevaerts kept Belgian colleagues informed in *Bruxelles Médicale*, where he warned that medicine should not be tainted by politics.[22] Ravina reported more critically for *La Presse Médicale* on the need to keep the issue of genocide to the fore, and how the prosecution was too detached from the atrocities. Drs Inbona and Berlioz of the Federation of Deported Physicians supported his criticisms.[23]

The Czechoslovak and Polish observers supported the Medical Trial with witnesses and evidence. The Poles facilitated the travel of the Ravensbrück 'Rabbits' as witnesses. The Polish observer Captain Acht obtained evidence for Alexander's resourceful assistant, Maryann Shelley. The Americans appealed on Czech radio, and to the Czech medical chambers, for victims and witnesses of human experiments to testify.[24] The Czech delegation under Bohuslav Ecer (who died imprisoned by the communists in May 1954) assisted the prosecution, 'especially in providing documentary material or witnesses.'[25] But the Cold War chill made itself felt, when the former Dachau internee, Blaha, a doctor and socialist politician, criticised the overwhelmingly American character of the Trial. Blaha attacked the Trial as without judges from countries that were 'most affected' such as France and the USSR, a criticism reflecting the continued demands for a second IMT.[26]

The onset of the Cold War meant the Soviets were icily detached from the Trial. The Soviet insistence at the IMT on 1 and 2 July 1946 that the Germans carried out the Katyn massacre of Polish army officers was loathed as a piece of tendentious propaganda. It provided a culminating point of tensions between the Soviets on the one side and the British and US prosecutors on the other. The Russians stalled on all requests for exchanges of information and handovers of suspects.[27] When Schreiber was requested for interrogation – and possible prosecution – as a follow-up to his evidence on germ warfare at the IMT, the Soviet authorities declined to make him available.[28] The onset of the Cold War meant that the French realigned with the Americans, whereas the Soviet critique of American research intensified.

Interrogations

The OCCWC conducted over 15,000 interrogations of over 2,250 individuals. Each defendant was interrogated between five and (in the case of Sievers) 20 times for the Medical Trial. The defendants with senior positions in the SS were interrogated most frequently: Karl Brandt 14 times; Rose in the same period only four times. The tone varied considerably. Alexander created bonds of confidence, while specialist interrogators could be aggressive in efforts to 'squeeze the juice' out of a subject. Interrogations focused on particular episodes to establish criminal responsibility or dealt with structural issues of organisation; others elicited a life story.

The interrogations changed in status. Until mid-1945 they were of military prisoners for intelligence purposes. Once Japan had capitulated, the Allies could pursue a war crimes agenda. The IMT trial team between August and October 1946 interrogated witnesses under oath, and from June 1946 Taylor established a dedicated team, who interrogated on a one-to-one basis in German.[29] Once the Medical Trial had been decided on and the indictments served in November, the interrogations became prosecution documents. They were mandatory for witnesses, as was pointed out to Germans who were unwilling to collaborate. The judges insisted that the interrogation was to be entirely voluntary for the defendants, who could have counsel present.[30] Affidavits had to be freely given.

The status of the interrogations remained ambiguous, as the accused were at first not informed that the interrogation record could be used as evidence against them. Karl Brandt was surprised when Alexander mentioned this in passing. The Austrian lawyer Steinbauer was meticulous in attending the interrogations of Beiglböck. Interrogators offered to improve the conditions of arrest to persuade subjects to co-operate.[31] The interrogation was in part preliminary skirmish, vindication, an earnest attempt to set the record straight, and in part psychological therapy and investigation.

Before indictment the interrogations were exploratory, and in the cases of Benzinger and Heinrich Rose led to release. The new set of prosecutors, McHaney and Hardy, joined forces with an experienced team of Nuremberg interrogators, who had detailed knowledge of the SS and its genocidal actions: Herbert Meyer, Joseph Maier, Benvenuto von Halle, Iwan Devries, Walter Rapp, Fred Rodell and Wartenberg were astute interrogators. They were well armed with documents and background information and rapidly made it clear that tales of innocence and exoneration were futile. The interrogators were bilingual and skilled in how to extract crucial information, which was formulated into an affidavit.[32] A guard and a translator were present. A few interrogations were in English (in the case of Blome, who had visited London in 1939), to make it easier when one of the senior prosecutors attended.[33]

In August 1946 the physiologist Ivy provided an agenda for pre-Trial interrogations. His menu included mustard gas, excision of ligaments, transplantation of tissues, poisoned bullets and the whereabouts of Rascher's collaborator Finke and Hirt. More broadly, he urged that experimental data and any pathological specimens be located, as well as details of techniques, what drugs were used, and their dosage. He wanted precise details of experimental procedures and what was done to the human subjects in terms of what tissue or bone was excised: 'How do they justify the fact that they used human subjects? A summary of these rationalizations will provide the keys to the arguments of the defence.'[34] The Ivy rules on human experiments for the ISC became central to the Trial.

Specialists assisting the prosecution came on board as interrogators only in mid-November 1946. Alexander sized up each defendant, and Mant and Thompson interrogated in the cases supported by the British, as did Bayle when there was a French connection. They followed Ivy's template to elicit issues of scientific significance. Alexander and Ivy interrogated Beiglböck and Schäfer to clarify technical aspects of the seawater purification method. It emerged that Schäfer's seawater purification method was ready in December 1943, but the air force medical experts spent a year equivocating on its merits and conducting human experiments.[35]

The verdict that the experiments were pseudo-science raised the issue of the sanity of the perpetrators. While the extreme rationalism of a calculating, manipulative and exploitative totalitarian state could provide one level of explanation, could the abusers of medicine have been psychopaths and sadists predisposed to murder? The 23 defendants were objects of sustained scrutiny from an informal grouping of psychiatrists who congregated at Nuremberg. Alexander on the prosecution side, and a number of observers – Thompson, François Bayle, Alexander Mitscherlich and Alice von Platen who shared a deep concern about therapy for mental disorders. They were aware of the historical significance of the Trial, and provided overviews of its origins and course so that psychiatric case history and the history of Nazi medicine fused.

Alexander was gentler in his questioning than the dedicated interrogation team. He did not place his subjects on oath for the psychological interrogations. He gave assurances that he would not use the replies in court.[36] The psychological interrogations followed a standard pattern. He asked about childhood and parents, siblings, behaviour and emotions as a child, leisure activities and youth groups, marriage, children, career choice and feelings about work, and investments. Why did the defendant become a doctor? Did they ever feel unjustly treated or contemplate suicide? He inquired about their attitudes to the Trial, whether it was significant for medical ethics, and on their opinions about the Hippocratic Oath. He also asked about illnesses (Becker-Freyseng explained he contracted pneumonia in the course of the oxygen experiments).

Alexander probed underlying values – money, foreign travel and what the defendant thought about Poles, gypsies and Jews and a range of nationalities (e.g. English, French, Italians, Spanish and Americans, Chinese and Japanese) and races; and also about their views on Nazi racial extermination. Rose condemned 'Negroes' as swindlers and medically irresponsible. Alexander explained to a suspicious Becker-Freyseng that he was asking him about his general attitude to 'gypsies' and not probing the seawater experiments.[37] Finally came questions about phobias and experiences with animals.

Alexander varied the agenda to elicit specific points – had Beiglböck known Kogon in Vienna? He spoke some Chinese to Rose.[38] He hoped to compile a 'neuropsychiatric-sociologic' analysis of war criminals.[39] He recognised that 'the material is fascinating from a social psychiatric point of view but I hope that these trials are instrumental in preventing something like this in the future'.[40] He aimed to identify the social psychological factors in the Nazi state, and the defendants generally appreciated the opportunity to vent their opinions.[41] Initially, the defendants welcomed Alexander's medical interest in their work, his evident intelligence and psychological understanding. But then they began to fear that Alexander understood them only too well with his combination of familiarity with German medicine and his psychological expertise. Responses became resentful, betraying a latent anti-Semitism, which could not openly find expression.

Alexander focused on the psychology of the medical criminal. He identified six types – from those with high-powered academic qualifications to the brutal concentration camp doctors. Between lay a spectrum with more limited intellectual powers.[42] This was a preliminary to probing each defendant's views about utilitarian ethics: was it right to sacrifice 20 lives in an experiment to save 2,000 lives? Rose endorsed such a ruthless act for significant and important scientific issues, but it was inadmissible as routine practice. He viewed it as a psychologically destructive situation for the experimenter.[43] Ruff similarly supported this view, citing George I's orders to use orphans for experiments.[44] While most said that German faculties did not rely on the Hippocratic Oath as part of medical training, Becker-Freyseng heard it referred to in the lecture of the dissident pathologist Büchner in 1941.[45] Schröder submitted a self-justificatory memo explaining how as an 'old officer' belonging to the pre-Nazi core of the army he encountered antagonism from the Nazis; but as military medical officer he felt a sense of duty to serve the welfare of his comrades rather than to the NSDAP.[46]

Alexander was interested in when the defendants lost their faith in a German victory, and whether they felt that Germans were superior. He developed an analysis of the social psychology of the SS as based on the central concept of '*Blutkitt*' or blood cement. An embittered Gebhardt explained that each SS officer had received a booklet on this, but he had only read it in the final week before his arrest when he was tending an ailing Himmler.[47] Hitler had been inspired by Genghis Khan's notions of blood loyalty, while unleashing devastation. The idea was that the criminal organisation could exploit personal insecurity and minor misdemeanours to secure complicity in atrocities. The result was a tight bonding among the SS doctors as well as a pathogenic ego. Alexander found this in his studies of the doctors, and SS officials Brack and Pohl.[48] He explored the idea that deaths were necessary to achieve greater strength, whether in medical research or in steeling themselves for killings. He built up psychological portraits of Hitler and Himmler, seeing the concepts of thanatology (the science of death) and ktenology (the science of killing) as rooted in the genocidal mentality of the Nazi leaders. He linked the human experiments to thanatolatry (delight in death).[49] Alexander dissected comments on Himmler's relations with the Raschers, noting Rascher and Himmler shared the personality trait of an obsessive attention to detail.[50] Like Taylor, Alexander was unwilling to accord the experiments genuine scientific status. He moved ever further away from seeing the medical researchers as engaged in normal science, but came to highlight their misdeeds as a macabre and sinister feature of the German war effort. He stressed the manipulative and murderous role of the SS, ideas of blood bonding and the strengthening of German endeavours by virtue of the sacrifice of lives. German medicine and the army were drawn into the criminal conspiracy by professional leaders like Gebhardt.[51] As he amassed evidence of 'all this madness', he ceased to view the medical defendants as colleagues, diagnosing them as defective personalities with a weak ego and strong superego.[52]

The psychopathology of SS medicine similarly preoccupied Bayle, who acted as liaison officer between the American and French war crimes authorities, and was on Thompson's ISC.[53] He requested permission to obtain handwriting specimens and take physical measurements. Justice Beals opposed this as potentially unfair to the defendants; Justice Sebring suggested, 'The French psychiatrist can sit in as a doctor in the course of the trials and can get information from that.' The judges decided that Bayle's psychological examinations would have to wait until the close of the trial, but by December he was collecting samples of handwriting and opinions from the defendants.[54] The assiduous Bayle established close relations with the prosecution, especially McHaney, and defendants, and even figured in the apologetic novel *Gasbrand*.[55]

Alexander initially found Bayle somewhat of an irritant: he recorded a 'Showdown re Bayle'. Yet relations improved when Bayle invited Alexander to address the International Commission in January 1947 in Paris.[56] The Americans came to value Bayle's presence as an insightful observer and supplier of documentation. He examined Karl Brandt no fewer than 20 times, taking handprints and assessing his athletic physique and lively character. He drew a contrast between Brandt and the monstrous brutality characterising other SS doctors.[57] Bayle's work found expression in two remarkable analyses: one on the psychology of the accused at the Medical Trial, the other in a study of the character and ethics of SS leaders.[58]

Mitscherlich stressed the need for the German delegation to have seats 'permitting an exact physiognomic study of the defendants'.[59] Alice von Platen noted that the first impression of their faces was that these were quite ordinary people and not psychopaths.[60] Mitscherlich gained permission to interview defendants with the psychoanalyst Wilhelm Küdemeyer in late February.[61] Lemkin also called for analysis of individual participants' mental health in relation to group personalities and cultures.[62] Taylor saw the possibilities of evaluating the defendants as objects of ethnology and sociology.[63]

The resourceful Thompson took up the plans of the US psychiatrists that a team of psychologists, psychometrists and psychiatrists report 'on the principal individuals accused at the Nuremberg Medical Trial as well as on individuals not available. Report of value for the future, on mechanism of Nazi ideology which is a concern of the present.'[64] He had long taken the view that the diseases of anti-Semitism and authoritarianism required appropriate antidotes. He hoped that psychological evaluations would throw light on the 'mechanism of Nazi ideology' as well as being of historical significance.

Thompson focused more on scientific questions when he interviewed Blome in early February 1947 about cancer research, its links to biological warfare and the Reich Research Council.[65] They considered science in an authoritarian state, whether it was possible to have a totally good man running such a state and its mass psychology. Thompson came to negative conclusions as to the quality of the German research, and viewed links between science and totalitarianism as 'a Suicidal belief'.[66]

When Thompson quizzed Blome about links between cancer and biological warfare,[67] Blome used this as an opportunity to establish his scientific credentials. He outlined a cancer morbidity survey, first conceived in 1936, involving 'the statistics of the patient as a whole in the entire country'. He explained that he published a special set of statistical reports in Mecklenburg. Similar surveys were for the mixed industrial/rural Saar, and Vienna and Merseburg (as industrial populations). Statistical data, collected over a five-year period, was abandoned in the cellars of the Posen anatomy department. Blome claimed to have examined heredity, and the toxicology of alcohol and nicotine, failing to mention that much of the research was at the initiative of his assistant, Carl Hermann Lasch.[68] Blome floated the idea of a national cancer scheme for all Germany, supported by the Reich Research Council, in which he administered a section for cancer. The scheme for a Greater German Cancer Institute was approved by Hitler, Bormann and Hess, but to be established after the war.[69] Blome developed rapid diagnosis for TB from the mid-1930s. He surveyed 150,000 persons in Mecklenburg, then Westfalen, Württemberg in 1941, and in the area of Posen. He established that there was a 2 per cent incidence of TB in a population, and that 1 per cent was unknown and infectious. He outlined a system of referral, and support for families.[70] His claims of an innovative preventive medicine contradicted the charge of shooting tuberculous Poles.

The aim between mid-August and early October 1946 was to assess criminality and the likelihood of obtaining a conviction. The pre-trial interrogations were conducted without any input from medical experts.

While medical scrutiny of the evidence might have eliminated some of the weaker cases (as against Schäfer and Pokorny), the prosecution team hunted for links to Himmler and the SS. Rose denounced the pre-trial interrogations as incompetent, and he vented his rage when the accusation was levied that as many as half a million were killed in the human experiments. Rose and Ruff indignantly pointed out that to be placed on trial was already an indication of guilt under German law. Once the defendants were formally charged on 11 November, the interrogations became court documents, where prosecution and defence staked out their positions. (See Table 6.)

A follow-up aim was to secure evidence for later trials. Rose gave information on IG Farben, Poppendick on the activities of officials in the SS Race and Settlement Office, and Genzken on other SS functionaries. At first Sievers was a rich source of information, adding much to what was contained in his diaries covering appointments and research contracts. He attempted to convince the prosecution that he was a member of a German resistance organisation *Geheime Organisation in Not*.[71] Once this was attacked in court, he refused point blank to allow an interrogation to proceed in the summer of 1947: he was angry that the prosecution had branded him a liar.

The interrogators tried to sort out the routines of who did what and under whom. Establishing the command structure was a preliminary for pinpointing the criminality of individual actions. The faction-ridden and feuding SS and NSDAP hierarchies made it difficult to see who was responsible for what. Hitler and Himmler had key roles, but the question was how did they delegate authority for the experiments and euthanasia? The euthanasia issue revolved around Hitler's command and administrative structure. Brack was asked about the euthanasia bureaucracy – how close was Karl Brandt to Bouhler, did Brandt have an office in the Führer's Chancellery, and how did the decision-making process work? The interrogators were puzzled by Karl Brandt's sweeping powers of command stemming from Hitler's order of 25 August 1944 – why then did Sievers say that the Ahnenerbe never received commands from him?[72] The role of medical decision-makers was scrutinised – notably the key position of the euthanasia psychiatrists Nitsche and Heyde, and whether Karl Brandt could override their diagnoses. It emerged that Bouhler (loyally aided by Brack and Globocnik) clashed with Bormann[73] who was the mainstay of Conti's support.[74]

Conti had his total powers over civilian health confirmed in the decree of 28 July 1942. He feuded with Blome (who used their arguments to deny that he was associated with euthanasia or human experiments) and with Karl Brandt, who looked to Speer for support. Schröder and Gebhardt expressed admiration for the pragmatic Brandt as restoring confidence in medical skills and tackling the problem of providing hospitals under war conditions.[75] The interrogators' efforts to reconstruct the Nazi hierarchies yielded spectacular displays of ambition, corruption, and personal intrigue.

Interrogators and defendants agreed that Himmler was pivotal for ordering medical experiments. The prosecution saw a direct chain of command, allegiance and internalisation of the macabre Nordic racial rituals and rites. Defendants responded that they resisted Nazi excesses and kept their distance from Nazi fanatics. Gebhardt – the closest to Himmler of the accused doctors – tried to vindicate his position by suggesting there were inherent tensions in the medical relationship with the SS: 'I was of the opinion the medical officer is first a doctor and second an officer, while Himmler was of the opinion he is first an officer and second a doctor.'[76] He accused Himmler of favouring nature therapy, citing his masseur Felix Kersten as evidence of his mysticism. Gebhardt protested that Himmler only consulted him when operative surgery was necessary.[77]

Sievers was anxious to deflect attention from his SS-Ahnenerbe organisation, and declared to the sceptical interrogators that Himmler operated primarily through Reichsarzt SS Grawitz and was advised by the ambitious Gebhardt and Rascher. He attempted to exonerate the Ahnenerbe, as lacking a medically qualified administrator.[78] Rudolf Brandt had to explain why it was so difficult for the SS to extricate Rascher from the Luftwaffe in 1942, with both organisations seeking to claim credit for the research.[79] Genzken and Mrugowsky were relentlessly asked about the command structure in SS medicine, as the prosecution attempted to establish responsibility for concentration camp research. In the end, both were interrogated together to iron out inconsistencies.[80]

The interrogators asked about contacts between the accused and SS functionaries. Sievers had to account for links to Bruno Beger, the SS anthropologist, and Adolf Eichmann.[81] There was constant probing of Himmler's activities, and the relations between Himmler and Hitler.[82] Alexander became fascinated with the demonic human vivisector Rascher as an extreme case of a psychologically perverted talent. Whereas the prosecution lawyers characterised Rascher as a fanatical pseudo-scientist, Alexander saw him as a talented but malevolently sadistic scientist. He collected other opinions on Rascher. Rostock viewed him as badly supervised, and generally blamed the experimental atrocities on Ding and Rascher. Alexander viewed the divorce of Rascher's parents as leaving him without a father's firm guidance.[83] Rascher's scientific rationality was poorly directed. Alexander similarly interpreted Himmler as gifted, hardworking but tragically misdirected: his fascination with death and the macabre resulted in lethal experiments and millions of racial killings.[84] Alexander reconstructed the Rascher network, which included the physiologists Holzlöhner and Finke, the prisoner doctors Blaha and Schneeweiss who had to conduct autopsies, Punzengruber as a prisoner chemist, and the prisoner assistant Neff, who had to select experimental subjects.

Alexander identified further networks among the doctors, as they transferred experimental subjects and body parts between the concentration camps: the persons killed for a Jewish skeleton collection at the Reich University Strassburg came from Auschwitz to Natzweiler, and organs of persons killed in freezing experiments were transferred from Dachau to Strassburg.[85] The freezing experiments involving 200 experimental subjects were a large series among the German human experiments. Of ca. 160 Russians who were victims, only one survived.

At the IMT prosecutors were faced with the enigma of Hess's amnesia, the truculent belligerence of Göring and the rabid prejudice of the anti-Semite Julius Streicher. In contrast, all the accused at the Medical Trial complied with the legal procedures. But they offered protestations of the harmlessness of the experiments, and of non-involvement, self-justification, and ignorance. The risk to the accused was that feigning excessive ignorance could backfire: the interrogators confronted the subject with accusations as to the role of the experiments in the murder of hundreds of thousands, incriminating evidence from other captives, and the accumulating stockpiles of documentation. To defuse such a highly charged situation, the accused calculated that revealing something substantive about German military medicine, weapons development or strategic operations could be advantageous. It might result in an appointment as a trusted consultant. But at the same time it was perilous to incriminate oneself or indeed to reveal knowledge of the full scope of what went on in such sensitive areas as germ warfare or aviation medical research.

Some captives treated their captors with a disarming frankness, while explaining that they were innocent of anything culpable. Sievers admitted to 20 experimental subjects for Hirt's mustard gas experiments, who were cured and who enjoyed considerable benefits, when there were 240 subjects and 60 deaths.[86] Oberheuser reduced the numbers of victims and deaths when she was first interrogated. Rose was careful not to admit that he experimented with dangerous insect vectors. The Allied interrogators were unwilling to take at face value the defendants' denials and justificatory stratagems. Both sides realised the advantages in establishing proven evidence, even if the Germans were privately contemptuous of the procedures. One tactic was to throw the interrogators off the scent by indicating the criminality of someone else. Rostock denied any knowledge of human experiments and was guarded concerning Karl Brandt; but suggested that Blome was a significant political appointee with strong Nazi contacts. Genzken similarly disclaimed any knowledge of human experiments, and was circumspect when it came to Mrugowsky. But both Genzken and Rostock unequivocally identified Gebhardt as Himmler's main scientific adviser – something that Gebhardt vehemently denied.[87]

Barr made an early start on Mrugowsky in mid-July 1946, and Rapp grilled Handloser for information on the SS hygiene services and the

Buchenwald human experiments. The position of the Waffen-SS emerged as controversial during the IMT – was it a normal fighting force under military command (in which case its medical services came under the wing of Handloser), or was it an extension of the SS as an ultra-Nazi elite? Handloser alleged that his responsibilities were purely nominal, whereas the accused SS officers saw their Waffen-SS status as a mitigating factor.[88] Genzken insisted that his first loyalty was to the Waffen-SS rather than to the general SS. He had to fend off attempts to appoint him as Reichsarzt-SS, and Himmler disliked him. He defended Handloser as only commanding Waffen-SS medical field units rather than being responsible for all Waffen-SS medicine.

Nine of the ten SS defendants joined the Waffen-SS.[89] Sievers was the exception. Brack, Genzken and Fischer claimed that Waffen-SS duties were a form of ethical protest, enabling them to extricate themselves from a criminal system. Gebhardt used the clinical crisis in Russia in 1942 as justifying the grotesque sulphonamide experiments. The prosecution was faced with medical defendants whose career profile in the Second World War showed the mobility between different theatres of military medicine, such as frontline duties, serving in concentration camps, normal research in university clinics and assignments to applied military medical research. Academic and administrative work and visits to concentration camps alternated with Eastern Front service: as in the cases of Fischer, Karl Brandt, Gebhardt (who explained his responsibility for surgical services on the Russian front from 1942), and Rostock, who was three months in 'Lublin bei Shitomir' in 1941.[90] Alexander took the view that Weltz had incriminated himself by visiting Dachau. Weltz maintained that he wished to see the camp for himself, and that he was taken in by its façade of order with the clear distinction between criminals and political prisoners.[91]

Concentration camp medical personnel tended to rotate positions rather than remain in one permanently.[92] Chains of command overlapped: when Mengele came to Auschwitz, he held the post of assistant to Otmar von Verschuer at the KWI for Anthropology Berlin. Mrugowsky held responsibility for SS medical research stations in Buchenwald and Auschwitz, although research personnel could also select on the Auschwitz ramp or give phenol injections. Hoven and Oberheuser were both camp doctors in concentration camps, and were in terms of medical hierarchies on the bottom rungs when they took on these appointments. Rascher was keen to move the freezing experiments from Dachau to Auschwitz where security was better and facilities were greater.[93] The mobility presented the investigators with a complex and at times confusing spectacle of doctors depicting themselves as outsiders from the SS state. Brack conceded a visit to Lublin in 1940 with Odilo Globocnik, although the transfer of T-4 personnel for the death camps of *Aktion Reinhardt* under Globocnik's command was not mentioned. Others insisted on their ignorance: Karl Brandt confessed to visiting Sachsenhausen in 1935 and Natzweiler once in 1944. Blome and Sievers were in Dachau in connection with Rascher's experiments, Schröder inspected the Dachau seawater drinking research, and Genzken and Poppendick also visited the camp. Weltz admitted to a single visit to Dachau out of curiosity. Only Mrugowsky admitted visiting Auschwitz. Rostock claimed never to have known what went on in a concentration camp.[94]

The defendants ranged widely in age from their late fifties to their early thirties. The group born between 1900 and 1909 stood out. These were mostly SS officers. They were deeply affected by the First World War, and this fired their interests in race and national regeneration. Karl Brandt had his schooling interrupted by expulsion from Alsace, and saw Nazism as redressing the injustice of the Versailles Treaty.[95] Mrugowsky lost his father in military action in 1914. They were typical of the generation of *Kriegsjugend*, who took a prime role in defending the nation's racial health.[96] (See Table 1.)

The interrogators took a top-down strategy. Hardy held Karl Brandt to be the chief culprit, and reconnoitred the evidence in conjunction with Berlin Document Center in July 1946. The Nuremberg prosecutors set about establishing Karl Brandt's administrative competence – culminating when Hitler vested in him overarching powers from August 1944. Then, on 19 October, the interrogator Rapp confronted Karl Brandt with the fact that thousands had been killed in human experiments. Brandt admitted to knowing about a small number of experiments in concentration camps, characterising the victims as 'volunteers' and as 'criminals condemned to death'. Brandt insisted that he only knew of the experiments of Gebhardt from his lecture in the Military Medical Academy in 1943, and that he heard from Mrugowsky in the autumn of 1944 about experiments on removing poisons from water. Hardy remained convinced that Karl Brandt was Hitler's main medical agent. As the interrogators tried to map the labyrinthine medical power structures, they were convinced that all threads of guilt would ultimately lead to the Führer's *Begleitarzt*.

From mid-August came interrogations of the SS officials Sievers and Rudolf Brandt (Himmler's stenographic secretary) to establish the command structure of the SS. Reconstructing the Nazi medical Behemoth was a precondition for establishing the guilt of the accused medical culprits. The interrogators' tactic was first to establish a factual biographical basis by discussing the defendant's career. The prosecution compiled biographical profiles to elucidate motives.

While the interrogators tried to reveal links to the mass killing of Jews, this was something that the interrogated prisoners strenuously denied. Rapp and Rodell were masters at using documented evidence to expose denials and excuses as lies. Brack insisted that euthanasia was a humane measure for the incurably ill and that he had pressed for its proper basis in law, and at first disclaimed any knowledge of the Holocaust. Brack was grilled intensively on 12 and 13 September, four times on the first day. Brack began cautiously retracting earlier information about the euthanasia programme. Faced by persistent denial from Brack as to complicity in mass X-ray sterilisation of Jews, Rodell made him first admit to the authenticity of his signature, and then confronted him with a signed document on the mass X-ray experiments on concentration camp inmates.[97] Even then, Brack presented a benign picture of Globocnik presiding over 'work camps' staffed by transferred T-4 personnel.[98] The reality was the extermination camps with carbon monoxide gas chambers of *Aktion Reinhardt*.[99] While Rapp triumphed over Brack, he found Blome impenetrably cunning and evasive.[100] The interrogations suggested that the Trial might yield mixed results.

Unlike the IMT, there was no trial as such of organisations. The prosecution assembled clusters of experimenters with an eye to their links with the Nazi state, Party, military and SS. The trial highlighted organisational links to the organised criminality of the SS, the German High Command and air force, and NSDAP, but took no interest in the nazification of professional organisations. The prosecution mustered defendants in vertical hierarchies from leading Nazis to their henchmen (and the woman) on the ground. Defendants fought this hierarchical concept. Blome was a Nazi political activist with a life-long loathing of communists and Jews. He had studied in Rostock under the bacteriologist Reiter, who from 1933 directed the Reich Health Office. He also had dark political links to Bormann. His book *Arzt im Kampf*, published in 1942, was a testimony to a career as a soldier, veteran anti-Semite and right-wing political agitator. By 1922 he was involved in three trials against the ultra-right, and was proud of shielding the murders of Walther Rathenau.[101] He saw the Trial as a way of settling old scores and turned the tensions between the NSDAP and SS to his advantage, distancing himself from the SS administrators and physicians. Blome condemned them as having betrayed the early Nazi ideals of a national struggle, and of the dedicated *völkisch* physician. He explained he had valiantly defended the independence of the Reich Physicians' Chamber against the SS and the machinations of his arch-enemy, the Reich Medical Führer, Conti.[102] He was only Conti's deputy within the NSDAP rather than in any matter connected to the state administration, and he claimed to have opposed euthanasia.[103] The interrogators accused Blome of pretending to have the innocence of a rabbit in the woods – concealing the true nature of his involvement in Nazi atrocities.[104] Alexander observed that his rise to a high position arose more from his political activism than academic attainments.[105]

The prosecution constructed its case primarily from documents, continuing the policy of Jackson at the IMT. Analysts combed through the US Documents Center collections, colour coding documents for each planned case, and producing crisp analyses of evidence known as SEAS (staff evidence analyses). Manfred Wolfson's masterly analysis of Verschuer and Mengele compiled in 1946 has been unsurpassed by 50 years of historical research.[106] The Americans understood that documents were authentic, whereas testimony could be fabricated, especially when whole groups conspired to argue that what had gone on was innocuous. The use of documents placed the burden on the defence to disprove authenticity or to prove that the prosecution was misinterpreting them. The incantation of documents into the court record dominated the proceedings. McHaney reflected, 'The documents were of their own making – and they came back to haunt them.'[107]

The mounting piles of documents necessitated establishing procedures for their consultation so that nothing should go astray and also to prevent deliberate destruction of evidence. The IMT prosecution had in fact lost the one surviving copy of the vital Wannsee Conference minutes (it was rediscovered in March 1947). Fortunately, all trial documents had been duplicated en masse. The defence was provided with a reading room. A small industry supported translation and duplication of the documents.

The Court established an archive on 21 February 1947, keeping track of original documents under conditions of high security. The archivist, Barbara Skinner Mandelaub, serviced the prosecution and defence, maintaining a definitive set of trial transcripts. The staff was subject to a rigorous efficiency rating to maintain output and quality of work. A requirement for an archives assistant was meticulous presentation and order, as 'These elements are of importance in preparation of records such as indexing, cataloguing and classifying of material which is to be of permanent value for legal and historical reference for generations to come'. Mandelaub increased the security and made sure that only archives staff retrieved or filed records, and the archives issued certified copies rather than originals so that nothing could be lost or destroyed.[108] The Trial Documents achieved what amounted to sacrosanct status, once they had gone through the stages of authentication and been presented to the Court, where their authenticity could be challenged. Culprits were to be convicted by the masses of paperwork that the desk-bound killers had generated.

In contrast to the American documents-based strategy, the British ran cases by locating witnesses able to testify against the prosecuted. The strategy of building up robust and incontrovertible testimony led to a number of British successes. At the trial of Bruno Tesch for supplying Zyklon B to Auschwitz enough of Tesch's employees came forward to testify against him.[109] By contrast, the American case against Gerhard Peters, the IG Farben official responsible for manufacture of Zyklon B, failed.

While documentary proof was considered to be of a higher order of reliability, affidavits from victims and selected witnesses developed the human side of the case. The prosecution contacted the 'Victims of

Fascism' organisation for witnesses to the medical experiments.[110] There were public appeals for testimonies on sterilisation and castration. Victims wrote spontaneously to the US prosecutors, particularly on the need to bring Mengele to trial.[111] However, the prosecution feared that their testimony could be challenged as exaggerated.

The prosecutors swung round to acknowledging the racial aims of the Nazi war machinery. Figures, whom the Allies originally targeted regarding chemical and biological warfare, emerged as perpetrators of a series of racial atrocities involving sterilisation and euthanasia. By way of contrast, interest faded in the role of scientists like the organic chemist Richard Kuhn. Mengele at first escaped attention as not involved in chemical warfare experiments. For him, and other racial biological researchers like Robert Ritter, the new Allied interest in racial atrocities meant either that they had avoided initial detection or that, given the time constraints of making snap decisions on whom to prosecute, constructing a case appeared too laborious.

Confronting 'victors' justice'

Germans viewed the prosecution as an official body, and expected it not to be partisan. Rostock pointed out that Germans did not understand the principle of remaining innocent till found guilty – in Germany pre-trial examining magistrates were required to establish probable guilt before a case came to trial.[112] The rapid pace of arrest, transfer to Nuremberg, interrogation and indictment meant that the accused were caught up in a legal whirlwind with only limited time to find exonerating evidence and formulate pleas. The autumn of 1946 was a period of self-justification, efforts to build up networks of support, and enforced introspection, at times tempered by self-pity. The more that publicity and press projected the image of the defendants as sadistic and fanatical SS doctors, the more the defendants' sense of outraged indignation grew.[113] Their elaborate pleas involved a mass of evidence and counter-accusations, which preoccupied the Nuremberg court for five months. The defendants can be divided into those who concealed and distorted, and those who saw nothing wrong in their conduct and faced the authorities with self-righteous indignation.

The defendants were kept in solitary confinement to avoid any possibility of collusion. Security measures intensified following Göring's suicide in October 1946. The isolation weighed heavily, so that the interrogation provided an opportunity for psychological release, reflection and dialogue.[114] The defendants deepened their sense of identity as representatives of a beleaguered profession and research corps. Once the indictment was served, the accused came out of isolation and encircled themselves with a protective shield of lawyers, and mustered extensive networks of sympathisers and supporters. Colleagues provided assistance in terms of information and extenuating evidence. They gave psychological reassurance by vouching that defendants had conducted themselves with duty and honour to the Fatherland and to the advancement of medicine. The Trial became a trial of German medicine's record under National Socialism. Defendants challenged the Allied claim to superior ethical values. The accusation that the Trial represented 'victors' justice' simmered on, accompanied by a sustained campaign to rehabilitate the record of German medicine.[115]

To the interrogations can be added diaries and notes kept by the prisoners. The IMT defendants had been allowed to write their memoirs as a self-legitimating and therapeutic exercise. In October 1946 the psychologist Gilbert judged a 1,100-page manuscript by the convicted Nazi Governor of Poland, Hans Frank 'of possible historical value', and other memoirs were passed to the defence counsel.[116] The prison authorities encouraged writing as occupational therapy. Life in the cell was conducive to introspection, and the prisoners were allowed an unrestricted supply of pencils and paper, although time weighed heavily. Rostock had a typewriter. While letters were rationed, the prisoners produced incessant memoranda on topics like euthanasia and human experiments. Rostock took the period in prison as an enforced sabbatical. He wrote a compendium of surgery and medical lectures (a radio report claimed this was evidence of his clear conscience).[117] Mrugowsky was historically minded. He wrote a history of the successful German anti-typhus measures during the war.[118] These were manuscripts written from memory and personal experience, but also making use of the document books produced for the trial. Karl Brandt's letters and papers from prison constituted a considerable *oeuvre*. Becker-Freyseng drafted a book on German Aviation Medicine, in which he glorified wartime medical experiments.[119] In all they were a very literary-minded set.

The reflections by the SS administrator Rudolf Brandt provide an example of his self-image at the time of the trial. He regarded himself as 'a decent German kept behind barbed wire'. He stressed duty above all, thrift, self-control and modesty, and a sober lifestyle without alcohol and tobacco – these virtues could regenerate Germany. His long internment was a crime against the rebuilding of Germany and of a united states of Europe, and against the progress of mankind. The Allies condemned wholesale everything to do with Nazism and the SS as involving crime and murder, without recognising what there was that was good and idealistic in their motives. After all the SS was itself an international organisation recruited from many nationalities.[120] Karl Brandt felt his conduct was fully justifiable, but that he faced false accusations and misunderstanding on the part of the Allies. The interrogation was an opportunity to correct their misconceptions.

Once the decision to prosecute was made, and the charges served on 5 November 1946, the interrogations became part of the judicial process, and defendants were allowed to have a defence lawyer present.[121] The defendants presented themselves not as grim sadists or psychopaths, but as academically educated doctors or administrators, who occupied positions of responsibility.[122] Psychologically, their moods swung between optimism and deep pessimism, and they vacillated between acquiescing in the trial and a deep antagonism to the Allies. The prosecution remained concerned by collective denunciations of the trial proceedings.

It was not until the third week of November that the defendants and the available pool of IMT lawyers agreed who should represent whom. The defendants were allowed a free choice, and a few brought in lawyers from outside. But the more seasoned proved effective. Sauter had a reputation as a capable defence lawyer, and he mounted a tactically astute case for Blome.[123] He had argued at the IMT that the defendants acted in accordance with the legal basis of the Nazi regime. For the provincial lawyers – such as Pribilla from Coburg – the Nuremberg tribunals offered legal challenges and human drama. Alfred Seidl persuaded Gebhardt, Oberheuser and Fischer to commission him, so that the Ravensbrück case was dealt with *en bloc*. Others had been assistant counsel at the IMT, but now took on a case for themselves. (See Table 7.)

The nationalism and antipathy to an alien American legal system meant that there were strong bonds of sympathy between counsel and client. Gustav Steinbauer was an Austrian lawyer who had represented the Gauleiter of Vienna and Reichskommissar of the Netherlands, Seyss-Inquart, at the IMT; he took on the case of the Viennese Beiglböck. Alexander conceded: 'The defense lawyers are quite smart too, especially the Austrian who is defending Dr Beiglböck from Vienna ...'[124] Hoven, who had spent some years in America, requested a US lawyer.[125] The judges agreed to accept an American lawyer on the defence side, but not to bring one over just for him. Hoven eventually settled for the belligerent Gawlik, who had conducted the IMT case for the SD security services.

In all, 14 out of 18 defence lawyers had experience of the high-profile IMT, where they defended individuals and organisations. The IMT verdicts on what was a criminal organisation were fundamental. The Tribunal also gave them valuable knowledge of court procedures, and of how to deal with the American authorities and to play the press and public to their advantage. Some German papers criticised the prominence of the American flag in court, and the unfamiliar American judicial procedures.[126]

Servatius had defended the SS. Seeing the professional advantages of acting for a high-profile defendant, he was interested in acting for Karl Brandt, whose defence was sure to receive great public attention, even though it was unlikely to succeed. Rudolf Merkel had represented the Gestapo at the IMT, where he refuted accusations of atrocities in the concentration camps;[127] he then represented the SS doctor, Genzken. Overall, the German lawyers were more experienced than the new wave of US prosecutors.

The judges recognised that lawyers with 'Nazi tendencies' acted in the IMT.[128] Twelve defence lawyers were NSDAP veterans, one had been an NSDAP candidate, another was an SA officer. Shared convictions strengthened bonds of trust with defendants, and they were belligerent and assertive. But the judges found ideological tirades and truculence in court distasteful.

Rudolf Brandt found his counsel aloof and disinterested.[129] His case appeared to be a lost cause. By way of contrast, while Servatius was at first cooly reserved, he was sensitive to Karl Brandt's wish to counter the US case by affirming that he acted in accordance with professional ethics. Pribilla was deferential to Rostock, the Berlin professor of surgery, referring to the case as 'our struggle'. He fawned over his client: 'I am fortunate to be defending such a great scientist as you.'[130] Together they played the judicial procedures to their maximum advantage. Pribilla recognised that meticulous attention to documentation was necessary, and left to Rostock the task of creating a winning impression in court. Rostock designed his plea as a critical and detached view of Nazism. Pribilla was characteristically flattering, that his astute and wily client 'might equally be the advisor to the prosecution'.[131]

Georg Fröschmann became a staunch supporter of Viktor Brack, whom he presented as an idealist. Rather than the punctilious organiser of pitiless mass murder, he believed that Brack was highly ethical and humane in his administering of the T-4 euthanasia organisation. He took to heart the tragedy of his six children about to be orphaned and felt that Brack must be saved at all costs.[132] This vulnerable miscreant merited pardon on humane grounds.

The defendants formed close relations to the legal assistants and secretaries. Rostock was charmed by Fräulein Unger ('liebe Ungerin'), the deputy of Servatius. When Rudolf Brandt thanked Fräulein Müller for bringing letters and stamps, other items may have been passed across. The assistants were adept at smuggling trifles in and out of prison and were valued go-betweens. The Germans were prone to flout rules whose legitimacy appeared questionable. Karl Brandt was observed destroying a concealed message.[133] The documents analyst Hedy Wachenheimer was once mistaken as an assistant who could smuggle out letters so avoiding prison censors.[134]

Each side became convinced that the other was not playing by the rules. The German defence felt strongly that the prosecution took advantage of the powers of military government and the security services to obstruct their case, not least by concealing evidence and witnesses. The prosecutors considered that the defence was ungracious in not appreciating a fair and equitable system, and considered the German lawyers to be grasping, obstructive and belligerent. The Allied authorities kept the defence under surveillance, and the prosecution saw no reason to refuse occasional offers of information. The defendants' tactics were revealed by the US Censorship Division, which passed letters from defence lawyers to the prosecution.[135] They revealed how the defence asked affidavits to be altered so as to omit information which could put a defendant in a bad light.[136] The British intercepted a phone call made by Servatius' office, dictating an affidavit for the Hamburg psychiatrist 'Buergerprinz' about how Karl Brandt saved lives during the war.[137] Servatius complained that he was harassed by British intelligence while gathering evidence in the Cologne area.[138]

The defence had even less time than the US prosecution to prepare their case, and applied on 3 December 1946 for postponement of the Trial. The lawyers complained about delays in translating documents. Fröschmann protested on 7 August 1947 that there was no guarantee that the defence counsels' final pleas would be translated in time for the judges' deliberations on the verdict.[139] The German lawyers nurtured a deep sense of grievance at procedural inequities.

The US government paid the defence lawyers, who had access to cheap meals, fuel and free cigarettes, although the quantities were in dispute. The lawyers considered that they were disadvantaged outside the courtroom: they petitioned the US trial authorities to provide better food and ease travel restrictions.[140] *In extremis* the lawyers threatened to strike. The German lawyers realised how the trial rules could be played to their advantage. They viewed the court with a mixture of hawk-eyed acumen to exploit loopholes, as well as contempt for an alien and less theorised judicial system. The Trial became a field of engagement with American justice, and pitted German national virtues and legal theory against what the defence tried to expose as the hollowness of US democracy.

The lawyers attacked the prosecution as duplicitous in revealing only documents, which were to the detriment of defendants. The defendants rarely had access to records in Allied custody until an incriminating selection was presented to the court. The defence had to construct its case from witnesses, the testimony of defendants and by spotting flaws in the evidence. The defence protested that they were severely handicapped in that access to foreign books and periodicals posed difficulties. Servatius petitioned to visit the British Museum library and the Bibliothèque nationale in Paris to gather material on 'euthanasia experiments on human beings, inmates of public hospitals or soldiers'.[141]

Servatius did much to build up a common front among the lawyers. He organised a protest, accusing the Allies of concealing the whereabouts of key witnesses, and harassing them. He maintained his combative vigour during appeal when some of the lawyers had given their mandates up as a lost cause. In February 1948 he arranged for a common collecting point for information and agreed on a shared strategy of appeal. Servatius published in *Die Zeit* on the legal failings of the Trial, and exploited critical public opinion against the Nuremberg proceedings.[142]

Family and friends

Most prisoners experienced the separation from their families with a sense of acute and protracted emotional pain. Rudolf Brandt had been so absorbed by doggedly taking shorthand for Himmler seven days a week and until 2 am every morning that he had hardly seen his family during the war.[143] Families had hitherto revolved around the father whose prestige had been boosted by National Socialist honours and promotion. Suddenly they simply did not know what was happening to the linchpin of their lives. The separation was exacerbated by the disrupted German postal service. Postal delays were compounded by the prison authorities scrutinising and translating letters. A card Conti was permitted to write on 5 August 1945, was received by his wife Elfriede on 12 December 1945. Only in mid-January 1946 did she realise that her husband had committed suicide on 9 October 1945 when her letters were returned.[144] Allowing family contact might have prevented suicide.

There were four solitary figures among the accused. The reticence of Herta Oberheuser, aged 35, added to the impression of cold cruelty in her dealings with prisoners. Her position as the sole woman prosecuted at the Nuremberg trials set her apart from the solidarity of the core group of the accused. She was described in a prosecution report as 'brunette, normal height, slender, looks good but not striking and gives the impression of a very middle-class woman'.[145] Fischer's tall, solid figure shielded Oberheuser as they sat in the prisoners' dock, so that his amputated arm was not immediately apparent.

Genzken was divorced and childless, and missed the SS cameraderie. Beiglböck separated from his wife in 1945. Pokorny was to make much of the family he had lost: his half-Jewish children had reached Great Britain on a children's transport in June 1939, and he claimed to have attempted to save his former wife, Lilly, from a concentration camp.[146] But he was not in contact with his estranged family during the Trial.

Rose had a dedicated supporter in his beloved 'marmot' Dr jur. Hella von Schulz. Before the Trial they mobilised a group of professors in tropical medicine for his support – the bacteriologists Kisskalt (professor of hygiene at Munich), Nauck (director of the Hamburg Tropical Institute), and Kikuth at Bayer-Leverkusen, and the chief aviation medical officer Schröder (who turned out to be a defendant).[147] Rose's request to the Rockefeller Foundation to vouch for him met an icy response. He expected release in the autumn of 1946 to supervise DDT spraying, rather than his arraignment.[148] Schröder protested to the ICRC against the detention of military medical officers.[149] The internment and prosecution of colleagues created rallying points. The sense of insecurity that they too would be implicated, combined with indignant pride at the tarnished reputation of German science.

Blome's wife, Bettina, collected materials on human experiments, and impressed on Sauter the defence lawyer the distinction between Blome and Conti.[150] Her intense interest in the trial culminated in a novelised account of the proceedings. She was ecstatic about her husband's defence as the highpoint of his career. She dangled psychological carrots: that making a good impression on the judges could win the return of a father for their children.[151] The lovers and mistresses gushed with passionate support.

Most defendants presented themselves as men of high morality with a model family life.[152] Karl Brandt had one son (the cause of Hitler's anger was that he brought his wife and son to safety), while Brack had six children. The defendants could send letters, initially just two a month until the ration was increased in January 1947. Only in July 1947 could wives and children visit – and then this was under strict conditions of observation, and visitors were separated by a glass partition. Mrugowsky had sent his family from besieged Berlin to their home town of Rathenow, which was in the Soviet zone. He never saw his four children again, although his wife twice illegally left the Russian zone to visit. For the younger of the accused, concerned parents gave support, while others benefited from colleagues and associates. Ruff's wife and children were supported by Josef Otto Zeuzen, who manufactured mobile pressure chambers used for the human experiments.[153] The defendants felt that their statements in court were their legacy to their children.[154]

A medical background was used as a defence. Beiglböck, Blome, and Becker-Freyseng had wives who were medically qualified, and Rostock had married an X-ray assistant. Beiglböck's lawyer stressed that his client came from a thoroughly medical family; he neglected to say that it was also thoroughly Nazi. Colleagues, former students and patients rallied to establish the prisoners' credentials as eminent scientists and decent citizens. Here the academics and senior clinicians (notably, Rostock and Gebhardt) had an advantage.[155] A massive number of colleagues mobilised, so that the Medical Trial reached deep into German medicine.

Nauck, the director of the Hamburg Institute of Tropical Medicine, rallied to support Rose to certify his achievements in preventive medicine and that his research was always purely scientific.[156] The irony was that as Nauck had been heavily engaged in genocidal aspects of epidemic control in the East – he had insisted on sealing the Warsaw Jewish Ghetto with a wall to prevent epidemics – such testimony was hardly objective.[157] Rose felt the decision to place him on trial as an ignominious shock. He considered that if the accusations had real substance, this should have been established during his 18 months' detention prior to the trial. As it was, he was confronted with a range of ill-formulated allegations confusing major crimes with routine medical activities. He resolutely counter-attacked American medicine. He saw that what was at stake was the propriety of clinical research. By predicting that medical research would be a major casualty of the trial, he was able to rally international support.[158]

Opening the Trial

Taylor proclaimed the proceedings were 'no mere murder trial'. By this he signalled that to judicially evaluate the German medical experiments required taking account of an additional and distinctive ethical and scientific dimension. Taylor's eloquent analysis of Nazi medicine as part of a regime of mass destruction was highly publicised. His address set the parameters not only of the Medical Trial but also of the whole trial programme. He was the manager of a vast legal machinery to produce justice, as well as its persuasive salesman: but he was not involved with the daily problems encountered in the courtroom. By July 1947 five courts were in simultaneous session and one trial had already been completed.

Taylor's remark that the trial was 'no mere murder trial' has been taken as indicative that the Trial would evaluate the grim series of Nazi medical experiments according to both legal and ethical criteria. Moreover, the judicial pronouncement of the Nuremberg Code at the close of the Trial drew out principles, raised in the courtroom by expert witnesses. There are three problems with this interpretation. Firstly, Taylor was absent from the prosecution stand. He received memoranda from Alexander and his prosecution team, at times he observed the proceedings, but he was preoccupied with the rest of the trials programme.[159] He heard the final pleas of the accused on 19 July, and the judgment and sentences a month later.[160] He liaised with external authorities, most importantly with the US Department of War, which financed the Trial, and with the British, French and Soviets.

Taylor requested evidence from the Soviets on points when the findings of the precursor IMT could not be substantiated. One unresolved issue was progress on the experimental germ warfare facilities at Nesselstedt near Posen. Taylor wrote: 'Our evidence is pretty weak on this point and I think the only approach to the Soviet Union is to ask them for further evidence on the Posen Institute and further elaboration upon the charge without indicating that we feel the said finding of fact is unsound.'[161] Taylor contacted the US ambassador to Moscow on 9 April 1947, requesting that the Soviet Political Advisor be contacted to see what help might be given, as 'Our staff has searched extensively through the record of trial of the IMT but has been unable to locate the evidence upon which the Tribunal found as a fact that Soviet prisoners of war were used in these experiments.'[162]

Taylor's stirring address marked a resumé of the pre-trial phase, when ethical issues were combined with establishing criminality during interrogations. He eloquently depicted the experiments as part of a common design while devaluing them as 'senseless and clumsy and of no real value to medicine as a healing art'. He made passing reference to Lemkin's novel concept of genocide and cited Alexander's view that the experiments as 'thanatology' meant they supplied the techniques of genocide. The Nazi experiments were 'devoted to methods of destroying or preventing life', especially to euthanasia and extermination methods, involving new lethal injections, new gases and poison bullets. These were views formulated prior to the Trial rather than worked out during it. Ivy's involvement in August 1946 indicates that ethical discussions shaped the decision to mount a trial primarily focused on human experiments.[163] But then the prosecution veered back to the clear-cut legal issues of conspiracy to murder, war crimes and crimes against humanity.

The protracted period of selection and definition of the case meant that it was only on 16 October 1946 that the press reported, 'Nazi Doctors on Trial Next'. The prosecution became aware that the German defence was going to argue that the human experiments were comparable to those in US penitentiaries. On 1 November 1946 Taylor cabled the War Department:

> re medical atrocities trial to begin last week in November. It is necessary that we have extensive paper on the history of medical experimentation on living human beings with particular emphasis on practice in U.S. Defendant Rose states that U.S. doctors have extensively experimented on inmates of penal institutions and asylums, especially with malaria. Any truth in this? If so, give us full facts. Do not limit to answering this question. Suggest you contact Dr A.C. Ivy, University of Illinois Medical School on this problem. Answer soonest.[164]

How the prosecution established Rose's position is an enigma. He arrived in Nuremberg on 5 October, and was interrogated only once by November, and then just to establish basic facts about his career. Rose was convinced that his letters to his supporter Hella von Schulz were intercepted.[165] Although letters were routinely read by censors, it was only on 27 November that the interrogator Rapp pointed out that the censors might not be fully informed and that copies should be made and passed to interrogators.[166] In Rose's case some copies of his prison letters survive.[167] Rose's strenuous attempts to gain release may have provided the prosecution with insight into the defence position. His defence kept issues associated with science and ethics, and the legal position during the war, to the forefront of the Trial.

On 16 November 1946 Marcus, the American War Crimes Chief, formulated a strategy linking ethics, the anti-state ideology of a free medical profession, and the Trial.[168] Taylor realised the need to refute the defendants' ethical excuses, but he was reluctant to concede points of law to ethical imponderables. He considered that a prosecution case based on ethical violations was vulnerable. Mant reported with characteristic precision that, 'there were no finer points of medical ethics under consideration, and the defendants are being tried under existing laws concerning murder'.[169] For Taylor, it remained above all a murder and war crimes trial.

When Kempner briefed Alexander in Washington on 13 November 1946, he predicted: 'Two main defence points ... expected: 1) experiments also in

US and other countries. Differences. 2) Under German medical, esp. military law, they were allowed to do so. It should be made clear that according to German law it was a crime.' Alexander arrived in Nuremberg on 18 November and rapidly reached the conclusion that the purpose of experiments was in fact to induce death and study the conditions under which it occurred. He recorded in his diary-logbook for 21 November 1946: 'Indictment read in Court. Afterwards: conversation with General Taylor, re Thanatology. "You have given the def[ence] too much credit, and Himmler too little".' He perceptively pointed out that relying on supposed links to a conspiracy instigated by Himmler was to be inadequate in explaining the rationales of the human experiments. Taylor considered that the political aspects of Nazi atrocities would override ethical complexities.[170] Alexander argued that the only scientific advance of the experiments was in killing techniques. Projects with a life-enhancing aim would veer the other way. He cited the example of Mrugowsky's research on impurities in vaccines as on the lethal effects of phenol injections: the experiments provided 'the techniques for genocide'.[171] Taylor interpreted Hitler as 'the focus of ultimate authority', but pointed out that Hitler derived his power from other influential men and groups – including doctors and the medical profession – who broadly shared Nazi ideals. Science thus boosted the German war effort and racism.

Taylor ensured that the Trials programme dealt with the key components of the Nazi war machine. He took a top-down model: the crucial issue was the interaction of industrialists, generals – and doctors – with Hitler and Himmler. Much effort was expended in reconstructing the hierarchies of command. Some defendants at the Medical Trial were there more by virtue of their office than actual evidence. Others were there because of stray items of correspondence with Himmler, rather than actual documentation on criminal medical activities. The accused were located within a corrupt system of public health, and their position led them to commit racial and ethical atrocities.

Despite Taylor's opening address fusing ethics with the criminal charges, the court proceedings indicated that this was primarily a murder trial. He insisted on the sufficiency of the criminal law taken over from the IMT. Thus the experiments violated charges of crimes against humanity, criminal conspiracy as well as murder and assault. His analysis was essentially political: that 'None of the victims of the atrocities perpetrated by the defendants were volunteers', and consent was not possible under Nazi dictatorship: 'In the tyranny that was Nazi Germany, no one could give such a consent to the medical agents of the state.' Taylor referred to 'crimes in the guise of scientific research', stressing the destructive objective of the experiments.[172] The experiments and other atrocities were part of a more general pattern of Nazi violence, mass killing generated by Nazi lawlessness and irrationality. Between Taylor's opening address and the Nuremberg Code at the conclusion of the Trial came a series of hard fought legal battles over the human experiments.

10
Science in Behemoth: The Human Experiments

The defendants

An international barrage of publicity, press releases and news reports reported the 'case against the 23 SS Doctors', and the malevolent medicine of 'Herr Dr Sadist'.[1] While the public gained the impression that the typical Nazi physician was a mass murderer, professional circles insisted that here were just a handful of medical miscreants. Between these extremes, the reality was no less chilling, but more complex. Three of the defendants were administrators, only seven of the 20 accused doctors were SS officers, and four were not Nazi Party members. They differed not only politically, but also in terms of their medical seniority and social background. The prosecution viewed the defendants in collective terms, and grouped them in a succession of charges. Behind this lurked the gargantuan state machinery of coercion and destruction. (See Table 1.)

The political and legal theorist Franz Neumann analysed the Nazi state as a lawless, annihilating 'Behemoth' – the Hebrew for a monstrous beast. Neumann's central concept was that of the 'Racial People', defined in part by eugenics and medicine. Alfred Cohn of the Rockefeller Institute for Medical Research supported the writing of *Behemoth* in 1942, at a time when he was embarking on sulpha drug trials on prison volunteers.[2] Neumann assisted Jackson and Taylor at the IMT as First Chief of Research.[3] Taylor found Neumann's exposition of the Nazi destruction of law illuminating, and observed that the same corrosive effects were evident on German medicine.[4] Medicine could be located within a general political theory of Nazism as destructive of the professions – their conduct, ethics and knowledge. Nazi medicine was to strengthen the racial basis of society. One symptom was the monopoly capitalism of IG Farben, which commissioned pharmaceutical research in Auschwitz and Buchenwald. The Behemoth analysis denigrated Nazi research as pseudo-science, and overlooked how the Nazis had maintained the institutions of normal science funding through the RFR and the traditional structures of academic and medical organisation. Behemoth became the blueprint for the Medical Trial.

The prosecutors depicted the medical atrocities less as scientific abuses and more as the products of a depraved political system. Each set of charges related to a fundamentally political analysis of the Nazi system of power, and of how medicine served Nazi military, racial and economic ends. Taylor and his team preferred to locate the miscreant doctors in the depraved politics of the Nazi Behemoth than to see them as guilty of violating any established principles of medical ethics, which had uncertain legal status. Taylor saw the Third Reich as in the grips of an 'unholy Trinity of Nazism, Militarism and economic imperialism'.[5] The Medical Trial was very much an exercise in contemporary history. The prosecution had to work historically, sifting through mountains of medical evidence and unravelling the structures of German medical research and its military applications. The Trial provided a detailed dissection of Nazi medicine.

Widely circulated photographs of the defendants captured them seated grimly in two rows in the same dock as the Nazi political leaders prosecuted at the IMT. The sole differences were that US soldiers replaced the 'snowdrop' helmeted military police, and that with 23 defendants rather than the 21 present at the IMT the back bench was now more crowded. The defence lawyer Merkel alleged that the seating of the accused represented their degree of culpability.[6]

The positioning of Karl Brandt in the seat once occupied by recalcitrant Air Marshal Göring underlined Brandt's putative key role, as Hitler had conferred on him co-responsibility for implementing euthanasia and overarching powers over civilian and military medicine. The prosecution accorded him prime responsibility by denoting the case as 'The United States vs. Karl Brandt et al.'. The judges were confronted by medical defendants in the front row wearing tunics bereft of insignia but indicating their military or SS status, whereas none did in the back row. In the furthest corner of the crowded second row were the junior and marginal figures.[7]

Closer inspection reveals striking dissimilarities of age, and professional and academic status. The eldest defendants at 62 were Handloser and Genzken, whereas the youngest, Fischer, was 35 when the Trial began. The seniority of Handloser and Schröder in the army and air force medical services, or of Genzken and Mrugowsky in the Waffen-SS medical services, was in contrast to the low-grade positions of Fischer, Oberheuser, Hoven and Schäfer. The prosecution identified the defendants as links in a chain stretching from Hitler and Himmler to the base executors of medical war crimes.

The careers of Rascher (born 1909), the two Verschuer assistants Mengele (born 1911) and Liebau (born 1911), and Gebhardt's assistant Fischer (born 1912) showed that academic ambition could be devastating. Rascher gained his MD in 1936 and Liebau and Mengele in 1938; Rascher and Mengele joined the NSDAP in 1937, and Liebau in 1938. While Liebau was an assistant of Verschuer he served with the Leibstandarte Adolf Hitler in the Ukraine, and then with the murderous Globocnik in northern Italy. All moved between the academic, military service with the Waffen-SS, and concentration camps. The administrators Rudolf Brandt and Sievers cemented the links to Himmler and Brack from the Führer's Chancellery to Hitler. (See Table 4.)

The opening of the Trial brought the full force of the accusations home to the defendants. Men (and one woman) who for the most part had only formal contact with each other, or engaged in rivalries, or had no contact at all were now accused of standing in a single conspiracy. The pressures resulted in a core group of defendants, and a few dissidents who placed the prime guilt on the core. Rudolf Brandt condemned Karl Brandt as aloof and 'hugely ambitious'.[8] Mrugowsky and Genzken (and the chief naval medical officer Fikentscher) identified Gebhardt as Himmler's chief medical adviser. Blome struck out independently, venting the antagonism of an SA veteran against the SS core. By way of contrast, a deep bond of loyalty remained between Rostock and his former assistant Karl Brandt, deriving from their time at the Bergmannsheil hospital in Bochum. Fischer, accused of a role in amputation experiments and ironically himself bereft of a limb, and the slight figure of Oberheuser, now sick and arousing adverse comment as the only woman, sat together beleaguered and bonded by common experiences in Ravensbrück.[9] Karl Brandt was on good terms with Handloser.[10] Gebhardt helped out by insisting that Karl Brandt had in practice only a middle-ranking and weak position. He protested that the Nazi rulers treated physicians as a relatively inferior group in the Third Reich.[11]

The Trial's landmark status for medicine under National Socialism makes it worth asking how representative a cross-section of the German medical profession were the defendants? Kater has shown that by 1937 over 40 per cent of German physicians were NSDAP members, and 7 per cent joined the SS. Only four defendants were not NSDAP members (Handloser, Schröder, Pokorny and Schäfer). Seven of the doctors were SS officers. Those on trial were more accentuated in their Nazi affiliations than even the 'average' German doctor, who was drawn more strongly to Nazism than other German professionals.[12] A strongly national culture shaped the mentality of the non-Nazi defendants. For example, Alexander found that Handloser was christened Siegfried as his musician father was a Wagner enthusiast.

Women were underrepresented in the Nuremberg dock, given that women made up 17 per cent of doctors registered with the Reich Physicians Chamber. Women were marginalised in German academic circles, and military medical research was a decidedly male sphere. Herta Oberheuser felt herself 'a woman in a difficult position'.[13] As the one woman on trial at Nuremberg, she felt unjustifiably vilified.[14] The Trial was a bitter reward, as she considered that she had been conscientious, dutiful, overcame financial hardships and had placed her career over any emotional attachments. She aspired to be at least a surgeon, if not a physiologist, and her Bonn MD meticulously analysed the effects of narcotics, that later would so often be denied to her victims.[15] Her move to Ravensbrück, and her elevation to Gebhardt's mighty orthopaedic clinic of Hohenlychen with its 1,000 beds were steps up the professional ladder. Her energetic participation in the experiments won her a place at this prestigious clinic. Fischer observed how 'Frau Dr. Oberheuser mainly came into the foreground because she was always ready to work and always to be found at the station, and that is how it came about that she took a greater part in the experiments.'[16] As she was acting under orders (the IMT judged this to be only a factor in mitigation but not a valid defence), she could not see why she should be culpable. She was a dermatologist like Blome and Pokorny – and this was one of the most marginal specialisms in the pecking order of German medicine. It offered scientific interest and a relatively secure existence, as sexually transmitted diseases and skin conditions were widespread in a pre-antibiotics age.

The religious belief of the defendants was mainly Protestant – 17 Protestants to six Roman Catholics. Looking closer, we find that some defendants had a mixed faith background. Oberheuser's father was Catholic although she was Protestant, and she felt relieved that the Nazi creed emancipated her from burdensome theological issues and divisions between Catholics and Protestants. Thirteen of the 23 accused left the Church, conforming to the Nazi surrogate of being a 'god believer'. But when Genzken wrote a treatise on his theosophical creed of harmonising polarities, Himmler prevented its publication. Mrugowsky felt drawn to the Reformation mystic Jakob Boehme and the nature philosophy of Alexander von Humboldt. His dour demeanour concealed a lively interest in holistic philosophy. Others continued as orthodox in their Christianity, as Handloser and Weltz, who annoyed the profane Rascher. The question of religion – pursued by Alexander in his psychological interrogations – was relevant to understanding the breakdown of inhibitions in becoming involved in gruesome experimentation. (See Table 2.)

All the physicians on trial had a middle-class background, their fathers being a mix of officials, professionals and landowners, while their mothers had the conventional role as housewives. Himmler's stenographer-secretary, Rudolf Brandt, studied law and economics, though his father was an artisan. The euthanasia administrator Brack made much of having a medical father, in order to stress his understanding of the profession's ethical norms. When asked by Alexander concerning investments and career aims, it emerged that defendants set less store in accumulating wealth or in enjoying a high degree of success, than in an ethic of service and duty. Hoven was exceptional in combining an unconventional lifestyle with a taste for escapist travel. He took this trait to extremes in the depravities of Buchenwald camp life, where he relished opportunities for sexual indulgence and exercising powers over life and death. By way of contrast, Oberheuser expressed distaste for the profligate lifestyle of her Ravensbrück colleague, Rosenthal. If things could have been otherwise without the loss of her family's wealth in the inflation, she longed for the tranquillity of a university research institute.

Just three of the accused doctors had parents who were physicians: Gebhardt, Mrugowsky and Beiglböck. Mrugowsky's father had been killed in military action in 1914, and Gebhardt's father (a Bavarian public health official and a friend of the Himmlers), and Beiglböck's father (an Austrian country doctor) were nationalist activists.[17] These cases show how a nationalist outlook was radicalised by the next generation. Most of those on trial did not come from traditional medical elite families, and lacked the socialisation of growing up in a medical family, from which they could have derived an ethical understanding of the patient.

Sievers was the sole accused who did not hold a university degree. He had worked in publishing, which was a respected para-academic career, and through this had come to administer the SS research organisation. The accused doctors attended every German university, with the exception of Königsberg in East Prussia. Some spent semesters in Innsbruck. Pokorny had studied from 1917 to 1922 at the German University in Prague, and had been assistant at the clinic for skin and venereal diseases in 1922–4.[18] The accused were products of the German university-based system of medical education with its values geared to the experimental basis of medicine. While some had scruples regarding the ethics of human experiments, none could remember ever having sworn anything like the Hippocratic Oath.

Some of the accused had worked overseas – Genzken as a ship's doctor, or the footloose Hoven in the United States. Karl Brandt had contemplated working with the inspirational Albert Schweitzer in French colonial Africa, until he found out that as an Alsatian he would be due for French military service.[19] Rose had the most extensive overseas experience having worked in China; he had also visited the USA. He was convinced of Germany's need to have its colonies restored.[20] International experience reinforced nationalist convictions.

The outcome of the First World War instilled the conviction that the sacrifice of life for the Fatherland was justified. The older defendants had seen military service, and the younger defendants were profoundly affected by the war. Karl Brandt's schooling was disrupted by expulsion from Alsace.[21] As a medical student Gerhard Rose wrote a tract *Krieg nach dem Kriege* calling for a rekindling of national cultural values, while repatriating prisoners of war. Rose turned words into actions by joining the Freikorps to fight for a German Upper Silesia, before joining the NSDAP.[22] He served as medical officer for the Condor Legion in Spain, before joining the Luftwaffe.[23]

The accused had a standard type of medical education and postgraduate training with junior clinical appointments at large clinics. Handloser and Schröder benefited from military medical training, which included exposure to experimental research in bacteriology. Of the 20 accused physicians, there were four surgeons (Karl Brandt, Fischer, Gebhardt and Rostock), three dermatologists (Blome, Pokorny and Oberheuser), four bacteriologists (Handloser, Mrugowsky, Rose and Schröder), a specialist in internal medicine (Beiglböck), a radiologist (Weltz), and two practitioners in general medicine (Genzken and the dubious Hoven). There were four pure specialists in medical research (chiefly the applied field of aviation medicine), Becker-Freyseng, Romberg, Ruff and Schäfer, and the medical geneticist and race expert, Poppendick.

Nazism radicalised the accused. As the physician was both *Volksgenosse* and scientific expert, the surgeons Karl Brandt, Gebhardt and Rostock gained influential administrative assignments. Their politics before 1933 varied between conservative nationalism (Oberheuser declared herself *Deutschnational*) and the ultra-right. Rose was an early NSDAP enthusiast. Brack was an SA recruit in 1923 before switching to the SS in 1929 with the low member's number of 180. Blome resisted the SS, because of prior loyalty to the SA. Ten of the 16 NSDAP doctors entered the NSDAP in 1933/4 (Becker-Freyseng, Beiglböck – albeit illegally as an Austrian – Fischer, Gebhardt and Romberg), or, more opportunistically, after the membership stop in 1937 (Hoven, Oberheuser, Rostock, Ruff, Weltz). Pokorny, although not an NSDAP member, had moved from Prague to the Sudeten German city of Komotau in 1939; as an ethnic German, he abandoned his Czechoslovak identity.

The prosecution saw Blome as corrupting humane professional values, because of his role in nazifying postgraduate medical education. By implanting notions of race and heredity, he created the demonic Ding/Schuler and Rascher.[24] Ivy suggested that the officers of the German Medical Association were 'accessories to the fact'.[25] But the prosecution's main thrust was to demonstrate the links between the SS and the corruption of research. Even though Sievers did not hold an academic post, he had an influential role in shaping the medical funding (along with Blome, Rostock and Karl Brandt) in the RFR, appointed to advise the German Research Society.[26] Ten had status at the University of Berlin, two were affiliated to the Munich medical faculty and one to Vienna University.[27] Although only the bacteriologist Rose held grants from the RFR, research funds came from the military and the Ahnenerbe. The compulsion to undertake unethical research had been absorbed in the military, the SS, and industrial conglomerates like IG Farben.

The academic standing of the accused was downplayed at the Trial. The Medical Faculty of Berlin had the lion's share of the accused. The Trial posed special problems for this flagship faculty. Rostock as dean of the

Faculty and NSDAP member was purged from the University on 13 July 1945. In October 1946 the pharmacologist Heubner was appointed dean when the faculty reopened. He encountered strengthening communism and allegations that he had insisted on seawater experiments in Dachau. Heubner became outraged that his conduct was impugned, and distinguished between genuine scientists and political opportunists.[28] He had protested against Blome's tenuous academic credentials in 1941.[29] In September 1947 the Berlin Faculty reflected that those condemned at Nuremberg were precisely those forced on the Faculty against its will.[30] Rostock impressed on the court his elite status and his sense of a strong stand on the divide between the ethical and unethical.[31] (See Table 3.)

The Vienna Medical Faculty had already dismissed Beiglböck as a long-standing 'illegal' Nazi Party member. No German university withdrew a properly awarded MD from any of those convicted at Nuremberg, although medical faculties had relentlessly annulled degrees of Jewish refugees under National Socialism. In effect, the strategy was to insulate German academic medicine by concurring with the prosecution that the convicted were pseudo-scientists and political fanatics.

Berlin University was implicated in the Ravensbrück trial with three defendants – Treite, Stumpfegger and Josef Koestler – from the SS-Lazarett Hohenlychen, where they took part in criminal experiments. Gebhardt supported their Habilitation in the Medical Faculty.[32] Gebhardt insisted that Himmler ordered Stumpfegger to undertake experimental bone transplantations for the benefit of wounded soldiers.[33] Despite his ambition to be a professor, Fischer avoided taking the Habilitation degree at the University of Berlin, as he considered that this would have involved further experiments. Fischer testified, 'five doctors had been habilitated before me in Berlin. I worked on some of these habilitations. This was the greatest opportunity we had.'[34] The carrying out of experiments to gain lecturers' posts motivated Mengele and Rascher to carry out gruesome research. Gebhardt used the Habilitation as a means of intimidating Rascher.[35]

In order to establish scientific credentials of the defendants, Alexander contacted the former dean of the Munich Medical Faculty, Alfred Wiskott (a paediatrician), who had been in post in 1943–45.[36] He explained the widespread consultation of referees for academic appointments; only Gebhardt, Rostock and Weltz were 'genuine members' of Faculties, and Handloser, Blome and Karl Brandt were just titular professors. He overlooked Rose with his research-based position at the RKI. Gebhardt revealingly claimed that Karl Brandt had a professorial title but was never a real professor, indicating tensions in Himmler's medical entourage.[37] The difficulty with Wiskott's view was that medical faculties had a tradition of appointing leading figures in military medicine and state health administration. Wiskott's faculty had as an honorary member Walter Schultze, a leading Bavarian Nazi medical official and crony of the Reich Physicians'

Führer Wagner. The university rector, the indologist and SS officer Walther Wüst, Ahnenerbe president and Himmler protégé, was Schultze's bitter enemy.[38] It meant that nazification of university faculties was held up by factionalism among the Nazis as well as the traditionalism of faculties with a long record of opposing state intervention.

The Nuremberg defendants were derided as pseudo-scientists, who held their position because of Nazi patronage. Rudolf Nissen, who had been assistant of the renowned surgeon Sauerbruch until forced to emigrate to Istanbul in 1933, condemned Gebhardt and Karl Brandt as incompetent surgeons – both were former students of Sauerbruch.[39] Gebhardt intervened on Sauerbruch's behalf with Hitler when Sauerbruch's son was arrested in connection with the bomb plot of 20 July 1944. Sauerbruch considered the trial unfairly singled out certain doctors, when others no less guilty were still in practice.[40] Rostock judged that Sauerbruch would do nothing to help him. Old rivalries meant most former Berlin colleagues could not be counted on to provide support, but he expected the medical historian Diepgen and the anatomist Hermann Stieve to remain loyal.[41] Rostock as former dean kept in touch with the Faculty secretary, the redoubtable Fräulein Dieterici, who was to reassure colleagues of his innocence. She bridged the Nazi and post-war periods, until her dismissal in June 1950 by the communist central administration.[42]

Experimental success offered the passport to a chair. In mid-1946 Ruff was nominated for a university chair in Tübingen.[43] Mrugowsky's ambition was to direct a university hygiene institute. Kater observes the dual process of the SS wanting to recruit academics into its ranks, and how Himmler backed candidates for university careers.[44] Human experiments advanced SS policy of gaining control of universities to transform them into ideological hotbeds to advance racial struggle. Rascher desperately wanted the accolade of a university Habilitation. This aspiration lay behind his murderous heart and altitude experiments. He hoped to gain the Habilitation at Munich or Kiel; Sievers of the Ahnenerbe thought the same at Rüdin's institute,[45] and then in the Munich Medical Faculty – but in July 1942 reported that the internist Alfred Schittenhelm and the University rector, Wüst – even though both SS officers – declined.[46] Rascher calculated that rather than reveal his murderous conduct to a Habilitation 'jury', he would do better by submitting another topic. This indefatigable practitioner of human butchery wrote up the Polygal blood anti-coagulant research in a 'fat ham' of a dissertation. He won praise from Himmler, who ordered a secret Habilitation examination – providing faculties with a welcome excuse to decline Rascher, as the Habilitation was traditionally a public examination.

Rascher's saga continued, as Blome (who had no Habilitation) tried to find a compliant university. Rascher proposed poison gas research in Dachau in August 1942, and Blome visited his experimental lair in August 1942 with the intention of grooming Rascher to undertake research on nerve gas.[47] The Marburg professor of hygiene, Wilhelm Pfannenstiel, consultant in hygiene for the Waffen-SS and a teacher of aviation medicine, was an ideal link person for the gruesome Rascher, not least because he observed trial gassings at Belzec. Pfannenstiel invited Rascher to Marburg in August 1942, because he hoped to continue his own experiments on nutrition and low pressure in Dachau. Sievers offered to finance a research position for Rascher through the Ahnenerbe. When the Marburg faculty rejected a secret appointment, Pfannenstiel had to decline the dubious honour of being 'Habil-Vater'. After sounding out Frankfurt University, where there was a willing examiner (von Diringshofen, the Director of the Aviation Medicine Institute), again the faculty insisted on the conventional public examination. Finally, Sievers arranged with the Ahnenerbe collaborator, Hirt, for Rascher's work to be submitted to Strassburg medical faculty on a secret basis in 1944.[48] Overall, Rascher's difficulties show that Himmler's powers reached an impasse when it came to academic procedures, thwarting academic preferment for his protégés. Alexander and Ruff considered that aspects of Rascher's research, although murderous, showed a demonic flair, but the Habilitation plans ground to a halt as Himmler withdrew favour from his unscrupulous protégé.[49] Himmler may well have sensed that Rascher was retreating into the ranks of the academics, and his imprisonment can be seen as a means of keeping him under the power of the SS.

The prosecution strategy

The courtroom proceedings were left to the prosecution lawyer McHaney and his assistants Hardy and Hochwald. McHaney commented in the final prosecution argument: 'the defendants are primarily on trial for the crime of murder...one should not lose sight of the true simplicity of this case'.[50] The prosecution dealt with the case as a series of experiments, rather than (as the IMT) a series of accusations against individuals. They relentlessly accumulated evidence that prisoners were forced to submit to experiments without consent.[51] The prosecutors released their case batch by batch – moving from one topic to another whenever evidence was ready or when a witness happened to appear.

The Trial began with the Dachau pressure and freezing experiments, as Alexander's expertise was to hand. Much of the case consisted of the presentation of documents, largely selected by analysts located in the dank cellars of the Berlin Document Center. Evidence from the British, the Austrian police and the Czechoslovak, French and Polish war crimes authorities arrived sporadically.[52] Court discussions revolved round the technical issue of admissibility of the evidence. At the end there was a protracted presentation of materials relating to euthanasia, taken over wholesale from the Hadamar Trial, which had concluded on 15 October 1945. Mant laconically observed that the Americans had simply run out of evidence regarding the human experiments.[53]

Behind the scenes Alexander worked tirelessly and to considerable effect: he evaluated evidence, suggested appropriate witnesses, and prepared questions for the lawyers to ask when cross-examining defendants and witnesses. His efforts did much to strengthen the prosecution's case. He recognised 'the emergency of the situation' and had to keep ahead of the pace of court proceedings.[54]

The prosecution's view was that what was on trial were not scientific experiments but their negation – a debased, nazified form of human torture. Himmler's interest in human experiments as supporting the war effort, while augmenting the SS's macabre stranglehold on power, was constantly hammered home. The freezing, high-altitude and seawater experiments exemplified these links. Ivy told the court how Himmler got hold of a *Life Magazine* story about research on desalination, and then hounded scientists with the demand for an effective device.[55] The prosecutors stressed the co-ordinating role of Gebhardt and Karl Brandt through the *Reichsforschungsrat* (involving Blome) and the Ahnenerbe (involving Sievers). The prosecution constructed vertical hierarchies from hands-on perpetrators like Hoven in Buchenwald, or Fischer in Ravensbrück up to senior medical figures like Karl Brandt, Rostock and Gebhardt. SS administrators – Brack, Rudolf Brandt and Sievers – were links to Himmler, Göring, Speer and Hitler.

The prosecutors located the coercive experiments within a genocidal context of the 'totalitarian' Nazi system of power. The experiments were construed as part of a broader pattern of aggressive militarism, coercion and mass murder. The prosecution constructed vertical chains of command for each cluster of experiments, so that senior figures in the army, Luftwaffe and SS medical services were linked to the SS. Himmler's medical advisers were said to be Gebhardt and the cardiologist Karl Fahrenkamp.[56] Himmler was a dark presence as the investigators derived the experiments from the Reichsführer's obsessions with survival at sea or in the face of the extreme cold of the Eastern Front, or mass sterilisation.

Totalitarianism as a political dynamic explained the lethal experimental research in the concentration camps. Kogon proved to be a formidable witness, as an authority on the Buchenwald vaccine experiments. He was a political scientist and journalist, whose Social Catholicism led to a critical attitude to the social injustice of the bureaucratic state, and to the desire to humanise science. Whereas Neumann provided a broad framework for understanding the complex structure of the Third Reich, Kogon had an insider's unique grasp of facts and events. He provided details of Ding's links to IG Farben and to its executives and researchers, and accused the company of making gigantic profits from vaccines and therapies.[57] Kogon linked the sinister SS state of the concentration camps to a criminal state.

The IMT charges had hinged on 'the acquiring of totalitarian control of Germany'.[58] Much effort was expended on reconstructing the command structure and in locating the place of medical organisations within the sprawling Nazi military and economic realm.

The prosecution focused on Karl Brandt's powers from 1942 onwards, as he came to represent the apogee of medical science in the service of a destructive political power. As a war crimes trial, the Medical Trial considered the *Gleichschaltung* and nazification of German medicine between 1933 and 1939 only as background. It paid no attention to the Nazis' centralisation of the sickness insurance system. The persecution and dismissals of Jewish physicians, dental surgeons and nurses were raised by defendants seeking to deny their involvement. Key figures in this first phase of the nazification of medicine, the bombastic Reich Physicians Führer Wagner and his arch-enemy, the SS public health official Arthur Gütt, were dead, and their successor Conti committed suicide. The prosecution scrutinised the centralisation of public health administration after 1939; the administrative structures and nature of medical research; human experiments; and sterilisation and euthanasia.

The prosecution had to make sense of the epic clash between Reich Health Führer Conti and his arch-enemy Karl Brandt as Plenipotentiary for Civilian and Military Medicine during 1942–4. Brandt was condemned as a young opportunist by the 'old' NSDAP activist Conti, who mobilised links to Bormann and Goebbels against Brandt and his ally Speer. The Führer favoured Brandt as a daring disciple, who was independent of the academic physicians, NSDAP activists, and SS. Hitler liked the genial Brandt for having the practical skills of a surgeon, who could operate on the decayed body of German academic medicine, and whose youthful energy challenged academic, professional and military medical hierarchies. Brandt was utterly mesmerised by the Fuhrer, who thrust on him massive responsibilities for euthanasia and raising the effectiveness of civilian and military medical services.

Blome as a veteran stormtrooper and Deputy Reich Medical Führer argued that he too was in opposition to Conti, although from a very different position to Brandt. This opened the way to analysing relations of public health to SS, with SS officers doubling as public health officials. Similarly, the SS's tentacles established a grip on the universities through the activities of Sievers, and through the Waffen-SS established a military medical role; above all, the SS shaped racial policy while unleashing the Holocaust.[59] Interrogators raised fundamental questions: Sievers was asked whether Karl Brandt influenced Hitler, or was it Hitler giving orders to Brandt?[60]

The prosecutors presented an impressive analysis of measures to combat infectious diseases, notably louse-borne typhus, as well as on tuberculosis and a range of other infections like hepatitis and malaria in areas under German occupation. They established the role of IG Farben, the SS and military medical research agencies in an effort to ascribe responsibility for human experiments. The appraisal of human experiments led to evaluation of what was going on in German physiology, anatomy and pharmacology – the rationales of German scientific medicine were thus exposed for analysis. The question was the extent that research had been racialised in terms of attitudes to clinical subjects, the research questions and the sponsors of research. Here plans for a second medical trial need to be recognised, even though this trial did not take place and Mengele slipped through the American net.

The concern with racial crimes, notably euthanasia, intensified interest in the culpability of Karl Brandt. He was the most senior figure in the Nazi medical administration in Allied custody; his convoluted career had ended in arrest and a death sentence on Hitler's orders on 17 April 1945. He was first interrogated for his knowledge of chemical and germ warfare experiments. He was an obvious candidate for trial, given that Hitler had conferred on him overarching medical powers, and because Hitler's euthanasia decree of 1939 identified him as one of the originators of euthanasia of the mentally ill and of children with congenital malformations. The prosecution suspected that Karl Brandt was a central figure in authorising human experiments.[61] While the prosecution failed to produce evidence for centralised research structures – indeed, the defence demonstrated that wartime German research remained highly fragmented – evidence mounted on how Brandt was involved in chemical weapons research at the Reich University Strassburg. Brandt pleaded that his responsibilities to the *Gemeinschaft* – by which he meant the disintegrating Germany at the close of the war – exercised a higher call than the physician's duty to the individual patient.[62]

The spotlight was on Karl Brandt as a member of the Führer's entourage since the mid-1930s. Until 1943/4 he was Hitler's surgeon in attendance (*Begleitarzt*) when the Führer travelled, to provide immediate assistance in the event of an accident. He gave the Führer occasional medical advice, and so was sometimes referred to as Hitler's personal physician, or 'Hitler's Doctor', even though in formal terms he never held such an office. The medical historian Fritz Redlich has observed: 'Dr Brandt the senior attending surgeon, never treated Hitler.'[63] Brandt eventually was relieved of this role.[64]

Brandt's athletic wife (a national swimming champion) was part of the Führer's social circle, and the alert and capable Brandt fitted in well in Hitler's entourage. He was in awe of Hitler, and was compliant and conscientious in carrying out medical assignments. He did not figure in Nazi medical politics until September 1939, when the Führer put him in charge of euthanasia. The logical choice might have been the rising star Conti, who symbolically operated one of the gas stopcocks in an experimental euthanasia gassing at the Brandenburg Psychiatric Hospital in 1940. Hitler saw Karl Brandt as able to check Conti's ambitions. Euthanasia became a central issue at the Medical Trial, which revealed significant links between the T-4 killing programme and the *Aktion Reinhardt* extermination camps of the Holocaust.

Karl Brandt unequivocally supported the killing of the severely physically and mentally disabled. He was entrusted to find an appropriate killing method, and concluded – probably on expert advice from chemical experts like the Würzburg professor and pioneer of Zyklon, Flury – that poison gas was more effective than the more medical solution of a lethal injection. He assumed the overarching symbolic role as the doctor entrusted to oversee the killings. At the Medical Trial he drew a distinction between rightful termination of life for the most severe cases of disability, and later abuses with the killing of milder cases and patients who were simply disliked by hospital staff as unruly or insolent.

Hitler ordered Karl Brandt to sort out the chaos of deteriorating medical resources and services after the bombing began. Again, Brandt was pitted against Conti. Hitler appointed Brandt as Plenipotentiary for Health and Medical Services on 28 July 1942, and he held the brief of co-ordinating military and civilian medicine. His powers were boosted in 1943 as General Commissioner for Health and Sanitation answerable directly to the Führer. This brought co-ordinating powers over medical research. The *Aktion Brandt* planned emergency hospitals on a functional, radiating design – although in the event only six were completed.[65] He was ordered to reach a settlement with the SS. In 1944 he became Special Commissioner for Chemical Warfare. But in the autumn of 1944 came a sudden reversal when Hitler condemned Brandt to death as defeatist. Brandt set out to show the court that his powers were actually far less than his grandiose titles suggested and that he knew of only a few involuntary human experiments. He disputed that the *Aktion Brandt* represented a covert means of continuing euthanasia by necessitating the killing of patients to clear wards for military personnel. The issue has remained controversial.

The prosecution suspected Karl Brandt was generally involved in human experiments. There was evidence of his support for hepatitis experiments by Dohmen in Sachsenhausen in 1943/4, and for mustard gas experiments in 1944. He supported nutritional experiments in concentration camps in 1943. He endorsed experiments on treatment of phosphor burns in September 1943, and ten inmates were made available in Sachsenhausen for tests in February 1944.[66] He energetically backed Bickenbach's experimental research, although here a motive may have been to challenge Hirt and the SS, in their research on chemical weapons. Brandt admitted that he visited Sachsenhausen in 1936 and Natzweiler in 1944, but there was otherwise no proof that he visited the camps. The question was whether his duties involved him in facilitating the experiments in the camps, and the holding of slave labour.[67]

While other defendants had links to Karl Brandt, and to the SS doctors Conti and Grawitz, his key co-ordinating role meant that he dealt with a small army of academics including Butenandt and Nachtsheim. The prosecution drew on CIOS and medical intelligence in distinguishing criminal from legitimate human research. But the assumption that the human experiments were 'pseudo-science' has blocked off evidence as to the broader impulses and extent of the human experiments. A few commentators saw the conduct in terms of 'pure science', but at the same time as deadly and immoral.[68] The prosecutors were less successful in the case against the Berlin professor of surgery Rostock, who assisted Karl Brandt with evaluating and co-ordinating medical research from 1943. The Allies suspected that medical research doubled as germ warfare research. Other members of Karl Brandt's staff, such as the chief naval doctor Fikentscher were interrogated, but only as witnesses. Rostock demonstrated that he had no direct responsibility for human experiments in the Office of Science and Research from the autumn of 1943. He held administrative responsibility for medical research, intensifying efforts to preserve Germany's research facilities and personnel.[69] Rose and Rostock attacked the central direction of German wartime science as a phantom, born one and a half years after the war. The reality was one of intrigue and factionalism.[70] Rostock's lawyer robustly condemned coercive human experiments in his final plea, although in reality Rostock was closer to his fellow surgeon Brandt on such matters.

The Milch trial was intended to deliver proof of such co-ordination for military medical ends. This opened on 20 December 1946, progressed rapidly and concluded on 2 April 1947 with judgment on 16–17 April. It was an important precedent for the Medical Trial, given the similar charges.[71] Six Medical Trial defendants testified for Milch. His defence was that he had no great interest in human experiments, as he was too senior given his rank of Air Field Marshal. The SS General Karl Wolff crucially affirmed that Milch did not personally know Rascher.[72]

The chief air force doctor Hippke was called as witness for Milch. He testified that Milch was against SS involvement in air force medicine. Behind this lay Hippke's sense that he was a convenient target, for if the blame for the Rascher experiments could be shifted onto him, this could exonerate some of the accused doctors.[73] Alexander gave evidence at the Milch Trial on 14 February 1947.

The section on aviation experiments was the least successful part of the Milch Case. Judge Robert Toms acquitted Milch on the count of responsibility for aviation medical experiments, thereby undermining the prosecution's notion of seamless hierarchies; by way of contrast, the slave labour charges were upheld. Despite Milch's letter of 20 May 1942 that cold experiments were desirable, the court endorsed the view that the phantom Rascher held prime responsibility. However, Judge Michael Musmanno took the view that Milch merited a death penalty, and submitted a dissenting report, stressing the pain and fatalities from the experiments, their criminality, and Milch's responsibility for them.[74] But the majority acquittal of Milch of responsibility for Rascher's experiments undermined the prosecution case regarding the Dachau aviation experiments at the Medical Trial. Milch was fortunate – his diary reveals him as a vicious and unrepentant anti-Semite, and a strongly committed Nazi, who would have regarded a few lives of racial undesirables well worth sacrificing in aviation experiments.[75] Musmanno went on to condemn the futility of the human experiments in the IG Farben trial. As judge in three cases, he was able to maintain a consistency as regards human experiments.

Eight of the accused at the Medical Trial linked clinical research and the Luftwaffe: Becker-Freyseng had membership of Strughold's Reich Aviation Ministry research institute, while serving as research administrator on the staff of the Air Chief Medical Officer from 1941. This meant he worked under the accused Schröder, and much aviation experimental research passed through his hands. Six others were involved in various aviation research groups. Ruff argued that he (in contrast to Strughold) had used self-experiments, which took him to extremes of epileptic cramps, so that he understood what was involved when he worked with volunteers. He pointed out that thousands of such experiments were undertaken, but Rascher stood out for his ruthless cruelty in experimenting on prisoners, and had Himmler's backing. Ruff insisted that he had experimented only on volunteers, who were well fed. He claimed that he took the pressure chamber away from Dachau to prevent its further use by Rascher. Witnesses were lacking, and McHaney and Ivy could not link the lethal experiments to Ruff's report. His testimony played on McHaney's lack of technical expertise.[76] Ivy pointed out that Ruff and Romberg were unwilling to treat prisoners as themselves, as they found the self-experiments painful and caused loss of consciousness. Indeed, their interest in Rascher's pathological findings demonstrated that they were willing collaborators with Rascher even though they knew murder was being committed.[77] In contrast, Ruff denounced US aviation medical experiments as causing casualties, and his counsel was confident of an acquittal.[78]

Weltz claimed that he was only nominally Rascher's commanding officer, and distanced himself from Rascher's experiments. Weltz testified that he only knew indirectly about Romberg's research at Dachau, and about the use of the pressure chamber in the camp.[79] He had been repulsed by Rascher having delivered his own father for imprisonment in Dachau, and described a growing conflict with Rascher, who took Himmler's view that anyone who disapproved of human experiments was a traitor.[80] Alexander was interested in Rascher's abusive treatment of his father. The threats made by Rascher against Romberg were mentioned in the closing address, and sent by Alexander to Judge Sebring and Ivy on 15 July 1947.

Romberg claimed that he resisted Rascher's intention to carry out fatal experiments. He objected because he was responsible for the group of 15 subjects, and forbade it. His aim was saving lives when parachuting from a great height. One experiment was to go up to maximum height for 10 seconds (the time that a pilot would take to bail out) and then descend at the rate of a parachute. Alexander judged this was an 'intelligent experiment'. In contrast, Rascher kept persons at maximum height for minutes thereby causing injury and death. Despite Himmler's telegram urging secrecy, Romberg discussed matters with Ruff and they decided to withdraw the pressure chamber.[81] The witness Hornung gave a chillingly precise description of how the freezing experiments were conducted, with Evipan anaesthesia being used. Rascher was reluctant even to use this, as he felt that any narcotic would distort the experimental results. Hornung felt Rascher's assistants Neff and Punzengruber were evil individuals.[82]

The selection of three non-medical SS administrators, Viktor Brack, Rudolf Brandt and Wolfram Sievers, served to cement the medical experiments and arising clinical abuses to Hitler and the SS. Brack worked in the Führer's Chancellery from where he administered Nazi euthanasia. He set up the administration overseeing the adult euthanasia killings, making sure that it was camouflaged by a series of pseudonyms. Brack adopted the name 'Jennerwein' as a cover when administering euthanasia. The T-4 official Reinhold Vorberg was a cousin.[83] Their role was highlighted at the American-run euthanasia trial concerning the killing centre of Hadamar. Brack presented himself as someone who constantly intervened to save lives: Otto Warburg, the biochemist who survived tenaciously in Berlin, testified to Brack's support.[84]

Rudolf Brandt joined Himmler's staff in 1936. He was adept at shorthand, and handled Himmler's voluminous correspondence. Matters concerning human experiments were here intertwined with a range of issues such as genocide, the building up of the SS as a fighting machine, and gaining control of the universities. The prosecution considered he took an organisational role and was therefore culpable, whereas he laconically saw himself as a mere cog in an administrative machine processing 40,000 letters in a year.[85] His lawyer, Kauffmann, had called the Auschwitz commandant Höss as witness at the IMT to prove Kaltenbrunner's distance from the killing process.[86] The strategy was to attempt to show his client was not involved in crucial decisions on human experiments.

Sievers was secretary of the Ahnenerbe Ancestral Research organisation, and of the Institute for Military Research.[87] He was an autodidact who had come to Himmler's notice because of his activities in Nordic circles. The Ahnenerbe at first mainly supported Germanic studies, notably prehistoric archaeology and the anthropological expedition of Ernst Schäfer to Tibet. When Himmler instigated the screening of the Tibet film in Dahlem in 1939, he used the occasion to develop the Ahnenerbe as a cultural and research organisation. The interrogators – alert to the economics of Behemoth – were interested in its Circle of Friends consisting of influential industrialists, who met continually between the early 1930s and 10 January 1945 and heard lectures on medical research. The Friends included Flick, Meyer of the Dresdener Bank, and Waldecker of the Reichsbank, while visitors included Otto Ambros of IG Farben, who used the opportunity to discuss with Himmler IG production in Auschwitz; the immensely powerful Pohl was the link between the donors and Himmler. The Ahnenerbe had departments for medical research and botany. When the anatomist Hirt demanded the bodies of 'Jewish Bolshevik criminals', Himmler obliged by ordering Jews from Auschwitz to be killed for the anthropological skeleton collection at the Reich University Strassburg. The prosecution saw Sievers as the link between the doctors and Himmler, processing reports and resourcing the experiments.[88] Sievers protested that he knew nothing about Auschwitz, had until then never encountered Eichmann, and that he could exert no influence on the order from Himmler that Hirt was to select 'the criminals' who were to be executed.[89]

Criminal experiments

The human experiments became the centrepiece of the Trial, arising from the diagnosis of medical research as at the heart of Nazi genocide. Taylor accused the defendants of complicity in the deaths of hundreds of thousands. He drew attention to the 'nameless dead', the cohorts of victims callously referred to as 'twenty Poles' or 'thirty Russians'. Servatius estimated the numbers killed in the experiments cited by the prosecution amounted to only 2,000 deaths.[90] Pohl, responsible for managing the SS's finances, provided a critical perspective on the experiments as uneconomic. He confirmed that he knew experiments of Schilling and Rascher, as well as of Clauberg, Heissmeyer, of the Madaus sterilisation drug and mustard gas, and that in 1942 he protested to Himmler against medical experiments as a waste of the labour force. He recollected that in autumn 1942 the medical officer in charge of concentration camps, Lolling, said that there were 30 to 40 series of human experiments. Pohl condemned these as criminal. But Pohl – in common with other Nazis seeking exoneration – gave an estimate that was far too low: he considered only 350–400 prisoners were involved.[91] The problem is that the Medical Trial focused on SS and military research. Experiments in clinics and prisoner of war camps remained in the shadows.[92]

The focus on numbers killed was also a distortion: many died after the experiments, either as a consequence of injuries or because the Nazis deliberately wished to kill them, so that they could not eventually testify. The killing at the Bullenhuser Damm is one such atrocity. The Nazis hunted down the Ravensbrück Rabbits with less success. Even those who survived to testify often had to live with debilitating injuries.

Blaha pointed out that while only seven died in malaria experiments, another 500 died from complications and indirect consequences. While Blaha may have exaggerated, it reflects an underlying situation of the vulnerability of the experimental subject, which can be confirmed from the hunt for Ravensbrück Rabbits in 1944–5 so that they should not testify. Blaha recognised that as a witness and victim, one could not understand the full extent of the events.[93] Yet survivors amassed considerable documentation, and the depositions and witness statements have still today not been assessed. Alexander was triumphant when he located Gerrit Hendrik Nales, a prisoner nurse at Dachau, who kept a list of 2,500 persons killed and noted who was killed by lethal experiments.[94] Historians would do better to listen to the direct testimony of survivors, immediate witnesses and those who first collected the testimonies. For many years the necessarily selective account by Mitscherlich, who was harried by the scientific establishment, was the standard source on the Trial rather than its rich underlying documentation.

The focus on experiments should not obscure how Nuremberg prosecutors cast their net widely by seeking evidence for a range of scientific atrocities. While they followed up the skeletising of murdered victims for anatomical collections, we know of approximately 100 Austrian Jews in Buchenwald in 1939–40 in addition to 90 such victims in Strassburg. They noted how scientists harvested about 1,000 brains from euthanasia victims, and experimental methods of sterilisation and fertility experiments, amounting to several hundred persons. Once we add the totality of victims from euthanasia (currently estimated at 400,000) and sterilisation (another 350,000 with an estimated 6,000 deaths), we can match Taylor's figure. Added to this should come groups persecuted on scientific grounds: the Sinti and Roma assessed as degenerate by psychologists and racial researchers of the Reich Health Office, the homosexuals incarcerated and experimented on at Buchenwald, and those stigmatised as hereditary criminals and anti-social. The evidence supports Taylor's accusation of hundreds of thousands of victims of 'atrocities committed in the name of medical research'. The divide erected between human experiments on the one side, and euthanasia and genocide on the other, appears artificial. The Trial generated evidence on links between the human experiments, euthanasia and genocide. While the prosecution was mistaken that the experiments were pilot studies for the Holocaust, there were still multiple links between the experiments and genocide. (See Table 8.)

The military and racial mobilisation of German medicine drove the different phases of German human experiments. The first phase of coercive experimentation in 1939–41 was neurological, mainly in clinics, and was linked to Nazi euthanasia measures. The neurologist Georg Schaltenbrand at Würzburg aimed to demonstrate the viral transmission of multiple sclerosis in 1940. The experiments involved injecting spinal fluid from humans to apes and from apes to humans. The victims were German, and included NSDAP members. Schaltenbrand deemed one cohort of 20 to be terminally ill and despatched them to a euthanasia institution. Between 1939 and 1944 anthropologists and racially minded psychologists investigated Jews and 'gypsies' (Sinti and Roma) – the subjects were invariably dispatched to camps for killing. There were large-scale experiments on sterilisation, mainly by X-rays, and gruesome experiments on human reproductive organs. Thirdly, between 1941 and 1943 came a wave of military experiments concerned with the management of war wounds, prevention of infectious diseases to protect the advancing troops, and aviation medicine. Finally, in 1944–45 medical researchers and concentration camp doctors preyed on children. Mengele investigated the inheritance of racial characteristics, and links between race and disease resistance.

The journalist Günther Schwarberg reconstructed the life histories of the 20 murdered children shipped from Auschwitz to Neuengamme for TB experiments.[95] What Schwarberg achieved in microcosm needs to be done for the totality of experiments. The overall numbers of experimental victims, let alone their identities, and the groups targeted – their nationality, religion, occupation, age and gender – and timing of the experiments have never been comprehensively assessed. Robert Proctor estimates that 'roughly 1,000 people died from the effects of human experimentation'.[96] He excludes the high proportion of what were initially non-fatal experiments, but whose victims were so weakened that they later died from infections and wounds, or from efforts to destroy the evidence. Many experiments involved small cohorts comparing groups of ten or twenty victims, but at times many hundreds of subjects were involved. Schilling used at least 1,000 for malaria research at Dachau, causing fatalities; the Buchenwald prisoner parasitologist, Alfred Balachowsky, reckoned that 600 prisoners died as reservoirs of different types of typhus pathogens, and the experiments involved not only trial vaccines and toxic drugs like Acridin and Rutenol, but also infecting batches of unvaccinated 'controls'.[97]

In 1940–41 experiments were at a low point, as research was disrupted by wholesale mobilisation. But from 1942 numbers of experiments climbed to reach a high point in 1944.[98] Even though the Reich was lost, an incentive was to continue experimental research. German scientists exploited unrivalled research opportunities on human captives, and faced by imminent defeat, science offered a passport to an uncertain future. The research became less strictly military (as for vaccines and aviation), and related to more general scientific problems. Children were a noted category of victim in 1944–5, as at Mengele's infamous twin camp in Auschwitz. The camps of Buchenwald, Dachau, Neuengamme and Ravensbrück were major experimental centres. Auschwitz became a major source of supply for experiments, as by Haagen at Strassburg and Heissmeyer at Neuengamme.

Telford Taylor observed how experiments were ordered on batches of prisoners referred to by numbers and nationalities. We do not know how many survived. It is artificial to separate experiments from 'atrocities committed in the name of medical research'. The numbers of medical victims rise when one includes the hundreds of 'experimental' X-ray sterilisations. Separating the human experiments from euthanasia is also artificial. Some victims were medically 'interesting cases', who were meticulously observed and then killed to order. Peiffer shows how 2,097 brains were with high probability examined by neuropathologists, and additionally 407 brains of murdered children were examined by Heinrich Gross in Vienna. Many brains were salvaged by scientists from the killing centres for permanent collections, and illustrated in articles and textbooks.[99] These forms of scientific atrocities need to be added to the minimum of several thousand victims of coercive research, and the result indeed bears out Taylor's contention of hundreds of thousands of victims of racial medicine.

A lethal Luftwaffe?

A cloud of suspicion hangs over the aviation experiments at the Medical Trial. The issues are firstly whether the US laundered aviation physiologists, by placing them on trial and acquitting them. The second issue is that the prosecutors targeted administrative personnel and lesser scientists rather than the leading physiologists Hermann Rein and Hubertus Strughold. On 16 September 1946 Taylor ordered a swoop of aviation physiologists at the American Air Force Aero Medical Center in Heidelberg. The physiologist Strughold directed 64 German scientists, preparing reports on wartime aviation medicine, and in research projects.[100] The Nuremberg prosecutors sent the commanding officer, Robert Benford, a lengthy list of German personnel involved in criminal medical research and euthanasia. The research administrator and aviation researcher Becker-Freyseng had been employed at Benford's Center since 15 October 1945, and the more junior Konrad Schäfer since 15 March 1946. They were arrested along with Ruff (Strughold's longstanding collaborator) and the senior air force medical officer Schröder. The fifth man arrested was Theodor Benzinger, who was carrying out animal experiments on freezing and blood circulation. The sixth was Heinrich Rose, who worked on the physiology of vision and (a sign of haste) was confused with Gerhard Rose. They were disconcerted by their abrupt shift from being privileged scientific collaborators to defendants at a trial with a death sentence as a real possibility.[101]

The aviation experiments exemplified how Himmler perverted normal research procedures, luring researchers with tales that the experimental subjects were volunteers, and that they would earn pardon and release. Romberg expected that the experiments would be on conscientious objectors, and that the Dachau convicts were volunteers. The accusation can here be levelled that the involved researchers were at best naïve and abrogated their responsibility of verifying the voluntary nature of the subjects' participation. Romberg insisted that he was involved in experiments on only 15 subjects and that none died.[102] He conducted many dangerous self-experiments and on volunteers in Berlin such as students, conscientious objectors, physicians and colleagues. He viewed animal experiments as only subsidiary.[103]

Weltz had been in custody at Dachau since September 1945. He was the third eldest of the accused, and had served in a balloon corps in the First World War. He had trained as a radiologist, and had only moved into aviation medicine research in 1938, when he built up a specialist department in Munich. Rein and Strughold dismissed Weltz as academically not at the forefront of the field. Alexander rated Weltz an amateur in aviation physiology, who failed to supervise Rascher properly.[104] Weltz maintained that his experiments on the effects of low temperatures only used animals. Various assistants and academics provided affidavits exonerating him, notably Büchner, chief of Luftwaffe pathology services (and later a militant critic of the Trial).[105] Although the Dachau interrogators earmarked Weltz for release, the US prosecutor Alderman on 19 August 1946 drew attention to his being on the IMT list for trial, and the Nuremberg investigators decided to proceed with the case.[106]

The prosecutors were keen to establish a chain of command that led from Göring and Milch down to Weltz. The defendants instead insisted that their orders came from within the medical hierarchy. Weltz maintained his orders originated from Hippke (as chief of Luftwaffe medical services), Hippke's assistant Albert Johann Anthony, and (the accused) Becker-Freyseng. The chain of command and the responsibility for issuing orders to undertake experiments were crucial. The prosecution argued that anyone who received reports on the deadly experiments was incriminated, and viewed Hippke and Milch as accessories to planned murder.

Defendants drew a distinction between legitimate experiments and the criminal atrocities ordered by the malign Himmler. Rascher was nominally under Weltz, who explained that the criminality of the research arose from the orders of Himmler. Weltz did not have to obey orders from Himmler, and when Rascher proposed joint experiments he declined. Weltz insisted he never received the report by the aggressive and ambitious Rascher, and outlined their tensions. One side of Rascher was civilised and cultured with a bent for music and anthroposophical ideas. But Weltz was disgusted by Rascher's bloodthirsty sadism of 'extreme experiments', and tried to transfer his renegade underling as quickly as possible. Weltz pointed out to Rascher that animal experiments on the effects of ascent had still to be completed, and Weltz saw no purpose in human experiments. As he knew that Ruff and Romberg had proposed human experiments on rapid descent, he passed Rascher over to them, believing that the resulting experiments would be safe, as self-experiments were initially envisaged. Weltz explained that he only knew by chance that the pressure chamber arrived for experiments at Dachau, and that there was just a perfunctory courtesy call from Romberg. He requested that Rascher be transferred from his institute by January 1942.[107] But the Nuremberg prosecutors produced signed orders from Himmler, ordering both Weltz and Rascher to carry out experiments. Weltz denied knowledge of these and pointed out that by the date of the order in May 1942, Rascher had been transferred to the Waffen-SS.[108]

Alexander found in Romberg an alert, competent and informative researcher, who explained how he had been lured into collaboration with Rascher between March and May 1942. He claimed that he knew at the time he had to extricate himself as rapidly as possible from Rascher's experimental killings. Romberg had wanted jointly to sign a report saying no deaths or lasting harm arose from their collaborative experiments, and that he had refused to participate in fatal experiments.[109] Alexander decided early on that the case against Romberg was lost.[110]

Benzinger rose to the challenge of making sure that his arrest did not lead to trial. He explained that he researched on height tolerance tests, as he was interested in the physiological possibilities of high-altitude flight.[111] He had been a member of the NSDAP from 1933, as well as of the SA, but his early political activism lapsed and he was primarily a researcher. From 1934–44 he ran the medical section of the Luftwaffe research centre at Rechlin.[112] Released after preliminary interrogation, the question arises whether Benzinger really should have escaped trial. He was linked to a group of aviation researchers involving the KWG biochemist Butenandt, whom Göring elected to the German Academy for Aviation Medicine in 1942. Butenandt organised experiments by Hillman, Ruhenstroh-Bauer and Ulrich Westphal on overcoming anoxia and cold. By releasing Benzinger, a series of scientific links was never followed up.[113]

Benzinger vehemently denied his culpability. He proved to be a good source of information about the research structures in the German air force. He explained to McHaney that he avoided contact with the experimenter Romberg at the Reich Air Ministry: 'I am not a timid man else ways but I decided to keep myself away from things like this and never to obtain any knowledge of experiments like this.' He denied that he was at the crucial meeting with Becker-Freyseng and Schäfer, when the experiments were agreed, protesting 'It doesn't agree with medical ethics … it is a crime'. The interrogators requested further information 'out of a sense of duty to the medical profession'.[114] Benzinger's research on the physiological basis for stratosphere aeroplanes interested American scientists. The US Air Forces, Wright Field circulated his report on this topic in October 1946 – just weeks after his detention and release.[115] Although announced as a trial defendant on 12 October 1946, he escaped the Nuremberg ordeal.[116] On 5 November he returned to the AAF Medical Center Heidelberg, and

from February 1947 he was directly employed under a US government contract.[117] He was fortunate to have regained his liberty, and he might have
made a better candidate for trial than Romberg. Hardy reflected that
Benzinger must have used the Rascher results, and that 'interrogations were
sloppy'.[118]

Ruff was director of the Department of Aviation Medicine at the German
Aviation Research Establishment in Berlin-Adlershof. He delegated
Romberg to take part in the experiments. He justified the altitude experiments as necessary, and as carried out on just a small number of volunteers
who all survived. He denied experiments were carried out on Roma. Ruff
had at first insisted that there had been no fatalities. Alexander elicited
from him how Rascher was a fanatical killer, using the pressure chamber
for executions, whereas Ruff insisted that the scientists and the air force
sanitary chief Hippke were guided by medical considerations. Ruff considered that the patient is always at a disadvantage, and could easily be
exploited. Alexander conceded to Ruff that he experimented on mental
patients but always with the consent of their guardians and the patient.
Ruff insisted that his series of high-altitude experiments in Dachau were on
volunteers.[119] In an unprecedented effort to maintain equity, the judges
invited Ruff and Romberg to leave the dock so that they could cross-
examine Ivy and advise their counsel on 13 June 1947 as to what questions
to ask.[120] Journalists relished the drama of the accused stepping into the
prosecutors' role.[121]

The Luftwaffe medical commander Hippke had insisted on Holzlöhner of
Kiel, Singer of Munich, and the pharmacologist Jarisch of Innsbruck being
involved in the experiments, claiming that he wanted to use physiologists
to restrain Rascher.[122] The aviation physiologist Becker-Freyseng approved
the experiments as part of his administrative remit covering all aviation
medical research. The questions remained whether Air Marshal Milch gave
approval to the responsibility of Hippke (who thanked Himmler on
19 February 1943 for allowing the cold experiments), and whether the
physiologist Strughold was briefed about these experiments. Himmler was
angry that Milch absented himself from Rascher's presentation, and that
Holzlöhner had taken credit for himself. Moreover, both researchers
complained how 'Christian ethical scruples' hindered the experiments.[123]

The accused internist Beiglböck worked under the renowned Eppinger at
the First Clinic for Internal Medicine at the University of Vienna.
Alexander remembered him as having been 'at the same class with me
during our first year in medical school ... He always was a Nazi roughneck,
and I guess he still is.'[124] Eduard Pernkopf (the Nazi responsible for an
anatomical atlas replete with swastikas) and Eppinger wanted to promote
Beiglböck for serving in Dachau and on the Eastern Front, while continuing to publish research. He was appointed on 23 June 1944 professor of
internal medicine at the time of the Dachau seawater experiments.

Bernhard Rust, the Minister of Education, praised Beiglböck's exemplary fulfilling of his duties, and his involvement gained him special protection of the Führer.[125] Aggressive experiments were a rung up the professional ladder.

Schröder had overall responsibility for aviation medical research and Becker-Freyseng had a key role as administrator in the office for aviation medicine. Since 1942 he had been concerned with making seawater drinkable. Becker-Freyseng encouraged Schäfer to work on this problem at the Strughold Institute for aviation medicine, and the result was the Wolfatit desalination method developed by IG Farben. But this competed with the method of the engineer Eduard Berka, developed at the Technical University Vienna, consisting of adding a tomato extract to seawater; the caramelising effect masked the salt taste, although the salt content remained. L. von Sirany conducted experiments on the Berka method in the Vienna Air Force hospital on German soldiers.[126]

The dispute reached the office of De Crinis, who brought in his former student friend from Graz and fellow illegal Nazi to adjudicate. The results convinced Eppinger of the viability of the Berka method, and Eppinger put Beiglböck's name forward to carry out the Dachau tests. The seniority of Eppinger and Heubner was crucial in reaching the decision to carry out the Dachau experiments. Eppinger argued that either the kidneys would adapt or the concentrated salts would be expelled in urine. The matter was referred to the eminent pharmacologist Heubner in Berlin, who said that the question could only be resolved by an experiment. Heubner later insisted that he did not intend this to be a coerced experiment in a concentration camp.[127] Strughold escaped complicity in that by late 1943 his department had been evacuated to Silesia. The Technical Office of the Luftwaffe (where Berka was a staff engineer) decided that the Wolfatit method was too costly in terms of raw materials, given that it used large quantities of silver nitrate. Its representatives accused the Luftwaffe medical officers Becker-Freyseng and Schröder of profiting from their connection to Schering. Schröder came across as a nervous, indignant and stressed figure when faced by his accusers, and made a poor public impression. Becker-Freyseng admitted that he chose the Dachau location in July 1944, because at this late stage in the war experiments could be done rapidly without fear of bombing. Forty experimental subjects aged between 20 and 30 were chosen as matched controls for pilots. Becker-Freyseng claimed (dubiously) that he ordered Beiglböck to conduct the experiments on a voluntary basis, and to terminate them when a subject refused or suffered adverse medical effects. The defendant Schäfer explained that he devised the purification method but was not involved directly in the concentration camps, of which he disapproved. Never a Nazi, his disengaged stance carried conviction. Schäfer was ordered to attend meetings when the Dachau experiments were decided on. He denied vehemently that he had either supported the decision to experiment in Dachau or was involved in any way with the experiments. He considered that the pre-trial interrogation should have determined his non-involvement.[128] The prosecutors had failed to see that he had merely devised a viable method of desalination, and had not taken part in either conducting or deciding on the Dachau experiments.

The 44 Roma victims were divided into five groups: one was given seawater treated by the Wolfatit silver-based desalination method; another, the Berka method using sugar to mask the taste; a third had fresh water only (the fortunate), a fourth seawater only; and a group was given no water at all, to see whether these fared better than the Berka group.[129] The Roma came from Buchenwald and Auschwitz and were selected from a larger group. The experiments took place from about 13 August to 3 September, and patients were kept under observation until 13 September 1944.[130]

Former Dachau prisoners said that fatalities resulted, not least because severely sick victims of human experiments were killed when they ended up in the camp hospital. Blaha pointed out that the number of deaths was fewer because food was smuggled into the 'gypsy block'.[131] A prisoner nurse, Joseph Vorlicek, told Alexander how the experimental victims made desperate efforts to find water in cleaning buckets, and that the infuriated Beiglböck threatened him with joining the ranks of the experimental victims in retribution.[132] Gerl claimed he saw two corpses of Roma victims, and that other emaciated victims contracted typhus.[133]

Beiglböck vehemently denied that there were fatalities, but Hardy accused him of giving a false impression of better conditions than actually existed and of concealing the identities of victims.[134] On 11 June 1947 Hardy asked Beiglböck about the two small notebooks: one was the laboratory book used in the experimental station, the other came from the laboratory of the entomological institute. He used these for entering results of chemical tests and compiled four fever charts on the experimental subjects. Beiglböck had kept these since he left Dachau in September 1944. Hardy pointed out that the covers with the names of the experimental subjects were now missing. Beiglböck's lawyer had shown the books to Alexander, Ivy and Hardy for their information but objected to their being used in court. Hardy pointed out that the research records were altered since the Trial started, and suggested a roundtable discussion to interpret the complicated red and blue pencil annotations.[135] Hardy requested that the Tribunal take custody of the booklets.[136] At first Beiglböck denied having recently made the annotations. Then he admitted that he erased material from the charts and notebooks in Nuremberg. He confessed that he marked these, as he felt that the court would take too negative a view of the subjects' extreme thirst. He tampered with his notes on weight loss and thirst, and with the charts to shorten the period of starvation.[137] Ivy, who researched on desalination for life rafts for the Naval Medical Research Institute in

1943, now waded into the proceedings.[138] He confirmed the attempt to destroy evidence. He had seen the chart in January giving individuals' names in the seawater experiments, and these were now missing.[139] He condemned the experiments as flawed, unnecessary, dangerous and inhumane.[140] The discovery of the fraudulent alterations seriously weakened Beiglböck's defence.

Infections

The leg wound and bone transplant experiments at Ravensbrück were linked to the pharmaceutical control of infections. Mant briefed the American prosecutor McHaney on the criminality of Gebhardt, who denied responsibility for the mutilating experiments at Ravensbrück.[141] The case fitted well into the overall portfolio of defendants, as Gebhardt was closely linked to Himmler, who preferred Gebhardt as medical adviser over Reichsarzt-SS Grawitz.[142] The physician-witness, Dr Maczka, grimly informed the court that none should have died if limbs would have been amputated at the right time.[143] On 20 December 1946 Alexander presented the scars and gashes of four victims to the Nuremberg court in a highly publicised episode. Their testimony proved that the German doctors refused to provide any reason for their gratuitous cruelty, thereby failing to obtain informed consent.

The two senior SS medical officers, who conducted infectious disease experiments, were the young and ambitious Mrugowsky, who directed the Hygiene Institute of the Waffen-SS, and the veteran, plodding Genzken, who directed the Waffen-SS sanitary services, and until August 1943 (when the SS medical services were reorganised) was Mrugowsky's commanding officer. Genzken's position was that the little research he carried out was 'only done as a soldier and under orders', and he resolutely denied knowledge of the human experiments. Genzken insisted that Grawitz maintained powers over research and planning in the SS, so that his own work was limited to organising field medical services.[144] He strenuously denied that he knew about human experiments or that he had authority over Mrugowsky's subordinate Ding. That Ding was located in Buchenwald meant that he came under Grawitz, but Ding maintained that it was Genzken who had sent him there.[145] Mrugowsky condemned Ding as a renegade, conducting deals behind his back with companies, and over his head with the more senior Grawitz. A debate flared as to whether the Waffen-SS was subject to military command – and thus its medical officers came under Handloser – or, as Himmler preferred, to the SS and thus to the ambitious Grawitz.[146] The chief military medical officer Handloser was accused of joining Himmler and Conti in authorising typhus vaccine experiments in 1941.[147]

Rose was an alert and pugnacious defender of German medical research. He denied involvement in the planning or execution of any typhus experiments.

He had criticised human experiments when – in his view – they were unnecessary, but undertook experiments on himself. He strongly favoured using DDT rather than Zyklon. He considered that the American deployment of DDT was a feat 'without any debasing of their ethics', refuting the claims of Handloser and Mrugowsky that vaccine experiments were necessary.[148]

The RKI supported population transfers from the East. At the third *Arbeitstagung Ost* of military medical specialists in November 1943 Rose discussed how whole village communities suffering from malaria were transplanted to the Reich.[149] The RKI report of 1943 – presented as part of his defence – showed him as mixing with Waffen-SS officers and engaged in geo-medical researches on the lines advocated by Mrugowsky. Rose documented that he used the malaria strain of tertiana known as 'Greece'. This was administered to 480 paralytics and schizophrenics in the year 1943; another five patients were infected with the Russian tertiana strain 'Odessa'. The report then dealt with experimental protective treatments of pharmaceuticals in psychiatric hospitals using Sontochin, and other drugs tested on psychiatric patients on behalf of IG Farben.[150]

Rose screened German settlers and Slovenian evacuees for tropical diseases. The resettlement camps became a focus of disease prevention, as well as of medical research, as when two different vaccines for scarlet fever were tested in the resettlement camps.[151] His theory of 'epidemiological anticipation' postulated that 'on the occasion of the re-settlement of large groups of people originating from rural areas, and their concentration under camp conditions' diseases which would have occurred over several years were compressed into a short period. Parallel data were cited for epidemic meningitis.[152] Rose benefited from the prosecution not having an expert in immunology or public health to hand, who could stand up to his justifications and counter-allegations. Alexander recognised that what was needed was someone with experience of wartime virus and rickettsia research 'able to point out the fundamental differences in manner of the performance of these experiments'.[153]

At a junior level, Waldemar Hoven, the Buchenwald camp doctor, was implicated in vaccine research and euthanasia killings in the camp. He appeared to be a significant researcher with his position as Deputy Director of the Buchenwald research station. But on scrutiny his academic qualifications were tenuous. He had difficulty in completing his studies in Freiburg. His state medical qualification attained in Munich on 17 October 1939 may have been due to NSDAP influence, or to a fortuitous war emergency provision. It was proved during the Trial that he obtained his doctorate fraudulently, as the camp Kapo for pathology carried out the research.[154]

Hoven relished the depravities of Buchenwald, where as assistant doctor from 1939, and camp doctor from 1942, he played on tensions between the administration and the antagonistic factions of criminal and political prisoners.[155] The most telling accusation against him was his giving phenol and water injections as part of euthanasia actions in the camp. Unconventional and impulsive, he enjoyed his power in the camp, and he was accused of having affairs with other officers' wives and of homosexuality. Arrested in 1943 by the SS for corruption, he claimed he sided with the camp resistance. He admitted killing hundreds of criminal prisoners in Buchenwald, and claimed he did this in consultation with a resistance committee of Jewish, Polish, Dutch, German and Czech prisoners.[156] The Trial dealt with his role in human experiments and euthanasia killings. As a doctor, he was a poor specimen; his dubious qualifications and limited abilities meant that he conformed in many ways to the stereotypical Nazi pseudo-scientist.

Hardy accused Hoven of having breached the Hippocratic Oath to protect life, and sweepingly condemned the German medical profession as itself guilty. A trial was necessary to restore the honour of the German medical profession. Hardy had Hoven placed in solitary confinement under close security watch (to prevent a suicide attempt), as he was deemed ready to make a statement 'as to all his activities involving the commission of murder'.[157] Hoven mustered prisoners willing to testify on his behalf that he had helped them; Kogon was the most influential of these.

Sievers explained links between Blome, Hirt, Rascher and Himmler in 1943/44 in research on a plant extract to treat cancer.[158] Rascher collaborated on cancer research with Ahnenerbe SS botanists to test plant extracts and vitamin E on cancers, and with the SS chemist Hans Tauböck and a camp doctor and nature therapist Hans Haferkamp on a means of early diagnosis.[159] Tauböck, an expert on secret inks designated by IG Farben, also worked on sterilisation for the SS.[160] Blome discussed tests on a herbal extract with Rascher. The intention was that Rascher should test this in Dachau in a special department of cancer research.[161] Experiments used cancer mice from IG Farben, contravening Rascher's wish to work on humans. Hirt claimed to have isolated living cancer cells in mice. Himmler wanted him to make human experiments, but Sievers insisted that the research was to be limited to mice.[162]

IG Farben was condemned for sponsoring pharmaceutical experiments in the concentration camps. The economic appetites of the depraving Behemoth provided a rationale. The researchers involved in military medical research cultivated links to pharmaceutical companies. The SS gained financially from such contracts just as it did from forced labour contracts. Individual researchers like Ding and Rascher lined their own pockets by testing drugs and vaccines. Pharmaceutical companies saw the camps as ideal experimental facilities at a time when animals were in short supply, and in any case human tests were scientifically desirable. For the war crimes analysts, the links between military medical research and the companies substantiated the axiom of the links between the economic and military arms of the Nazi Behemoth.

Rascher had in 1942 tried to introduce a blood styptic, but had been abrasively criticised by Gebhardt as lacking clinical experience.[163] In 1943 a Jewish Dachau prisoner, Robert Feix, who knew about pectin manufacture, suggested that a pectin-based substance could be taken in the form of tablets as a blood anti-coagulant. Rascher saw this as a means of gaining academic distinction, as well as making his fortune. Rascher reported the discovery to Sievers. He alerted Himmler, who ordered its development for military use.[164] At this point Himmler agreed – and Blome concurred – to request the Führer's Chancellery to declare Feix a half-Jew and release him, so as to develop the production process. Despite Bormann's opposition, Blome obtained favourable pedigrees for Feix from the eugenicists, Eugen Fischer and Verschuer of the KWI for Anthropology. Rascher and the Dachau researcher, Kurt Plötner then set to work on 'Polygal 10' with backing from Gebhardt and Pohl. The RFR supported the Polygal experiments in Dachau. The human experiments by Rascher and Haferkamp were published in the *Münchener Medizinischer Wochenschrift* on 28 January 1944 as research from the 'Institute for Military Scientific Research (Dr Rascher)' in Dachau. This demonstrated remarkable openness about human experiments, indicating Rascher's desperation to achieve scientific recognition. It earned him a reprimand from the SS for a breach of security. The RFR intended to shift Polygal production to Waischenfeld near Bayreuth. As the German armies retreated, an alternative plant on the Bodensee was planned.[165] Polygal showed the power of the SS to gain support from a wide range of scientists for a chimeral scheme.

Scientific counter-attack

Defendants were keen to describe the scientific significance of their work. At the same time they argued that they were not culpable for the worst atrocities, shifting responsibility to SS doctors like Grawitz, Conti, Ding/Schuler and Rascher, who were no longer alive, and that SS administrators like Bouhler were responsible for the horrors of euthanasia. At its bluntest the defence accused the prosecution of falsifying evidence. The defence lawyers refused to accept the authenticity of crucial documents. Mrugowsky tellingly pointed out that the Ding diary contained inconsistencies: events were mentioned, which occurred later than the date they were referred to in the 'Diary'.[166]

A second line of defence was to argue that the evidence for a conspiracy was a fabrication. For the prosecution the charge was important, as interlinking representatives of organisations already found to be criminal. Reiter, Handloser and Mrugowsky denied being at the conference of Conti and the RKI president Gildemeister, which called for typhus vaccine experiments on humans.[167]

A third line was that the defendants were conscientious scientists. Gebhardt explained to his Allied captors that Himmler really favoured homoeopathy, and thus treated him with contempt. Karl Brandt was bemused by Hitler's predilection for the quackery of Morell. The defendants hoped that their scientific credentials would divide them from the pseudo-science of the Nazi fanatics. Rostock's lawyer, Pribilla, took the view that the Americans would not prosecute an academic except when he occupied a leading position in the state.[168] But he found that the denazified rump of the Berlin medical faculty was mostly unwilling to provide support. The surgeon Sauerbruch held aloof, although Rostock shielded Sauerbruch from being investigated for having approved the funding of unethical human experiments through the RFR. He had greater success in working with the formidable Fräulein Dieterici to drum up support from senior figures in German medicine; some were drawn into the Trial because of the prosecution accusation that they were informed about the results of human experiments.[169] Rostock's colleagues were outraged that they were considered as compromised, and Heubner swung round from being critical of Nazism to insisting the experiments were necessary and justified.

Rostock favoured settling the issues through expert medical witnesses.[170] The defendants mobilised academic networks to endorse the legitimacy of their research and their academic credentials. The mobilising of colleagues meant that wide sectors of the German medical profession became concerned about the Trial as tarnishing reputations of individuals and of German medicine as a whole. Servatius recruited a panel of 14 scientific experts to exonerate Karl Brandt as having been criminally involved in euthanasia and human experiments, including Rodenwaldt to comment on malaria, Flury on mustard gas, Rein on aviation experiments, and Gutzeit (in detention in Nuremberg) for typhus and jaundice experiments.[171] Rein had been reluctant to assist the prosecution.[172]

Some of those rallying to the cause of the accused had themselves been under some pressure from the Nazi establishment. The inherent nature of the system was to use a misdemeanour to bind the miscreant to the regime. Better still for defendants was when character witnesses had been forced to emigrate (in that their testimony would carry more weight), or were from one of the Allied powers. Rose appealed to tropical medicine experts from around the world, including the parasitologist Brumpt in France and Kenneth Mellanby and other pre-war contacts in the Society for Tropical Medicine in Britain to testify that he was no racist. E. Payne confirmed that: 'Rose was always a Gentleman and never made any exceptions because of race or religion (naming two or three Jewish patients with whom Rose was friendly or had helped.) Sturton says that this fact is true and will vouch for that. She will send such a statement and this can be kept in reserve, should someone try and twist: making out that R would have done something against races or religions (now called "humanity").'[173] Many German academics sensed that if a former colleague was found guilty then they too could be implicated. While some felt it opportune to feign ignorance or illness, there was a high level of solidarity for interned colleagues. They looked forward to the dawning of an era of normal science with rather less controls and interference than under Nazism.

Former colleagues endorsed the good character and morality of the defendants as conscientious clinicians. More problematic was the obligation to testify on issues, which could incriminate. The IG Farben manager, Demnitz, initially denied all knowledge of human experiments; he was struggling to maintain the autonomy of the Behringwerke by presenting himself as having resisted Nazism. Mrugowsky's lawyer Flemming refreshed Demnitz's memory by intimating his legal powers to force a witness to testify – a veiled threat of having him arrested and held indefinitely in Nuremberg.[174]

The German defence turned out to be dogged and protracted. Whereas the prosecution conducted the case by accusing groups of defendants of culpability for different atrocities, the defence limited the responsibilities and complicity of each defendant. Their lawyers made much of the disparate nature of the positions that they had occupied, and sought to break down the links that the prosecution made to establish a genocidal and militarist SS-state. They did this by demonstrating a lack of knowledge of and power over what went on, and also to argue that the prosecution fundamentally misunderstood the nature of the experiments.

The defence efficiently and effectively refuted the conspiracy charge. Seidl was sharp on weaknesses in this charge at the IMT, when the charge was thrown out as untenable.[175] The operation was repeated at the Doctors' Trial. Each defendant testified that s/he had not known most who were charged, or if they did know them often it was only to a very limited extent. The defence succeeded in demonstrating that they were not a centrally directed phalanx. Gebhardt maintained that his cluster of Fischer and Oberheuser was separate from bacteriologists or aviation experimentalists, and that he really was an outsider in the medical power struggles.[176] Accused as being the linchpin of the Nazi medical system, Karl Brandt stated that although he had known Handloser reasonably well when his office was evacuated to the Beelitz sanatoria on the outskirts of Berlin in 1945, he had not known the other defendants. The prosecution gathered correspondence to try to refute the denials, but with scant success. The charge collapsed.

The pugnacious Rose led a counter-attack, which went to the heart of the prosecution's case. He joined Karl Brandt in accusing US medicine of carrying out human experiments under coercion. Rose and Brandt were inspired by a story in the forces edition of *Life Magazine* of June 1945 on malaria experiments in the penitentiary of Stateville, Illinois – the story was intended to convey heroism and self-sacrifice on the American side.[177] The defence found much ammunition in *Life*, *Time*, and *Reader's Digest*, which extolled the achievements of US wartime medical research as a glamorous and heroic activity.[178] Brandt cited Allied opinions of using Nazi prisoners for therapy and clinical research.[179] The defence's attempt to turn the tables on the prosecution was expressed by the Latin adage *tu quoque* (lit. and you too), a tactic already used at the IMT.[180] The defence hit on a type of experimentation, which was widespread in the US.

Ivy was a stalwart campaigner for experimental medicine based on animal and human experiments. He fervently believed in the righteousness of prison experiments and readily justified them as a crucial stage in medical innovation. The prison experiment was in his eyes an act of redemption. The defence lawyer Servatius challenged Ivy regarding whether the prisoner-research subjects in the United States were as idealistic as a radio report claimed.[181] While the Germans had hit on a vulnerable point in the American charges, Ivy's ethical fervour prompted him to insist on the voluntary nature of the experiments.[182]

Surviving prisoners' autobiographies have been variously interpreted. Moreno suggests that the experiments were genuinely voluntary, and that there was much genuine idealism.[183] By way of contrast, Hornblum demonstrates that the penitentiary experiments involved coercion, and that between the 1930s and 1960s American medical researchers saw the controlled penitentiary environment as a unique experimental resource.[184] The most charitable explanation is that Ivy considered that research ethics could gain more from winning the case than to concede that the penitentiary experiments were unethical. He gave the Tribunal the impression that the consent procedure was more formalised than it really was. Whatever the reality about the treatment of prisoners, the equation of an American penitentiary or even an internment camp with a Nazi concentration camp was grotesquely tendentious.

Rose and Karl Brandt set out to document that coercive human experimentation was widespread. Brandt cited British experimental research on German infants with birth defects. Brandt used a similar defence to Rose that Allied scientists were experimenting without consent and with state support. The Cambridge Professor of Experimental Medicine, Robert McCance (who had been involved in seawater drinking experiments), and his collaborator, Elsie Widdowson, were carrying out research for the Medical Research Council (MRC) in Wuppertal.[185] In June 1946 McCance had a request circulated by the British zonal administration to German doctors for information when terminally sick babies with meningocoeles (an abnormality of the spinal cord) or other abnormalities were born, so that he could test renal function by blood and urine tests. McCance selected terminally ill babies as he was unsure whether the tests were safe. He failed to ask for parental permission.[186]

Servatius was informed about this when on a trip to Cologne, and he cited the request in court on 17 April and 26 June 1947. But he encountered objections from the prosecution, because he had no copy of McCance's circular to enter into the court record. The Tribunal allowed him to submit 22 questions to McCance, including whether the children's lives were endangered and whether he had the parents' consent, but denied Karl Brandt's request to summon McCance in person.[187] The incident caused the British authorities and McCance considerable embarrassment. McCance defensively insisted that a 'test' was not an experiment, as Karl Brandt ingeniously lambasted the blood and urine tests as an example of Allied human experimentation.[188] Genzken also accused the British of testing a 'flu vaccine on Germans imprisoned at Neuengamme.[189] The MRC was concerned by the similarities of German and British experiments, which it sponsored, noting its nutritional experiments on vitamin-deficient diets, and parallel German research.[190]

Rose's counter-attack was observed by General Clay, the US Military Governor, on 18 April. His stance can be compared to when the IMT defence lawyers condemned the bombing of Dresden and Hiroshima as Allied atrocities. The argument was that the US was as guilty as the Germans, and American morality was hollow and false. Other iniquities were historically more remote, but resonated strongly with the German public. Seidl (defending the Hohenlychen trio) was obsessed with the injustice of the Treaty of Versailles.

Rose claimed that research in concentration camps did not differ from experiments in US penitentiaries and on colonial peoples. He attacked the Americans as little more than bank robbers, kidnappers and gangsters. They were opportunistic in bringing the case against patriotic German scientists, who were doing no more than their duty to state and society, but could not be held responsible for general politics. It was a political decision whether to experiment on political prisoners. There were cases in other countries when beggars were used (against a small payment) as in Italy, or on 'coolies' in the Canal Zone. The US volunteers were offered improvements on their normal prison routine. Rose considered it was better to execute in the form of a generally beneficial experiment than merely to take life. He defended the position that the experimental subjects were under a death sentence, and that in Dachau they were promised a pardon.[191]

Rose pointed out that on three occasions he had performed experiments on himself. He had tested the safety of vaccines by using a double dose on himself, and on several occasions had infected himself.[192] He also criticised Ding/Schuler's experiments. He explained that he visited Buchenwald shortly after March 1942, and that he went to Conti to explain that the Buchenwald experiments did not conform to normal bacteriological procedures based on animal experiments. Conti considered Ding's experiments justified by the typhus emergency in the East, and the threat this posed to the Reich. Rose was adamant that no more knowledge could be expected from these Buchenwald vaccine trials than what was already known at the time. He claimed that Mrugowsky supported him, when he raised the objections at the meeting of the consultant military medical specialists.

A difficulty was that key individuals Conti, Gildemeister and Ding were no longer alive, but Rose affirmed that there were corroborating witnesses.[193]

Rose denied involvement in Schilling's malaria experiments at Dachau by having supplied some cultures. He argued this was standard practice and did not make him responsible for the conduct of the experiments.[194] He vehemently objected to the accusation of coercive vaccine testing. He conceded how, on 29 September 1943, he had recommended the testing of Ipsen's 'Copenhagen' mouse liver vaccine. He wrote to his former RKI colleague, Haagen at the Reich University Strassburg on 13 December 1943 urging him to test this vaccine.[195] Rose was medical consultant to the Luftwaffe and Haagen was a reserve officer, but Rose denied that Haagen was under his command. Richard Haas, the virologist, required some prompting to admit that he remembered Rose's report having reached him while he was in charge of the Behring Institute Lemberg. But he had not received further details from Rose, and doubted whether he ever received a sample batch. He lamented how he had to evacuate his splendid Institute, and much was lost.[196] Olga Eyer, secretary at the Strassburg hygiene institute, testified that Haagen collaborated with Dohmen in 1944 on hepatitis research, and that there had been fatalities.

Rose rallied support from German colleagues in tropical medicine.[197] Nauck, the director of the Hamburg Tropical Institute, gave extensive help supplying documents and publications.[198] Rose claimed that at his final posting to his Luftwaffe fever department at the Pfafferode asylum, he made sure that the clinician had the right of veto over the experimentalist. Yet for all his denials, it appears that he experimented with dangerous anopheles strains of mosquitoes.[199]

The prosecution had to contend with the German defence that they were experimenting on prisoners condemned to death. This extended the traditional rights of doctors to use bodies of the executed for teaching and research. The accused claimed that they were told that concentration camp prisoners were legitimately condemned to death, imputing a tenuous legality to Nazism that did not exist. Although some prisoners were earmarked for death, this was an arbitrary decision by an unjust regime rather than in any way resulting from due process of law. Prisoners were picked at random, and the excuse that the subjects were being given 'a sporting chance' by earning release was also a fiction, except in a few cases whom Rascher wanted to use as helpers. Alexander pointed out that experimental subjects were killed in autopsies when their bodies were cut open while still alive.[200]

Initially, Karl Brandt pointed out that he had neither knowledge of nor administrative competence over the human experiments. His defence countered accusations of criminal human experiments on hepatitis and chemical warfare by arguing that these were legitimate and non-fatal. He shifted his strategy in February 1947, when his third book of defence documents included extensive American materials on human experiments and on their ethical justification. He rejected that the US penitentiary experiments on malaria were voluntary, and attacked the US hepatitis experiments.[201] He went on to cite Allied opinion on using convicted Nazis for medical research. Servatius presented a critique of Ivy's contention that US experiments were voluntary.[202]

Karl Brandt insisted that the health of humanity justified human experiments. He argued for a strictly medical form of euthanasia, and he shared with Brack the view that euthanasia was ethically justified. Brack introduced documents concerning euthanasia discussions in Britain in the 1930s, and abuses in US psychiatry.[203] The SS official Werner Best obligingly testified as to the legality of euthanasia on the basis of Hitler's decree.[204] Brandt's self-portrayal was as a courageous idealist. He was among those seeking recognition as *Täter aus Überzeugung* (perpetrators by conviction), which under German law could be recognised as mitigating. This position meant that he did not express regret or condemn wholesale euthanasia or involuntary human experiments in the war. But he denied connections to murderous human experiments, as alleged by the prosecution. His defence was that he only knew of the benign experiments rather than those that were dangerous and lethal.[205] McHaney was well briefed by the prosecution team. His cross-examination of Karl Brandt 'really shook him to his foundations' by establishing that he was in fact far more knowing of experiments, such as those at Ravensbrück.[206]

Otto Bickenbach, who collaborated with Haagen and Hirt in poison gas experiments and phosgene experiments at Natzweiler, testified that Karl Brandt criticised unwarranted experimentation and that he interceded with Conti on this point. Bickenbach wished to interview Karl Brandt to establish that Hirt conveyed the orders of Himmler for him to carry out the experiments. Brandt insisted that he only knew of the phosgene experiments, although he denied complicity in the four deaths. Bickenbach had told Brandt that such experiments were contrary to medical ethics, and asked him to intervene with Himmler, but that Brandt did so without result. Hirt – backed by Himmler – insisted on further experiments at Natzweiler as a result of which four Roma subjects died.[207]

In the interrogations of 1945 Gebhardt argued for the scientific value of the wound experiments in the saving of lives of military casualties. He presented himself as a non-ideological clinician, contemptuous of the futile experiments of Rascher at Dachau. Gebhardt shifted his position. He pleaded that animal experiments always had primacy. But in experimental work on infectious diseases, animal experiments had severe limitations as there were immunological differences between humans and animals. He recognised that animal experiments were fundamental for surgery. He drew a distinction between his own research, and research ordered by the state in time of war, particularly as self-experiments were banned during the war.

His plea was thus the classic one of the accused Nazi – that he was acting under orders – but transposed to the sphere of medicine. Similarly, Gebhardt argued that confidence between physician and patient was paramount – unless the physician was acting under State orders.[208] Here, he insisted that his actions did not contravene any law in force at the time.[209]

Under interrogation in October 1946, Gebhardt alleged that Fischer had carried out the experiments under orders from Himmler. Gebhardt claimed that he was not able to carry out the order as he lacked pharmacological expertise.[210] Gebhardt maintained he was a conscientious and humane physician, who on occasions assisted persons who were Jews or half-Jews. Defendants stressed how they were dedicated and humane physicians, and produced testimonials on their capacity for dedicated, selfless service. It rapidly became clear that good doctoring was incidental to the Trial, and that to show this amounted to no alibi but was the equivalent of providing a character witness. The defence had to face the prosecution's mounting evidence of murderous experiments. In subsequent testimony, Gebhardt elaborated his views on human experiments in war and peace in extensive memoranda on the topic.[211] But he became increasingly desperate, as evidence was amassed against him. The camp commandant Suhren explained that he had tried to prevent further experiments as disrupting camp discipline, but that the imperious Gebhardt overrode him.[212] Gebhardt emerged as inextricably caught up in the SS and Waffen-SS, and with the human experiments as part of building up Hohenlychen as a major SS medical research centre.[213]

Handloser came from the experimental tradition in German medicine, and in 1913–14 was involved in pioneering experiments on blood pressure and altitude to determine suitability for airship crews.[214] His defence insisted on the priority of 'uniformity and synchronized operation of all medical capacities in the East', not least, as his memo of May 1942 stated, because of 'the vastness of the Eastern space, the cultural standards of the country and of its population' and 'the permanently lurking dangers of the introduction of epidemics from the Asiatic territories'.[215] While prioritising research on infectious diseases, his stance was that he only endorsed experiments which were safe, as in the case of Dohmen's epidemic jaundice experiments.[216] Handloser strenuously distanced himself from the SS, although the prosecution accused him of having authority over Waffen-SS medicine.[217]

The defence rejected the prosecution's view that there was a concerted conspiracy by the army and SS to perpetrate massive programmes of human experiments. They dented, disrupted and delayed the prosecution's case. They insisted on the legitimacy of their research work under war conditions. But the prosecutors were astonished that however dedicated they were as physicians, for the most part they were complicit in murderous science. The prosecution case was robust when it came to major sets of experiments at Ravensbrück, Dachau and Natzweiler. It all added up to what Alexander and Thompson felt was the madness of the Nuremberg situation, in having to understand the rationales of cataclysmic crimes, and whether mundane, earthly justice was adequate to the task.[218] They viewed the Germans as depraved psychological specimens.

Allied scientists and physicians divided over the legitimacy of the German wartime scientific investigations and the worth of the Trial. The most sustained controversy was carried out in the pages of *The Lancet* and *British Medical Journal*. The outspoken entomologist, Kenneth Mellanby was accredited as correspondent of the *British Medical Journal*. He went to Germany to hunt for medical documents, to sift through FIAT documents, and to attend the trials of doctors and scientists at Nuremberg and Dachau. He wanted to salvage material of scientific value, irrespective of the circumstances in which it had been obtained. Mellanby's position as observer at the Medical Trial was to defend human experiments. He had himself requested permission to infect consenting volunteers during the war, first with typhus and then with scrub typhus, but permission was refused by the MRC.[219] Lord Horder, the King's Physician, recommended that the experimental results be restricted to military use.[220]

Whether the German experiments were scientifically justified was a constant bone of contention. Mellanby provided the defence with support, which was a powerful source of rebuttal of the prosecution's accusations. He felt that the malariologist Schilling was unjustly accused of causing a high number of deaths when he used tertian malaria. He informed Dr Fritz (the defence lawyer of Rose) that the malaria experiments of Schilling were 'carried out carefully and with a reasonable regard for the safety of his subjects', pointing out that tertian malaria was benign. But he conceded that 'by using subjects in a concentration camp, even though he himself behaved reasonably, he was condoning the bestialities which were going on in the same institution'.[221]

While Mellanby roundly condemned Nazism and the sadism of coerced experiments, he identified a legitimate scientific component in some of the contentious experiments: 'Where they are bestial and unfit for [the] public it also happens that they are no use to medical science either.' He naively believed that Schilling had obtained prisoners' consent at Dachau. He commended the Buchenwald typhus experiments of Ding as a 'useful evaluation of the various vaccines', overlooking that around 1,200 Russian, Roma and French prisoners were killed in the experiments. Mellanby strongly believed that Rose, whom he knew personally, was innocent, as was Handloser. The defence counsel of Becker-Freyseng cited Mellanby's extenuating opinion in his document book.[222]

The *Daily Telegraph* reported on 11 February 1947 that some medical experiments on inmates in concentration camps could be of great value to medical science. It quoted Mellanby's belief that they should be made available to genuine investigators. The editorial in *The Lancet* sparked off a major discussion: Mellanby and Sidney M. Hilton of Preston Hall Hospital, Maidstone favoured publication of the results of the concentration camp experiments. Hilton recommended a special commission sift through the results of the Nazi experiments. Ivy took an intermediate view that no German work of any scientific value occurred, so that it was pointless to seek to defend any individuals; T.B. Layton (ear, nose and throat surgeon of Guy's Hospital, London) condemned any use of criminally obtained material. Layton had taken over the hospital at Belsen for UNRAA. He prescribed that data should be destroyed 'uncopied and unrecorded'.[223] One experimental victim, Father Leo Michalowski, a Polish priest and former inmate of Dachau (where 540 priests died), favoured publication of the results, but with a warning that the data had been obtained through horrific cruelty.[224]

Beiglböck called on the British physiologist W.C.S. Ladell to testify on the basis of his *Lancet* article on seawater requirements of the shipwrecked.[225] Ladell had worked for the British Committee on the Care of Shipwrecked Personnel, as had Mellanby and McCance, experimenting on the effects of drinking seawater. Steinbauer argued that the research of Ladell and his client were equivalent.[226] Ladell pointed out that he only experimented with 'free and willing volunteers', and that he preceded the experiments with tests on himself or his collaborators; he was careful to stop short before any real danger to health or life occurred.[227] Alexander was determined that his evidence should contrast legitimately conducted and criminal experiments, and demonstrate the mental anguish and dangers to diseases and death in a concentration camp.[228]

The prosecution argued that the Berka method was lethal as it only masked the saline taste, while not extracting salt. Alexander argued that the Schäfer method was superior in removing chemically all salt. This was a method similar to the British Permutit desalination process, which was adapted by Ivy in the US. The Dachau seawater experiments caused psychological distress and would be criminal in Britain.[229] To refute this, Beiglböck's lawyer advertised on 30 June 1947 for experimental subjects. He gathered testimonies of good treatment in the seawater tests from a few former test subjects and witnesses. The statements came too late for the judges to take these into account.[230]

The prosecution case rested on the view that Nazi administration was a centralised hierarchy. The accused stressed fragmentation. The vast chart of a streamlined administration became complicated by rivalries and by the Nazi chaos of polycratic authority. Gebhardt complained that he attempted to unify the disparate military medical services of the army, air force and navy.[231] Karl Brandt failed in his unifying endeavours as regards medical services and research.

Rudolf Brandt, who dealt with Himmler's correspondence on human experiments, was accused of having a coordinating role. Whereas Gebhardt depicted Rudolf Brandt as 'an extremely unimportant typewriter', Blome condemned him as having masterminded the human experiments. Rudolf Brandt adopted the stance of being Himmler's cipher rather than an instigator of policies. He expressed regret at having signed orders to conduct human experiments, and condemned experiments.[232] His lawyer pleaded not that he was innocent but that there were mitigating circumstances.[233]

The most senior defendants argued that they were not informed about the human experiments. They stressed that experiments were unauthorised by persons higher up the chain of command: Gebhardt accused Stumpfegger of conducting irresponsible experiments on the blood styptic, Polygal. Gebhardt interpreted Himmler's and Hitler's belated confidence in Stumpfegger as a means of disposing of Karl Brandt.[234] Less convincingly, Genzken explained that he had no wish to become the chief SS physician, preferring to keep his distance from the malevolent Himmler.[235]

Defendants exonerated themselves by explaining that they subjected themselves to self-experiments. Romberg and Ruff did this in their aviation research, although Alexander pointed out that they only went up to 12,000 metres and their coerced subjects were sent to 20,000 metres.[236] Rose experimented on himself with dysentery toxin and had infected himself accidentally with cholera, scrub typhus, dengue, dysentery, malaria, blackwater fever, skin tuberculosis and DDT suffering atrophy in his right leg.[237] Karl Brandt said that in the Military Academy practically every officer underwent self-experiments using mustard gas.[238] Beiglböck claimed that he proposed seawater experiments on military personnel, instead of on 'such unreliable and irresponsible subjects' as the Dachau convicts. The blame was shifted onto Himmler.

The defendants complained the Americans withheld documentation, while the prosecution accused the Germans of destroying incriminating evidence. Beiglböck was exposed red-handed. In other cases circumstances were to blame: the petitions of parents of severely malformed babies, or in Gebhardt's case the assassination of Heydrich as legitimation of the bone transplant and sulpha treatments at Ravensbrück. The Ding/Schuler diary has remained contested.[239] The underlings argued that they could not defy the lethal orders from above. Rarely were they willing to take responsibility for their actions. The structure of authority was laboriously elucidated. Sievers argued that it was really Wüst who was in charge of the Ahnenerbe, while he was only an administrative subordinate with no autonomy. He felt caught between the demands of scientists and the political orders of Himmler. He explained that the best that he could do was to involve the RFR, so as to ensure that the research had at least scientific validity. Mrugowsky argued that his Waffen-SS Hygiene Institute was concerned with combating epidemics, and that the underling Ding acted on his own account. Blome highlighted the power struggle against Conti. Poppendick was chief of the medical department at the *SS Rasse- und Siedlungs Hauptamt* (which had a major role in population displacements and the Holocaust), and subordinated to the leading SS physician Grawitz from 1939 as chief of his personal office. He argued that his responsibilities as assistant of Grawitz needed to be precisely defined, and that it was wrong to see him as a proxy for Grawitz. He denied any link to typhus and virus research at Buchenwald.[240]

The Germans attacked the American case as resting on a series of misunderstandings. They accused the prosecution of mistranslating German medical terminology. One instance concerned diseases like typhus (in not translating the German term *Typhus* as typhoid), and another the distinction between an experiment and a clinical trial. Phrases were hard fought. The bellicose Rose contested the court record on the basis of his bacteriological expertise. Behind the acrimony over mistranslations lay the effort to prove that the Germans were engaged on programmes of legitimate medical research, and that the Allies had wilfully misconstrued these. Both sides interpreted the human experiments in a wider context. The defence insisted the experiments were for the most part necessary tests of innovative and potentially beneficial therapies. The prosecution insisted that the experiments were part of wider patterns of mass killing, aggressive war and genocide.

The grim debates generated a sense of gallows humour. Alexander conveyed all the difficulties of mounting the case in a satirical sketch of the proceedings. The drama focused on Gebhardt's counsel, Seidl. As 'Seidl' was the term for a beer jug, he was 'Dr Pint'. A servile Dr Servatius provided assistance. The parody revolved around the issues of who was whose gang boss. Alexander poked fun at his own role – he failed to prove that a dead man was really dead. The German lawyers introduced evidence from *Time* magazine, and constantly objected.[241] Explaining Nazi human experiments was far more difficult than expected – indeed, it was so hard to secure a conviction for what seemed so obviously criminal, that the experience had its moments of frustration and absurdity. The humour underlined the problems in establishing how the medical Behemoth functioned.

11
The Medical Delegation

Medical dissent

When the Trial opened, a delegation of six medical observers appeared. Its head, the psychiatrist Alexander Mitscherlich, explained that the German universities of all zones and Chambers of Physicians had appointed them to report on the Trial to the German medical press and to compile a summary of the proceedings.[1] Unlike the IMT, Taylor allowed Germans in the visitors' gallery. The 60 places were generally empty apart from two steadfast medical observers, who were gripped by the horrific evidence.[2] The delegation saw its task as providing information and impressions about the Trial. Mitscherlich produced articles, and co-authored a pamphlet *Das Diktat der Menschenverachtung* – literally, 'The Order to Despise Humanity'.[3] While the German medical profession expected the delegation to draw a sharp line between the medical atrocities, and the conscientious German doctor, Mitscherlich highlighted the issue of modern science as inherently unethical and destructive.

Mitscherlich had been a prisoner of the Nazis in Nuremberg. He stood for a distinctive form of socialism, which drew on a nationalist critique of modernity. His personal history provides insight into his social concerns, historical interests and renegade nationalism. In the summer of 1945 he was minister in the short-lived American zonal region of Mittelrhein-Saar, but was suddenly arrested with the eminent political scientist Alfred Weber. They found that they shared a critical understanding of modern mass society and its implications for individual psychology in the technicised state. They condemned a military occupation that had restored the old order rather than releasing creative forces of change, and they published on 'free socialism'.[4] A quotation from Weber on the inhumane psychological consequences of totalitarianism introduced the book on the Trial.

Mitscherlich approached the task of presenting a digest of the voluminous Medical Trial documents as an exercise in contemporary history. He had studied history for three years in Munich. The nationalist historian

Karl Alexander von Müller declined to supervise his planned dissertation on Luther's reception in the nineteenth century, as he had embarked on this with a Jewish supervisor. Mitscherlich switched to medical studies, as he realised that he was both academically and politically blocked because of his affiliations with radical nationalists. He was linked to the movement resisting modernity led by the charismatic Ernst Jünger and to the national bolshevism of Ernst Niekisch. Caught up in nationalist cliques, who were none the less critical of Hitler, Mitscherlich was arrested in 1937.[5]

Mitscherlich was critical of experimental medicine. He believed that history was the basis of human existence. Disease had to be understood as an existential problem. He rejected the crass determinism of Social Darwinism as denying the essential subjectivity of the sick person. Psychosomatic medicine was a crucial interface between mind and body.[6] The Trial raised the issue of the cultural legacy of mechanistic biology, and individual psychopathology. By linking reductionist and experimental physiology with eugenics, euthanasia and Nazi atrocities, the Trial offered an opportunity for a critical historical sociology of medicine. His aim was a fundamental reform of scientific values. This broader agenda explains why he became obsessed with assembling a full documentation and providing social and psychological analysis of the mass of trial materials.

Viktor von Weizsäcker, formerly professor of neurology at Breslau and now on the Heidelberg faculty, had shielded Mitscherlich during the Third Reich after his release from prison in 1938. Weizsäcker had outlined a German anthropological medicine, which focused on the subjective feeling of pain and sickness, so standing in sharp contrast to a medicine based on racial hygiene and genetics.[7] He too was critical of Social Darwinism and mechanistic biology, and yet his holism had affinities of Nazi ideology, as he defined health in terms of productive work. He collaborated with the holistic Austrian psychiatrist, SS officer and euthanasia advocate Prinz Auersperg.[8] Weizsäcker's Breslau institute received the brains and spinal cords of child euthanasia victims.[9]

Weizsäcker's attacks on Nazi research were a covert form of self-castigation. His brother, the Secretary of State Ernst von Weizsäcker was under arrest; despite his claims of resisting Hitler, he was convicted at the Nuremberg Ministries trial for endorsing the deportations of Jews from Paris and Rome to Auschwitz.[10] Viktor von Weizsäcker confronted the trials with a sense that as a German doctor he was implicated and that he had to take a stance on the issues. He provided Mitscherlich with steadfast support, yet he established contact with the accused. He sent his pamphlet on euthanasia to Rostock, who considered that this exonerated the defendants from the vilification to which they had been subjected.[11] Weizsäcker stressed that information on the Trial and its background belonged to the formative historical forces of modern society and that it would provide a basis to define how medicine would be constituted in post-war society.[12]

The delegation was to provide documentation, and refute any idea of the collective guilt of the medical profession.[13] Mitscherlich envisaged a delegation to collect documents, conduct physiognomic studies of the defendants' psychology and organise publicity. The delegation initially consisted of Fred Mielke, a medical student from Heidelberg; the doctors Benz, Koch, Spahmer and Jensen, and Alice von Platen, an assistant of Weizsäcker, who had been perturbed by euthanasia first as assistant physician at Brandenburg and then as a country doctor near the euthanasia killing centre of Hartheim in Lower Austria during the war. Her command of English was fluent as she had grown up in England, and had a liberal education at Kurt Hahn's school at Schloss Salem.[14] It rapidly became clear that the Commission would take a critical position on mainstream scientific medicine and violate the conventions of German collegiality, which demanded that colleagues did not criticise one another in public. Mitscherlich's helpers were reduced to the dedicated Mielke and von Platen.

The 'German Medical Commission to the American Military Tribunal' represented the Chambers of Doctors of the American, British and French zones as a result of an agreement at 2 November 1946. The Chambers were based at Bad Nauheim where the William G. Kerckhoff-Stiftung was a noted centre of aviation physiology. The Chambers wanted to refute any notion of collective guilt of the medical profession, fearing that the imminent trial could be tendentiously extended to discrediting the German medical profession and German medical science in the eyes of the world.[15] The intention was to demonstrate the innocence and ignorance of Nazi crimes of the overwhelming majority of German physicians.[16] But events would turn out very differently. When asked by Karl Oelemann, chair of the Hessen Chamber of Physicians, to form the delegation, Mitscherlich laid down as a condition that he wanted it to represent German medical faculties including those in the Russian zone, and also state public health authorities.[17]

Mitscherlich vexed the traditionally independent-minded German medical faculties, because of his view that leading professors were involved with concentration camp research. The matter was complicated by the emerging East-West split. Oelemann conducted inconclusive negotiations with the medical chamber of the Russian zone.[18] The prestigious Berlin Faculty favoured Mitscherlich having observer status at the Trial, but not that he should inspect the prosecution documents.[19] The faculties of Giessen and Marburg, newly reopened in the US zone, supported the delegation.[20] Not all faculties wanted to be represented. Faculties at Mainz and Würzburg considered themselves too depleted to endorse the mission, and Göttingen had doubts as to Mitscherlich's competence.[21] Other faculties expected Mitscherlich to vindicate German medical research under National Socialism. The small, belligerent and scarcely denazified medical faculty at Freiburg numbered among its ranks the aviation pathologist Büchner and the aviation physiologist Benzinger, who had been taken for interrogation to Nuremberg. The Freiburg Faculty prejudged the issue by pronouncing that the Trial had unpleasant implications for science and medicine, even though 'the German physician and scientist knew nothing of the atrocities'.[22]

After tortuous negotiations to secure support from the majority of medical faculties, in December 1946 Mitscherlich announced the German delegation represented 'German Universities of all zones and Chambers of Physicians'.[23] He mentioned the Russian zone universities of Berlin and Leipzig, when approaching the court authorities. He stressed the involvement of German doctors in the horrific crimes and the need for the profession to reformulate its values.[24] In February 1947 he described himself as representing all universities of the former Reich (in fact, an extension as this would include 'lost' territories). He also represented the Medical Chambers of the Western zones, which met in June 1947 to consider his report.[25]

The newly reopened Medical Faculty at Halle in the Russian zone thought it right that students consider the evidence presented at the Trial. The local communist official (Otto Halle) insisted that all medical students submit written statements after the judgment.[26] The dean, Werner Budde, drew a sharp line between the small number of Nazi doctors and the majority of the German profession, and opposed written statements by the students as this would undermine public confidence in the profession. He wanted a special lecture by an appropriate medical expert to present an 'objective' analysis of the Trial.[27] The students asked the court to supply documents, which they could study.[28] On 10 September 1947 the Ministry of Sachsen-Anhalt insisted on written statements from all students. The Faculty decided that it would refer the question to the students' representative organisation – in the hope that the students would reject the Ministry's demand.[29]

The Students' Council at Halle formed a study group on the Nuremberg Doctors' Trial. Medical and law students wanted to travel to Nuremberg and inspect the documents there. The students found that they were unable to cross the zonal border, and so in the end had to settle for the dean's lecture series.[30] Even though Mitscherlich's dedicated collaborator Mielke was also a student, the delegation remained isolated, whereas it might have drawn support from a younger generation of students.[31]

German communists publicised the atrocities revealed at the IMT and remained broadly supportive of the Medical Trial, whereas the Soviet Military Administration was already distancing itself from the American trials that might otherwise have taken place in Berlin.[32] The Medical Trial offered the opportunity of exposing the iniquities of a reactionary bourgeois profession. Havemann as Director of the KWG (albeit only in the

Russian zone) published an indignant article on 'Humans as Experimental Animals'. He pointed out that the difference between legitimate clinical trials and the criminal experiments was that the victims were deliberately infected.[33]

Mitscherlich's representation of universities in the Russian zone meant that the nascent German administration in the pre-DDR period gave support. Emilian Ritter von Skramlik, the physiologist and dean of the Jena Medical Faculty, congratulated Mitscherlich on *Das Diktat*, and promised to circulate it locally. Skramlik had cared for survivors of Buchenwald, and soon found himself in conflict with the communist authorities over penicillin and food aid from an émigré physician in the US.[34] The University of Rostock dismissed the aviation physiologist Anthony, its newly appointed professor of internal medicine; he died in August 1947 in mysterious circumstances. Here, the Faculty was preoccupied with political pressures, and the Trial had no apparent impact.[35] Mitscherlich was invited by the physicist Robert Rompe (of the education administration) and the left liberal Theodor Brugsch (vice president of the Department for Higher Education and Science) to lecture to university teachers about the Trial in April 1947.[36] By way of contrast the Berlin Faculty endorsed the record of its dean, Heubner, in response to his being mentioned in connection with seawater experiments.[37] Under fire from conservative West German colleagues, Mitscherlich recognised that his links to the Russian zone had become an embarrassment. Left-liberal academics found themselves displaced by heightened ideological polarities. Mitscherlich's embarrassment increased when he was asked by a denazification panel in East Berlin to supply Trial evidence against Sauerbruch in November 1947.[38]

Mitscherlich was an outsider from the German medical establishment not only politically, but also because he was developing psychotherapy and psychosomatic medicine. He was at this time Privatdozent in neurology. Mitscherlich, supported by Weizsäcker, tried to establish psychotherapy as part of mainstream medicine. But Kurt Schneider, dean of the Heidelberg Medical Faculty, was antipathetic. The philosopher and physician Karl Jaspers defended Mitscherlich in his clash with Büchner, but did not share a reverence for Freud, and judged the critique of Nazi medicine as disruptive. Although Mitscherlich achieved his immediate aim of establishing a university clinic for psychosomatic medicine, his involvement with the Trial was to be at considerable professional cost. The German medical establishment exacted revenge on this 'fouler of the nest'.[39]

Having been imprisoned in 1937–8, Mitscherlich did not wish to lock himself into the role as a full-time courtroom observer. He divided his time between teaching in Heidelberg and psychoanalytic work in Switzerland, and appeared only sporadically amidst the ruins of Nuremberg. Mielke and Platen endured freezing and makeshift billets, and felt isolated.[40] Mitscherlich returned to Nuremberg for the opening of the defence case in late January 1947.[41] Mielke and the insubordinate but attentive Platen gathered the grim Trial documents. Mitscherlich found mastering the massive documentation time-consuming, and spent over a year with Mielke on their hoard of documents. He then had the wearing task of defending their work in the face of a massive medical onslaught.[42]

Mitscherlich highlighted the academic eminence of several of the defendants. Their mistake was to adopt utilitarian aims and abandon the Hippocratic principle that humane concern with healing the sick individual must be a doctor's sole motive. This article went down well with the defence.[43] Alice von Platen contacted Servatius, who was still hopeful that the commission would turn out to be an ally, while she gained insight into the euthanasia issue and documents.[44] Mitscherlich received documents from all the defence lawyers, and was in touch with Becker-Freyseng's wife, a doctor who also lived in Heidelberg.[45] But a rift with the defence soon emerged.

The delegation's contacts with the prosecution remained slight, although Mitscherlich had established their official status with the American authorities. When visiting Heidelberg in 1945, Alexander first learned of Mitscherlich as a psychotherapist with links to the Göring Institute for Psychotherapy in Berlin.[46] The delegation promptly contacted Alexander and Kogon, who shared their insights into SS psychology.[47] The Americans decided to give the delegation equivalent status to the press. Alexander made a room available for the analysis of documents.[48] The Americans were gratified and surprised to see the positive presentation of their case.

Mitscherlich marked the start of the defence case with the comment that defendants presented a depressing picture with protestations of ignorance of Nazi atrocities, and they lacked any sense of critical responsibility in blindly carrying out orders. After 30 days of trial hearings (in February 1947) he proposed a sequence of short papers consisting of five sets of documents on such topics as typhus, pressure chamber experiments and euthanasia, to be published in the *Deutsche medizinische Wochenschrift*.[49] Its editor took the view that as a journal for communicating the results of medical research to the clinician and general practitioner, it would have to restrict its coverage to scientific opinions on the Nuremberg evidence.[50] He conceded that it was necessary for the German profession to re-conceptualise the ethical basis and methods of medicine as a result of the trial.[51] Mitscherlich ideally wanted a special issue devoted to trial documentation.[52] He submitted a paper on the problem of human experiments and eugenics, but then withdrew it as he wished to incorporate it into the collection of documents.[53]

The priority was publication of the trial documents to inform the profession about what many had no real wish to know. *Das Diktat der Menschenverachtung* was issued on 3 April 1947 in an edition of 30,000 copies.[54] In this stark presentation of Trial evidence, Mitscherlich pointed to the dangers of a mechanistic view of disease and how service to the state caused a betrayal of the physician's ethic to relieve suffering. The medical student, Jürgen Peiffer, took its message to heart as pointing to the need for a Christian renewal of medical values.[55] He retained this conscientious attitude in his career as brain pathologist and historian of euthanasia, but such a thoughtful response was the prerogative of a minority.

In 1949 *Das Diktat* was republished with American commentaries by Alexander and Ivy as *Doctors of Infamy*. The German edition's subtitle stressed how the Trial concerned German science as a whole, as coming from 'The German Medical Commission ... a Documentation against 23 SS Physicians and German Scientists'.[56] By way of contrast, the publication *Wissenschaft ohne Menschlichkeit* mentioned only the West German Physicians' Chambers. The role of the medical delegation in representing the universities was discreetly omitted, having been killed off by a combination of revisionist nationalism and Cold War politics.

Mitscherlich wanted to evoke a sense of responsibility and atonement, and a reappraisal of scientific values. His diagnosis was that a free profession was suppressed by a manipulative state, and by militarist and National Socialist ideologies. An aggressive search for truth was combined with servile obedience to a dictatorship. The result was unprecedented evil and cruelty, with the doctor transformed into the licensed murderer.[57] He hoped that demonstrating connections to the wider scientific community would result in a process of soul-searching and questioning of the aims of medicine. The irony was that his books were criticised and suppressed by a manipulative profession.

By May 1947 only the Medical Chamber in the British zone ordered copies of the trial documentation.[58] Mitscherlich and Mielke were convinced that the matter was urgent.[59] Their book won some praise from university rectors, and elicited perfunctory and polite acknowledgements – as the universities distanced themselves from National Socialism while rehiring Nazi doctors like the eugenicist Fritz Lenz in Göttingen. *Das Diktat* provoked heated criticism in medical associations and the universities.[60] Mitscherlich exposed how medical faculties were bound by conservative ideas of rank and status, as his endorsement of the Trial was taken as an attack on the professorial elite.

Counter-attack

In June 1947 Mitscherlich reported to the West German Medical Chambers. Several doctors attacked him for not having published on the Trial in a professional journal, and complained that they were criticised in public but had no firsthand knowledge of the Trial. They reprimanded him for besmirching their profession's reputation in *Das Diktat*. They demanded that the edition be bought up, and circulated to physicians rather than the public. Although Oelemann commended Mitscherlich's efforts as contributing to the international rehabilitation of German medicine, leading figures among the German medical profession and faculties were seething with animosity.[61]

Four leading medical professors spearheaded the attack on the documents. They drew a sharp line between mainstream medical research and the guilty few. They were uncompromising in defending the value of medical research, seeing the guilty as having lapsed from an ethically and intellectually inviolable tradition of German medical science, which had flourished despite of and in opposition to Nazism. Not only was the book personally defamatory, but it brought the whole German medical profession into disrepute. They accused Mitscherlich of irresponsibility and of violating the code of conduct that colleagues should not publicly criticise each other.[62] Because Mitscherlich represented all German medical faculties, his critics demanded that the medical faculty of Heidelberg reprimand him.[63] Mitscherlich realised that medical professors wanted to obliterate the career of a critical young Privatdozent for violating the code of slavish obedience to superiors.[64]

The sharpest attack came from the pathologist Büchner of Freiburg. Although Mitscherlich pointed to the dangers of the closeness of science and the militarised state, Büchner defended the ethical parameters of his work as aviation pathologist and brain anatomist. He was outraged at being portrayed as not having voiced any criticisms of the human experiments at the Nuremberg meeting on survival at sea. He insisted he had publicly opposed euthanasia (although he was silent about links to Hallervorden and Spatz who had researched on the brains of euthanasia victims). On 26 April 1947 Büchner sued Mitscherlich, Mielke and the publisher Lambert Schneider. He insisted that his lecture on 'The Hippocratic Oath' – delivered in November 1941 in a series on 'the nation's health – criticised Nazi euthanasia, and that he incited students to demonstrate against the euthanasia propaganda film *Ich klage an*.[65] He was ordered to be present at the Nuremberg military medical conference 'Sea and Winter Distress' on 26–27 October 1942, and the lecture by Holzlöhner and Rascher was unannounced. Büchner claimed that Rascher and Holzlöhner never divulged where the experiments took place – only that they were on criminals condemned to death. He could not take part in discussions the next day, as he had paratyphoid; he was treated by Weltz, to whom he uttered his opposition and condemnation of the experiments. A public protest was impossible because of the secret status of the experiments, although Jaspers pointed out that objections could be raised in closed discussion.[66]

Mitscherlich pointed out that these criticisms were made only later in private discussions among Becker-Freyseng, Weltz and then with Hippke as the head of German air force medicine. Büchner denied that his presence at the conference meant he was complicit in the experiments; his affidavit for the defence expressed the view that Weltz wished to distance himself from the human experiments. He vehemently denied complicity in or knowledge of Haagen's hepatitis experiments, having merely been in contact with him over hepatitis-infected mouse livers.[67]

Mitscherlich felt it was unjust to prosecute him for citing from authenticated documents, which had been presented in court. He appealed to Jaspers for advice as to how to balance moral and legal issues, as having already confronted the question of German guilt.[68] Büchner's stance was upheld by German courts in April and July 1947.[69] Legal action blocked the circulation of Mitscherlich's documentation in the French zone from May 1947, although the US zonal authorities publicised the book.[70] The injunction began a process of German courts undermining the Trial. In the event Büchner agreed to a slip being inserted in *Das Diktat* stating that he was not involved in the planning or execution of human experiments; he had only dissected mouse livers for Haagen, and he had protested against human experiments.[71]

The pharmacologist Heubner and the surgeon Sauerbruch joined the offensive. These were two of Berlin's most distinguished medical figures. Heubner pointed out that he had merely endorsed the view of Eppinger that experiments were necessary.[72] He insisted that they did not know seawater experiments were to be carried out on prisoners in a concentration camp. He overcame initial hesitations as it would have meant disagreeing with Eppinger as a senior colleague. He had accepted the probity of the experiments given that they were supervised by Beiglböck as Eppinger's assistant, and felt reluctant as a civilian to challenge the military medical establishment. Heubner hid behind rank, status and collegiality to justify his ethical lapse.[73] Heubner said that he heard Gebhardt lecture on the sulphonamide experiments on prisoners, but he believed that the prisoners had volunteered and that, despite his distaste at the findings, he had no opportunity to express criticism. Heubner was outraged that Mitscherlich connected him to the Ravensbrück sulphonamide experiments by virtue of not having objected to their report; this was at the Arbeitstagung Ost in May 1943 at the Military Medical Academy Berlin. Mitscherlich eventually dropped this point in the 1975 edition of his book, although he retained the details on Heubner's complicity in the seawater experiments.[74]

When Heubner made his initial criticisms he was professor in the Soviet zone, and his tone was more moderate than Büchner's belligerent indignation.[75] His position hardened by summer 1947, as he began legal action with Sauerbruch. Mitscherlich pointed out that he was faithfully citing court records, and it would be inappropriate to doctor them. The German lawyers insisted that the court records were not an accurate reflection of events. On 29 July 1947 Heubner obtained an injunction from the Berlin court preventing distribution of Mitscherlich's documentation unless changes were made. This Mitscherlich considered impossible. Heubner insisted that the documentation was defamatory.[76] Heubner's influence increased, when he switched affiliation to the new Free University of Berlin in the American zone. The acquittals of a number of the defendants meant that he called for a fundamental revision of Mitscherlich's documentation. The publisher replied that as the edition was sold out, the court decision was redundant, and in the event of a second edition, fundamental changes would be made. Mitscherlich and the lawyer for Heubner and Sauerbruch quarrelled for a year over the extent of these changes.[77]

Büchner lobbied the medical faculty of Freiburg to take action against Mitscherlich. He tried to shame and marginalise Mitscherlich by complaining to the University of Heidelberg. He mobilised the German medical faculties – after all, Mitscherlich was their representative. Büchner lobbied state ministries for higher education, so that none should offer Mitscherlich a professorial appointment. The Freiburg Faculty attacked Mitscherlich's view that no one protested against human experiments as defamatory to the medical profession. It claimed that Büchner, Rein and Strughold had all energetically opposed the human experiments.[78] In Heidelberg a commission of Friedrich Weber, the psychologist Witte and – remarkably – the physiologist Strughold (who held the chair of physiology from November 1946) sat in judgment on Mitscherlich.[79] The attack on the reputation of the conscientious and unbending Mitscherlich was vengeful: it meant that Privatdozent Mitscherlich was never to enter the hallowed ranks of the German medical professoriate.

The Göttingen University magazine became the arena for controversy about the validity of the Trial evidence between Mitscherlich on the one side and Rein, Heubner and Sauerbruch on the other. In an article on science and inhumanity Rein attacked Mitscherlich as irresponsibly defaming senior academics.[80] He repeated the by now standard exculpation – that the atrocities were the work of a small group of pathological individuals, and that mainstream German science had resisted the perversions of fanatical Nazis. Rein had criticised Rascher's experiments as poor science. As Rector of Göttingen he held a key academic position for science in West Germany. Alert to wider currents of revisionism, he praised the ideals of autonomy of science as advocated by the Society for Freedom of Science, an Anglo-American organisation attacking social planning. At this juncture Rein sought to draw a sharp line demarcating his research from lethal experiments or – in the case of the vision research of Gothilft von Studnitz – experiments on subjects who were to be executed.[81] He tried to persuade Weizsäcker to censure Mitscherlich, although Weizsäcker responded that he fully endorsed Mitscherlich's critique of the morality of natural science.[82]

Mitscherlich was convinced that what was at issue were two different views of medical research, and prepared a polemic on inhumane science. He conceded that the personal psychology of a few of the convicted was abnormal. But most were average types. How was it that they had been drawn into criminal politics and destructive science? Real issues about the nature of medicine were at stake. At root was an immoral and unethical natural science, which examined organs but never saw the whole person.[83] Mitscherlich claimed the Trial proved the necessity of his programme of medical reform. Heubner and Sauerbruch rallied to Büchner in further articles and proceedings aimed to silence the errant Mitscherlich.[84]

Analysing Nazi euthanasia

Mitscherlich's assistant, Alice von Platen, felt that the grim atmosphere of the court proceedings did not appear in the transcripts. Observing the defendants, she felt, 'The horror was not only the documents but in the very closed faces, no feeling, expression of justification and in the tone of the voice. To us they were not a group.'[85] Mitscherlich was irritated when she decided to produce her own analysis of Nazi euthanasia.[86] He accepted that her work ran in parallel with his, and commissioned research on the problem of human experiments over the previous 40 years.[87] Her analysis dealt with the conflict between the conscience of the researcher and state authority, and she too saw events in the political paradigm of 'totalitarianism'.[88]

She did not limit her studies to Nuremberg, and consulted documents of the Hadamar euthanasia trial. The Eichberg euthanasia trial opened on 2 December 1946, dealing with ill treatment and killings in an institution designated as a collecting point for condemned patients. Friedrich Mennecke was sentenced to death on 21 December 1946. Three further death sentences were the outcome of the Kalmenhof Trial. A second Hadamar Trial was run by the German authorities, when nurses were prosecuted for complicity in 1,000 deaths. The trial concluded on 21 March 1947 with far lighter sentences, the heaviest being eight years; 14 accused were acquitted.[89] Von Platen consulted the evidence from the satellite trials. She saw euthanasia more in terms of terror and bureaucracy than of medical behaviour.[90]

Platen had worked at the Landesanstalt Brandenburg-Potsdam under the Nazi psychiatrist Hans Heinze, who later joined the panel deciding on euthanasia. Distressed by the abusive and demeaning treatment of patients, Platen opted out in 1935 when she went to Italy, and during the war worked in Austria as a country doctor. She drew attention to how every practising doctor must have known of the euthanasia killings and the extent of compliance with compulsory sterilisation. By October 1947 she completed her pioneering text on the Nazi euthanasia atrocities. She rightly recognised the Medical Trial as pioneering in demonstrating the link between euthanasia and Nazi genocide. Here the testimonies of Karl Brandt, Hoven and Brack were crucial. She added documents on euthanasia in what was now the French zone; Kogon provided access to documents on the Frankfurt euthanasia trial.[91]

Platen was curious about what made Karl Brandt so naïve about Hitler, and felt irrationalism underpinned atrocities.[92] She discussed with the prosecution the psychology of medical brutality in the conduct of Gebhardt and Oberheuser.[93] Although Platen's origins were Protestant, she was sympathetic to Kogon's social Catholicism, not least because she felt that the Catholic Church had maintained a greater distance from Nazism. Both Kogon and Platen shared the view that anti-Nazi Germans should have had a role in the trials.[94] They were perturbed by the lack of remorse on the part of the defendants. Kogon reprinted the poem of Werner Bergengruen, which was circulated by Leo Alexander to the defendants: that the spurned Jew, mental patient and slave labourer had now returned in judgment.[95]

Mitscherlich was more secular in outlook and was annoyed by Platen's contact with Kogon, pointing out that she had been employed to work under him. He therefore wanted his own book to be published first.[96] Platen no longer had Mitscherlich's support for qualifying in psychoanalysis. (Coincidentally they both received their teaching analysis from Michael Balint, who pioneered the humanising of physician-patient relations). She completed her succinct and thoughtful analysis, *Die Tötung der Geisteskranker in Deutschland,* at Kogon's house near Frankfurt, and it appeared in 1948 in the *Verlag der Frankfurter Hefte,* which Kogon co-directed.[97] Its subtitle stated that it was issued by the German Medical Delegation at the American military court, directed by Alexander Mitscherlich, and the book was dedicated to Viktor von Weizsäcker. The edition of 3,000 was pioneering in its analysis of Nazi euthanasia, and warned against voluntary euthanasia under medical direction.[98]

Whereas Mitscherlich pointed to the dangers of mechanistic science, Platen saw state collectivism as menacing. It was all too easy for the state and insurance-bound physicians to take a view that lives were valueless and costly. She noted how Karl Brandt described the medical commission on euthanasia as a state body, and how Brack criticised a physician who recommended too often that euthanasia was inappropriate.[99] Judge Beals established that the doctor saw himself in an executive role rather than carrying full responsibility for the decision to carry out euthanasia – even though the killing of individual patients emanated from a doctor's orders.[100] That holding institutions for patients on the route to the death chambers were really 'observation stations' was a cruel fiction.[101] Competition between ruthless and radical administrators exerted pressure on the psychiatrists to comply.[102] Brack was a noted enforcer of euthanasia measures, terrorising doctors and nurses to continue killing procedures.[103]

Platen pleaded for the doctor to stand on the side of the patient, taking a cue from Weizsäcker's notions of the solidarity of doctor and patient.[104] Science failed in its obligations to sustain lives of individual patients, and instead let itself become a destructive agent for the racial state. Doctors lacked self-confidence in their critical judgement.[105] Biology was a pseudo-religion for the Nazis, and Nazi decrees had a destructive psychological effect on those who executed them.[106] The Catholic Church had rejected euthanasia and the racial measures of the Nazis, and stood against the rising tide of racial hygiene, which paved the way for euthanasia.[107] Platen's analysis of the psychology of the perpetrators portrayed science as a destructive agency. Mitscherlich's *Wissenschaft ohne Menschlichkeit* also adopted this theme when it appeared the following year. Although Mitscherlich and the Trial observers earned enmity of their rank-and-file colleagues, there were also perceptive statements made in their support. The Heidelberg lawyer, Gustav Radbruch praised *Das Diktat*. The perpetrators were not pathological types but psychologically normal, physicians who acted in response to the Nazi subversion of moral values.[108]

Public reactions

The reactions triggered by *Das Diktat* divided between replicating the US line on the Trial and outright rejection of the validity of the trial as an effort to purge German medicine of miscreants. The responses have to be seen in a broader context of the growing political hostility to war crimes trials. The poet and journalist Stephen Spender (a close friend of John Thompson) recorded that there was disbelief that the actions of the commandant Josef Kramer were humanly possible. Spender heard that: 'The majority of Germans believed the trials were a put-up job, and that they were only being prolonged because the accused had so much to be said on their side and we could not suppress the evidence.'[109] The poet T.S. Eliot went even further. After visiting Germany, he expressed disgust at the war crimes trials, and lobbied for their cessation.[110]

Taylor's staff gave interviews to counter critical publicity and gain public endorsement of the Trial. Alexander condemned the experiments as either useless or as yielding results that could have been obtained by other means.[111] The German lawyers claimed that the trials were an unfair exercise in 'victors' justice'. Other opinion formers were close associates of the accused, who justified the conduct of the defendants. But they had a hard task, because the view of doctors as Nazi torturers caught on. Bettina Blome-Ewerbeck wrote to her beloved Kurt that many do not trust the doctor any more. The evidence disturbed many in the German medical profession. She noted how shaken the eminent surgeon Gustav von Bergmann was to learn what Gebhardt had done, and that country doctors were also perturbed by the revelations. She reflected that: 'You were right in Alt Rehse to see the doctor as having a care of souls and not to treat people as machines.'[112]

The Munich orthopaedist Hohmann condemned the experiments as not true science – as science requires a moral content – but as dilettantism. Hohmann attributed the dehumanisation of experimental subjects to materialism, observing that the defendants lacked goodness, truth and human love. The defendants violated their true medical calling in the service of a totalitarian state.[113] Similar public expressions of disgust with medicine were reported by the representatives of the German professional chambers in June 1947. Mielke argued that it was necessary to re-establish the confidence of the patient in the physician.[114] The critics of *Das Diktat* argued the irresponsible circulation of Trial documents destroyed the bond of confidence rather than the Nazi atrocities.

Rostock became a crystallisation point for a new wave of apologetics, and set out to challenge Mitscherlich. Radio Baden praised his integrity; he was only accused as the nominal head of a research department, where experiments were systematically registered; he had never personally conducted experiments or was involved in Nazi propaganda. His air of innocence and honesty reflected to the benefit of his fellow accused.[115]

The Bavarian journalist Süskind gave insightful commentaries for Radio Munich, and the *Süddeutsche Zeitung*. He described how through the sound system, observers were treated to the 'New World Symphony', which as a fusion of the Czech and American evoked the encounter of the Old and New Worlds at the Trial. For this was an international event as Germans were to be judged by non-Germans. Süskind picked up the issue of how the medical and biological researcher had come into conflict with the physician. Scientists had taken on the corrupting role of administrators and become tools of a depraved and corrupting state. He stressed that the 20 accused physicians were wholly untypical of the rank-and-file physician. He highlighted how Rostock showed that some defendants were elite figures, who took a strong line on the divide between what was ethical and unethical.[116] Rostock came to occupy a central position in attempting to persuade the press to take the part of the defence.[117] The ability of the defendants to find an echo among German intellectuals and the medical profession grew with the mounting campaign against the legitimacy of the trials.

12
A Eugenics Trial?

The charge of genocide

Whether the Medical Trial dealt adequately with race, eugenics and genocide has been a matter of debate. Alexander saw the experiments as driven by an annihilatory creed of race and extermination. Their aim was to induce death, and experiments on poison gas were a test bed for the Final Solution. The genocide operation required experts in physiology, statistics, public health and genetics to weed out racial undesirables.

The Behemoth template gave the prosecution a functionalist explanation: that the experiments served the utilitarian ends of using surplus human material to satiate the needs of the predatory military and industrialists. Eugenics was subsumed into the totalitarian apparatus of racial policy and conquest. The Behemoth template stressed not so much the Nazi racial ideology and practice as Nazism's link to militarism and the German industrialists as the mainsprings of the medical crimes and lawlessness. These corrosive links explained the decay of science into irrational and sadistic fanaticism. Race was very much a presence at Nuremberg, but explained in terms of socio-political functions and its murderous outcome in motivating genocide.

Alexander Mitscherlich unequivocally portrayed the Trial as confronting the issues of eugenics and medicine. Yet Holocaust historians have criticised the Medical Trial for ignoring the genocidal role of physicians in racial atrocities and above all in the selections at Auschwitz. This criticism has been raised in a more nuanced form against the IMT. The charge of crimes against humanity was substituted for genocide. The failure was legal, but at a rarefied diplomatic level: the victims of acts of terror and murder were defined as citizens of Allied nations rather than as groups victimised for ethnic and racial reasons. Thus – so the argument runs – crimes against Jews or Roma had no place in the Allied accusations.[1]

The IMT revealed for the first time the full horror of Nazi racial crimes to a shocked world.[2] This Trial presented the first reliable estimates of the mass murder of 5.7 million Jews – a figure later revised to six million. The figures were based on the Nazi exploitation of census statistics and population registration.[3] This raised the issue of the genocidal role of the bureaucrats, statisticians and disinfection experts. The French witness, Marie Claude Vaillant-Couturier, told a hushed court of gassings at Auschwitz, and the Soviet evidence included the murder of over three million prisoners of war.[4] The concept of crimes against humanity was introduced to deal specifically with the lawlessness of the Nazi state and covered 'persecutions on political, racial or religious grounds', including extermination, torture and other inhuman acts. While the definition excluded localised outbursts of race hatred, it included wholesale, nationwide campaigns 'to expel, to degrade, to enslave, or to exterminate large groups of the civilian population'.[5] The case against the SS as a criminal organisation included not only human experiments but also the role of the Reich Commissioner for German Nationality, the *Volksdeutsche Mittelstelle* and the SS Race and Settlement Office under Otto Hofmann.[6] Although racial atrocities were an element of all charges, it was above all crimes against humanity, which came to mean crimes against Jews.[7]

Former prosecuting lawyers argued that the Nuremberg trials provided definitive evidence of the Holocaust.[8] However, Bower and, more recently, Bloxham and Marrus suggest that the Allies all too often shied away from the Holocaust. Non-Jews were the main witnesses at the IMT, as there was concern that Jewish evidence was biased.[9] Marrus sees the IMT as dealing with the murder of European Jews in a sporadic and uneven manner. But he is categorical in criticising the Medical Trial as failing to tackle Nazi eugenics.[10]

Bloxham and Marrus draw their argument from evidence at a high political level, rather than war crimes investigators working in conjunction with victims in Germany. For those at the grass roots of the evidence of criminality, the issue is more one of whether they received political support and resources for investigating racial atrocities than the turning of a blind eye to the Holocaust. Yet the war crimes authorities did not ignore Auschwitz; medical atrocities there were investigated in some detail. Strong cases were made for the criminality of sterilisation and euthanasia. The plans for a Second Medical Trial at Nuremberg suggest that the American authorities continued to take a serious view of genocide. The selection of witnesses at the Medical Trial included Roma and Jewish victims.

Neither the American nor British war crimes authorities were wholly dismissive of the charge of genocide. The British tried hard with limited means to investigate atrocities in Auschwitz. That the Allied investigators at first gave more attention to Clauberg's brutal experiments in reproductive medicine than to Mengele suggests that the strategic priority of medical exploitation skewed the data on war criminals. But by the time of the Medical Trial many war crimes prosecutors understood that Mengele had conducted gruesome crimes in the name of medical research.

The Chief Prosecutor at the IMT, Jackson, appointed Raphael Lemkin as legal consultant. The US war crimes section under 'Mickey' Marcus – a pronounced sympathiser of Jewish causes – retained him. Lemkin's interest in the vulnerability of ethnic minorities went back to when as a child in the new Polish state, he sensed how endangered they were to coercive state power. Christians in Iraq or Jews in the new Central European states were equally vulnerable. He drew the conclusion that minorities required special protection.[11] His expertise in international law prompted him to invent the term genocide to define the systematic annihilation of an ethnic group.

Lemkin had escaped from the German occupation of Poland to Lithuania, and he reached the United States in 1941. He was convinced that Jews and Roma were targeted for obliteration, having discovered that the Nazis had decreed that hiding Jews and Roma would incur the death penalty. Legal and administrative evidence for a war of racial extermination culminated in his pioneering *Axis Rule in Occupied Europe*, published in 1944. He showed that law in countries under German occupation was 'bereft of all moral content and of respect for human rights'.[12]

Lemkin linked mass murder to biological and medical strategies to eradicate ethnic groups and cultures. His analysis of the biology of annihilation was especially significant for war crimes prosecutions of medical atrocities. He predicted that the German attempt to change 'the balance of biological forces' would result in 'a stunted post-war generation', whereas Germany would still enjoy 'numerical, physical and economic superiority'. He condemned preventing births as a means of physically destroying any ethnic, racial or religious group, and the forcible transfer of children to another group – both key features of Nazi racial and population policy.[13] In January 1947 Lemkin recommended a special trial on the abduction of women into prostitution by the SS.[14] Lemkin highlighted the racial dimensions of family and population policy, in which medical experts were massively involved.

Lemkin's contacts with Jewish leaders in London in August 1945 led to inserting the charge of genocide in the Nuremberg indictment.[15] His encyclopaedic concept of genocide covered 'not only the destruction of the Jews and Poles but also gypsies', thereby reflecting the diversity of the ethnic background of the victims of the human experiments.[16] He saw the genocidal destruction of the Jews included human experiments, and this occurred within a broader historical and sociological framework. On 28 September 1945 Lemkin sent a memo to Telford Taylor stating that the defendants at Nuremberg were responsible for cultural and physical genocide by lowering the victim groups' birth rate, and by policies of starvation and subjugation to unhealthy conditions.[17] He criticised the concept of 'crimes against humanity' as tied to proving a conspiracy to commit medical atrocities, which for legal purposes were shown to amount to hands-on murder.[18]

After this promising start, Lemkin became disillusioned with the IMT, because the judges took the view that genocide was only punishable when linked to the waging of aggressive war. He condemned the judges for not going beyond the constraints of a military tribunal and dealing with Nazi atrocities in peacetime conditions – this would have covered compulsory sterilisation (especially illegal measures such as against the mulatto Rhineland children) from 1933, and numerous acts of persecution against gypsies, Jews and other racial undesirables. The legal quibbling over the punctuation of the Allied agreement excluded pre-war Nazi crimes and the generic phrasing of 'crimes against humanity'. The upshot was to weaken the preventive value of the crime of genocide, as lawyers found it difficult to define aggression.[19]

Lemkin continued to campaign obsessively for recognition of genocide, and for the Nuremberg Trials to tackle issues concerned with racial annihilation.[20] The prosecution staff remained preoccupied by the implementation of the Final Solution. On 26 April 1946 Walter Rapp suggested action be taken against the manufacturers and instigators of mass killing with Zyklon B.[21] Lemkin contacted Marcus about 'The participation of German Industrialists and Bankers in Genocide', citing the manufacture of Zyklon B, as well as the industrial organisation of the killing installations in an effort to bridge the military-industrial paradigm with that of genocide.[22]

In the spring and summer of 1946 Lemkin contacted Thompson's initiative on medical war crimes. He liaised with the French war crimes authorities, and in July attended the UNWCC, remaining in touch with its research officer Lieut.-Colonel H.H. Wade.[23] These informal links have to be set against the legal parameters of the Trials. By the time of the Medical Trial, the polarity between crimes against humanity and genocide was apparent. The UNWCC was critical of Lemkin's concept of genocide. The Chair insisted on the US concept of crimes against humanity, and that war crimes were primarily of a military nature. In December 1946 the UNWCC pedantically objected to the etymology of genocide with a Greek beginning and a Latin ending.[24] By then Lemkin had already triumphed at the General Assembly of the nascent United Nations. He lobbied delegates (notably from Cuba, India and Panama) to the UN Economic and Social Council to condemn genocide as a crime under international law on 11 December 1946 – just two days after the Medical Trial opened.[25] Lemkin was keen to point out that the Nuremberg Trials were no substitute for a Genocide Declaration, in that the Declaration was agreed by nations as equals to prevent and prosecute genocide. But he remained interested in the extent that the Nuremberg trials were genocide trials.

Eugenics as genocide

Given the ambivalence over genocide, the question arises: were the Americans and British reluctant to place German eugenicists on trial? Although the rabid anti-Semitic propagandist and *Gauleiter* of Nuremberg,

Streicher, was successfully prosecuted at the IMT, there was a lack of attention to the ideological rationales of the Final Solution. The Allied prosecutors were similarly cautious when it came to the Nazi appropriation of anthropology and eugenics. The ranks of embarrassing figures included Alexis Carrel, the Rockefeller Institute of Medical Research scion who had returned to build up medical research in Vichy France, and whose popular book *Man the Unknown* advocated euthanasia and sterilisation while condemning democracy.

Former French prisoners provided numerous testimonies against the eugenicist Edwin Katzenellenbogen, who held American citizenship and had been a eugenic activist in New Jersey. He became a prisoner doctor at Buchenwald, where he was loathed for his callous cruelty. Katzenellenbogen was discreetly absent from the Nuremberg proceedings. Alexander considered that he was trying to cover up his role in the hospital charnel house at Buchenwald, and recommended that he be turned over to the team investigating the camp.[26] Katzenellenbogen's eventual defence made much of the links to American medical scientists. He proved to be an obdurate and stubborn defendant, who earned a prison sentence.

Sterilisation laws with a eugenic rationale remained in force in several US states, as well as in the Scandinavian countries. Karl Brandt taunted the prosecution by including a key text of American Nordic racism in his defence documents: Madison Grant's *The Passing of the Great Race*, published in New York in 1923, prescribed the extermination of the insane as the scum of society. Translated by Lehmann Verlag, it suggested that the United States rather than Germany had the foremost role in applying eugenics.[27] Brandt mustered documents on the pioneering role of US sterilisation measures and the spread of sterilisation to Scandinavia from 1929.[28] Although the prosecution could not portray Brandt as a racial theoretician, he held prime responsibility for the implementation of euthanasia.

The German sterilisation legislation of 1933 was profoundly influenced by psychiatric genetics. Sterilisation was required for nine supposed diseases: hereditary feeblemindedness, schizophrenia, manic depression, hereditary epilepsy, Huntington's chorea, hereditary blindness and deafness, hereditary malformations and severe alcoholism. Although race was according to the text of the law not in itself grounds for sterilisation, in practice some of the sterilisation courts proceeded harshly against Jews. In 1945 each zonal authority had to adjudicate whether the law was specifically Nazi, or was normal and legitimate medical practice. While all zonal administrations prevented further sterilisations, the legal status of the law differed in each zone.

Alexander was well qualified to evaluate the scientific claims of eugenicists, as he had been a signatory of an American Neurological Society report of 1936 on eugenic sterilisation.[29] This committee argued for a more rigorously scientific approach to eugenic sterilisation, and warned that harsh

measures could have little effect on eradicating genetic diseases in a population. At Nuremberg Alexander pressed for investigations of the criminality of Otmar von Verschuer (Director of the KWI for Anthropology from
1942) because of his links to Mengele in Auschwitz. After Nuremberg
Alexander appears to have hardened his opposition to eugenics in any
form, including abortion. Alexander's contention – that the experiments as
thanatology were part of the Nazi endeavours in racial eradication – was
supported by evidence concerning sterilisation and euthanasia. He recognised the 'general framework of genocidal activity'. The selection of racial
inferiors linked the experimental programme to the Holocaust.[30]

Alexander was on friendly terms with Lemkin. They had met at Duke
University, North Carolina in 1941.[31] He promptly consulted 'old Lemkin'
in Washington on 12 November 1946, when he was given a batch of 'genocidal material'. Lemkin had just decided to resign from the US War
Department to devote himself fully to securing the UN Genocide
Convention.[32] Lemkin's view was that the medical case dealt with how
genocide was implemented scientifically.[33] He argued, 'by attaching the
stigma of genocide to the acts of the defendants we will effectively preclude
them from invoking the plea that their experiments were scientific'.[34] He
classed the Nazi doctors as 'professional genocidists', who developed
methods for realising 'the most hideous genocide program the world has
ever witnessed'. He included human experiments and the prevention of
births, especially coerced sterilisation, as momentous chapters added by the
Nazis to the long and still unfolding history of genocide. Medicine provided psychological conditioning for the killers 'who regard their odious
task as they would fighting a plague'.[35]

Lemkin was pleased that Taylor used the term 'genocide' in his opening
address, and felt that the Medical Trial was potentially a classic genocide
case.[36] But he was disappointed that Taylor and the US prosecutors continued to use the generic accusation of crimes against humanity rather than
the more specific charge of genocide as the basis for the medical case – and
indeed for ensuing cases. For this meant a missed opportunity to insert
genocide into international law and so prevent its recurrence. Taylor
insisted that: Crimes against Humanity as endorsed by Control Council
Law No. 10 covered 'persecutions on political, racial or religious grounds'.
As Taylor explained in Paris on 25 April 1947, it was not a question of isolated acts of murder or violence but wholesale, nationwide campaigns 'to
make life intolerable for, to expel, to degrade, to enslave, or to exterminate
large numbers of the civilian population'. One limitation was that the
authority of the military courts could not cover crimes committed prior to
1939. It meant that Nazi medical abuses such as sterilisation, whether legal
or illegal, before then could not be tried; nor could any feature of Nazi
racial policy until the onslaught against Poland. Fortuitously, the wave of
human experiments and euthanasia coincided with the declaration of war

in 1939. The court also took the view that crimes had to be committed against non-Germans; crimes against Germans being left to the jurisdiction of German courts.[37] Lemkin saw the weakness of tying crimes against humanity closely to war crimes. However, he continued to be esteemed as an authority on Nazi mass murder. In February 1947 Judge Beals requested that the US War Department provide him with a copy of *Axis Rule in Occupied Europe*.[38] Lemkin was convinced that the human experiments were designed to test mass killing techniques. This was a point on which he and Alexander were in agreement. But as the defence case opened, the defendants vehemently denied that the medical and poison gas experiments were pilot exercises in mass annihilation.

Lemkin wanted the utilisation of Zyklon prosecuted, and when the Trial was over the World Jewish Congress was concerned that the planners of genocide were not being prosecuted.[39] Lemkin kept a critical eye on the proceedings. On 10 January 1947 his memo on 'the Importance of the Genocide Concept for the Doctors Case' argued that 'The doctors case has nothing to do with the misuse of human life for science'. He warned that the Trial should remain based on the war crimes concept, rather than on violations of medical ethics involving the relationship between doctor and patient. An individual scientist might stray from the canons of professional ethics – but 'By attaching the stigma of genocide to the acts of the defendants, we will preclude them from effectively and justly from invoking the plea that their experiments were scientific ...' For murder should not be camouflaged as science.[40] He hoped that the prosecution would develop its case on the basis of the genocide concept.[41] He criticised Alexander's term thanatology, as he favoured a geopolitical and 'genopolitical' approach (the latter concept betrayed a biologistic strain in his analysis) to Nazi medical war crimes.[42] Lemkin considered that geo-medicine had a key role in generating murderous medical research, necessitating the need to link medicine to plans for racial expansion.

Alexander dissected the brutalised mentality behind the killings as 'geno-suicide' in a psychological analysis, contrasting to Lemkin's legal and historical approach. He interpreted Himmler's belief in the hardening effects of the killings – declared at Posen in 1943 – in terms of a thanatolatrous death cult as a destructive-aggressive urge.[43] Alexander pinpointed SS concepts such as the self-selection of leaders and the glorification of heroic death as regenerative. He attributed Himmler's cyanide capsule to Rascher. What interested Alexander was how Himmler inculcated a murderous ideology of exulting in the death of others and of oneself.[44] Alexander diagnosed the medical will to supplant the old leadership with 'a new medical-biologic-scientific super-dictator' who could destroy even Hitler.[45]

When preparing the Medical Trial, the interrogators hunted for evidence of genocidal activities by the SS. The SS Ahnenerbe organisation was a prime target, given its interest in Germanic archaeology and medical research. The interrogators delved into Sievers' links with Karl Brandt, the anatomist Hirt and the SS official Eichmann with respect to obtaining bodies from Auschwitz for the Reich University Strassburg anatomical institute.[46] The prosecution looked for evidence of medical killing. They passed over Sievers' concerns with racial anthropology – his diary noted meetings with the KWI anthropologist and SS officer Wolfgang Abel, his assistant Bruno Beger who examined Soviet prisoners and measured skulls in Auschwitz, and the statistician of the Final Solution, Richard Korherr. Although Sievers was condemned for his involvement in the killing of Jews for the Strassburg skeleton collection, the eugenic and anthropological concerns of the Ahnenerbe were marginal to the hands-on murder strategy of the Trial.[47]

The prosecution interrogated Jewish survivors of Auschwitz for evidence on Mengele and other doctors committing human experiments.[48] The British interest in medical war crimes led to scrutiny of the gynaecologist Clauberg's gruesome fertility experiments on Jewish women (especially from Greece) in Auschwitz. The British saw that they could embarrass the Soviets over their secrecy regarding Clauberg, as a major war criminal. (The Russians in fact sentenced Clauberg to 25 years' imprisonment, although they repatriated him to West Germany in 1955). Taylor outlined the links between euthanasia and genocide, and quoted a letter from Brack to Himmler on the links between T-4 and providing personnel for the 'special mission' of Globocnik, who directed the extermination camps of the *Aktion Reinhardt*.[49] In March 1947 Kempner, preparing for the Ministries Trial, discovered the Wannsee protocol – Taylor was aghast at the calculated murder.[50] Nazi race experts defined not only Jews, but also 'gypsies' as meriting total eradication. Other 'races' like Slavs were defined as inferior and earmarked for exterminatory measures. Medical expertise was essential to maintain the fitness of higher races by eliminating the mentally ill and the severely disabled, and preventing reproduction of carriers of inherited diseases. In the press the Trial was presented as a genocide trial, while court proceedings dealt with mass murder.[51]

Racial victims as witnesses

Nuremberg became a public focus for victims willing to testify: a sterilisation victim wrote: 'my sister has just informed me that the wireless stations have requested to report all who were compulsorily castrated during the Nazi regime. I am one of these victims. The marks of the operation still to be seen will prove this.'[52] Their evidence was important to understand the conduct of the experiments, to ascertain the identity and numbers of the victims, how they were treated, and to find out who was involved. The victims countered a general pattern of denial of defendants' visits to concentration camps.

The experiments varied as regards the racial identity of the victims. Survivors of Rascher's Dachau experiments made it clear that the selection depended on age and physique.[53] Auschwitz figured at the Medical Trial as a locus for human experiments, sterilisation and euthanasia, and a source of experimental victims. Defendants claimed they helped individual Jews facing persecution. Himmler's masseur, Kersten, testified on behalf of Rudolf Brandt that he 'always helped' Kersten's efforts to free imprisoned victims.[54] But it was difficult to deny virulent anti-Semitism. Mrugowsky had written an anti-Semitic tract and had a longstanding track record of NSDAP and SS activism. The majority of the accused had worked in or had at least visited concentration camps. But all vehemently denied any role in the mass murder of Jews. Karl Brandt was at pains to point this out.[55] Blome similarly sought to extricate himself from the accusation that he had contributed to genocide. His proposed questions for the leading Nazi medical activist, Richard Dingeldey, included 'Wie stand Blome zum Judenmord?' Dingeldey's response on 19 January 1947 was that Blome supported the plan to lobby senior party officials to remove unpleasant aspects of NSDAP as represented by the rabid anti-Semite Julius Streicher and to restore a moral tone. He confirmed that Blome never discussed the liquidation of Jews with him.[56]

One key problem was to identify survivors. Beiglböck altered names of seawater experiment victims to stop prosecutors locating witnesses, though he found witnesses to testify on his behalf. Beiglböck's lawyer advertised on 30 June 1947 for experimental subjects and obtained six testimonies of good treatment in seawater tests from the Austrians George and Raimund Papai, Nettbach, Pillwein and Xaver Reinhart, and the French witness, Jean Senes. There were numerous categories of victims, and Catholics, Roma and Jews gave evidence. One of Alexander's tasks was 'getting the right witnesses selected'. He described the time-consuming interrogations as 'separating the wheat from the chaff' in finding witnesses who could testify against the defendants.[57]

The reading of documents, and contests over authenticity and validation, were punctuated by chilling testimony from witnesses. Presentation of victims of experiments had a strong impact. The court was moved by the evidence of the Catholic priest Leo Michalowski, and four of the Ravensbrück 'Rabbits' selected as witnesses for their intelligence. Alexander stressed that the medical student Maria Kusmierczuk and the pharmacy student Jadwiga Dzido were bright and attractive young women; they were not 'sub-human'.[58] Judge Sebring remembered them (incorrectly) as the only survivors of the 74 Polish Ravensbrück experimental victims.[59] While the gratuitous injuries inflicted on the educated might make a strong impact on the Tribunal, the prosecution argued that the uneducated Roma seawater victims experienced greater anguish.[60]

The defence was cautious in challenging stunning evidence of human butchery, and courageous protest, evasion and resistance. Victims stated whether they had volunteered. The victim Ferdinand Holl (who testified about Natzweiler) was asked 'Did you consider this a normal medical experiment?', to which he replied: 'We were not allowed to think in the camp.' From the start it was established how only exceptionally the experimental subjects were volunteers, few received anaesthetics or proper aftercare, and in the rare cases when food and cigarettes were offered as inducements, this was offset by the coercive regime of the camps.[61]

The seawater drinking experiments were conducted on 44 Roma, who were being detained as 'asocials'. Beiglböck claimed that he was experimenting on petty criminals whose sentences would be mitigated. He took a racial view of his experimental subjects as congenital social parasites.[62] His evidence was sensationally countered by the German Roma Karl Höllenreiner: his liver had been punctured and he was forced to drink putrid yellow water. He had lost his child, sister and both of her children at Auschwitz. When Höllenreiner was asked to identify Beiglböck, he sprang into the dock and punched the defendant.

Alexander reflected on the tension generating the violence. Having worked on the effect of stress on aircrew during bombing missions, he diagnosed that the witness had suffered from 'an anxiety-tension reaction of a type similar to what combat veterans who have been through severe crushing-threatening experiences in the course of their combat duty during the war. This anxiety-tension which builds up and is again revived by memories of the experiences, create a potential of aggressiveness, which is bound to burst out in emotional stress.' Alexander pointed out that he had been cross-examined mercilessly and told not to be evasive, 'as gypsies usually are'. Höllenreiner was given a sentence of 90 days for assault (although he was soon released on parole).[63]

The dramatic incident demonstrated the problems of a victim confronting the perpetrator. A great gulf separated the victims who had the physical and psychological scars of atrocious experiences from the Germans' concerns with drugs, vaccines and military efficiency. The challenge was to link the ordeals inflicted on camp prisoners with the specific charges made against the accused. The witness gave the prosecution the opportunity to identify the 'nameless dead'. Alexander located a Dutch witness, Johann Broers, who was a medical student and prisoner assistant to Haagen, and Gerrit Nales, who helped define the case as a murder case by giving details of Haagen's individual victims.[64]

The fact that Mr Nales knew exactly the names, dates and places of birth in regard to most of Haagen's victims added an air of definiteness which helped the judges to find themselves on familiar ground, namely on the ground of murder cases of definate specific persons. After each name he was asked whether he saw the corpses and he said 'yes' and in most cases... . The great definiteness was impressive and a sort of relief from the nameless murdered masses which we dealt with on so many other occasions. [65]

McHaney highlighted 'the extent to which the patients suffered', and how sterilisation was used to commit the crime of genocide.[66] Initially prisoners volunteered for additional food and were told that experiments were harmless. As time went on victims were increasingly coerced; at the end of the Third Reich they were centrally selected from other camps. The experiments met with protest, sabotage and resistance. An affidavit by the philosophically perceptive prisoner-researcher Ludwik Fleck indicted the low level of scientific competence of the bacteriologist Ding/Schuler. The reflections of the century's most perceptive critics of experimental science were telling evidence of German scientific incompetence.[67]

McHaney saw what happened on the ground in the concentration camps as less significant than the experiments with their high-level authorisation. He later reflected, 'I was not interested in that selection of who was to be cremated.'[68] The Nuremberg prosecutors had their eye on the devastation inflicted by the German military, and the human experiments appeared as the significant characteristics of the Nazi Behemoth, whereas they saw the grim slaughter in the concentration camps as essentially a matter of routine. But they did not ignore the racial master plan of resettlement and extermination.

The Trial evidence showed the accused were involved in the resettlement policy. Sievers had headed the commission for the transfer of Germans from South Tyrol in 1940–43, removing 70,000 to Germany. Rose's defence documentation revealed his experiments were to assist the *Generalplan Ost*, and shows how he was involved in screening repatriated ethnic Germans from Romania.[69] Rose developed his theory of 'epidemiological anticipation' on the basis of observations of 'the re-settlement of large groups of people originating from rural areas, and their concentration under camp conditions'.[70] The materials relating to the *Generalplan Ost* were held over to the Eighth Trial concerning the Race and Settlement Office and other racial organisations.[71]

Sterilisation and genetic criminals

The Allies at first took a hard line against sterilisation. But a few trial prosecutions meant this foundered on the rocks of legal complexities. Lemkin drew a distinction between 'sterotechnics' by which he meant sterilisation and abortion as genocidal techniques, and 'sterology' as a humane science of birth control.[72] He argued that sterilisation was crucial in pointing to the need for a law on genocide, as the killing of the unborn could not be prosecuted under the charge of mass murder. Sterilisation was a constant point of reference in war crimes investigations, but (as Lemkin feared), the zonal administrations equivocated over the criminality of sterilisation.

War crimes prosecutions in the Soviet zone were based on crimes against humanity. On 12–14 November 1946 a sterilisation trial took place in Schwerin (the administrative centre of Mecklenburg in the Russian zone) involving the deputy chair of the local hereditary health tribunal, the director of the health office, a medical officer, a member of the sterilisation tribunal and medical director of a local hospital. In all, seven doctors were prosecuted for crimes against humanity.[73] The matter was by no means clear-cut as, after initial conviction, the sentences were reduced or quashed by a higher court, and the case dragged on. In the British zone administrative directives on sterilisation were shifting, and a German medical lobby argued for the legality of sterilisations as a medical rather than a racial measure.[74] The war crimes prosecutors became reluctant to prosecute abuses before the outbreak of war in September 1939, or to tackle cases when the victims were German citizens. While the sterilisation law was abolished in the US and Soviet zones, the sterilisation tribunals were replaced by administrative courts in the British zone in 1947. Here individuals could appeal against sterilisation, although the majority of verdicts were upheld.[75]

Perpetrators of sterilisation research were held in custody, war crimes authorities collected evidence and then nothing happened. A notable case of non-prosecution was that of Hans Reiter, director of the Reich Health Office, who was interrogated at Nuremberg on 19 December 1946. Reiter insisted – falsely – that the Reich Health Office had nothing to do with euthanasia or eugenic legislation. Under Reiter came a Department for Hereditary Biology, where Robert Ritter and his assistants Eva Justin and Sophie Erhardt undertook mass surveys of Roma and vagrants. Reiter explained that his institute was concerned only with sterilisation among civilians rather than in the camps. His evidence linked the psychiatric eugenicist, Rüdin to the medical official and euthanasia administrator, Herbert Linden who assisted Brack. Legal proceedings against Ritter were suspended in 1950.[76]

Due partly to Lang's well-aimed denunciations, the sterilisation expert Rüdin was interned from December 1945 to the decisive month of August 1946.[77] Lang accused Rüdin, his Austrian colleagues Albert Harrasser and Friedrich Stumpfl, and some German radiologists and officials of developing X-ray sterilisation, and first experimenting in 1940. The Russian X-ray geneticist, Timoféef-Ressovsky, could corroborate his deposition, as they had discussed the matter in January 1941.[78] Although the legal status of sterilisation made it difficult to prosecute, Rüdin was on the margins of medical criminality. He had opened negotiations with the SS-Ahnenerbe to provide special research facilities at his Institute, supported aviation medical research in Munich, and he was prepared to exploit euthanasia for brain research.[79] He was on the list of medical criminals screened by Hardy in September 1946. He remained unprosecuted, although his period of detention was punitive.

During November 1946 the American prosecutors at Nuremberg collected details of drugs used for procreation and sterilisation, as well as of doctors involved in sterilisation.[80] A Staff Evidence Analysis included 26 letters from people sterilised through verdicts of the hereditary health court; among the reasons were the remark by a 16-year-old girl – 'What comes after the Third Reich? – the Fourth', or for being part-Jewish.[81] The staff analysts prepared batches of letters of victims sterilised by the hereditary health courts.[82] The prosecution collected evidence from victims of sterilisation. Many victims and witnesses spontaneously contacted the Nuremberg prosecutors. French associations of survivors pressed their government to forward evidence, and insisted on the case as being one of genocide affecting all who lost their lives in the camps.[83]

The prosecution made efforts to locate victims of sterilisation. A former Spanish Republican officer testified that his sterilisation in April 1942 while a prisoner in Dachau was punitive. Alexander reassured him that the operation was not as damaging as feared.[84] The interrogator Rodell recollected the agony and sickness of a Jewish survivor of the sterilisations in Auschwitz:

He could only stand for ten minutes and then could only sit for 20 minutes and would have to lie down. I was there when the doctors examined him. He was jet black from the waist to the thighs and his insides were burned.[85]

Alexander took up the case in November 1946, providing the OCCWC with a deposition and photos of the injuries:

Emotionally this man was deeply hurt and humiliated by his mutilation. He has not yet been able to tell even his own sisters about it. Although all this happened through no fault of his own, he feels deeply ashamed about his castration. He is afraid that his increasing gain of weight and loss of male characteristics are bound to ultimately give him away for the wreck which he has become. He feels that he has no future and has nothing to live for and has had no real life so far, and nothing to really live for ahead of him. At times his thought and emotions overcome him and he begins to cry when talking about what has happened to him.

When he heard over the radio that the people responsible for the German medical atrocities are going to be tried, he decided that it was his duty to come here and to testify although he is afraid that, especially if his name is printed in newspapers, his sisters might find out about his condition that way. However, he feels that it is his duty to be helpful in bringing those responsible for the atrocities, to which he and others have been subjected, to justice.

It appears that he is one of 100 young Jewish boys who were castrated for no reason other than to confirm the fact that they had been sterilised by sufficient Xray radiation as if Xray burns which resulted from a fifteen minute exposure were not enough to prove that point. A great many of his fellow sufferers have in the meantime developed cancer of the irradiated skin. While his skin is severely induarated no evidence of cancer is yet discernible.

When he gave evidence to the Tribunal, the judges requested that his identity be kept anonymous.[86]

Given that the Nazis flagrantly flouted the provisions of the sterilisation legislation of July 1933, it laid them open to charges of illegal sterilisation. The prisoner doctor Robert Levy gave evidence about experimental operations in Auschwitz.[87] He was a French citizen but served with the German army in the First World War for two years, and in the French army in 1939–41. He was arrested by the Gestapo in Limoges on 12 May 1943 for anti-German political propaganda. He worked in the central hospital Auschwitz, and then in Mauthausen. As chief surgeon in the in Auschwitz Revier (camp hospital), he found patients had been used for sterilisation experiments, their testicles were extracted or they were exposed to X-rays.

'Dr Levy is very certain that few of the patients so treated can be alive ... most of the patients were very unhappy and psychologically broken by the sterilisation ... more serious cases developed into Xray cancer.'[88]

The French collected testimony about medical atrocities in the camp of Struthof-Natzweiler. A deposition of 25 May 1945 to the French War Crimes authorities described the preparations of human testicles by the anatomist Hirt at Strasbourg, and 54 slides of arrested spermatogenesis (the specialism of Hirt's assistant Kiesselbach) derived from experiments at Struthof:

> A Dr Stive [actually Stieve] had the nerve to publish in a German magazine of 1944 observations made of Haemorrhages produced in women by bad menstruation. These experiments were conducted on normally menstruating female prisoners who were told that they were about to be shot, thus producing an internal haemorrhage which was studied by Dr Stive. [89]

Stieve and Hirt's assistant, Kiesselbach, enjoyed successful post-war careers.

The American investigations of the KWI for Anthropology in the US zone in Berlin, and the KWI for Psychiatry, were at first perfunctory.[90] The UN

War Crimes Commission and OSS (Office of Strategic Services) Wanted List of 12 May 1945 included Rüdin and Verschuer, the directors of these two institutes.[91] A British FIAT evaluation of 'SS Medical Research' of 4 September 1945 mentioned 'Mangerlay or (Mengle)' in connection with Auschwitz gassings. Another listing of concentration camp medical personnel correctly described Mengele as 'SS Camp Doctor Birkenau', but the focus on the camp obscured the link to Verschuer.[92] The investigators overlooked the importance of the funding records of the *Deutsche Forschungsgemeinschaft* and the *Reichsforschungsrat*. The Mengele-Verschuer axis might have compromised other eminent scientists, notably the biochemists Abderhalden and Butenandt for research on whether there was a racial basis to protein reactions.[93] Abderhalden did not dare to intervene publicly on behalf of Becker-Freyseng, for fear of damaging his position in Switzerland.[94] As long as the scientists maintained that they resisted the cancer of Nazism, their post-war careers remained viable.[95]

The Czechoslovak biologist and historian of genetics, Hugo Iltis, suggested that '"Scientific" representatives of the German race theory should be dealt with as War Criminals'.[96] An OSS report of 30 May 1946 similarly took a hostile view to research on heredity, because it was linked to racial research. The racial hygienists connected to the KWI for Psychiatry, as well as Ulrich Ploetz – son of Alfred Ploetz who coined the term Rassenhygiene – and Verschuer were cited as war criminals. Verschuer was characterised as 'a man who remained in the background, but in spite of all, one of those principally responsible for the theory of racial hygiene and executions by gas'.[97] The report made no mention of Mengele and the Auschwitz twin research. Verschuer remained under suspicion for advocating 'Perpetuation of racial research', and in July-August 1946 the CIC (Counter Intelligence Corps) placed him under house arrest.[98] The war crimes investigators looked to science as providing a lethal paradigm for genocide.

The survival of the KWI for Anthropology was due to the geneticist Nachtsheim's assertion that he was untainted by Nazism. He was not an NSDAP member. The Allies overlooked that he joined the hereditary pathology project of Verschuer, experimented on epileptic children and had no hesitation in collaborating with a pathologist working for Gebhardt in the SS sanatorium of Hohenlychen.[99] The Allies characterised the Institute in October 1946 as: 'Closely related to Nazi race theories'. But after the war, it claimed to research on inherited diseases of blood, nerves, sense organs and skeleton, and 'psychophysical development of man'.[100] These could be seen as useful tasks at a time of conspicuous shortage of medical resources, or as eugenics without an explicit racial component. War crimes investigators paid too little attention to Verschuer's multifarious contacts to NSDAP racial experts and organisations.[101]

The prosecution might have used Sievers' links to the SS anthropologist Abel and Rüdin to mount a wider case against the KW anthropologists and psychiatrists. The KWI eugenicists opportunistically exploited links to the SS and party political organisations. But in 1945, the psychologist Kurt Gottschaldt, in the Russian zone, and the geneticists Lenz and Nachtsheim sought legitimation from the Allied zonal authorities, while sniping at their former colleagues. They secured their academic survival by demonstrating that they were untainted by past scientific crimes. Verschuer had not been able to take an academic position at Frankfurt at the time of the Trial, but remained free. Nachtsheim obligingly concluded that using body parts of Holocaust victims for scientific research was legitimate. Notables on the Berlin Medical Faculty, particularly Heubner and Sauerbruch, took a belligerent position to exonerate themselves, while undermining the Trial's legitimacy.[102]

The rising interest in medical war crimes in the summer of 1946 meant that Verschuer came under renewed scrutiny. Initially, FIAT called for his arrest.[103] After a brief period of house arrest, he was kept under observation while the Nuremberg prosecutors collected evidence; he finally came under consideration as a defendant in a second Medical Trial. Auschwitz loomed large at the Medical Trial when it came to X-ray sterilisation experiments, Clauberg's damaging of reproductive organs and the selection of Jews as euthanasia victims. Brack claimed (dubiously) that he recommended X-ray sterilisation to Himmler so that the lives of Jews would be saved.[104] He denied involvement in the mass killing of Jews.[105] His wife testified on contacts with Globocnik.[106] Mitscherlich thus classed the X-ray experiments as a form of 'Genocidium'. [107]

On 3 May 1946 Robert Havemann, who had been entrusted with overseeing the KWG by the authorities in the Russian sector of Berlin, denounced the KWI Director Verschuer for research on eyes from Roma in Auschwitz.[108] Verschuer's former colleague, Gottschaldt, now set on a career in East Berlin, had supplied the chilling details. Havemann had good links with concentration camp survivors through the association 'Victims of Fascism', and gathered evidence from Auschwitz survivors.[109] He supplied the OCCWC documents analyst Wolfson and the lawyer Kempner with details of Mengele's links to Verschuer.[110]

Wolfson and his documents team trawled through the cellars of the Berlin Document Center, and in September 1946 sent his findings to Nuremberg. Mengele's name was prominent in a list of criminal suspects for the 'medical experiment case'. Depositions from victims and observers of the human experiments at Auschwitz had identified Mengele and the fertility researcher Clauberg as prime culprits.[111] The Nuremberg interrogators kept a bank of witnesses concerning Auschwitz medical atrocities. These included Siegfried Liebau of Mrugowsky's Waffen-SS Hygiene Institute, and assistant to Verschuer at the KWI for Anthropology from December 1942 to October 1943; Liebau served with SS medical corps in the Ukraine, before joining the murderous Globocnik in Trieste, where many former T-4 personnel also served. His career linked eugenics, bacteriology and euthanasia, and illustrated the continuous movement between the academic and the SS. Wolfson drew Alexander's attention to Liebau in Nuremberg, and to correspondence between Magnussen and Gottschaldt, concluding that Verschuer should be charged with war crimes.[112] Even after the Medical Trial, the prosecutors continued to collect witness statements: on 8 October 1947 the interrogator von Halle took a statement from Mengele's pathologist, Nikolas Nyiszli.[113]

The expectation and perception of the Medical Trial was that it was a trial of German eugenicists, even though mainstream eugenics as represented by those involved in German eugenic propaganda and public health measures were not prosecuted. The issue arises whether the Americans were reluctant to put Nazi sterilisation advocates on trial, as Karl Brandt had highlighted American sterilisation legislation. The decision to prosecute X-ray sterilisation linked eugenics to a savagely cruel atrocity, which occurred during the war, and had a large number of foreign nationals as victims. The prosecution here felt on legally safe ground, and could avoid potential difficulties from the defence about US sterilisation and the legal status of sterilisation conducted under the German legislation of July 1933.

Prosecutions concerning the German sterilisation measures were difficult, although by no means impossible. Unlike euthanasia, the first phase of sterilisations was carried out under an enacted law, even though the sterilisation courts did not always respect legal niceties in proceedings against Jews. The Nazi Physicians League's leaders agitated for vindictively sterilising the half-caste Rhineland children in 1937 in a covert and demonstrably illegal manner.[114] This action was supported by the veteran racial anthropologist Eugen Fischer, who remained untroubled by war crimes prosecutions. Fischer had been involved in racial propaganda for Rosenberg's Ostministerium and adjudicated on cases of racial ancestry.

The issues of sterilisation and euthanasia linked clinical medicine to genocide. The prosecutors found many cases of sterilisations in the concentration camps. Sometimes the purpose was experimental, sometimes punitive, and at times just the wish to sterilise racial undesirables. Thus the prosecutors began to deal with overlooked aspects of sterilisation directly related to genocide. That the victims of sterilisations in the camps were mainly Jews and Roma highlighted the genocidal rationale, while undermining the defence that sterilisation was simply a public health measure. The prosecution accused Pokorny, Karl Brandt and Poppendick of responsibility for experimental research on methods of mass sterilisation. Blome cunningly argued that his superior, the bullish Reichsärzteführer Wagner, opposed sterilisation, and this led to suspension of the system of tribunals during the war.[115] He failed to mention that Wagner's views were dictated by the view that the sterilisation law was inadequately racist, and that Wagner lobbied for euthanasia killings. Evidence that Oberheuser was involved in forced sterilisations of the healthy and on children from eight years old, was overlooked at the Medical Trial, as were her euthanasia killings with phenol injections.[116]

The central role of biological research and race at the Trial emerged in the prosecution of Poppendick. One of the longest serving doctors in the SS, he took a course in genetics at the KWI for Anthropology from November 1934 to July 1935, when he was trained by the racial hygienists Eugen Fischer and Fritz Lenz. He moved to the SS Office for Hereditary Health and Population Policy in 1935–6, where he dealt with marriage applications by the SS. In 1941 he headed the medical department of the Sippenamt of the Race and Settlement Office of the SS. In September 1943 he became head of the Personal Office of the Reichsarzt SS Grawitz, who had pressed for human experiments. He was also accused of complicity in sterilisation research.[117]

The defence characterised Poppendick as 'a follower of the Lenz school, whose attitude with respect to heredity and racial hygiene was not exactly in accord with party doctrine'.[118] He claimed that he was not a member of a hereditary health court, that he had no role as adviser on population policy, and approved of neither compulsion nor extermination.[119] Thus the case against him did not stretch to the twin genetics researches of Verschuer and Mengele, and routine sterilisation was not prosecuted. He was the subordinate of Reich SS Doctor Grawitz from 1939, but he argued that his responsibilities as head of Grawitz's office did not mean that he was Grawitz's deputy. He explained that Grawitz disapproved of his adoption of Lenz's views, as Lenz also did not support compulsion and extermination.[120] Although at a major office for racial policy, and deputy of a top SS doctor who authorised human experiments, Poppendick contested any guilt. The SS dental officer Blaschke explained that Grawitz was marginalised, had few actual duties and that as staff members under his command the positions of Mrugowsky and Gebhardt were purely nominal.[121] Poppendick successfully disproved any link to typhus and virus research at Buchenwald.[122]

While Poppendick was an obscure racial hygienist, Lenz was renowned as a pioneer of genetics on a Mendelian basis. Lenz's praise of the Nazis in 1930 as the party most favourable to eugenics has earned him the characterisation of being an archetypal Nazi eugenicist. Lenz alleged that Hitler in Landsberg had been influenced by his writings, and he advised the SS on their hereditary screening. He commended the health services and marriage screening for the SS as a Nordic elite, involving genealogies to prove sound racial qualities. This represented what Lenz called *Aufartung*, or racial regeneration, and he was in contact with Walter Darré, the Nazi agriculturalist when the marriage code was drawn up.[123] Lenz favoured emigration of the mulatto Rhineland children (who in the event were forcibly sterilised), and suggested that there be a Reich Commissioner for

Population Policy (this was never instituted). Karl Brandt cited in his defence that Lenz favoured the killing of severely malformed or 'idiotic' newborn children, and Lenz sat on the committee for a euthanasia law, which never materialised. He opposed the conferment of a professorial title on Mrugowsky during the war, so angering elements within the SS but the reason was Lenz's opposition to advocates of geo-medicine (which at the Trial included Mrugowsky and Rose). At the insistence of the public health official Gütt, Lenz joined the NSDAP in 1937, but in 1943 was at loggerheads with Himmler over the racial value of children born out of marriage, and was critical of the Lebensborn's efforts on behalf of single mothers.[124] The 1936 edition of the textbook on human heredity by Baur, Eugen Fischer and Lenz was cited as evidence that Lenz did not believe in compulsion. After the war he insisted that he was not an anti-Semite and that he opposed compulsory sterilisation.

Lenz had the distinction of having held the first teaching post in racial hygiene in Munich from 1923, and between 1933 and 1945 he was director of the Institute for Racial Hygiene at the University of Berlin. He was also departmental director at the KWI for Anthropology. The medical faculty at Göttingen commended Lenz to the British Control Commission in January 1946 as untainted by Nazism, and that he had controversies with the Nazi leaders Darré and Rosenberg. The Commission was more interested in control of strategic science than resolving the criminality of eugenics. On 1 November 1946 he was appointed professor of human genetics in the medical faculty of Göttingen. Lenz stressed his British contacts throughout the 1930s, and a number of right-wing eugenicists vouched for him including the Dane Tage Kemp, and the American Paul Popenoe.[125] Nachtsheim provided a denazification reference in August 1946, although he privately condemned Lenz for his Nordic racism.[126]

In the whitewashing cycle of exonerating 'Persil certificates', Lenz provided an affidavit on 26 January 1947 that Poppendick was benign in his views and critical of Himmler. He presented himself as an opponent of the 'fanatical antisemitism' of the Nazis (although this was not a rejection of anti-Semitism wholesale), and affirmed that Poppendick had taken no part in the measures to exterminate Jews.[127] Poppendick was referred to as 'Poppendiek' at the IMT where he had been arraigned for low temperature and pressure experiments at Dachau.[128] Poppendick explained that in the Race and Settlement Office his work in screening SS marriage applications arose from the concerns of the Reich Peasant Führer Walter Darré with the need to have a hereditarily healthy population settled in the rural east. Poppendick's work as the senior physician in the Office between 1941 and the autumn of 1944 was oriented to hereditary health of the SS and intended marriage partners, and the encouragement of 'child-rich' families. There were some SS villages near to Berlin, and he claimed that his medical work ran parallel to genealogical and statistical studies. Himmler ordered that the SS genealogical work continue through the war, although Poppendick became increasingly involved with personnel issues as a result of the requirement to have SS doctors in frontline service.[129] He denied that the Office was concerned with settlement in the East and with exterminating Jews, Poles and Russians, or that the Office worked on the Jewish question or the existence of a register of Jews.[130] But under psychological interrogation, he showed how he still thought in general categories of race, and regarded 'gypsies' and 'negroes' as being without culture.[131] (Similarly, Blome's mindset was deeply racist: he blamed the extermination of the Jews on their preponderance after the First World War and remained an unrepentant Nazi.[132]) The defence argued that Poppendick was an isolated figure with idiosyncratic views on genetics.[133] His extenuations were a precursor for the defence of others from the SS Race and Settlement Office, who pleaded that the Office was primarily concerned with the marriage applications of the SS, and that during the war its tasks were welfare-oriented.[134]

The defence insisted that treatment of infertility was confused with sterilisation.[135] Poppendick had referred infertile SS wives to Clauberg for treatment, but denied involvement in Clauberg's Auschwitz experiments. Poppendick was accused of supporting the sex hormone therapy, conducted by the Danish doctor Carl Peter Vaernet (originally Jensen) at Buchenwald. The 'therapy' drew on the KWI biochemist Butenandt's discovery of the chemical constitution of sex hormones.[136] The SS supported Vaernet's research on an artificial sex gland. Although based in Prague, Vaernet experimented at Buchenwald with the aim of 'curing' homosexuality with hormones and glandular implants, causing the death of his victims.[137] The British interned Vaernet in 1945 and handed him over to the Danish authorities; the Danish Medical Association sent an affidavit on Vaernet to the Danish authorities in May 1945, but Vaernet was allowed into Sweden and a Nazi escape network made possible his escape to Argentina in late 1946 or 1947.[138]

The case against Pokorny appeared to have a strong basis, because of a documented link to Himmler concerning mass sterilisation. In July 1945 the use of herbs for sterilisation caused concern to US war crimes investigators.[139] Herbal medicines appeared to match Himmler's penchant for pseudo-science. The prosecutors at the IMT accused Pokorny of having proposed to Himmler in 1941 the use of the Brazilian herb *Caladium seguinum* (also known as *Dieffenbachia seguine*) for mass sterilisation. His arraignment at the Medical Trial was a consequence.

Pokorny (like Blome and Oberheuser) was a dermatologist. His application to join the NSDAP was rejected, because he had been married to the Jewish radiologist, Lilly Pokorna-Weilova. He told the court of his feeble attempts to shield her from deportation. To his accusers Pokorny insisted on his identity as a Czech, left-wing doctor. As the Czechoslovaks expelled him at the end of the war as a Sudeten German, he appeared at the Trial as a German physician. Pokorny's identity in terms of nationality and politics remained controversial, and evidence of the actual use of the herbal extract in mass sterilisation proved elusive.

Pokorny had based his letter to Himmler on claims by Gerhard Madaus of having sterilised mice with the herb.[140] In March 1947 Alexander found a paper by the pharmacologist Friedrich Koch as well as a more popular article, arousing the suspicion that clinical trials of herbal sterilisation were carried out on prisoners.[141] Madaus, who held an MD from Bonn, was a celebrated homoeopathic pharmaceutical manufacturer. However, he had died in 1942, and as the company was based in Saxony in the Soviet zone investigations were difficult. The case highlighted how Himmler's favouring of herbal and homeopathic medicines was rooted firmly in his racial mentality. The prosecutors found that an IG Farben botanist, Tauböck, had been ordered to research on the sterilising effects of plant extracts, as a result of Madaus's interest in the drug for increasing sexual potency, while reducing fertility. He explained that the SS wished to develop the drug as a means of mass sterilisation of mental defectives and racial undesirables.[142] It was overlooked that studies were made on the effects of herbs on Roma at Lappenbach in Austria. But the prosecution hoped that the Pokorny case could represent abuses of human sterilisation in Nazi Germany, and took the view that his explanations were an incredible fantasy.[143]

The defence of Pokorny and Poppendick contested links drawn by the prosecution to sterilisation, despite the formidable and distressing documentation. Verschuer escaped mention in the Trial record. The Trial did not serve as a comprehensive analysis of Nazi sterilisation and eugenics. To do so it would have been necessary to examine a far wider range of racialised health care, as well as the activities of leading anthropologists, geneticists and clinicians. But the Trial did examine intersections between eugenics and genocide. It therefore drew the prosecution away from science and towards the worst atrocities.

Alexander recognised a series of weak points in the prosecution case and became preoccupied by omission of the evidence of racial atrocities at Auschwitz. On 31 December 1946, he wrote excitedly home:

Fifi darling, The mad old whirl is going on and more and more war crimes are unfolding. It sometimes seems as if the Nazis had taken special pains in making practically every nightmare come true. Some new evidence has come in where two doctors in Berlin, one a man and the other a woman, collected eyes of different colour. It seems that the concentration camps were combed for people who had slightly differently colored eyes. That means people whose one eye had a slightly different color than the other. Who ever was unlucky enough to possess such a pair of slightly unequal eyes had them cut out and was killed, the eyes being sent to Berlin. This is the carrying out into reality of an old gruesome German fairy tale which is included in the Tales of Hoffmann, where Dr Coppelius posing as a sandman comes at night and cuts out children's eyes when they are tired. The grim part of the story is that Doctors von Verschuer and Magnussen in Berlin did prefer children and particularly twins. There is no end to this nightmare, at least 23 are being tried now and, I trust, the others will follow later.[144]

The quest for the eye specimens deepened Alexander's insight into the psychology of the perpetrators of medical horrors. His letter introduces a perplexing sequence of events: the disappearance of Mengele, the survival of his mentor, the geneticist Verschuer, and a Spallanzani, like Nachtsheim, overseeing the conversion of racial anthropology into human genetics. As in Hoffmann's tale *Der Sandmann,* dark forces have a tenacious existence. Clearly, Alexander thought that Verschuer and Magnussen were candidates for a second doctors' trial, which appeared imminent amidst the frenetic searches for human vivisectors. By mid-1947 this prospect faded with the onset of the Cold War and the weariness of the investigators. The evidence suggests that Alexander initiated investigations, but could not locate the incriminating collection. He referred to the eyes being sent to an unknown destination in Berlin.[145] On 23 December 1948 he denounced Verschuer in the anniversary discourse of the New York Academy of Medicine on 'Science under Dictatorship'. After discussing the (mainly) Jewish skeleton collection of Hirt at Strassburg, Alexander commented: 'A few scientists were more fortunate in destroying evidence. A collection of eyes from identical twins with heterochromic iris was traced from Auschwitz concentration camp to the laboratory of Professor Dr von Verschuer in Berlin, but the corpora delicti were never found.'[146] Although Magnussen was interrogated, no prosecution resulted.[147]

A denazification application by Verschuer led to a CIC interrogation on 25 July 1946. Verschuer was denazified as a Nazi supporter (*Mitläufer*) on 9 November 1946, while continuing to press for a new institute of hereditary pathology in the US zone on the basis of his researches into the genetics of TB. He remained on a list of medical suspects 'who may be included in the medical experiment case'. He was characterised on 7 September 1946 as 'One of the principal persons responsible for Race Purification theory and for executions by gas. Was appointed confidential agency of the Party to Institute for Purification of the Race in Francfort.' The suspicions prompted a dossier to be compiled by Wolfson: 'von Verschuer is accused of experimentation on living human beings in concentration camps, particularly Auschwitz. He is accused of being a Nazi activist.'[148]

Wolfson's strategy was to locate Verschuer within SS networks. He outlined links to Mengele and also to Heinrich Schade, another former assistant of Verschuer (and NSDAP member since 1931 who served at the Racial Political Office), and Siegfried Liebau, who was briefly seconded to the KWI from the SS Sanitäts-Amt. Wolfson added for good measure connections between Verschuer and Reichsgesundheitsführer Conti.[149] The Research Analyst, Miss H. Wachenheimer, of the Office of the Chief of Counsel for War Crimes, Berlin Branch asked the Berlin Document Center for the personal files of both Verschuer and Mengele, indicating how their names were twinned.[150] On 7 November 1946 Wolfson drew on evidence provided by Gottschaldt that Verschuer had been 'informed of the detailed setup as it existed in Auschwitz'.[151] Not unreasonably, Wolfson recommended that Magnussen, a known Nazi activist, be arrested and interrogated. Verschuer counterattacked that the denunciations derived from communists.[152] The report was Alexander's source of information on the heterochromic eye atrocities.[153] On 7 January 1947 Wolfson contacted Alexander concerning Liebau, as an assistant of Verschuer and SS officer, and Magnussen as 'important figures in the war crimes which Verschuer can be charged with'.[154]

Wolfson persisted in his accusations: on 12 February 1947 he placed Mengele at the head of a table of Auschwitz officers.[155] Attempts were made to have the denazification verdict revoked. On 13 May 1947, Verschuer spoke of Mengele's excellent relations with his patients in Auschwitz.[156] The strategy was to analyse Verschuer's links to the SS rather than to reconstruct the research programmes, which drew on human materials from the concentration camps. The matter of the blood protein research was not fully investigated and prosecuted at the Medical Trial, as this RFR project was neither commissioned by the SS, nor directly ordered by Himmler or by the Ahnenerbe. The extent that Butenandt was informed by Verschuer about the provenance of the blood analysed at his Institute remains a matter of conjecture.[157]

Verschuer counter-attacked that Havemann's evidence against him was provided by the communist sympathiser, Gottschaldt. Havemann had in any case lost the fight to control the KWG as head of the Berlin *Amt für Volksbildung*. Verschuer denounced him as a KPD member prior to 1933, who was now established in the Soviet zone. Verschuer consistently pressed home the point that those discrediting him were 'Communist agents'.[158] Verschuer, although a Nazi Party member since 1940, lacked the criminal profile of Mengele as an SS officer; Mengele's main duties were as camp doctor, preventing the spread of infectious diseases to the Germans (this included gassing complete huts and killing the gypsy camp), and in undertaking selections of new arrivals for the gas chambers. Given that his research was officially recognised as a sphere of activity, which went beyond his duties as a camp doctor, it was easy to overlook the linkage between the KWI for Anthropology and Auschwitz. The prosecution's approach was that the SS-sponsored experiments were not so much medical research as murder. Medical witnesses repeatedly cited Mengele's researches, and Alexander obtained copies of depositions collected by the British.[159]

The opening of the Medical Trial prompted survivors to offer evidence against Mengele. His prisoner assistant, Gisela Perl, contacted the Office of the US Chief of Counsel at Nuremberg on 11 January 1947, offering to testify against 'Mengerle', as news reports of his capture had appeared in Vienna and Budapest in December 1946. However, the US authorities conceded responsibility for the trial of the Auschwitz atrocities to the Polish government.[160] Perl's exchange with US war crimes investigators ended on 19 January 1948, when Telford Taylor made the extraordinary claim that 'Mengerle' was dead – something that had been known since October 1946.[161] This was probably based on the false claims of Mengele's wife, when investigators went to Mengele's home town of Günzburg. The fateful exchange with Perl has received a number of interpretations. Mengele was in fact in hiding none too far away from the Augsburg US Counter-Intelligence Center until 1948. That Taylor was misinformed on so serious an issue raises disturbing questions as to the efficiency of US investigations of medical criminals. At the very least, it exposes Allied war crimes intelligence as rigid and bureaucratic, and as relying on hearsay rather than on proven evidence.

Alexander recognised how atrocities at Auschwitz could reinforce the prosecution's case at the Medical Trial. On 16 April 1947 he drew McHaney's attention to the diary of the Auschwitz camp doctor, Josef Kremer. Alexander collected further reports on Auschwitz experiments while in the Netherlands during March 1947.[162] The US Trials Program was still not fully worked out in mid-March 1947, when it was noted that 'several [medical] persons have been located who would have been defendants had they been apprehended soon enough', the most important being Hippke, who was in charge of German air force medicine when the cold and pressure experiments were carried out.[163] Milch had called on Hippke to testify in his defence, but he was hauled off for intensive interrogation when he arrived in Nuremberg.[164] Against any effort to prepare for a second Medical Trial, some British and US circles thought that one exemplary trial of medical crimes was adequate, and by the summer of 1947 it was clear that the Nuremberg programme would not stretch to a second Doctors' Trial.

KWG researchers remained cautious. When the defence lawyer of Schröder and Becker-Freyseng asked the biochemist Butenandt in April 1947 to exonerate research on the hormone-based project Haemopoetine as a styptic, he delayed his reply until 4 July 1947 when the defence had concluded its case. Butenandt explained that he had approached Hippke with a research project as a way of conducting research under the difficult wartime circumstances. He thereby recovered an assistant from frontline duties, as he believed a substance preventing bleeding could be achieved. (He overlooked that the assistant in question conducted pressure chamber experiments with children, and that research was conducted in his institute using blood plasma obtained from Auschwitz.)[165] Butenandt was above all keen to play down his link to Verschuer: the substrate research was carried out by one of Verschuer's staff, Günther Hillmann, in Butenandt's residual Berlin Institute.[166]

The evidence against Sievers and Rudolf Brandt exposed another KWI link that was not followed through. Racial anthropological examinations of Soviet POWs were carried out by Abel, assisted by the Race and Settlement Office and supported by the Ahnenerbe.[167] Sievers explained that the 115 victims for the skeleton and skull collection of Hirt in Strassburg were selected from Auschwitz by the anthropologists Beger and Hans Fleischhacker.[168]

The escape of Mengele, the non-prosecution of Verschuer and suspicions surrounding Butenandt reveal defects of co-ordination in the operation of the war crimes machinery, as well as a lack of interest in the criminality of science at prestigious institutions like the KWIs. Magnussen had completed her doctorate under the KWG zoologist Alfred Kühn, before joining Nachtsheim. Verschuer's attack on his detractors as communists carried increasing weight, because the location of the atrocities in Auschwitz could have meant that – as in the case of Mengele's colleague Hans Münch – a handover to the Poles was involved just when Poland was going over to communism.

With the US coming round to the British position on the need to revive German science, a transfer of the human geneticist Verschuer to the Lysenkoist communists was hardly a viable option, and US hostility to Verschuer melted with the onset of the Cold War. Stalinist support for Lysenko in 1948 did wonders for the position of Nachtsheim in Berlin.[169] The line was drawn between legitimate science and extremist fanaticism, and prospects for a second Medical Trial receded in the midst of rising East-West tensions.

The Lysenkoist onslaught on genetics meant reluctance in the West to place German geneticists on trial.[170] German scientists pleaded that they were oppressed by Nazism; they shrugged off the suggestion that science served the German war effort. Indulgent Allied scientists resumed relations with German colleagues as if nothing had happened. The plant geneticist Cyril Darlington, who toured Germany in mid-1945 for BIOS, was positive in his view of German biology. But he loathed the Soviets for liquidating the brilliant plant geneticist Vavilov. The bio-statistician R.A. Fisher promptly resumed contacts with von Verschuer, broken off only late in 1939 after Verschuer lectured at the Royal Society. The Soviets made offers to the bio-chemist Otto Warburg, who had survived the Nazi persecution. The Soviets also courted Nachtsheim, who in turn sought rehabilitation of other tainted geneticists. He wanted to keep Verschuer marginalised as a suspect, as this gave him a free hand to develop scientific initiatives in Berlin. But at the same time, too sweeping a condemnation of Nazi genetic crimes could pull down the whole pack of biologists into the mire of criminality.

13
Euthanasia

Taylor's opening address added euthanasia to the chilling catalogue of atrocities and murder of 'hundreds of thousands of human beings'. This programme involved 'the execution of the aged, insane, incurably ill, or deformed children and other persons by gas, lethal injections ... in nursing homes, hospitals and asylums.' Such persons were condemned as 'useless eaters, as burdens to the German war machine' (the defendants constantly denied using these expressions). The deaths encompassed children's euthanasia between October 1939 and April 1945 with approximately 5,000 child deaths, the T-4 programme of special killing centres between early 1940 and August 1941 when 70,273 adults and juveniles were killed, and the programme code-named 14-f-13 from April 1941 to 1944 with an estimated 50,000 concentration camp prisoners killed. The prosecution cited the *Aktion Globocnik* in the killing of Jews. The killings of POWs and forced workers from the East were identified as a distinct phase of euthanasia.[1] The historian Suess suggests that there is no basis for associating the *Aktion Brandt* to clear hospital beds from August 1943 to the end of 1944 with euthanasia; however, killings continued throughout the war.[2]

Taylor presented Karl Brandt as a key figure in Hitler's euthanasia programme. Three other accused were implicated: Hoven for the transfer of Jewish inmates from Buchenwald to the killing centre of Bernburg, Brack as running an office in the Führer's Chancellery (KdF), and Blome as deputy Reichsärzteführer. Brack and Blome were first in line as assistants to Bouhler and Conti, respectively. This chain of command was contested at the Trial. Blome stressed that he and the Reichsärztekammer were not involved in the administration of euthanasia – unlike Karl Brandt, who with Bouhler liaised with Hitler and formulated policy;[3] Brack was 'the manager of euthanasia for adults and children', directing an office in the Führer's Chancellery under Bouhler;[4] finally, there was the peripheral and subordinate Hoven. In the event, Brack and Karl Brandt emerged as central figures in Nazi euthanasia.

Taylor saw euthanasia as a pretext for the killing of Jews and other undesirables. He gave the figure of 275,000 victims (deriving from the IMT).[5]
What the Trial failed fully to do was to establish the origins of euthanasia, as
the account of Karl Brandt was not adequately scrutinised. Reichsärzteführer
Wagner objected to the sterilisation law as inadequately racial. By 1937
medical officials were promoting a census of psychiatric hospitals and
requiring registration of newborn children with deformities. Hitler radicalised the racial health measures to ensure a 'final solution of the social
problem, and of the burdens of the past'. Euthanasia had origins in German
racial hygiene, and in the 1920 tract of the lawyer Karl Binding and the neurologist Alfred Hoche, advocating the destruction of '*lebensunwerten Leben*'
(life unworthy of life). A circle of radical Nazi doctors close to Hitler pressed
the Führer to implement euthanasia. Certainly Wagner supported the
protests of Nazi Party members about being taken to a Hereditary Health
Tribunal. Could this group have picked up international medical discussions? Alexis Carrel, a longstanding member of the Rockefeller Institute for
Medical Research in New York, published the bestseller *Man the Unknown*
in 1935, when he recommended that criminals be executed in 'small
euthanasic institutions supplied with proper gases'.[6]

The strategy of the Medical Trial prosecution was to reconstruct vertical
chains of command. Euthanasia fitted this hierarchical schema well: here
was a demonstration of the mass destruction emanating from the totalitarian state. Hitler's secret decree authorising euthanasia was symbolically
dated 1 September 1939 to coincide with the German invasion of Poland.
This was useful for the prosecution in demonstrating how euthanasia was a
war crime, and so within the military court's sphere of competence.
Overall, the strategy was to attack Nazi medicine as inefficient, unscientific
and monumentally destructive.

The defendants and their lawyers legitimised euthanasia as the genuine
relief of the suffering of the incurably ill. Karl Brandt cited the petition of
parents of a severely disabled child to the Führer, requesting that their
severely handicapped newborn be killed. This made the point that the Nazi
leaders were responding to a popular wish; Brandt dated the incident to
1938. The medical historian Udo Benzenhöfer has established that child 'K'
was in fact Gerhard Herbert Kretzschmar who was born on 20 February
1939. Hitler sent Brandt to visit the child in July 1939, when he was in the
'care' of the Leipzig professor of paediatrics, Werner Catel. Baby
Kretzschmar died later that month.[7] Karl Brandt's confusion as to the date
gave the impression that his visit was before the euthanasia policy was
settled; but Hitler and his circle of physicians and bureaucrats had reached
the decision to kill malformed newborn by spring 1939.[8] Brack similarly
claimed that he first heard of euthanasia only at the end of September or
early in October 1939, and recollected hearing Hitler mention petitions
from parents. He appears to have been involved in establishing the

euthanasia machinery from early 1939. The Nazi leadership sought legitimacy for euthanasia by claiming that it was in response to popular demand. They also alluded to the miserable existence of long-stay patients in other countries.[9]

A second defence was that the decisions to kill were made by doctors adjudicating individual cases. This position was based on the view that clinical judgement relied on the personal integrity of the physician, and that a court had neither the knowledge nor right to question the clinician. Karl Brandt affirmed: 'Every individual doctor was responsible for what he did in the course of those measures which led to Euthanasia. Each doctor was absolutely responsible for his judgement.'[10] Brack even claimed that as the son of a doctor his rationale was fundamentally medical.[11] Again, this defence relied on a fiction: the doctors in the T-4 central office made the decision to kill on the basis of hastily compiled and often inaccurate medical questionnaires. Alexander prepared questions for Karl Brandt for McHaney to attack medical rationales for euthanasia.[12] Cross-examination revealed how the system of medical decision-making operated: although several doctors could be involved, face-to-face evaluation of the victim was not part of the system.

Brack's position emerged as ambivalent – was he an obedient functionary or imbued with a perverted set of ideals of medical ethics? Tensions surfaced between the medical panels and the administrators. Brack observed: 'I am not a doctor, and I could not criticise an expert in any way concerning his medical work in the questionnaires concerning euthanasia.'[13] The sphere of the scientific expert was to remain inviolable. At the Trial there was outrage that medical credentials were challenged. The asylum director Hermann Pfannmüller objected to judicial probing: 'I am a doctor confronted with a lawyer, and our points of view are completely divergent.'[14] He had arranged public tours of his asylum, Eglfing-Haar, to convince the public that his patients were mere 'lumps of meat'. He explained to the military tribunal how he had worked tirelessly over long hours to select patients for death. (He earned a five-year sentence for his murderous endeavours in March 1951.)[15]

Brack cultivated psychiatrists who were keen to develop research out of euthanasia; his relations with the Heidelberg psychiatrist Carl Schneider were cordial.[16] Brack and Karl Brandt insisted that they had never heard of the 14-f-13 programme until their trial. They denied that the killing of the infirm, asocials, those incapable of work and Jews had anything to do with euthanasia proper. Their defence was that what was intended was genuine medical care.[17] Brack did nut accept such coarse expressions as '*nutzloser Esser*' (useless gobblers). He considered that the prosecuting lawyer Hochwald tried to impose an interpretation on him of routine mass extermination, running contrary to his individualist point of view. Here, his defence accorded with that of Blome, who declared his aim as Deputy

Reich Physicians Führer was to secure clinical autonomy against the centralising incursions of Conti. Bouhler feared Conti would extend euthanasia beyond the incurably mentally ill – though Brack's justification could not be verified, as Bouhler was no longer alive. Blome had to contest a deposition of Conti made before his suicide, incriminating him.[18] The defence attempted to confuse Nazi mass murder with the killing of a terminally ill person who had freely expressed a will to die. In effect, the defence shifted guilt onto the deceased.

The charge of common design or conspiracy to impose euthanasia could not be proved. The accused showed that most neither collaborated with nor knew each other. Blome, Karl Brandt, Brack and Hoven denied any close connections. Thus Blome stated that he dealt with Brack at only a single meeting in 1941, when Conti ruled that the Reich Physicians had no jurisdiction over euthanasia. Blome affirmed: 'I did not talk to Hitler during the whole war. I never spoke with Himmler and Bouhler about the euthanasia program, and with Himmler when the so-called euthanasia action had already stopped.'[19] Similarly, Brack testified to having very occasional contact with Karl Brandt.[20] The judges accepted that there had not been a concerted conspiracy to impose euthanasia, but they did not dispute the prosecution evidence of a series of planned euthanasia programmes.

The prosecution argued that euthanasia was a preliminary to genocide. Again, Brack denied anti-Semitism, and involvement in the killing of Jews.[21] He explained that he joined the Waffen-SS in 1942 to distance himself from a regime with which he was increasingly disillusioned. Karl Brandt argued that the euthanasia programme was medical in its aims, but that the Aktion 14-f-13 (in which he disclaimed involvement) was racial and political. A cloud of suspicion hung over Karl Brandt, whose devotion to Hitler had made him a willing medical overseer of euthanasia. Even when the T-4 programme ceased, Brandt had ordered that patients be transferred by the same transport organisation which had conducted psychiatric patients to their death, in order to clear beds for the injured from Allied bombing.[22] The prosecution believed that he took orders from Himmler to convert the T-4 organisation for killing Jews and asocials, while Brandt denied links between euthanasia and the concentration camps. He insisted that he was daggers drawn with Himmler and the SS, while the prosecution cited his steady promotions through the SS hierarchy.[23] Blome claimed to have adopted a critical position towards the Nazi regime by 1939, citing his memoir *Arzt im Kampf* as evidence. Hoven was under arrest for a considerable period at Buchenwald. Several of those on trial claimed that they in fact came to resist Hitler.

Courtroom statements yielded information on events and insight into the character of the perpetrators; but testimonies could be highly misleading when there were no corroborating documents. A deep and at times perplexing contrast emerges between the brutality of euthanasia as actually implemented and the retrospective moral legitimations the accused presented. Blome succeeded in gaining his acquittal: he conceded that he had been a Nazi activist in medical organisations; he believed in euthanasia as an ideal, but disputed the legitimacy of the Führer's order of September 1939. Blome's stance was that he had since 1940 opposed euthanasia as practised by the Nazis, and that his opposition was apparent in his *Arzt im Kampf*, when he praised the idealism of the early Nazi movement.[24] Alexander reflected with satisfaction on Blome's critical position: 'We ought to have Blome on the bench, he realizes things were not merely unethical, but criminal!'[25]

Brack, Karl Brandt and Hoven failed to convince the court. Brandt denied his involvement in racial euthanasia killings after 1941. He had argued for a strictly medical form of euthanasia. He was deeply religious – although a non-church member, he was humane and an admirer of the Alsatian medical missionary Albert Schweitzer.[26] He cited how Pastor Bodelschwingh at the Bethel institution recognised his idealistic motives, indicating a common understanding and mutual respect with the evangelical church.[27] Brack similarly claimed a deep belief in humane ethics: he believed that euthanasia was accepted throughout the world; only unjust killing was against religion.[28]

The defence claimed that what was intended was the release from suffering of incurable and severely disabled patients whose lives were not worth living (*lebensunwert*). Karl Brandt and Brack affirmed their conviction that what was intended was genuine euthanasia, no different from the practices advocated in other countries. They cited a range of literary justifications of euthanasia, including essays by the Austrian socialist and (Jewish) eugenicist Julius Tandler, and extracts from Carrel's *Man the Unknown*. These were deliberately chosen as by recognised progressives, Tandler having been persecuted by Austrian nationalists, and Carrel as a former Rockefeller Institute scientist. Brack's lawyer, Fröschmann, was impressed with his client's high ethics and humanity. Brack was portrayed as an idealist advocate of euthanasia in the sense of seeking the release from torment of the severely disabled. American evidence for euthanasia was cited to give the impression that Nazi euthanasia amounted to no more than a legitimate moral position.[29] The defence, however, failed to confront the euthanasia killings as relentlessly carried out until 1945, which neither Karl Brandt nor Brack criticised. Brandt insisted on psychiatric hospitals continuing to provide the Reich authorities with reports on patients – information which could be used to order a patient's death.[30]

Karl Brandt denied that he held overarching medical powers, that these extended to euthanasia and genocide, and that he was involved – or had knowledge of – human experiments. He drew a distinction between euthanasia as ethically and medically justified in individual cases, and the mass killings of psychiatric patients. He stressed his distance from the SS's medical policies. He tried to counter the effect of Brack having heavily implicated him in the administration of children's euthanasia; Brack also accused him of a crucial role in the transition from the T-4, centralised phase of euthanasia to what he claimed to have believed were Jewish slave labour camps but turned out to be the extermination camps of *Aktion Reinhardt*. The prosecution insisted that it could only have been with Brandt's 'personal knowledge and consent' that doctors like Eberle and Horst Schumann were drafted in from Auschwitz.[31] The prosecution placed Brandt in the midst of genocide.

The prosecution used interrogations to prove that Brandt was involved in the medical administration of euthanasia from the beginning. Hermann Boehm, the NS Physicians League expert on racial hygiene, Referent for hereditary biology in the NSDAP Hauptamt für Volksgesundheit, and from 1943 professor at Giessen for Erb- und Rassenpflege, joined the ranks of those interrogated at Nuremberg. Boehm was an early convert to Nazism. Since his time as a medical student in Munich before the First World War, he combined his interests in pathology with racial hygiene. In March 1925 he joined the NSDAP and had the prestige of the low membership number of 120. Boehm took a lead in propaganda for hereditary health at the Nazi medical training centre of Alt-Rehse in the depths of rural Mecklenburg, where he also had an institute for hereditary biology and enjoyed the support of Blome. The interrogator did not regard twin research as in any way problematic, but Meyer was angered by Boehm's assertion that doctors were powerless in the Nazi state after 1940.[32] By stressing his critical position towards euthanasia, he was able to exculpate himself, and emphasise that Karl Brandt and Brack (under the alias 'Jennerwein') were in charge of the T-4 euthanasia programme.[33]

Boehm testified as to the division of labour between Brack and Brandt. In the autumn of 1940 a medical colleague, Kurt Klare, was perturbed by protests from victims' families, the way relatives were informed and the use of bogus diagnoses. Boehm wanted to replace the subterfuge with consent of the families of the euthanasia victims and to secure legislation. He complained to Bormann, who turned down any direct approach to Hitler and instead recommended a meeting with Karl Brandt (whom Bormann actually loathed) as he was responsible for the euthanasia measures. Brandt explained to Boehm how euthanasia was to provide a 'gentle death' and release from 'unbearable suffering'. Boehm testified that Karl Brandt was also critical of aspects of the programme, but that Hitler demanded secrecy. Brandt struck Boehm as young, weak and insecure, and that he did not know how to change procedures. Boehm provided the prosecution with a deposition that Brandt actively ran the medical side of the euthanasia arrangements.[34]

Brack testified that he too wanted to bring euthanasia out into the open, so he commissioned a doctor and author, Hellmuth Unger, to prepare a film script to win over public support. Unger was a press officer for the Nazi Doctors League, author of a novel promoting euthanasia, *Sendung und Gewissen*, and of popular medical histories of Robert Koch and Behring and of the development of the sleeping sickness drug Germanin. He glorified the medical researcher as empowered to take liberties with life.[35] Unger joined the circle of doctors around Hitler, who persuaded the Führer to adopt euthanasia at a point in time, which the Medical Trial left undetermined. Brack supported Unger to provide a screenplay for the notorious feature film *Ich klage an* ('I accuse'), which romanticised euthanasia as a matter of personal tragedy: the storyline was how a professor's glamorous wife contracted multiple sclerosis. Brack believed that the film provided the best conclusion for his testimony in its appeal to public sympathy for the killing of the incurably ill.[36] Karl Brandt had also requested the film to prove to the tribunal that euthanasia was ethical and humane, as well as the films *Life Unworthy of Life* and *Existence without Life*.[37]

Brack claimed that he supported euthanasia out of ethical conviction. Euthanasia differed fundamentally from *Völkermord* (genocide) as unleashed by Himmler and Globocnik (a revealing reference, as Brack had associated with the latter when T-4 personnel were transferred to the *Aktion Reinhardt* extermination camps). He stressed that he was from a respectable and well-educated home – his father was a doctor who had assisted Frau Himmler at birth, and that this connection brought him an administrative post. He wished to help mentally ill concentration camp prisoners and give them access to medical care, and that Bouhler advised him to send psychiatrists to the camps. Brack claimed that he did not even know of the secretive T-4 organisation. This is incredible given Brack's important role in its administration. Brack hoped that he could convey an atomised view of the state bureaucracy in which functionaries carried out assigned tasks without any broader understanding of the administration or its aims.[38]

Alexander was concerned not only with the responsible individuals but also with the psychological conditions underlying the genocidal atrocities. Schoolbooks of 1935–36 illustrated 'the subtle way in which the taboo against mass killing was broken down in school children' by asking them to make calculations using inflated figures. It revealed how the German psychology was prepared to lie when it served a purpose, but only to inferiors.[39] Nazi euthanasia was 'a preliminary step towards genocide', and 'German doctors involved in the euthanasia program were sent to the eastern occupied countries to assist in the mass extermination of the Jews'.[40] The prosecution saw euthanasia as the cruel culmination of racial medicine.

14
Experiments and Ethics

Why a Code?

At the close of the Trial on 19 August 1947, the presiding judge promulgated guidelines for 'permissible medical experiments'. These principles have been known since March 1960 as the Nuremberg Code. At the time they appeared more as guidance to prevent a recurrence of the gruesome abuses in Nazi medical research, rather than as general ethical and legal principles. The Code's origins, motives and significance are matters of controversy. The judges said that the guidelines assisted them in 'determining criminal culpability and punishment'.[1] At the same time, the Code marked the culmination of developments, which went beyond the immediate business of a military tribunal. The ethical concerns ran parallel to the Trial, occasionally intruding into court proceedings, and the final guidelines summarised discussions on experimental procedures.[2]

The judges agreed that the Nazi experiments came under the Tribunal's authority to prosecute crimes against humanity and war crimes: 'These experiments were not the isolated and casual acts of individual doctors and researchers working solely on their own responsibility, but were the product of co-ordinated policy making and planning at high governmental, military and Nazi Party levels, conducted as an integral part of the total war effort.'[3] Taylor argued – and the judges agreed – that well-established laws concerning murder, manslaughter, assault and battery, as well as the laws of war, were the primary basis of the Trial. Although the charge of conspiracy failed, Taylor held firm that 'the crimes which these doctors committed were not a thing apart by themselves but contributed merely a part of the crime committed by the Nazi state'.[4]

The question thus arises: why did the Trial judges feel compelled to outline a set of ethical and moral principles governing clinical experiments? To answer this requires locating the Medical Trial within a broader ethical discourse on Nazi medical atrocities. The court only sporadically engaged with ethical issues, and the ethical discourse was distinct from the

Trial proper. Nor was the Code an intrinsic part of the sentencing. In contrast, Katz and Shuster see the Code as formulated by the judges in response to the courtroom testimony of experts, witnesses and victims.[5] While there certainly was 'the clash of opinion and standards' in court, the ethical issues can be seen as driven by the momentum of the discourse on medical war crimes preceding the Trial.

Taylor considered that the judges – and especially Judge Sebring with his 'probing mind' – were 'primarily responsible for the famous ten principles'. Sebring, a Florida Baptist, had an active interest in judicial ethics and in the responsibility of the doctor to the public: 'Sebring dealt with all questions related to medical ethics', and appeared to Taylor as the most intellectually able on the bench.[6] Whereas Judge Biddle's private diary provides detailed insight into the IMT verdicts, no record of judicial deliberations has surfaced for the Medical Tribunal. But in reminiscences and lectures to students Sebring confirmed that 'during the course of the 150 page opinion and judgment I wrote the following for the court respecting what I thought should be valid limitations on human experiments even though conducted on concentration camp inmates'. Sebring then gave the text of the Code and its preamble. His reflections confirm that it was his task to deal with the experiments when delivering judgment, and that he drew together expert opinions to formulate the Code.[7]

Sebring's contribution can be linked to how medical experts at the Trial took a formative role in the evolution of the Code. Ivy's work was recognised by medical ethicists in the 1960s. Alexander published on his role in the mid-1970s. The bioethicist Michael Grodin has identified Ivy and especially Alexander as the prime instigators of the principles, which Sebring then drew on. Jon Harkness has shifted the emphasis back to Ivy's involvement in the AMA's terse guidelines publicised before the Trial, while adding a sinister twist to the story: he suggested that the judges connived at Ivy's duplicity in misrepresenting ethics controls on the Stateville malaria experiments. Thus, ironically, the Code was based on unethical misrepresentation. Harkness placed his important findings before President Clinton's Advisory Committee on Human Radiation Experiments. He raised the matter with Taylor and published in the special issue of *JAMA*, marking the 50th anniversary of the Trial in 1996.[8]

For the most part, medical ethicists have analysed only published materials, and did not seek rough sketches or examine the mountains of manuscript materials generated by the Trial. The origins of the Code lie further back than the AMA's brief principles. In fact, the scheme of a regulatory Code was conceived in the summer of 1946, when it prompted the decision to hold the Medical Trial. The ethics of clinical research was on the agenda during the preparatory interrogations in autumn 1946.[9] Ivy outlined a basis for the principle of informed consent to the ISC on 1 August 1946. He recommended voluntary consent and that experimental subjects be 'informed of the hazards, if any'. His requirements raised informed consent as an issue, although the substance of the Code was to evolve substantially during the Medical Trial.

Ivy's Rules of Experimentation on Human Subjects of August 1946 focused the prosecutors' attention on what constituted criminal research. The war crimes staff weighed the issue of whether the Nazi experiments were crimes, and a suitable matter for trial. The prosecutors drew a sharp contrast with the probity of American research – pointing out the Office of Scientific Research and Development required accident insurance. Neither German law nor the Hague Conventions on combat allowed experimentation without consent. The AMA's Rules for Animal Experimentation showed that the Nazis gave far less consideration to the human victims of experiments with respect to pain, suffering and hygiene than to the standard care received by laboratory animals. Moreover, the experiments interfered with both the happiness and desires of an individual, and any results would thereby be morally debased. There was no single instance ever justifying the use of non-volunteer subjects: conversely, it was always possible to secure volunteers without coercion. Forced participation would vitiate both the method and results. Furthermore, experimenting without their consent would cause patients to lose confidence in their physicians, and beyond this would result in an amoral profession and society: good has to be achieved by moral means. Nazi physicians performed experiments because of their lack of regard for vulnerable, poor and underprivileged patients, denigrated as 'useless'. The transfer of the rationales of euthanasia to genocide and the resulting killings provided 'a rationalised licence to use human beings of alleged subhuman grade as experimental material'. The ideological spurs of the master race and *Lebensraum* accelerated these steps.[10]

False start

The impulse to formulate ethical requirements goes back to inter-Allied evaluations of the scientific quality of German wartime science, and – in the final analysis – to the demands of surviving victims on the need to protect the experimental subject. German scientists were self-righteous and sought to establish their ethical probity as predating Nazism. They claimed that German regulations of 1931 on physicians' obligations to seek consent for new therapies meant that research was conducted on an ethically sound basis during the Third Reich.[11]

Fierce public controversies over patients as human guinea pigs reached a highpoint with the BCG disaster at Lübeck in 1930, when 77 children died following administration of contaminated TB vaccine. This had an impact on Reich Health Council guidelines for human experiments. Christian Bonah has pointed out that what were at first intended as guidelines for researchers became a code regulating clinical therapy in hospitals. The Reich Ministry of the Interior issued 'Regulations on New Therapy and Human Experimentation' on 28 February 1931, and a revised version on 11 June 1931. The provisions included informing the subject: 'Innovative therapy may be carried out only after the subject or his legal representative has unambiguously consented to the procedure in the light of relevant information being provided in advance.'[12] A final section sought to eliminate abuses in research: this required consent and the primacy of animal experiments, while experiments on children and the dying were denounced as unethical.[13]

That the 1931 regulations were legally binding until 1945 was refuted at the Medical Trial, and there is no evidence that the consent procedures were routinely enforced. By the time Ivy discussed the 1931 guidelines in court on 13 June 1947, other formulations of consent had taken shape.[14] The regulations assumed a mythical status, when after the war leading German scientists asserted that they had adhered scrupulously to them. The Heidelberg pharmacologist Eichholtz sent the regulations to Mitscherlich (the prospective trial observer) on 22 November 1946.[15] (Eichholtz had in fact urged human experiments on the Plötner-Rascher blood styptic in 1944.[16]) German pharmacologists launched a pre-emptive strike just prior to the opening of the Trial. They petitioned Sir Henry Dale, the President of the Royal Society, that the German medical profession was under merciless legal attack. In order to prove how physicians behaved humanely and conscientiously, Eugen Rost (also of Heidelberg) forwarded to Dale the text of the 1931 code on human experiments. He sent this at the suggestion of the pharmacologist Heubner, who advised on seawater experiments. Dale respected Heubner as an opponent of National Socialism (he had forced the Nazi chemist Druckrey to leave his institute and had been perturbed by the dismissals and deportations of Jews). Rost asserted in a clumsy allusion to British liberties that this 'Magna Carta' of German physicians had been in force since 1931 and throughout the Nazi period.[17]

Dale replied positively to Rost that he understood 'the desire of yourself and my other friends ... that British colleagues should understand the completeness with which general medical opinion in Germany would desire to repudiate, and to disassociate itself from, practices such as those for which certain German medical men now stand accused before international Courts of Enquiry'.[18] Nevertheless, he held back from following Rost's request to publish the German experimental code. Dale, an inveterate Germanophile since his studies with Paul Ehrlich, was a member of the British Committee on Medical War Crimes, which had been convened in September 1946 to adjudicate on the ethics and scientific value of the German human experiments. The chairman Lord Moran insisted that there be no publications prior to the issue of a final report, and so Dale registered Rost's point without publicising it.[19]

Criminality and medical ethics

Scientific intelligence officers prioritised the ethical analysis of Nazi experiments. Thompson lobbied to co-ordinate scientific intelligence and war crimes investigations. The investigation of nerve and poison gases in early 1946 produced evidence of ethical violations. The British officer Edmund Tilley pressed for evidence of human experiments concerning Tabun in a sequence of interrogations. At first, the interned German chemists and doctors feigned ignorance: but gradually Hörlein of IG Farben and Wirth admitted to animal and then human experiments, albeit self-experiments and on student volunteers. What they feared was punishment for unethical experiments on 'involuntary "human guinea pigs"'. There were parallel tests with mustard gases.[20]

At the FIAT meeting of 15 May 1946 of British, French and US representatives, the key issue was 'unethical experiments on living human beings'. The bacteriologist Lépine supported 'moral condemnation of the unethical practice of German scientists', and 'that this should be done by the representatives of the scientific bodies of the four Powers'. The chairman, Sydney Smith, went further, calling for 'a meeting at which the practice of this criminal activity could publicly be condemned by representatives of science from all countries'.[21] The aim was international guidelines on human experiments. Smith, an expert on ballistics and firearms injuries, endorsed the twin demands of trials of medical miscreants and ethical evaluation: 'I am of the opinion that many experiments have been carried out on human beings without their consent, that the experiments show as far as our information goes, inadequate planning, crudity of technique, gross indifference to the value of human life and callous disregard of human suffering.'[22] Smith accused the German human experiments of violating the ethical canon of consent of the experimental subject. By the summer of 1946, the strategy had crystallised of a dual ethical and war crimes approach to coercive experiments. Ethical issues underlay forensic investigations, interviews with victims and interrogations. Somerhough explained that 'a case involving gross breaches of medical ethics should be given all the publicity available so as to engender the interest of other powers in cases of a similar nature'.[23] He supported Mant's investigations of survivors of Ravensbrück with the aim of a Medical Experiments Trial under British jurisdiction.[24]

When the AMA nominated Ivy to the embryonic medical war crimes commission in mid-May 1946, his mission was fuelled by moral and scientific convictions. He believed in the time-honoured medical custom that a researcher should undergo self-experiments before recruiting volunteers: 'In order to get the corpsmen interested in serving or "volunteering" as "assistants" or subjects in the earlier experiments at the Institute, I served as a subject in a trial experiment. This was true of the dilute seawater tests, the Goetz water tests and straight seawater test. This, however, has always been my policy in laboratory work.'[25] Alexander similarly took part in clinical experiments 'both as experimental subject and as experimenter', and Thompson was a subject in decompression chamber experiments at some personal risk.[26]

Ivy was 'a man with strange religious twists'.[27] His sense of mission to inculcate sound morals among medical students and physicians was apparent in his incessant public speaking in favour of temperance and Christian conduct. He relentlessly advocated ethical conduct, while defending the experimental basis of medicine. He saw the opportunity for using his AMA credentials to get quiet legitimation for animal experiments. In 1946 Ajax Carlson, Ivy's Chicago mentor, was taking steps to found the National Society for Medical Research to counter critics of animal experiments. The aviation physiologists Bronk, as Chief of the Division of Aviation Medicine, and Fulton gave support.[28] The time was ripe to campaign for the legitimacy of animal experiments. In January 1946 Ivy commended to Fulton an article in *Pageant Magazine* favouring animal experiments.[29] Ivy linked the delicate operation of establishing legitimate use of animals in medical research to providing an ethical basis for experiments on humans.[30] His ethical guidelines stressed that animal experimentation was to safeguard humans from harm. His embryonic Code of August 1, 1946 enshrined animal experiments, as absolutely necessary for protecting human life. In November 1947 he addressed the Central Association of Science and Mathematics Teachers on the need for physiologists and physicians to oppose anti-vivisectionists, to safeguard medical progress in the interests of human welfare.[31] When he testified at Nuremberg he argued that animal experiments rendered the hazardous experiments of Ruff, Romberg and Rascher unnecessary.[32]

The demand for ethically based clinical research found support in Britain – where physiologists were well organised to counter anti-vivisection – and among the leaders of French medical science. Some were victims of the concentration camps, and the *Association Nationale des médecins deportés* kept a vigilant eye on war crimes verdicts; others had soldiered on as research scientists under German occupation. The bacteriologists from the Pasteur Institute, who pressed ahead with the French branch of the scientific commission, had close British and US links. Pierre Lépine, who had been dismissed by Laval because of his American sympathies, was in London in May 1945, assisting the revival of French research.[33] Lépine and Legroux convened the inter-Allied conference at the Pasteur Institute, instead of Wiesbaden.[34] The Paris location had a symbolic significance: the heirs of Pasteur were to sit in judgment over the misdeeds perpetrated in the tradition of Robert Koch as an icon of German medicine. The presence of an ethical tribunal enhanced the Institute's reputation as a beneficiary to mankind during post-war reconstruction. The Institute had endured a difficult war, faced by the choices of collaboration or resistance.[35] Some German medical criminals like Ding/Schuler were trained in Paris, and the concentration camp experiments involved the RKI in Berlin.

The ISC minutes underlined the sense of the historic significance of the location. 'The Chairman [John Thompson], voicing the feeling of all members, expressed his gratitude to the Pasteur Institute for its hospitality in supplying accommodation for the meeting and expressed his gratitude for the deep honour done to each member in allowing him to meet in a room formerly and frequently used by Louis Pasteur.' Ivy venerated Pasteur as a heroic genius of Napoleonic grandeur.[36] The bacteriologist Legroux headed the French scientific commission, and liaised with the UNWCC. He facilitated American investigations of the German vaccine experiments in Strasbourg.[37] The process of peer review would judge the Nazi research according to the procedures of science.

As a preliminary to the Paris meeting, Ivy met Thompson on 29 July 1946 at the FIAT offices, Hoechst: 'I found that Dr. Thompson held views very similar to those I had formulated relative to the problem of war crimes of a medical nature.'[38] Thompson concluded that the combination of the demonic psychology of the perpetrators and the precepts of science detached from religious faith generated the unparalleled medical crimes. He sent the draft of a code for vetting by Canadian Catholic experts in canon law.[39]

Thompson was perturbed that the British and Americans were also conducting unethical human experiments.[40] He contemplated medical teaching based on students conducting self-experiments rather than animal experiments, convinced that almost all physiology can be taught from the human being.[41] He made available to Ivy extensive documentation on human experiments, and they discussed the general problem of the need for a regulatory framework. Despite their shared concerns, a polarity emerged between Ivy's permissive view on animal experiments and Thompson's efforts to minimise these. The momentous meeting at the Pasteur Institute on 31 July/1 August 1946 pursued the issue of experimentation 'without the subjects' consent and with complete disregard of their human rights'.[42]

Ivy participated at the Pasteur Institute meeting as 'Special Consultant, Secretary of War, War Crimes Branch (US)'. He shouldered US interest in violations of medical ethics. 'Doctor Ivy warned that unless appropriate care is taken the publicity associated with the trial of the experimenters in question, and also the publicity which is bound to be attached to the official report of this meeting, may so stir public opinion against the use of humans in any experimental manner whatsoever that a hindrance will therefore result to the progress of science.' He cautioned that the knowledge that human beings were extensively used in experimental work could entice people in other countries to engage on human experimentation.[43]

Ivy's demarcation between legitimate clinical research and criminal abuses stressed that animal research was a fundamental prerequisite for clinical research: 'Therefore, Dr Ivy felt that some broad principles should be formulated by this meeting enunciating the criteria for the use of humans as subjects in experimental work ... it was agreed that each member should carefully consider Dr Ivy's proposals and suggest any changes which he may see fit to do so at the next meeting. On the basis of this the criteria believed to be necessary would then finally be formulated.'

Ivy's draft code contained the germ of the principle of providing experimental subjects with information on hazards:

'OUTLINE OF PRINCIPLES AND RULES OF EXPERIMENTATION ON HUMAN SUBJECTS
I. Consent of the subject is required; i.e. only volunteers should be used.
(a) The volunteers before giving their consent, should be told of the hazards, if any.
(b) Insurance against an accident should be provided, if it is possible to secure it.
II. The experiment to be performed should be so designed and based on the results of animal experimentation, that the anticipated results will justify the performance of the experiment; that is, the experiment must be useful and be as such to yield results for the good of society.
III. The experiment should be conducted
(a) So as to avoid unnecessary physical and mental suffering and injury, and
(b) by scientifically qualified persons
(c) The experiment should not be conducted if there is a prior reason to believe that death or disabling injury will occur.'[44]

The requirement that volunteers should be told of hazards before giving their consent represented an important step towards informed consent.

An annotation on the agenda of the bacteriologist Lépine at the first ISC meeting indicated that it was intended to discuss Ivy's code more fully on 16 October 1946. The US delegate at the follow-up meeting was the Nuremberg prosecutor Alexander G. Hardy. He assured the Commission that it was expected that by the next meeting the American Committee would be in existence.[45] The Commission at first provided the Nuremberg prosecution with background on medical atrocities, but it lost relevance when, during September 1946, the British and US authorities collaborated on preparations for the Medical Trial. Ivy's draft Code shaped the Trial agenda, as dealing with 'the principal doctors who engaged on involuntary experimentation on human beings'.[46]

Ivy obtained scientific information from interned Germans, and briefed the US war crimes lawyers. He recommended that they should deploy a physician-scientist 'in interrogating witnesses and potential defendants, as in all medical cases', and this led to the appointment of Alexander.[47] Ivy met Taylor's staff to discuss: 'Were the experiments necessary? Were they properly designed? Were the results of any value? Is it legal or ethical to experiment on human subjects? Since some of the victims were condemned prisoners, was it not proper to use such prisoners in experiments? Some of the attorneys knew we had performed a great number of experiments on ourselves and on volunteer human subjects during the war ... So the attorneys were somewhat confused regarding the ethical and legal aspects as well as the scientific aspects of the question.'[48] The Code was then reformulated in the light of the forthcoming Trial with appendices on the 'Experiments' as crimes, and an explanation of why Nazi physicians performed them.[49] Ivy's ethical agenda remained an undercurrent at the Trial, which sporadically erupted in courtroom discussions of experimental ethics.

At the conference on 6 August 1946 with Taylor, Ivy suggested, 'caution should be exercised in the release of publicity on the medical trials so that it would not jeopardise ethical experimentation'. Taylor introduced Ivy to John M. Anspacher, who was in charge of public relations, to organise publicity to emphasise the difference between ethical and unethical experimentation. He suggested that it would be advisable for the AMA and BMA to provide rules for ethical experimentation, to clarify legal aspects of euthanasia, and the probable arguments of the defence. Ivy supplied statements regarding ethics and a list of questions for questioning witnesses and potential defendants. His report to the Department of War indicates that ethical discussions shaped the decision to mount the Trial as primarily focused on human experiments rather than other areas of Nazi medical killing.[50]

Ivy prescribed a dual strategy of obtaining 'as much scientific data as possible', and that 'caution should be exercised in the release of publicity on the medical trials so that it would not jeopardise ethical experimentation'.[51] In accordance with Anspacher's priority, an American press release from Nuremberg emphasised the 'inhuman experimentation program', and the 'thousands of experiments' which violated 'ethical rules for human experimentation'.[52] On 7 November 1946 Marcus, the energetic Chief of the US War Crimes Office, contacted Ivy concerning the need to ensure that the AMA House of Delegates approve a code of experimentation. He hoped the AMA would send an observer to Nuremberg.[53]

Hardy, recently returned from the ISC meeting, showed its effect when he confronted the Buchenwald camp doctor Hoven with the implications of the Hippocratic Oath on 22 October 1946:

Q. When you became a doctor and were given your degree you took the oath of Hippocrates. You stated when you took that oath that you would do everything in your power to preserve life.

A. Yes

Q. Doctor you breached that oath. You may have breached the oath because of orders from above but, nevertheless, you breached the oath of a doctor. The medical profession in Germany, as you know has sunk to a depth that is disgraceful, not only to you as a German doctor, but to American doctors and doctors of other nations. It is something which will take a thousand years to wipe out. The medical profession in Germany has done things that have never been heard of before ... In this trial we are going to bring it to light so it will never happen again, so that other men so sincere in their profession won't allow such a thing as the German Reich to destroy their belief.[54]

The interrogator Meyer accused the aviation medical specialist Schröder of violating the binding promise of a physician and that it was ethically wrong to have approved experiments at Dachau. Schröder defended the seawater experiments as non-fatal and as contributing to the war effort. As preliminary tests were carried out on volunteers and colleagues, he believed the seawater drinking experiments to be safe.[55] This was an early salvo in what was to be a sustained exchange of views on the ethics of experiments.

Faced by the German defence that the American research was itself coercive, the prosecution intensified its interest in ethics. Taylor reflected, 'Curiously enough, we were educated in large part by our opponents. We had ample opportunity to interrogate, and in the course of interrogating doctors, of whom some were sophisticated and very able physicians, we began to realise the kind of problems we would be up against when presenting the case.'[56] On 1 November 1946 Taylor cabled the War Department for academic reinforcements: 'It is necessary that we have extensive paper on the history of medical experimentation on living human beings with particular emphasis on practice in U.S. Defendant Rose states that U.S. doctors have extensively experimented on inmates of penal institutions and asylums, especially with malaria. Any truth in this. If so give us full facts. Do not limit paper to answering this question. Suggest you contact A.C. Ivy, University of Illinois Medical School on this problem. Answer soonest.'[57]

The prosecution rapidly realised that it would have to contrast ethical and non-ethical procedures for human experiments. Alexander contacted Abraham Myerson for details of routines in Boston State Hospital, and then leading American medical institutions for evidence on human experimentation. These included the Commissioner for Mental Health of Massachusetts, and the Rockefeller Institute of Medical Research for procedures for human experiments in hospitals. He wanted details of permission from next of kin and guardians, as well as on rules covering experiments on normal control subjects.[58] The idea was to elicit a collective reaction from leaders of American medicine.

On 23 November *JAMA* published its editorial 'The Brutalities of Nazi Physicians'. Taylor's request and Rose's counter-attack spurred Ivy to find additional support for his code. He submitted a set of rules to the Judicial Council of the AMA, requiring that: 'Before volunteering the subjects have been informed of the hazards, if any.' He elaborated criteria allowing the pursuit of experiments as 'the method for doing good', and argued that the Hippocratic precepts of benefiting the sick, not giving any deadly medicine and of a duty of confidentiality to the patient 'cannot be maintained if experimentation on human subjects without their consent is condoned'.[59] The references to Hippocratic duty to the individual patient, and the need to provide information appear not to have found favour.

The AMA's Judicial Council recommended to the House of Delegates 'Principles of Medical Ethics'. These watered-down principles demanded far less than Ivy's stipulation of information on potential hazards:

1. The voluntary consent of the individual on whom the experiment must be performed must be obtained.
2. The danger of each experiment must be previously investigated by animal experimentation.
3. The experiment must be performed under proper medical protection and management.[60]

These rules were in small type in *JAMA*, and were likely to be overlooked by readers. A regime of discretionary controls by the physician replaced Ivy's postulates of informing the subject of the hazards, the good of society; avoiding suffering, injury and disability. Apart from his rudimentary notion of voluntary consent, Ivy secured approval for animal experiments as an absolute prerequisite for clinical research. In court at Nuremberg in June 1947, Ivy repeated his original principles and cited AMA's restricted requirements.[61]

Thompson continued to develop an ethically informed evaluation of medical war crimes. Already seconded by the RCAF to the RAF, by the RAF to FIAT, and back to the Canadian Department of External Affairs, he was now seconded once again to British military service to assemble evidence on coercive experiments. Thompson's multiple missions created an administrative tangle, while giving him considerable latitude to follow his ethical and spiritual concerns. Somerhough stressed the significance of his inter-Allied medical links.[62] Thompson had embarked on an ambitious mission to establish an International University to be sited in Germany. Intellectual leaders, theologians and psychologists were to define the values of Western civilisation as a bulwark against the advancing tide of communist materialism. He accepted the post of secretary of the ISC on the British payroll, while continuing a broader international agenda of rallying Centre-Right intellectuals like the philosophers Jacques Maritain and Isaiah Berlin. He embarked on an arduous agenda of collecting and classifying records of German human experiments, visiting the Documents Centers in Nuremberg, Augsburg and Heidelberg.[63]

The investigating commission was directed by Thompson with Mant and Bayle. The hope was that a US officer (ideally Alexander) would also come on board. In October 1946 the Prime Minister, Clement Attlee, appointed the devious and ambitious Lord Moran, Churchill's physician – rather than the committed and capable Smith – to lead the delegation of British medical scientists to evaluate the German human medical experiments.[64] The Anglo-French attempts to launch an international commission on medical war crimes looked to expert authority of a panel of scientists. Ivy hoped that the conclave of scientists would promulgate a Code on permissible human experiments, and he continued to press for US participation.

Ivy, Alexander and Taylor supported ethical evaluation of human experiments by the ISC, whereas the US State and War Departments declined to support any general evaluation. The War Department objected that there were no funds for publishing a report to inform the scientific and medical world about the Nazi misuse of human beings. Its position that 'personnel now assisting in the medical trial cannot be used for any such purpose' blocked the development of the ISC as a tri-Allied authority. On 17 December 1946 Secretary of War Robert P. Patterson sought advice from the State Department on whether a report should be written 'informing the scientific and medical world at large of the type of experimentation conducted by the Nazis, and (2) of condemning extra-legally the misuses of human beings in alleged experiments of a medical nature'. However, Taylor and the US government felt that such a task was more suited to UNESCO or the UN – a decision ultimately conceded by Ivy.[65] By the spring of 1947 Ivy and Taylor had accepted that they would have to work to achieve promulgation of a Code on legitimate experimentation within the parameters of the Trial. While Ivy regretted that the full ISC never materialised, it increased pressure on the judges to make an ethical declaration.[66]

Expectations ran high that the Trial would make a contribution to medical ethics. The German Union of Physicians and German faculties of medicine approached the Office of the Chief of Counsel to evaluate the Trial records from an ethical, scientific and humane point of view, 'As conclusions of widespread influence must be drawn from these trials'.[67] Medical scientists realised that a wholesale denunciation of Nazi medicine might jeopardise their own position. The War Department commissioned Ivy to spearhead an ethical strategy at Nuremberg, as a check to Rose's attack on American research. The prosecution enlisted the support of Morris Fishbein, the veteran editor of *JAMA*, as part of its strategy of publicising the ethical status of the American wartime experiments.[68] Although AMA nominated Ivy as medical adviser to the Nuremberg prosecution, he preferred the detached status of an expert witness (he had similarly declined military rank during the war). The plan was for him to come to Nuremberg armed with a History of Medical Experimentation on Humans, the *JAMA* editorial, and a revised code of ethics.[69] Ivy's editorial on the brutalities of Nazi physicians appeared in *JAMA* and this was forwarded by Damon Gunn of the War Department to Taylor, just prior to the opening of the Trial.[70]

Ivy feared that Nuremberg could turn into a trial of American medicine unless a more sophisticated strategy was pursued than that of the condemnation of human experiments as war crimes and murder. Alexander had similar qualms on the way to Nuremberg, when he heard in Washington that he was to be the expert for the prosecution: 'don't tell [Telford Taylor] that the indictment is sloppily drawn'. Alexander set out to rescue the case by collecting evidence that human experiments on US prisoners and the mentally ill were voluntary – a point to be endorsed by Ivy, who fronted the Green Committee on the Illinois penitentiary experiments.[71]

The Trial observer, Mitscherlich, also signalled the importance of ethics, as he was concerned with the human component in doctor-patient relations. He felt that it would be a mistake for physicians to distance themselves from the Trial. In fact, every doctor needed to recognise what happens when the individual suffering human being becomes an object or a case – '*einen Fall*'.[72] Mitscherlich discussed ethical dimensions in early February 1947 with 'the official observers of American, Canadian and French universities', meaning Ivy, Thompson and Bayle. Mitscherlich argued that an ethical response from the German side should take place no later than the end of the Trial. He believed (echoing Thompson) that given the large numbers of '*Mitwisser*' in German universities, it would be wrong to see the Trial as limited to judging a handful of criminals.[73] The Trial was one not just of German medical science, but of the dubious ethics of unbridled medical experimentation.

15
Formulating the Code

Defensive struggles: mind, body and spirit

The German defence accused the Allies of conducting human experiments and devising grim scientific weapons. The prosecution asked each defendant about what principles should be followed when conducting human experiments.[1] The accused stressed that they were highly ethical in their overall outlook, and that they had a firm sense of how ethically to conduct their research. The judges allowed the defendants considerable latitude to express their views on ethics, and this culminated in the cross-examination of Ivy by Ruff, Rose and Beiglböck. Taylor considered this a very unusual thing, but it showed the judges' recognition of the complexity of the case – 'Curiously enough, we were educated in large part by our opponents.'[2]

The judges' ethical concerns prompted questioning of Karl Brandt about the importance and non-importance of the experiments. They were curious whether the prosecution had overdone its condemnation of the poor quality of the scientific work. Brandt felt that some experiments were indeed useless but that others were justified by the military emergency.[3] Judge Sebring posed the question whether the skeleton collection or the shooting of tubercular Poles had any military justification. Karl Brandt conceded there was no military rationale for the skeleton collection, and roundly condemned shooting the tuberculous: 'I see no justification because a person is sick or suffering, or because he can no longer work, to kill him, no matter what his nationality is or what his age is.'[4] Judge Beals confronted Brandt with the fact that he was speaking only about 'a physical danger to life' rather than 'serious physical injury'. Brandt responded that it was important 'that in all the experiments one must make it clear to the subject what the experiment is about and what results may be expected'.[5] Here Brandt reinforced the emerging notion of informed consent.

Brandt drew attention to the practical difficulties of obtaining monkeys for experiments on chemical warfare, and that Swiss currency reserves had to be drawn on.[6] He insisted that human experiments were justified for diseases when there was no clear animal transmission model, and at times of war to avoid greater loss of life, but that the experiments should be on as small a scale as possible. His awareness of the evolving Code is shown by his comment: 'It will probably be necessary to settle these questions basically, probably on an international basis ... every state is guilty.'[7] These sentiments reveal the belief that state power was potentially genocidal and inhumane: any antidote had to come either through an international body of physicians, or a judiciary untrammelled by dependence on any state interests.

The onslaught on American science focused on coercive experimentation and on research into weapons of mass destruction. Karl Brandt and Gebhardt attacked the criminality of all involved in the research, manufacture and dropping of atomic bombs on Hiroshima and Nagasaki. If the Allies claimed that atomic weapons were justified by the war, why not also the human experiments on chemical weapons and sulphonamides, which had a strategic rationale?[8] They did not go as far as Milch, who ruminated on the bomb as a Jewish weapon, because of the refugee scientists involved in its development.[9]

The defence plea that the accused acted out of conscience stood against the charge of perpetrating genocidal experiments. Initially, Karl Brandt's defence denied either knowledge of or administrative competence over the experiments; when it came to the hepatitis and chemical warfare experiments he claimed that these were legitimate and non-fatal. The prosecution withdrew the charge on hepatitis, although chemical warfare stood. The defence changed from emphasising how the accused were dedicated physicians to their ethical rationales, once the ethical parameters of the Trial became clear. Brandt's Third Document Book contained extensive American material on human experiments and on their ethical justification. He marshalled evidence in support of euthanasia going back to Thomas More's *Utopia*.[10] Brandt's view was that securing the future health of humanity justified human experiments. Becker-Freyseng applied the argument that experiments should be necessary.[11] Brandt's self-portrayal was as a courageous idealist. His defence was that he knew only of benign and beneficent experiments, and not those that were dangerous and lethal.[12]

Rose documented how he criticised Ding's experiments, and the consternation at his view that the experiments were unnecessary. The Soviets supplied the prosecution with documents found at the RKI on Rose's approach to Himmler through Rudolf Brandt. McHaney felt this was a momentous triumph over the key orchestrator of the German ethical defence.[13] Despite the failure of his case, Rose's moral position continued to exert influence. He condemned as unrealistic American views that it was possible to inform a volunteer about the risks: the subject could not have the necessary medical knowledge, and experimenters would be tempted to play down the risks. He considered that neither prisoners nor conscientious objectors, who were subject to stigmatising public criticism, could be said to freely volunteer. His solution was to place full responsibility on the physician carrying out the experiments. He saw no problem with conducting an experiment on any person sentenced to death, as all means of execution were cruel and this form of death could yield positive benefits.[14]

The defendants adopted a range of ethical positions. Ruff was belligerent. He complained that it was unfair to be stigmatised as an SS doctor, Alexander replied that the aim of the Trial was to determine the ethical basis for human experiments.[15] Ruff argued that American training practices and experiments were inhumane because they caused multiple deaths, whereas there were none in Germany.[16] The elderly air force doctor, Oskar Schröder, appealed to Christian ethics. He confessed to Alexander that he was shocked by the revelations at the Trial, which he regarded as absolutely necessary in distinguishing medicine administered with the best of intentions from criminality. Schröder was – like Alexander's father – an ENT specialist, and they built up a collegial rapport. Alexander asked McHaney to give Schröder the opportunity to express his misgivings to the court.[17] Schröder's demeanour was 'angelic' and his defence counsel elicited a direct and personally engaged response to the experiments. But McHaney exposed his participation in the Natzweiler experiments.[18] The journalist Süskind expressed scepticism not of Schröder's sense of rectitude, but of the metaphysical depth of his Christianity. Rudolf Brandt's lawyer appealed to Christian ethics. He argued that a combination of natural law and Christian ethics justified the conduct of the accused, and that the experiments were motivated by the desire to save lives.[19] Rostock's fawning legal adviser, Pribilla, stressed that it was necessary for his client to establish his scientific credentials, while distancing himself decisively from Nazi atrocities.[20] Alexander considered that by the end of January 1947 only one defendant – Fischer – had expressed remorse.[21] Alexander arranged for suitable poems to be sent 'to those of the accused whose moral sense we would like to awaken'. One, 'The Final Epiphany', was about a prisoner who then returned as a judge.[22] Alexander felt this strategy had positive results with Schröder and Blome.

Blome's defence papers show him developing a position inimical to his colleagues'. At first, he took the justificatory stance of the defence by heroising the role of medical researchers from Leonardo da Vinci to Walter Reed's human experiments on yellow fever in 1911.[23] Blome's wife, Bettina, advised him to condemn outright all the human experiments that were carried out. She complained that his being imprisoned with the other accused meant that he had lost his sense of values. But it was now the case that many feared consulting a doctor in case they were used for experimental purposes. He owed it to the German medical profession to condemn the ruthless experiments.[24]

Blome duly revised his position to legitimate the rank-and-file physician. He suggested that the physician assumed a religious role under Nazism and had roles as confessor and therapist. The German physician's moral obligations meant that the majority of doctors did not support coerced human experiments.[25] He conceded that human experiments were necessary to alleviate human misery and to cope with the damage to life and limb during the war – or (here Blome indulged in self-justification) for defensive measures against biological warfare. Many self-sacrificing doctors experimented on themselves. He suggested that an expert medical commission should authorise all experiments. It should seek to establish that:

1. the experimenter is qualified
2. the experiment is scientifically justified
3. the numbers and type of experiment are appropriate
4. the subject be protected
5. the research should be supervised
6. prisoners should always be volunteers and receive a reduction of sentence or an amnesty (this justified the US prison experiments while condemning those of the Germans). Political prisoners and prisoners of war should not be subjected to experiments
7. volunteers should be used whenever possible
8. children and the mentally ill can be experimented on with permission from their guardian, but never when pain or danger is involved.

He latched on to the issue raised by Alexander of the autonomy of the subject and the responsibility of the doctor, but offered the subject only information as to hazards.[26] Ethics provided Blome with an opportunity for vindicating his own and the German medical profession's past conduct – indeed, he claimed to have steered the profession on a moral course through the final phase of the war. By establishing a responsible attitude to the place of research in medicine, he castigated defendants with links to the SS as amoral psychopaths.[27] Bettina saw the proceedings in terms of individual spirituality and encouraged her husband to take his justificatory stance that the physician had a spiritual duty in society.[28] Blome's casuistry cleared the way for his acquittal.

The Trial was not just a calling to account of German science, medicine and administration, it was a test of the individual's psychological resilience. The prisoners' diaries yield insight into the trial as a spiritual and emotional test. Pokorny scrawled in old German script in pencil on thin paper a daily record of his emotions.[29] He was sickened by the evidence of euthanasia, the trial proceedings and the physical circumstances of captivity. Captivity meant time for introspection and self-justification. Alexander found that while most defendants conceded that the Trial was necessary to prevent medical abuses in the future, they resented the charges and felt unjustly accused.

Given that Himmler railed against 'Christian medical circles' for opposing human experiments, the defence stressed religious beliefs and concern
to act morally and disinterestedly, as demonstrating the ethical sense of the
defendants.[30] While Fischer and Schröder were conscience-stricken, there is
no evidence of any defendant being overwhelmed by an intense wave of
guilt or the need to make a momentous confession. Defendants explained
the ideals which had lured them to Nazism in the first place, and secondly
to participate (even if reluctantly) in coercive experimentation. Justice
Sebring felt that the atrocities and crimes were 'the end products of an
immoral and unChristian philosophy of teaching and thinking'. He was
shocked by the evidence of schoolbook calculations on war gases.[31]

The first seven of the accused to join the Nazi Party had a Protestant
background. The six Roman Catholics were in the minority. Handloser
stressed his steadfast commitment to Roman Catholicism to underline
his ethical regard for human life and his opposition to Nazi ideology.[32]
Pokorny affirmed that he had returned to the Roman Catholic Church at
the beginning of 1945. The doomed Brack similarly resumed his
Catholicism.[33] The loquacious Gebhardt stressed his Roman Catholic faith
and the importance of religion as a stabilising basis for post-war Germany,
not least as a corrective to American commercialism.[34] Hoven had left the
Protestant Church in 1925, but had recently converted to Catholicism.[35]

Most of the Protestants were less forthcoming about their faith. Fischer
and Weltz (whom Rascher loathed for his 'black' religious convictions)
returned to their original Protestantism.[36] Others were far less conventional.
Genzken had a longstanding interest in theosophy and had even submitted
a theosophical treatise to Himmler.[37] Sievers (like Handloser the son of a
church musician) found solace in the medieval mystic philosophy of
Meister Ekkehart and poetry.[38] Oberheuser was guarded about her unconventional religious views, suspecting that these might be used to discredit
her.[39] Most found belief in organised Christianity a solace and vindication
of their ethical and moral outlook.

Two of the most vigorous defenders of the German medical record were
not religious – at least not in any conventionally Christian sense. Rose was
an affirmed atheist. Karl Brandt explained his moral commitments as a
physician as justifying his actions, and did not seek refuge in any belated
return to Christianity. Fischer explained that the prevailing ideology at
Hohenlychen was not typical of the SS, but more of enlightened scientific
circles. He characterised it as a combination of Darwinism and Hegelian
ideas of the state as the realisation of the divine. Gebhardt had refused to
allow an SS Weltanschauung unit to be stationed at Hohenlychen.[40]

Beneath Mrugowsky's rigid demeanour lay a deep interest in the German
tradition of spiritual enthusiasts for nature, dating from the Reformation
divine Jakob Boehme.[41] He cited his edition of the ethics of the romantic
physician Hufeland in his defence. The holistic philosopher of biology

Adolf Meyer-Abich declared, 'How could someone who has studied Humboldt and Hufeland have knowingly transgressed the laws of international medical ethics?'[42] Meyer-Abich explained that Mrugowsky was a pupil of the physiologist Emil Abderhalden, who edited the journal *Ethik* from the Leopoldina Academy in Halle.[43] The defence calculated that commitments to ethics would distinguish the accused from the unscrupulous immorality of the Nazi leadership. Mrugowsky had an intense, private spirituality, finding solace in the writings of the philosopher-physician Jaspers on the problem of collective guilt; but to the court and outside world he appeared resolutely defiant.[44]

The prosecution's onslaught built up group solidarity among the accused. Their code of loyalty was that a defendant could plead extenuating circumstances, but implicating a fellow accused amounted to betrayal. The bonds of loyalty strengthened between Karl Brandt, Rostock and Rose. There was a circumspect aura of mutual respect between the former rivals Karl Brandt and Gebhardt, and the SS doctors closed ranks. Hoven and Pokorny rallied to this core group.

Blome turned accuser: his mixture of self-justification and attack on SS medicine earned the contempt of his fellow defendants. The final pleas caused bitter antagonisms to peak as he criticised SS medicine. Blome complained of curses and threats from his co-defendants: Brack was outraged, and Gebhardt, Mrugowsky, Poppendick and Hoven ('der schlimmste Hitzer') were determined to exact revenge.[45] Mrugowsky condemned Blome as a shameless liar to Gebhardt ('Da ist ganz dick drin gelogen'), because Blome was in fact involved with the concentration camps and with Rascher. Gebhardt felt wounded by Blome identifying him closely with Himmler.[46]

Gebhardt took the opportunity to establish his ethical and scientific credentials by elaborating on experiments.[47] Although animal experiments always had primacy, they had severe limitations due to immunological differences between humans and animals. Animal experiments were fundamental for surgery. He drew a distinction between his own research and research ordered by the state in time of war. He pleaded that he was acting under orders. Self-experimentation, he alleged (and there was much reference to mythical decrees), was banned during the war. He argued that confidence between physician and patient was paramount – unless the physician was acting under state orders. Ivy responded by pointing out that 'the State can never assume the moral responsibility of the physician to his patient', and that in war the doctor has a duty of care to all, whether friend or foe.[48]

Fischer had, since his capture in July 1945, been open and informative about the experiments. He readily admitted his role in carrying these out, but insisted that his superior Gebhardt had given him no choice.[49] Fischer regretted that generalised conclusions were not drawn from clinical observations, but that Grawitz had insisted on the experiments. Alexander asked him why he did not refuse point blank to carry them out. Fischer felt he had to carry out orders from Gebhardt as his medical superior and military commander. But that he then abandoned his ambition of a career in academic medicine.[50]

Beiglböck under cross-examination similarly insisted on animal experiments having to precede human experiments. He undertook the self-experiment of drinking 500 ml of seawater for $4\frac{1}{2}$ days.[51] He felt that a layman could voluntarily decide on participation in human experiments and that a prisoner had the option to volunteer providing there were no reprisals if he refused. He insisted that as an army officer he had no powers to intervene in anything to do with the SS's administration of the concentration camp.[52] Nor was he prepared to allow any of his subjects to leave the experiment voluntarily if the thirst became unbearable.

Witnesses contributed substantially to the discourse on ethics. Wolfgang Lutz was a military medical officer, who refused to experiment. Alexander prepared the question: 'Why did you refuse Dr Weltz's offer? Dr Lutz replied: "Because I didn't have the brutality in me." I then said: "I take it that this means that it was made obvious to you that brutality would be required for this job." Dr Lutz answered: "If one takes a human being, and experiments on him, *why*, the thing might suddenly start talking to one. I found it already difficult to kill dogs in an experiment let alone a human being."'[53]

Hans Luxenburger was a neurologist at the KWI for Psychiatry until 1941, and then in the Luftwaffe. He co-authored with the aviation medicine specialist Hans-Erich Halbach a study on 'Experiments on Human Beings as Viewed in World Literature'.[54] Luxenburger was approved as expert witness for the defence, and their documentation was completed in mid-April 1947.[55] He had left Rüdin's Institute as a result of SS disapproval of his Roman Catholicism, although he remained a committed eugenicist. He was Halbach's commanding officer at the Science and Research Instruction Group at the Luftwaffe Medical Academy. (Halbach also swore an affidavit confirming Becker-Freyseng's lack of involvement in human experiments for infectious diseases.[56]) Their overview dealt mainly with American experiments. The first paper cited was on experiments on hepatitis using human volunteers, published by *JAMA* in 1945.[57] Luxenburger was disposed to condone German wartime research, as the medical historian Werner Leibbrand was critical. Luxenburger had worked alongside with Becker-Freyseng; he had encountered Haagen while evaluating Luftwaffe research during the war; and helped Rose to meet the advancing US troops, to safeguard his experimental wards at Pfafferode.

Luxenburger and Halbach observed that experiments and interventionist therapies were not regulated by law. In war, dangerous experiments were justified (a point also made by Karl Brandt, Gebhardt and Mrugowsky).[58] They stressed the dangers of Goldberger's pellagra research, yellow fever experiments in Cuba and the Canal Zone, and experiments on endurance:

Ladell on thirst, and Dell and Forbes' experiments on a schizophrenic subjected to freezing. Luxenburger took the view that in oxygen deficiency experiments, the lethal stage was readily preventable.[59]

They assumed most experiments were on volunteers, with the exception of those on children and the mentally ill. They identified nine sets of voluntary experiments on convicts, as well as one set of toxicological experiments on 11 men sentenced to death. Experiments using plague, leprosy and typhus carried a high risk. Luxenburger pointed out that convicts could be used under conditions which could be standardised and controlled.[60] Malaria experiments would have been impossible on free citizens, but some were connected with the offer of remission or rewards.[61]

Luxenburger and Halbach stressed:

1. the responsibility and integrity of the researcher
2. the experiment should be necessary – in terms of scientific innovation and therapeutic advance (although difficult to define)
3. animal experiments and self-experiments as preliminaries wherever possible (animal experiments were recognised as not always appropriate; self-experiments were left out of the Code)
4. the result could never justify the experiment.
5. the need to have a necessary number of subjects for statistical significance.

They saw that experiments varied greatly in terms of numbers of victims, citing cases of between 1 and 800 subjects. The spectrum ranged from the very dangerous to the harmless, making it difficult to establish a rule. Their overview was overtaken by the defendants' dramatic cross-examination of Ivy earlier in June, when they asked him about American deaths.[62] Luxenburger was not called to testify, but his documentation was presented to the court along with Kisskalt's 'Theory and Practice of Medical Research' as part of Becker-Freyseng's defence. It provided both a critique of Alexander's principles and Ivy's testimony, and was available to the judges as a defence appraisal of research ethics.[63]

The accused defended the rationales for their actions in the context of war.[64] The judges recognised that defendants had a valid voice in countering the prosecution's damning critique of the experiments as scientifically defective, unnecessary and blueprints for mass murder, and that their fall from grace rendered their views all the more interesting. The defence contributed substantively to the emergent codification of research procedures. Introducing the Nuremberg Code, the judges commented, 'all agree' on basic legal and ethical principles governing research. Who the 'all' could be has puzzled later commentators, who have focused on ethical highlights rather than analysing the Trial proceedings as a whole and in broader context.[65] The accuracy of defence testimony was open to challenge. But defendants and their supporters had a clear voice in the making of the Code.

Ivy and ethics

Ivy's prime motive was to secure a new legal standard for human experiments. In 1964, when congratulating Irving Ladimer, Roger Newman and W.J. Curran, professor of legal medicine at Boston University, on their volume on 'Clinical Investigation in Medicine', Ivy reminisced: 'I accepted the invitation to serve at the Nuernberg trials only because I had in mind the objective of placing human beings as subjects in a medical experiment, so that these conditions would become the international common law on the subject. Otherwise I would have had nothing to do with the nasty and obnoxious business. I believe in prevention, not a "punitive cure".'[66]

Ivy returned to Nuremberg in June 1947 to refute the evidence of Leibbrand that American prison experiments were unethical. The prosecution elicited from Ivy a definition of an ethically correct position, when conducting dangerous experiments. Hardy framed the questions as 'problems that will confront scientists all over the World', as well as to counter the defence allegation that US pressure chamber experiments killed six American airmen (something Ivy absolutely denied, although he cited the case of a parachutist killed in a free fall experiment). Sebring, concerned that the questions were objectionable, pointed out 'that this Tribunal will, in its opinion, answer that question in such a way scientists in the future will have some landmark to guide them'.[67] When he asked how far the AMA principles corresponded with 'the principles of the medical profession over the civilised world generally', Ivy confirmed that they were identical, citing the German principles of 1931 as confirmation.[68]

Ivy felt that 'the Principles and rules as set forth by the American Medical Association' corresponded with his recommendations. When Hardy asked about the basis on which the American Medical Association adopted those rules, Ivy answered, 'I submitted to them a report of certain experiments which had been performed on human subjects along with my conclusions as to what the principles of ethics should be for use of human beings as subjects in medical experiments. I asked the association to give me a statement regarding the principles of medical ethics and what the American Medical Association had to say regarding the use of human beings in medical experiments.' The toned down *JAMA* Code was presented as generally corresponding to American practice.[69] Ivy fudged the issue by invoking the Hippocratic Oath, which he interpreted as endorsing an experimental approach to medicine while protecting the patient. The prosecution lawyers were still confident in securing wholesale conviction of the German experiments by proving links to Himmler and the SS. While Taylor's prosecuting lawyers were convinced that the laws against murder and assault were sufficient to gain convictions, Ivy and Alexander sought to show a lapse from the Hippocratic principle of 'do no harm' to patients.

The defence challenged Ivy's credentials as a bacteriologist, and objected to him giving any opinion on the proper conduct of infectious disease experiments. Sauter pointed out that so far the court had objected in almost all the cases to discussion of the general problem of the ethics of human experiments, as a matter to be dealt with after the Trial.[70] Ivy responded that he had advised AMA and chaired the Committee appointed by Governor Green of Illinois on prison experiments. He defended malaria experiments in the Illinois penitentiary of Stateville as ethically correct, based on a putative report of the committee on the ethics of the experiments.[71] The experiments had been publicised in *Life* on 4 June 1945, along with details of experiments on 800 convicts at the New Jersey State Reformatory. Karl Brandt introduced these as evidence.[72] The article in *Time* on 'Conscientious Objectors as Guinea Pigs' was rejected by the court as evidence.[73] The defence attacked American medicine as recklessly experimentalist.

Blome praised Paul de Kruif's *Microbe Hunters*, an American popular history of medical research dating from 1927, for the view that 'You must kill men to save them'. De Kruif (a bacteriologist turned writer) justified dangerous experiments, proving how a disease is transmitted. He glorified Walter Reed, who infected volunteers with yellow fever to prove its transmission by mosquitoes, and Ehrlich for his discovery of Salvarsan. He weighed the deaths of a few from adverse reactions against thousands of cures.[74] De Kruif provided the basis for Alexander's question whether it was right to sacrifice five individuals to save the lives of many. Blome's response was that Reed had shown that there were medical problems, which can only be resolved by human experiments. But he let slip a racist observation that some lives were of less value than others.[75] Overall, Alexander found this popularisation full of misstatements and sensationalising exaggerations of the heroism of laboratory workers – remembering how de Kruif had once come to Vienna and interviewed the psychiatrist Wagner Jauregg, the pioneer of malaria fever therapy. Alexander ranked de Kruif as the Louella Parsons of medicine.[76]

Ivy endorsed the tradition of human experiments of Reed, Strong and Goldberger, insisting that the experiments were voluntary and legitimate.[77] Ivy suggested to the Illinois Governor Dwight Green to convene a committee to advise on the conditions under which prisoners could take part in experiments, and whether any reduction in sentence could be granted. Sauter and Servatius questioned Ivy for how long he had been concerned with the historical basis of medical ethics. Ivy claimed the origins of the committee went back to December 1946 (certainly he was alerted to Rose's accusations by then), although Harkness deduces that the formation of the Green Committee was a response to Ivy's hearing Leibbrand's testimony in January 1947. Certainly, Governor Green issued invitations to serve only on 13 March 1947. Ivy contacted all members on 21 April informing them that he was studying the issues involved so as to have a 'factual basis' for a meeting.[78] In court on 13 June 1947, Ivy – correctly – described himself as Chairman of the Green Committee.[79] He tried to arrange a meeting of the Green Committee only on 30 June, but then suggested that circulation of his draft report might be sufficient.[80]

Under cross-examination Ivy explained that he had prepared the report as a rebuttal witness at the Trial, in response to the German defence attorneys' attempt to equate American prison experiments with those in the concentration camps.[81] He admitted having experimented on conscientious objectors.[82] But he defended these experiments, when Servatius pointed out that the objectors were not at full liberty. He considered the use of prisoners was ethical, although the defence objected that some were on the 'racially inferior' (as Hispanics in Cuba or Filipinos experimented on by Strong in 1905–11), and they were sceptical of Ivy's numbers of prisoners condemned to death.[83] Rose sought to discredit Ivy's competence for judging the ethics of the infectious disease experiments by exposing his ignorance of numerous experiments involving deliberate infection, for example an American experiment on a subject whose legs had been amputated.[84] Rose's verdict was that Ivy ignored the danger of the experiment, as his sole concern was whether the subject gave consent.[85]

The most controversial part of Ivy's rebuttal was on the ethics of experiments on prisoners. Ivy reflected that 'The American malaria experiments with 800 or more prisoners were absolutely justified, scientifically, legally and ethically even though they bring with them danger to human life.'[86] Under cross-examination by Rose, Ivy conceded that the American experiments used the more dangerous malaria *tropica* strain, whereas Schilling experimented with *Plasmodium vivax* in Dachau.[87] Harkness observes that Ivy answered only in the first person, so obscuring that the committee had never actually met.[88] Ivy circulated the report to committee members, although they did not meet until November 1947. Ivy's report made reference to the AMA code rather than the Nuremberg guidelines, but added that scientists should make self-experiments when death or disability might occur.[89] The report was publicised in *JAMA*. In 1952 the AMA condemned the participation of prisoners involved in violent crimes or treason in experiments.[90] Ivy's concerns with ethics ran deep, and his extensive human experiments appear to have been based on consent, however perfunctory. He was convinced of the ethical probity of American wartime research. His tactics betrayed an excessive zeal to present an unblemished image of American medical research. His enthusiasm would later be his undoing.

Ivy was not willing to concede that experimental medicine was inherently inhumane. This stands in contrast to Werner Leibbrand, a psychiatrist and medical historian – the author of a celebrated study on Romantic medicine. Leibbrand came to Nuremberg in 1943 to work in the neurological clinic, and in May 1946 was appointed at nearby Erlangen.[91] He was active in

Catholic intellectual circles, had links to the 'Red Band' resistance group, refused to notify the authorities of patients with inherited diseases to save them from sterilisation and euthanasia, and went underground to protect his Jewish wife. He was a witness who could speak with genuine authority.[92] He established the centrality of Hippocratic precepts at the trial. In 1939 Leibbrand published 'The Divine Rod of Aesculapius' – an erudite treatise demonstrating the religious basis of medicine. He argued that Hippocrates was misinterpreted as a pioneer of scientific observation in medicine, and was instead to be understood in religious terms.[93]

Leibbrand advised the prosecution on the biographies and publications of the accused.[94] Speaking to a journalist on 9 December 1946 on the 'SS Doctors' Trial', he commented that the Trial exposed how medicine had been used under National Socialism to make people sick. He explained that the demonic events derived from the purely biological thought of the later nineteenth century.[95] This statement, and a treatise on patient rights in psychiatry, suggested that Leibbrand could turn out to be a problematic witness. He gave evidence for the prosecution on 27 January 1947 on German medical ethics with respect to experimentation on animals and human beings.[96] Servatius manoeuvred Leibbrand to criticise experimental medical research for relegating the patient to a mere object. This shift to identifying ethical flaws in science rather than seeing science as corrupted by Nazism was unacceptable to Ivy, who introduced his criteria for legitimate experimentation into the trial proceedings. Leibbrand testified that American research was ethically dubious, because prisoners were in a forced situation. The Trial had to tackle the conditions under which risky experimentation was ethically permissible.[97]

The Hippocratic Oath became a rallying point for both sides. There had been considerable emphasis on the Oath as a universal and traditional requirement, initiating students into medicine. A widely held view of Hippocrates was that he was the father of modern medicine and that the Oath was commonly subscribed to by all graduating students. All this was far from the case. The émigré medical historian Ludwig Edelstein argued that the Oath derived from an esoteric religious sect.[98] The text of the Oath was so opaque as to admit a variety of interpretations on euthanasia, abortion and experiments. Both the defendants and prosecution experts claimed their views as in accordance with the Hippocratic Oath.

The Nazis developed ideas of a Hippocratic medicine to counter the excesses of rationalism in medicine, which they attributed to a calculating Jewish spirit. The homoeopathy of Madaus had been part of the Nazi project of a new biological medicine.[99] The defendants felt confident in asserting that they were the heirs of Hippocrates. At the Milch Trial, the prosecution cross-examined Hippke about Rascher's experiments: 'Did you take the Hippocratic Oath' with its principle 'Abstain from whatever is deleterious and mischievous', and 'Did you ever hear of Dr Rascher, Dr Ruff or any other Luftwaffe-doctor getting into a tank of cold water and staying in it three hours?' Hippke's defence was that although he was concerned that subjects would die, he viewed them as 'people who had been condemned to death, who, in this case, had been given a chance to prove their loyalty to the Reich'.[100] Judge Phillips asked Hippke about the voluntary nature of experiments and whether any of the prisoners were actually pardoned. The Presiding Judge Toms asked about the conditions of volunteers. Hippke replied that he was interested that their physical condition should be 'normal average', because if really sick people were used, 'the result of these experiments would then be fallacious'.[101]

Luxenburger and Halbach demonstrated that human experiments went back to Hippocrates and Galen.[102] But Blome pointed out that none of the accused ever took the Hippocratic Oath. He praised Hippocrates as a pioneer of modern scientific medicine, but said that it was inappropriate to follow the Oath dogmatically.[103] Romberg agreed that modifications to the Oath were necessary.[104] The witness Boehm confirmed that no German doctor ever took the Hippocratic Oath.[105] Alexander discussed the Oath when interrogating Becker-Freyseng, who expressed the wish that the medical ethics should gain legal force. Alexander learned from Becker-Freyseng about Büchner's courageous lecture of November 1941 on the validity of the Hippocratic Oath, which he saw as a still valid statement on the scientific approach to disease.[106] Becker-Freyseng defended the right to experiment on the condemned for the sake of humanity. By way of contrast, he considered that the physician was often faced by cases of persons with terminal illnesses, and here relentless therapy was pointless when death was inevitable.[107]

Ivy argued that Hippocrates affirmed the sacrosanct principle of respect for life and the human rights of the experimental subject. His Commencement Address on 'Basic Principles. The Significance of the Moral Philosophy of Medicine' to the University of Nebraska College of Medicine on 22 March 1947 came between his two sessions as expert witness at Nuremberg. He stressed Hippocratic medicine as laying down the principles of the scientific, technical and moral philosophy of medicine. He considered that AMA's 'Principles of Medical Ethics' were in harmony with the Hippocratic Oath. By way of contrast, 'the atrocities committed by a small group of Nazi SS-physicians' in forcing human beings to serve as experimental subjects arose from dedication to the state and politics. Nazism had no place for 'the ethics of medicine which teaches that the physician-patient relationship is a holy and individual matter'.[108] Ivy condemned the state for usurping the moral responsibility of a physician for his patient or experimental subject.[109] He praised the Oath as 'the Golden Rule of the medical profession', providing the individual physician with moral guidance.[110]

Ivy reaffirmed his position in an address to high school teachers given in November 1947: medicine was grounded in the Hippocratic method of experimental procedures, and the Hippocratic Oath commanded respect and reverence for life. The individual physician's duty is to the welfare of the patient, and the profession 'is to serve humanity by maintaining life and postponing death'. Ivy cited the atrocities 'committed by a small group of Nazi-SS physicians' – he estimated these as variously numbering 70, 100 or (in *Doctors of Infamy*) 200 – fewer than Thompson's estimate of several hundred. The crimes arose from a disregard for the value of life. Science lacked freedom of discussion and Nazi fanatics controlled the medical profession.[111] 'Truly, the only solution for the problem of the misuse of scientific discoveries is the development of a good society.'[112] Ivy reiterated his aim of a Code on human experimentation at a War Department meeting of 9 April 1947. He still favoured US participation at the ISC, because 'the attitude of scientists towards the evidence collected for the medical trial of Nuremberg is somewhat different from that of the prosecutors'. He supported an international code of ethics on experimentation on humans as 'a guide for the future'.[113]

Ivy was recalled as a witness on 12 June 1947 because of his 'expert standing as a physiologist and experimenter'.[114] Prior to his departure, he addressed the Federation of State Medical Boards. He referred back to his Paris principles as demonstrating 'the conditions under which human beings were used as experimental subjects in the U.S.A. during the War'.[115] Linking ethics to politics, he stressed the profession's duty to resist any form of totalitarianism, not least because socialised medicine would lead to ethical violations.[116] This theme would loom large in the aftermath of the Trial, as physicians mobilised against state medicine. The prosecutor McHaney in his closing argument on 14 July 1947 restricted himself to Ivy's three research rules concerning consent without coercion and informing the subject of the hazards. Ivy's rules, formulated for the ISC in July 1946, shaped the prosecution's case.[117]

Alexander's ethics

Alexander reinforced the prosecution with his encyclopaedic scientific interest and ethical, philosophical and psychological insight. In contrast to Ivy's view of the Trial as 'nasty and obnoxious business', Alexander worked with intense dedication to refute each twist and turn of the defence. He was interested in the defendants' group psychology – why some did experiments and others, such as (he believed) Strughold, refrained. He attempted to ensure that the Trial was 'no mere murder trial'. Among his guidance notes was a memo of 23 December 1946 on 'Countering the Defence that Germans were experimenting on Prisoners Condemned to Death'; he concluded that experimental wards were a death mill from which few emerged alive. He showed that Himmler's concession of April 1942 that prisoners should be pardoned after undergoing a potentially lethal experiment, was never acted on. 'Our evidence is that they [i.e. the victims] were picked at random, that none of them was asked whether he was willing, that none of them signed any written agreement ... That should explode the German claim that these experiments were voluntary, quite apart from the fact that a concentration camp was certainly no setting for anything voluntary.'[118]

Alexander agreed with Lemkin that the legal basis of the Trial – the prosecution of war crimes as crimes against humanity – was too narrow. He coined the term 'thanatology' to demonstrate the lethal aims of the experiments as a perversion of medicine and saw the relevance of genocide to human experiments. Thanatology was the scientific counterpart of genocide, as it meant 'idolatrous delight in death', and echoed Jung's concept of *thanatos* as a death wish. The Nazi experiments were 'devoted to methods of destroying or preventing life'.[119]

Alexander's analysis rapidly took shape on arrival at Nuremberg:

> I have delved into that mass of material, and have finally grasped its meaning and have come out with an appraisal that makes sense. It is thanatology pure and simple, and it is the techniques of genocide. Thanatology is a word I have coined: thanatos in Greek means death. 'Genocide' is the 'murder of the people', a word coined by that old Pole Lemkin. I shall send you the carbon of an appraisal of the whole problem which I have been writing for General Taylor, the Chief Prosecutor; and which I may publish independently, if the War Department approve.[120]

The only scientific advance of the experiments was in killing techniques. Mrugowsky's research on impurities in vaccines and experiments on the lethal effects of phenol injections provided 'the techniques for genocide'.[121] Mant assessed the thesis of how the declared aim of the experiments obscured studies of the onset of death – he questioned whether the trials on making seawater drinkable were really a study of the time it would take to kill.[122] Alexander redrafted the memo, sent to General Taylor on 5 December 1946, on 'Thanatology as a Scientific Technique of Genocide'. He argued that 'The German research therefore definitely constitutes an advance in destructive methodology', 'as killing methods for a criminal state' and as 'an aggressive weapon of war'.[123] Lemkin did not endorse the concept of thanatology: 'The word is too all-embracing. His point that this case must be viewed in the Geopolitic and Genopolitic light is certainly correct and when the International Trial fell down they missed a lot when they missed Haushofer.'[124] Lemkin favoured an approach based on the Nazi expansion of *Lebensraum* and the geopolitical theory of Karl Haushofer, rather than the minutiae of doctor-patient relations.

Alexander and Lemkin agreed that the experiments were designed to develop the science of killing. In January 1947 Lemkin introduced two further variants to cover techniques for outright killings and abortions on the one hand, and sterilisations and castrations on the other. He called the first 'ktenotechnics' (from the Greek *ktenos* meaning murder) and the second 'sterotechnics' (from the Greek meaning infertile, or *steirosis* meaning infertility).[125] Alexander responded by using the term 'ktenology' to describe the science of medical killing. He wrote on 'Ktenology as a Scientific Technique of Genocide', and concluded: 'The frightful body of new methods of killing ... constitute a formidable body of new and dangerous knowledge, useful to criminals everywhere, and to a criminal state if another one is permitted to establish itself again, so as to constitute a new branch, a destructive perversion of medicine worthy of a new name, for which the term ktenology is herewith suggested. This ktenological medicine supplied the technical methods for genocide, a policy of the German Third Reich which would not have been carried out without the active participation of its medical scientists.'[126] Unlike thanatology – which was taken up by *Newsweek* – ktenology was short-lived as a term.[127]

Alexander focused on experimental codes between 3 and 7 December 1946, just prior to the Trial, and then between 20 and 26 January 1947. On 3 December 1946 he outlined his paper on the proper conduct of human experiments:

> *Plan:* Ethical and non-ethical experimentation in human beings: the crucial experiment (Pettenkofer) – the scouting experiment – the model experiment with physico-chemical systems – the theoretic thinking through. In which way were the German experiments non-crucial experiments, inadequately prepared; therefore inaccurate and misleading (example: high altitude) and unnecessary (example: sea water).[128]

On 6 December he again noted: 'Worked on ethical and non-ethical experimentation in human beings ... Completed ethical and non-ethical exp. on human beings.'[129]

This text outlined the conditions for 'permissible experimentation by a doctor'. As in Ivy's draft Code of 31 July 1946, Alexander required the consent and voluntary participation of the experimental subject. While Ivy required the experiment to be useful, Alexander preferred a more generalised viewpoint, that the experiment should not be unnecessary; both concurred that results should be for the good of society. This overlap suggests that Alexander took Ivy's report as a basis for his views.[130] Alexander amplified the concept of consent, as based on a proven understanding of the exact nature and consequences of the experiment. A doctor or medical student was most likely to have the capacity for full understanding. The degree of risk was justified by the importance of the experiment and the readiness of the experimenter to risk his own life. Overall, Alexander produced a more rigorous set of requirements than either Ivy or the minimalist AMA code.[131]

These criteria highlighted the lethal side to the German experiments. Alexander submitted his paper on ktenology to Taylor and to the ISC in January 1947. Lord Moran was hostile to Alexander's endeavours, resenting that the ethical impetus had shifted to the Medical Trial. Alexander pointed out 'that running throughout all the experimental work, one could see a "fine red thread" of measures designed either to kill or sterilize. On the basis of data collected by Doctor Alexander at Nuremberg, he had prepared two papers for submission to the American medical press.'[132] Moran pounced on Alexander for briefing *Harpers* and *Time Magazine* journalists on Nazi medical crimes, arguing that 'the effective function of the Commission would be undermined by releases either to the medical or lay press before the final pronouncement made by the Commission itself'.[133] Given that the US was not a member of the Commission, Moran's arrogant attempt to exert authority was futile. He overlooked the fact that Alexander had gone through the proper military channels to gain authorisation for publication, and that informing the public was ethically desirable.

In mid-January 1947 the prosecution case was drawing to a close, and Alexander returned to the issue of the rationales for human experiments and 'enlightened consent'. On 21 and 22 January Alexander met Ivy to discuss the issue of ktenology.[134] Ivy returned to Chicago 'with the recommendation that an international, legalized Code of ethics should be published on the use of human beings as experimental subjects'.[135] On 24 January Alexander recorded: 'Sent off Ktenology article. Finished the additions to the article re ethical and unethical experimentation.' On the following day he 'Worked on Ethics article'. On 28 January he outlined his six rules: 'Evening: discussion with Mitscherlich, later with Wing Commander Thompson', showing that the German medical delegation and the ISC were part of the ethical discourse surrounding the Trial.[136] Alexander's crucial input to the Nuremberg Code occurred in the two weeks after the Paris meeting and culminated in the affidavit of 25 January. The conversations with Ivy, Mitscherlich and Thompson were formative.

Alexander's six rules for permissible experiments were now in the form of a prescriptive Code. How is it that a later date, 15 April 1947, has been ascribed to the Code? The formulation was incorporated in Alexander's January papers on ktenology: 'Ethical and Non-Ethical Experiments on Human Beings – General Ethical, Medico-Legal and Scientific Considerations in Connection with the Vivisectionists' Trials Before the Military Tribunal in Germany'. The military authorities approved these papers for circulation only in mid-March 1947, after checking the legal and historical evidence. The six-point formulation of conditions for permissible experiments represented a substantial advance on the position of the AMA, which disregarded the requirement of voluntary consent.[137]

Alexander was in the Netherlands collecting evidence between 12 March and 14 April 1947. His Logbook for 15 April recorded that on his return to

Nuremberg he immediately 'Worked on medical experimentation', and he then submitted memoranda to Taylor. He concluded with a formulation of the Code in six sections. The sequence of events indicates that Alexander had settled his views on the topic of an experimental code by 25 January 1947, which was then embedded in the discursive papers of mid-March.[138]

Alexander drew on his psychological expertise, when he defined what constituted 'enlightened consent'. His criteria were 'legally valid voluntary consent of the experimental subject' requiring 'a. The absence of duress. b. Sufficient disclosure on the part of the experimenter and sufficient understanding of the exact nature and consequences of the experiment for which he volunteers, to permit an enlightened consent on the part of the experimental subject.' Alexander's overall criteria were: 'The nature and purpose of the experiment must be humanitarian, with the ultimate aim to cure, treat, or prevent illness and not concerned with the methods of killing or sterilization (ktenology).'[139] The elaborate judicial statement on 'voluntary consent' in the final Code was substantially due to Alexander.

Ivy and Alexander both claimed authorship of the eventual Code. Alexander stated in 1966: 'In order to define conditions under which medical experimentation on human beings is ethically and legally permissible, I prepared a memorandum entitled, "Ethical and Non-Ethical Experimentation on Human Beings", which was submitted to the United States Chief of Council for War Crimes and the Court on April 15, 1947. With additions derived from Dr. Andrew C. Ivy's testimony of June 12, 13 and 14, 1947, this memorandum became the basis of the so-called Nuernberg Code incorporated in the judgment.'[140] This represented Alexander's fullest statement of a sustained engagement with the problem since November 1946. Alexander and Ivy consulted together to pull together their testimony.[141] As Ivy later reflected: 'The Judges and I were determined that something of preventative nature had to come out of the "Trial of the Medical Atrocities".'[142] As witness for the tribunal, Ivy had access to the judges.

Alexander's paper 'Ethical and non-ethical experiments on human beings' developed Ivy's criteria. He defined voluntary consent more fully – as the absence of duress, and '[1] ... Sufficient disclosure on the part of the experimenter and sufficient understanding on the part of the experimental subject of the exact nature and consequences of the experiment for which he volunteers to permit an enlightened consent. A mentally ill patient should have the consent of next of kin or their legal guardian, and where possible should give his own consent.' His principles required:

2. experiments should be humanitarian 'with the ultimate aim to cure, treat or prevent illness, and not concerned with killing or sterilization.

3. No experiment is permissible when there is the probability that death or disabling injury of the experimental subject will occur.

4. A high degree of skill and care of the experimenting physician is required.

5. The degree of risk taken should never exceed that determined by the humanitarian importance of the problem. Ethically permissible to perform experiments involving significant risks only if not accessible by other means and if he is willing to risk his own life.

6. ...the experiment must be such as to yield results for the good of society and not be random and unnecessary in nature.'[143]

Finally, there was a move to protect the research subject. Luxenburger and Halbach raised the problem of experiments on children and the mentally ill on 23 April 1947. While this was after Alexander had drawn up his memo, Luxenburger's evidence was placed before the court.[144] Alexander had included special provisions to protect mentally ill patients, requiring where possible the consent of the patient in addition to the next of kin or guardian. This provision lapsed from the eventual Code. (See Table 12.)

The judges' role

Taylor attributed the Code to Judge Sebring, who was interested in the broader ethical outcome of the Trial.[145] The questions arise whether one can single out Sebring, and whether it is right that the judges took the initiative in formulating the Code. The Trial protocols show that all the judges intervened to seek clarification about experimental procedures. Beals kept an alert eye on ethical issues. Judicial intervention was massive at the IMT, and the judges were more restrained at the more smoothly running Medical Trial. The tribunal questioned Rostock on 24 February about wartime medical research, Rose on 27 February about research on conscientious objectors and about Jews as victims on 9 May. The interest of the judges in experimental procedures stands out. Beals passed the testimony of Leibbrand and transcripts of trial proceedings to Fishbein, the editor of *JAMA*, for editorial comment in March 1947.[146] Faced with Rose's testimony, the judges requested an 'adviser to the court', and Alexander proposed Ivy.[147] Sebring questioned Ivy on 12 June about the voluntary nature of experiments, and differences in results from human and animal experiments.[148] On 13 June Sebring clarified the conditions under which conscientious objectors acted as human subjects.[149] On 11 June Sebring asked what was involved in a liver puncture;[150] and on 17 June he clarified the circumstances concerning Eppinger's dismissal of three Jewish assistants from his clinic in March 1938.[151] Sebring's keen interest in Ivy's evidence lends support to Taylor's view, although the alert probings by Beals should be noted.

Shuster argues that the judges shifted the focus away from the physician to the research subject. She suggests that the requirement for informed consent was 'new, comprehensive, and absolute'. Certainly, what was novel was the right to withdraw from the experiment. Ivy had required far less when he called for informing the subject of potential hazards. But asked by Judge Sebring about the seawater experiments, he explained that when he conducted similar experiments, he permitted 'the volunteer subjects to withdraw from the experiment whenever they desired to do so'.[152] Ivy reflected that if subjects were in any way coerced there would be cheating and unreliable results.[153] The view that the Code 'grew out of the Trial itself' omits the formative preliminary period, the crucial inter-Allied discussions and the protests of the victims.[154]

Key issues remain the contact between the judges and the medical experts at the Trial, and the discussions between the judges. Alexander provided objective background information in his capacity as an expert. He sent Sebring evidence on the German psychology, and on 'the perversion of superego values'.[155] Ivy later provided crucial information to Ladimar:

> I did not reveal my objective to anyone except Messers Haney [*sic*] and Hardy until January, 1947. And, at that time I was invited by a member of the Tribunal (late in January, 1947) to have lunch with them. At luncheon the Judges indicated that the prosecution had not made a case since the defense was arguing that medical scientists had done to prisoners and 'Conscientious objectors' in the U.S.A. the same thing and under the same conditions that the Nazi physicians had done to their prisoners. This opening was the first opportunity that I had to inform the Tribunal of my objective and the reasons for it. I also told them that I had outlined the rules for the use of prisoners at Stateville, Illinois, and that I had used and was using conscientious objectors. I was then asked by the Presiding Judge if I would return in June, 1947, as a rebuttal witness, since it was not proper at that stage of the Trial to give direct testimony.[156]

Ivy nurtured the judicial interest in the Code 'so that something of a preventative nature had to come out of the "Trial of the Medical Atrocities"'.[157]

While Ivy's retrospective comments came when his own ethics were heavily criticised for his support for a natural cancer therapy, they can be seen to tally with the Trial documentation. Although in March 1947 Ivy failed in his attempt to persuade the War Department to support a scientific report on medical atrocities based on the Nuremberg evidence, he remained committed to his aim of the Code as 'a guide for the future':

> He thinks that an international commission could finish its report within six months, that Dr Alexander, now at Nuremberg could do the necessary 'leg work' by July 1 and that thereafter an American committee of scientists could finish its work in this country.[158]

The issue remained whether scientists could deliver a more effective verdict on Nazi medical crimes than the adversarial legal procedures. The judges' initiative made the point that scientists remained subject to law.

Internationalising the Code

Judge Beals stressed how 'certain types of medical experiments on human beings, when kept within reasonably well-defined bounds, conform to the ethics of the medical profession generally'. His verdict left the question open whether the criminal experiments could be tied to Nazism. Although the judges supplanted the ISC, there was no obvious way to disseminate the Code in international medical science and professional organisations. Taylor criticised the ISC as slow and that it 'has not carried out any significant investigation of its own but has relied mainly on documents unearthed by American and British war crimes investigators'.[159] But the necessarily aloof Nuremberg judges were far removed from medical circles in Germany let alone in the wider world.

Thompson continued the work of scientific evaluation. He was in Nuremberg to sort and microfilm German medical documents in February and again in August 1947. He considered that scientists could achieve a great deal more than the Trial in an ethical appraisal of all German wartime research. He conveyed his misgivings about the Trial to the Canadian official, Lester Pearson, who was supporting Thompson on his 'special mission':

> The intention of the ISC (WC) is to gather all evidence of German exper-
> imental work carried out in an unethical manner on human beings, and
> as representative scientific bodies, to
>   i)   pass judgment on the value of the scientific results obtained,
>   ii) condemn, in the name of science, the prosecution of such experi-
>   ments, and finally,
>   iii) lay down some definition of what may be termed a justifiable
>   experiment where a human being is being used as a subject.
> It will be realized that some or much of the data which is available will
> never appear in court when the experiments have been carried out by
> someone who is now dead or has not been apprehended; furthermore,
> legal authorities are not in a position to judge the validity of an experi-
> ment as scientists themselves are better able to do. And finally, their pro-
> nouncements will carry only a juridical condemnation by the scientific
> bodies, which may have more influence in determining the future
> course of experimental work on human beings. This much is being
> written on this particular point since some lay people are known to have
> expressed an opinion that all has been taken care of by the trials in the
> American or British Zones, in which some doctors are being accused.
> This is entirely to have missed the purpose of the ISC (WC).[160]

Thompson continued to liaise with Alexander and Ivy, conducted interrogations and collected evidence. Once the US drive to promulgate an ethical

code through the Trial became clear, the British became concerned less with ethics than with demonstrating German scientific incompetence. Smith noted on his 'Visit to Germany, re War Crimes and Medicine' about medical experiments on Jews in concentration camps: 'With the amount of material at their disposal and their complete freedom from the restraints imposed by ordinary ethics quite apart from the stringent prohibitions on medical ethics a great deal of valuable information about human experiments on human beings might have been obtained.'[161]

A barrier to rendering the ISC a genuinely international body was its restricted membership. The Danes tried to participate: V. Fenger of the Danish Medical Association went to Paris as an observer, but was dismayed that Legroux stressed *La Patrie* over ethical evaluation. He criticised the ISC for not dealing with euthanasia. He pointed out that unless there was an ethical declaration, physicians in any future war might lapse into similar abuses. He asked about the possibilities of a comprehensive evaluation 'shedding light on the German "scientific" crimes in all their forms'. He was opposed to only physicians evaluating the evidence of Nazi atrocities in secret, as the findings merited full publicity: 'Will this material, if obtained, eventually be given to the world at large, or perhaps in some special form or other be laid before the medical profession of all countries?'[162] The ISC was never to reconvene after January 1947. Instead, the British focused on the scientific quality of the German human experiments. Although poor science and inhumane ethics appeared to be linked, ethical guidelines remained necessary.

One way forward was through a revival of the Hippocratic precepts. Alexander and Ivy frequently cited the Hippocratic notion of the doctor's duty of care for a patient. But Hippocratic ideas (inherently opaque given the problems of translation and interpreting the semi-mythical Hippocrates) became subsumed in the political ideology of totalitarianism. Medical opposition to interference in the doctor-patient relationship meant that – in Ivy's words, 'We must oppose any political theory which would regiment the profession under a totalitarian authority or insidiously strangle its independence.'[163] An editorial in the *British Medical Journal* diagnosed the problem as political: 'the surrender, in fact, of the individual conscience to the mass mind of the totalitarian State'.[164] Fishbein, the editor of *JAMA*, took a similar tack. He linked the evidence on compulsory sickness insurance to the deterioration of the ethics of the German medical profession.[165] Physicians referred to the abuses of Nazi medicine as a rallying cry against the socialisation of medical services. The autonomy of science reflected a situation of doctors opposing central state planning and the welfare state. The scales of justice were heavily tilted by the weight of Cold War requirements for strategically relevant clinical research, and by professional defence of the status of the individual practitioner.

Despite its unpromising inception, international developments rein-forced the Code. The International Declaration of Human Rights developed parallel to the Code in discussions overseen by UNESCO. In March 1947 special guarantees included freedom from preventable disease, the right to health, and the rights of children, the disabled and the aged. The statement promulgated 'The Right to Live', and 'The Right to the Protection of Health', marking a symbolic refutation of the Nazi devastation.[166]

The medical profession became vocal on the world stage, claiming to speak paternalistically for the patient, while seeking to guarantee profes-sional freedoms. Its agenda was to counteract post-war enthusiasm for social planning and socialised welfare. The World Medical Association (WMA) was founded in September 1946 as a non-governmental organisa-tion, representing national physicians' associations.[167] The WMA was heavily influenced by the BMA. Its joint secretary was the NHS opponent, the avuncular British wartime Radio Doctor and opponent of the nascent NHS Charles Hill, who as BMA Secretary, was a diehard opponent of the socialisation of medicine through concern for doctor-patient relations.[168] The co-secretary, Pierre Paul Cibrie, a Paris-based physician who had been active in the resistance, was keen to see the ethical response to Nazi medical atrocities. Its offices were at the New York Academy of Medicine. The Soviets did not join, and the WMA excluded Germany and Japan. In June 1947 the BMA submitted a resolution on war crimes and medicine to the WMA, which met shortly after the Nuremberg verdicts to invite the German profession to acknowledge and condemn the participation of German doctors in acts of cruelty and oppression, and to expel the guilty.[169] It diagnosed that the corruption of medicine arose from its becoming 'an instrument in the hands of the state to be applied in any way desired by its rulers'. The BMA's view conveniently absolved physicians and their professional organisations from primary guilt for Nazi medical atroci-ties.[170] In December 1947 the BMA published the UNWCC's dossier on German medical war crimes. The BMA's report 'on breaches of medical ethics' was compiled from the materials of the UNWCC research officer Col. Wade who had liaised with Thompson. While the UNWCC/BMA report was a response to the failure of the ISC to issue a report, it fulfilled some of the original ISC aims. These reports concluded with a recapitula-tion of the Nuremberg guidelines, which began to achieve a wider impact.[171] In May 1949 the WMA approached the World Health Organisation for support for an international code of research ethics.[172]

Thompson's interest in the moral regeneration of Western civilisation, Alexander's in forensic psychology and Ivy's in safeguarding animal exper-iments meant that the Allied medical investigators focused on a Code for very different reasons. But their interaction was fluid and fruitful, overcom-ing the crabby arrogance of Moran and bureaucratic divisions. The scheme for a Code arose from Thompson's scientific branch of FIAT where he became concerned with medical war crimes. The first formulation of a Code should be credited to Ivy, who insisted on the voluntary status of the experimental subject, and then convinced the prosecution and the judges of the necessity of regulatory principles. Alexander amplified the obligations of the medical experimenter, defined what constitutes consent, and insisted on humane aims. While the Medical Trial judges recognised the rights of the experimental subject, they were picking up a theme already articulated by survivors of experiments in Auschwitz, Dachau and Ravensbrück. Although it appeared that the Code at Nuremberg was generated by the judicial procedures, the codification of medical ethics was in the throes of repeated reformulation since the end of the war. While unveiled to the public as a coherent set of principles in August 1947, the different interests in the origins of the Code can now be identified. The questions remain, did the mission to legitimate clinical research render the Code too permissive in what it condoned, and too weak in the provision of safeguards for the patient?

Part III
Aftermath

16
Cold War Medicine

Judgment

Between the final personal statements of the defendants on 19 July 1947 and judgment on 19 August 1947, prison conditions were relaxed, visits were allowed and prisoners socialised freely. There was doom-laden suspense. The defendants joked grimly about the impending 'Prize Giving'.[1] Rostock, despite favourable press responses to his testimony, predicted prison sentences for himself, Weltz and Blome, and the acquittal of Becker-Freyseng. He wrote letters in the event of his being condemned – ruefully reflecting that his career and life would be at an end.[2] Judgment Day meant either liberty, the prospect of continuing imprisonment or – as a final act of a devastating war – a death sentence carried out by the military authorities. While judgment was a matter of personal vindication, if the court was unconvinced by the defendants' justifications, then the judicial process was itself flawed.

Most defendants considered themselves to be ordinary scientists and physicians caught up in seismic political shifts of war. The legacy of the war was that the state recognised an obligation to fund medical research on an unprecedented scale. While the Trial was predicated on autonomy and liberty, the defence argument that the state could legitimate dangerous research as in the national interest gained chilling relevance. The Cold War had profound repercussions for medicine. The problem of atomic fallout led to defence-related studies in genetics and cancer studies. Researchers wanted resources to explore fundamental problems in the human metabolism, heredity and the causation of disease without onerous responsibilities to human subjects.

Cold War medicine was characterised by state funding of medical research, while ideals of liberty attenuated the socialisation of medicine. The thrust of these developments was to exonerate the individual physician for committing medical crimes and place the blame on totalitarian politics for exploiting medicine. A handful of courageous medical philosophers warned that the epistemological demands of scientific medicine could themselves be destructive of human life. The Trial observers Mitscherlich and Thompson and the medical witness Leibbrand found themselves lone voices against the onward march of human experimentation and quantification of clinical trials.[3] Those involved in scrutinising the Nazi experiments found themselves marginalised. The Americans recruited defendants for Operation Paperclip, and the German medical establishment was poised to rehabilitate the defendants, while vilifying supporters of the Trial. The danger was that the Medical Trial legitimised Cold War exploitation of medical research, rather than safeguarding patient rights.

After 139 days in court the judges had a firm sense of the overall picture of Nazi genocide and medical crimes. Judge Sebring reflected on the task of establishing individual responsibility: 'When I first sat as a judge, I looked at these shabby little men [the defendants] sitting there looking just like the rest of us. Then the prosecution began to put on its evidence. It was all so clear, so one sided, I began to doubt the evidence. I thought it couldn't be. People don't act like that in a civilised world. If they hadn't confessed their guilt on the stand, I don't think I would have believed it even after it was all over.'[4] Sebring's passing reference to 'the civilised world' raises the question as to the values of the Western world in the confrontation with communism, materialism and the spectral totalitarian state. The Medical Trial was predicated on a view that the Soviet judicial system was tainted, and that international justice administered by America would demonstrate the higher values of law and humanity. There was a growing sense that core values of Western civilisation required definition.

The question arises whether the verdicts were in any way compromised, if not by US strategic requirements, then by a new ethics designed to sustain Western freedom during the Cold War. Alexander hoped for death sentences on all the defendants with the exception of Schäfer, Rostock and Pokorny.[5] The trial observer Mielke reflected that the acquittals were those most needed by American aviation medicine.[6] The American pursuit of a dual policy of scientific exploitation and war crimes prosecutions raised the possibility that having lifted four scientists from the Aero Medical Center at Heidelberg, these researchers could now be legitimately returned to pursue their allotted tasks. The suspicion persists that the aviation medicine verdicts were compromised;[7] but no evidence has ever surfaced of undue influence being exerted on the judges.

The Medical Trial judges were careful to cite extracts from documents and testimony as the basis for their verdicts. They weighed problematic evidence in terms of the judicial principle of evidence for guilt being 'beyond reasonable doubt'. The operation of separating guilt from innocence required a clear understanding of actions, motives, and the medical and administrative hierarchies. The judges carried out their task with surgical skill in cutting through swollen tissues of accusation on the one side, and malignant lies and obfuscations on the other. Guilt was contingent on the circumstances of Nazi racial and military medical assignments as providing a framework of power within which actions occurred. The judges discounted all circumstantial rationalisations, by insisting on firm evidence.[8] The criteria of reasonable doubt meant that the judges could be left with residual suspicions, even when conceding that a defendant was innocent of the charges.

Judgment began with promulgating the guidelines on human experiments. The judges presented a succinct analysis of the wartime German medical services. They outlined how medicine succumbed to inhumane racial ethics. They dealt with the legal basis of the charges, rejecting for the most part that of 'Common Design, or Conspiracy', confirming a ruling of 14 July 1947 that conspiracy was not a substantive offence.[9] The charge of crimes against humanity remained crucial in condemning the involuntary experiments. The judges rejected any defence relating to the legality of human experiments under German law. They took the view that under international law coerced human experiments on citizens of another country was a crime against humanity. They stressed that the experiments were conducted without consent, and involved 'murders, brutalities, cruelties, tortures and other inhumane acts'. They rejected the argument of Gebhardt and Rose that a state could order experiments on those condemned to death.[10] The argument of superior orders was also dismissed, as the scientists initiated the experiments.[11]

Karl Brandt was first in line: he was pronounced guilty of overall responsibility for, and receiving information about, human experiments with sulphonamide, hepatitis and mustard gas. But the evidence did not show 'beyond reasonable doubt' his involvement in freezing, malaria, bone, muscle and nerve regeneration/transplantation, seawater, sterilisation and typhus experiments, although he knew in general of experiments in concentration camps.[12] The experiments for which he was guilty involved large numbers of experimental subjects, but relatively few deaths; even so, many survivors had crippling injuries. Most importantly, he was held to be criminally responsible for euthanasia involving deaths of non-German nationals. Here there was a paradox, because Karl Brandt's prime responsibility was for the T-4 euthanasia measures to which German citizens mainly fell victim. There were indeed a probable 200,000 non-German victims of Nazi euthanasia, but only by stretching Brandt's remit of responsibility to all euthanasia could he be seen as responsible for the later phases of 'wild' killings. No evidence was offered that he played a part in the transfer of T-4 staff to extermination camps. But as an SS member, he was guilty of having a senior rank in an organisation found by the IMT to have committed war crimes. Ironically, he was at loggerheads with many in the SS leadership.

The judges could find no mitigating circumstances and passed a verdict of guilty. Brandt's sweeping powers were found to be less than comprehensive.

But the judges took the view that because Hitler had invested him with prime responsibility for health and sanitary provision at a time of total war, he carried a substantial burden of criminal responsibility. The sentence confirmed medicine as a lethal element of Nazi power. Brandt felt aggrieved that the Tribunal had accepted neither his defence that there was a medical basis for euthanasia, nor that his efforts to organise emergency medical facilities in 1944 were genuinely effective operations, rather than camouflage for further euthanasia killings. In the eyes of the court he was the linchpin of the Nazi racial abuses of medicine, whereas he felt he had shouldered honestly the burdens of medicine in confronting severe disabilities and the disintegration of the final years of the war.

Next in line of seniority came Handloser, who held overall responsibility for the performance of military medical officers. He was found guilty of hearing reports of cold and freezing experiments. The meeting on typhus vaccines of 29 December 1941 was found to be incriminating, despite the efforts of Handloser's medical subordinate, Willibald Scholtz, to offer extenuating evidence.[13] The evidence of Ding/Schuler's diary, which Kogon had delivered to the OCCWC, implicated Handloser in authorising typhus research in Buchenwald. Although the judges agreed that the entries were not written on the date in question, they considered that the diary had probative value, as Ding signed every entry.[14] Citing the opinion of the US Supreme Court on the Japanese commander, Yamashita, the judges considered that a military commander was responsible for preventing violations committed against civilians. Here again, Handloser was culpable, and pronounced guilty.[15] The verdict stood against German military claims of immunity from war crimes.

Schröder (whose immediate superior was Handloser) was found guilty of having aided and abetted and taken part in medical experiments on non-German nationals against their consent. He was found innocent of the hepatitis and freezing experiments. The guilty verdict reinforced the view of military culpability.

The professor of surgery, Rostock, had knowledge of human experiments, but – despite his responsibility for co-ordinating wartime research – the judges realised that he did not have authority over them. To his apparent surprise, he was acquitted.[16] This marked an important distinction. Other scientists, such as the physiologists Rein and Strughold, could have had knowledge of (but not authority over) experiments, conducted by the German forces or SS. By taking this line, the judges opened the way for a distinction between the criminality resulting from German military and political organisations, while scientists could be untainted.

The judges accepted as plausible Blome's explanation that he attempted to subvert the plan to exterminate tuberculous Poles.[17] Even though Blome may have intended biological warfare experiments at Posen, the court found that there was no evidence of these actually being carried out.[18]

Although there was a 'strong suspicion' of Blome's support for the Polygal tests undertaken by Rascher, the judges found this was not sustained 'beyond a reasonable doubt'.[19] Thus the one defendant who had a senior position in Nazi physicians' organisations was acquitted. Unless a defendant was involved in an organised hierarchy – either to perpetrate criminal experiments or as a member of an organisation pronounced criminal – taking an ideological and organisational role in the placing of medicine on a Nazi basis was not in itself deemed criminal.

The judges considered the aviation medical researchers Romberg, Ruff and Weltz as a group. The prosecution maintained that Weltz had suggested and organised the high-altitude experiments, and instigated the collaboration of Rascher with Ruff and Romberg. Weltz had wanted volunteer subjects for a Luftwaffe-approved research assignment on the rescue of aviators at high altitudes.[20] Rascher approached Himmler for authorisation of the cold and pressure experiments at Dachau, and Weltz recruited Ruff and Romberg. The judges accepted that they expected that prisoners condemned to death would be rewarded by clemency. Friction arose between Weltz and Rascher. As Rascher used his links to Himmler to assert independence, he was transferred from Weltz's command.[21] The judges recognised that Romberg was closely involved in the experiments with Rascher, and took the view that 'the issue of the guilt or innocence of these defendants is close'.[22] But they decided to accept the defendants' explanations as truthful. The judges found the three accused not guilty of the air pressure experiments, and that Weltz was also innocent of the freezing experiments.

When it came to the seawater drinking experiments, the judges wholly exonerated Schäfer, who had devised a seawater purification method. His was the one unequivocal verdict of not guilty. Becker-Freyseng, Beiglböck and the Luftwaffe medical officer Schröder were pronounced guilty. Although Schäfer had attended the crucial conference of April 1944 when the seawater drinking experiments were decided on, the judges exonerated him from criminal involvement as he had merely devised a relatively effective desalination process. But Beiglböck conducted the experiments under coercive conditions, causing pain and suffering. Although the judges could not find evidence of deaths, they condemned his attempts to destroy and alter evidence, and pronounced him guilty.

The cases against Gebhardt, Fischer and Oberheuser lacked supporting documentation. But a strong set of corroborating testimony from victims and witnesses meant that the prosecution case was unassailable. The judges found no mitigating circumstances, and found Gebhardt and his assistants guilty.

Poppendick was guilty by virtue of his senior position in the Race and Settlement Office of the SS, but the judges found no evidence of responsibility for specific crimes. By way of contrast, Pokorny held no position of responsibility in the party or state hierarchy, but was a practising dermatologist. The judges did not believe his defence that he wished to sabotage the sterilisation programme. But they considered that no attempt was made to carry out the herbal sterilisation programme in the concentration camps. They acquitted Pokorny, 'not because of the defence tendered but in spite of it'.[23]

Overall, the judgment drew a fine line between war crimes and politically sanctioned racial atrocities on the one hand, and legitimate scientific inquiry on the other. The judges pronounced Schäfer as innocent without expressing any reservations, whereas the six others acquitted were all in some way tainted. The judges found greater criminality in the organised structures of the SS and the military than among the scientists. The verdicts drew a distinction between criminal atrocities inflicted by the German military and Nazi authorities, and the legitimate aims of the researcher. This divide was to remain axiomatic in the recasting of medicine for the Cold War.

The sentences

On the following day, 20 August 1947, the court pronounced sentences. The SS officers received the heaviest sentences: Brack, Karl Brandt, Rudolf Brandt, Gebhardt, Hoven, Mrugowsky and Sievers were sentenced to death by hanging. Two more SS officers, Fischer and Genzken, as well as the high-ranking military medical officers Handloser, Rose and Schröder, received life sentences. The judges sentenced Becker-Freyseng and Oberheuser to 20 years, Beiglböck to 15 years, and the SS doctor Poppendick to 10 years. A higher proportion at this trial (over a third) received the death penalty than the 10 per cent in the American Nuremberg trials overall. The total of death sentences was 24 out of 185 tried. Given that 35 were acquitted overall, the proportion of acquittals was also higher for Case One.[24]

The convicted were viewed less as medical miscreants than as vestiges of a genocidal regime. Alexander delivered his verdict on the accused at the Custer Institute on the far reaches of Long Island, New York in mid-August 1947. He analysed them as specimens of a dynamic totalitarian personality, which justified destruction. The problem was to ascertain how far Germany was infected, and to institute educational remedies.[25] Thompson took a similar view, considering that Germany needed a therapy based on humane spiritual values. He devoted his energies to establishing the German UNESCO programme, focusing on initiatives for education, youth and social science.

The convicted were immediately transferred to Landsberg prison, which had once housed Hitler. They had a limited period to file appeals for clemency. Time was to show that a life sentence meant far less than life, as these were reduced to 15 years, and parole followed. While the death sentences were upheld, the prison sentences became subject to the contingencies of the Cold War and to American policies of seeking to placate conservative German nationalists.

The German medical faculties neither acknowledged any guilt of academics, nor did they move against the condemned as individuals. In Nazi Germany many emigrated doctors and dissidents had their degrees annulled, and after the war these were rarely restored. Universities were reluctant to respond to the implications of the Medical Trial and abolish the degrees of the convicted. The medical faculty of Freiburg overlooked that Karl Brandt was a graduate. But outside pressure from concentration camp survivors prompted the Faculty to cancel Hoven's MD degree.[26] This was on the basis of its having been fraudulently obtained in 1943, because the research was carried out by concentration camp prisoners. Hoven's criminal misconduct did not prompt its annulment.[27]

Weltz lost his position as associate professor at the University of Munich under denazification regulations on 22 November 1945. The denazification tribunal of July 1948 confirmed Weltz was a Nazi sympathiser. He secured the restoration of his position on the basis of an amnesty of 8 March 1948, and the dean of the Medical Faculty endorsed his scientific achievements in aviation physiology.[28]

Beiglböck retained his professorial title. He did not venture to return to his native Lower Austria or to Vienna, as these were in the Soviet zone and the Austrian police left open the case files against him. He kept his German nationality rather than reverting to Austrian citizenship.[29]

The US military governor General Clay, who had attended the court on 18 April 1947, became the focus of clemency appeals.[30] While broadly supportive of the war crimes programme, he was reluctant to intervene in judicial procedures. He received clemency petitions, as there was no court of appeal. The US Supreme Court decided by five votes to three that it was not competent to hear appeals. President Truman also disclaimed competence for receiving appeals. Servatius considered an appeal to the United Nations. The lack of an appellate authority was a flaw and weakened the authority of the Trials.[31] Clay felt the burden of the hundreds awaiting execution. But he could not be swayed to alter the Medical Trial verdicts, rejecting the appeals for review of sentence and for clemency on 22 November 1947. But he ordered a stay of execution pending further appeals.[32] It was a period of frenetic lobbying on behalf of the defendants for clemency, but to no avail.

The acquittal of the wily Rostock was greeted as a triumph, and he was much congratulated by colleagues.[33] He still cultivated the prosecutors – as he explained to Karl Brandt – 'at our former residence'; their support was necessary for denazification. He then intended to go on the offensive to assist the convicted, to whom he felt obliged for support during the Trial. He found that Hardy and McHaney along with Bayle were disposed to help; only Hochwald was icily cold.[34] Rostock was the linchpin of a revisionist network, which attacked the legal basis and judgment of the Trial. He helped establish the 'Central Office for the Defence in the Physicians' Trial', which circulated a 'Compendium on Human Experiments as Carried

Out in Other Countries' as a refutation of Mitscherlich and Mielke.[35] Rostock lobbied the press to explain how the Trial was fundamentally unfair.[36] Rostock had the support of the historian of medicine, Diepgen (a former Berlin colleague), and maintained contact with wives and some lawyers (complaining that others simply went cold on the matter).[37] Rostock found allies among his professorial colleagues in his campaign, including Heubner and Rein. He agitated for the pardon of his close friend Karl Brandt, who hoped Rostock could intervene with Judge Crawford to support clemency for him and Beiglböck.[38]

Pokorny settled in Landsberg; he and Rostock (who obtained a clinical post in nearby Bayreuth) maintained solidarity with the condemned and imprisoned, visiting, writing, sending magazines and lobbying for clemency.[39] The devoted 'Ungerin' vividly described the scene in Landsberg to Rostock: she found her 'Röschen' [i.e. Rose] wandering about cleaning windows, while Genzken and Karlchen [i.e. Karl Brandt] languished in the prison hospital. The condemned were outraged and depressed by their fate.[40] Rostock (conspiratorially codenamed 'Vati') energetically campaigned to drum up support for his protégé Karl Brandt, and Servatius, codenamed 'Pan', agreed in February 1948 to co-ordinate a joint appeal and drum up sympathy with a newspaper article in *Die Zeit*.[41] The agitation to overturn the death sentences meant that the clemency appeals were rallying points for nationally-minded German physicians, scientists and intellectuals. They viewed the condemned as conscientious colleagues, who had erred under pressure from an oppressive regime.

'I am a Physician', Karl Brandt's emotive defence, resonated among medical colleagues. While convinced of his obligation to obey the Führer's orders, he felt that he had conscientiously fulfilled his obligations towards life and humanity.[42] The renowned surgeon Sauerbruch, who had distanced himself during the Trial for fear of being found culpable, now threw his authority into the petitions. He felt indebted to Gebhardt for interceding with Hitler on his own and (more importantly) his son's behalf at the time of the 1944 bomb plot.[43] Despite his bleak prospects, Karl Brandt mustered an impressive array of supporters.[44] Professorial colleagues and surgeons rallied to his cause. Twenty-six supporters included the Berlin surgeons Domrich and Sauerbruch, as well as the renowned pathologist Robert Roesle, the pharmacologist Heubner, the gynaecologist Stoeckel, and the historian of medicine Diepgen. Other noted petitioners included Tönnis (by then in Bochum where Brandt had once been), as well as noted physiologists, pathologists and surgeons.[45] General Clay was again not to be swayed:

> Regardless of what inner convictions Dr Brandt may have held, he was directly responsible for much of the suffering and death caused to the unfortunate concentration camp victims chosen to be used as subjects in

brutal medical experiments. In justice to these persons who underwent torture and death, I am unable to grant clemency in this case.[46]

Conservative Republicans and German nationalists joined forces to attack the legality and political justifications for the trials. The American lawyer Earl J. Carroll, representing the industrialist Flick, lobbied President Truman and General Clay to roundly condemn the Nuremberg trials. He litigated on behalf of Krupp to challenge the jurisdiction of the Nuremberg Tribunals.[47] Carroll alleged infiltration of the support staff at Nuremberg by émigré German communists seeking revenge.[48] These criticisms peaked with the trial of 24 IG Farben officers.[49] Servatius found Carroll a useful ally in the campaign to discredit Taylor.[50]

Rudolf Brandt's lawyer used the clemency appeal to condemn the long drawn-out trial procedures and the consequent disinterest of the German people.[51] Karl Brandt petitioned – unsuccessfully – for his execution to constitute a human experiment. This would have placed him in the category of a condemned experimental subject and vindicate the moral commitments outlined in the Trial.[52]

In desperation Emmy Hoven petitioned President Truman in February 1948, insisting on her husband's non-involvement in human experiments and euthanasia.[53] Several former Buchenwald prisoners, including Kogon and his comrade Werner Hilpert (Minister President of Hessen), pleaded for a pardon. Kogon also believed that Sievers had genuinely worked for the opposition.[54] He warned that Germans could not heap all responsibility on a few demonic figures like Rascher. Other former Buchenwald prisoners petitioned Military Governor Clay.[55] The Archbishop of Freiburg appealed for clemency for Hoven, and condemned the injustice of the trials. Hoven's American links were reflected in a petition from Hendrika Young of Boston, citing Carroll's opinion that the court's procedures were flawed.[56]

On 11 May 1948 the execution order sealed the fate of the condemned.[57] On 18 May the Americans refused to release Karl Brandt to the French, who requested his transfer for interrogation on Bickenbach's opposing mustard gas experiments.[58] The prisoners were allowed a final visit from their families. Some of the children would not or could not make the final visit.

The executions took place on 2 June 1948. The prisoners gave a short speech on the gallows. Sievers cryptically greeted the *Heimat* in the name of God, and Brack wished for God to give peace to the world.[59] Most condemned the injustice of the executions: they considered themselves victims of an unjust trial and convictions. Karl Brandt defiantly condemned his execution as 'nothing but political revenge' by a nation whose responsibility for Hiroshima and Nagasaki conferred an eternal mark of Cain: 'It is no shame to stand on this scaffold. I served my fatherland as others before me.'[60] The *New York Times* reported that Karl Brandt 'was the only one who scorned religious aid on the scaffold'. When he refused to end his speech, the hood was dropped over his head in mid-sentence.[61] Mrugowsky remained defiant, declaring that he was unjustly executed as a German soldier. Gebhardt had a Catholic priest in attendance while he mounted the gallows.[62] Only Rudolf Brandt affirmed that he faced death without hatred and vengeance. The US military paper *Stars and Stripes* reported that: 'Seven Nazi doctors hanged for wartime atrocities ... They paid an eye for 1,000 eyes and a tooth for 10,000 teeth. In a chilling rain they died unfrightened.'[63]

The condemned were photographed on the scaffold, and then lying on sawdust in their coffin. Hanging had been used at British insistence for the culpable at the IMT, but was not a usual form of execution in Germany or the US (as much depended on whether the length of rope was adjusted to the weight of the victim). Justice had to be seen to be done; only the photos on the scaffold were released to the press with the image of the guards removed. The bodies were returned to the families for burial.

The French authorities considered confiscation of Hoven's property, but concluded that the tribunal never ordered this.[64] Sievers' family was eventually granted a pension by the Bonn government. Later, some children of the executed received assistance from *Die stille Hilfe*, the secret SS members aid organisation run by the erstwhile euthanasia administrator Allers. But despite their private tragedy, the orphaned grew up quietly in West or East Germany, the children often entering the professions. A contrast can be drawn to the children of the high-profile Nazi leaders for whom the burdens were greater, because of their parents' public prominence.[65]

Attention turned to the prisoners serving out their sentences in Landsberg, as German conservatives stepped up their opposition to the Nuremberg trials. The appeals reflected the defensive mentality of the scientific community, closing ranks to defend German science. Heubner and Rein protested against the public presentation of the Trial, while Rose continued the campaign against the sentences. Rose was incensed that he was condemned for his knowledge of human experiments rather than participation in these. He indignantly asked why others with equivalent knowledge had been acquitted. The network of Rose supporters meant he received a stream of visitors and correspondence.

In the meantime the whole trial programme came under political strain. By June 1948 the Cold War was fully evident as the Soviets cut access to Berlin.[66] The campaign intensified for an amnesty for war criminals, supported by Heidelberg jurists and former Nuremberg defence counsel.[67] The Bonn government, the German churches and charities took up the agitation to abrogate the injustice of Nuremberg and Landsberg. Clay as High Commissioner remained impervious to the agitation, but it gained concessions from Clay's civilian successor, John McCloy, who organised a review panel.[68]

The Cold War chill bit ever deeper into the Nuremberg trials: midwestern conservative judges turned out to be extraordinarily indulgent towards the defendants, whom they viewed as patriotic soldiers and industrialists. Although the prosecution produced detailed evidence on genocide and war crimes, the accused were acquitted or given light sentences. The trial of industrialists failed in its prosecution of the supply of Zyklon B. The agriculturist Konrad Meyer was acquitted in March 1948 for drafting the genocidal *Generalplan Ost*, which combined science and racial policy.[69]

Demands intensified for clemency. Ivy was asked in 1950 to support appeals for clemency for Schröder, Becker-Freyseng and Beiglböck. He was inclined to the view that these researchers were decent men forced to be involved in the criminal acts. He sounded out the prosecutor Hardy, who realised that a concerted German attempt to undermine the legality of the Trials was underway. While recognising that the prison governor should be able to remit sentences, Hardy defended the convictions, believing that 'To say no one was responsible for these dastardly crimes is to say they never were committed.'[70]

As the pressure for German rearmament built up during the early 1950s, its promoters argued that clemency to war criminals was necessary to ensure West Germany's goodwill towards the West during the Cold War.[71] High Commissioner McCloy reviewed sentences of 79 of the Landsberg prisoners on 31 January 1951.[72] Although he upheld five death sentences, he indulgently cut many sentences. He commuted Genzken's and the ailing Handloser's life sentences to 20 years, and Fischer's life sentence to 15 years; Rose's sentence was commuted to 15 years, as was Schröder's.[73] Oberheuser and Becker-Freyseng's sentences were reduced from 20 to 10 years, and Beiglböck's from 15 to 10 years. When Schröder was paroled on 1 April 1954, it was on condition that he should not take part in any activities in the field of medicine or pharmaceuticals, and not to associate with his co-defendants. The Americans took the view that those condemned had 'lost all rights and privileges of a doctor'.[74] Only Fischer observed the condition, the recalcitrant Rose was cautious about opening a practice, but Beiglböck, Oberheuser and Poppendick flagrantly violated the restriction in the British zone.

The whole prosecution staff was now under attack. Behind the scenes Hardy and Drexel Sprecher expressed their concerns.[75] Steinbauer gathered evidence from the unprosecuted brain researchers Hallervorden and Spatz to discredit Alexander for duplicity and false accusations against Eppinger.[76] Taylor was a resolute spokesman for the Nuremberg Trials. He condemned McCloy's commutations as 'a terrible mistake'. While McCloy denied he was swayed by political considerations, Taylor diagnosed 'a wholly changed political outlook'. He complained that McCloy damaged wholesale the Nuremberg proceedings, as well as the integrity of those involved in the legal procedures in individual cases.[77] The acquittals undermined Nuremberg's function in cleansing Germany of its past misdeeds, so that democracy should not be burdened by Nazi criminality.

German conservatives preferred to emphasise national continuities. This in turn meant that West German justice was flawed and sporadic. Scientists closed ranks. Heubner joined Butenandt and two further KWI directors to exonerate Verschuer from colluding in medical murders with Mengele.[78] By 1953 numbers of prosecutions had dropped dramatically in both Germanies, and there was a rash of pardons and releases for the convicted.[79] Those among the Hadamar staff who were convicted were pardoned during the 1950s when an era of collective amnesia concerning medical crimes dawned, and the death penalty was abolished in West Germany for the convenience of Nazi perpetrators. Although the German Democratic Republic took the IMT as the basis for Nazi prosecutions, the notorious culprit Heissmeyer (the Hohenlychen physician responsible for the Bullenhuser Damm atrocity when 20 children were murdered) and euthanasia doctors, such as Rosemarie Albrecht at Jena, were allowed to remain at large. The GDR was ever more reluctant in its pursuit of medical miscreants, wishing to accuse the West of harbouring Nazis while suppressing evidence of past crimes.[80] Nationalists and conservatives vociferously campaigned for the release of the convicted and burying war crimes allegations.

The Medical Trial did not pave the way for further Allied prosecutions of medical atrocities on any large scale. The Americans and British had frozen war crimes prosecutions and extraditions of suspects to victim countries by 1948, as they were deemed to be a stumbling block in the emergence of a Western coalition against communism.[81] The trials' approach – to pinpoint the guilt of individual miscreants within Nazi ideological and bureaucratic structures – was ditched as politically inopportune.[82] Liberty as a political value came to mean freedom from war crimes prosecutions, and burying the whole issue of medical atrocities under National Socialism. This in turn would serve to liberalise medical research at a time when defence requirements saw a need for human experiments on nerve gases and radiation. The ending of the Nuremberg Medical Trial was to open the way for human experiments and clinical trials.

The Allies set a time limit on further prosecutions, and the German judicial authorities were at best sluggish and at worst acted only under intense pressure. A conservative ethos dominated German medicine, as it rejected the opportunity to draw a line under the past and reform itself for a democratic future. The effects of Nazism were downplayed in terms of the numbers of perpetrators and victims of medical war crimes. Nazism was used to excuse the involvement of doctors in human experiments, and few ventured to remind the profession that there had been widespread knowledge of the coercive experiments.[83] The German physician preferred traditional structures and to venerate past medical glories. The ethos of German history of medicine reflected this with a conservative agenda, which was slow to take a critical interest in Nazi medical atrocities. The lessons of Nuremberg were ignored.

'Paperclip'

The British and Americans were concerned that the Soviets were recruiting German scientific personnel for weapons research on nuclear fission, radioactivity and aeronautics. The Western Allies screened German science for its offensive potential while talent-spotting scientists for strategic projects. The US employed in all 765 German and Austrian scientists and engineers for military research under the 'Paperclip' scheme. It was at this time that US authorities gave the Japanese bacteriologists of Unit 731, responsible for the horrific experiments in biological warfare, immunity from prosecution. The interest of the Soviets in the atrocities and in gleaning intelligence data, as well as the appetite of US chemical warfare specialists in the data on human experiments, hardened American determination to shield Takeo Ishii, the key scientific perpetrator of mass murder. By March 1948 American policies to shield the Japanese military scientists were in place.[84]

Just as the Trial commenced, *Life Magazine* ran the story 'Nazi Brains Help U.S. German Scientists are revealed as Army researchers'.[85] The accusation has been repeatedly levied against the Medical Trial for whitewashing scientists to allow their transfer for US defence contracts. While the American military certainly wished to exploit some defendants' expertise, the available evidence indicates that the prosecution had complete discretion in selecting defendants, and judicial independence was strictly upheld. Strughold was vetted by the US war crimes experts, but exonerated. He arrived in the US in August 1947 just when the Medical Trial had drawn to a close.

Strughold co-ordinated the projects of German aviation medical experts at the Aero Medical Center in Heidelberg. The Center closed on 15 March 1947; the researches were published by the US Air Force in 1950.[86] The English-language compendium included research by Strughold, Rein[87] and Heubner[88] (all accused of having knowledge of the criminal experiments), as well as by Benzinger,[89] who had narrowly missed prosecution, and the defendants Ruff[90] (acquitted), and – remarkably – the guilty Becker-Freyseng[91] and Schröder.[92] The policy of exploitation gained momentum.

The end of the Trial cleared the transfer of German scientists to the United States. Strughold recruited 'Paperclip' specialists. Benzinger, former head of the German Aero Medical Laboratory at Rechlin, had been arrested at the US Aero Medical Center Heidelberg. A cohort of 'Paperclip' scientists was decided on at the time of his arrest.[93] He was released after interrogation in Nuremberg and, after working at the Aero Medical Center from 1 October 1945 to 5 February 1947, was transferred to the US. He admitted membership of the Nazi Party and SA, although his record stated that politics waned in favour of science, and he quarrelled with the medical advisers of Milch.[94] Benzinger declined Butenandt's efforts to lure him back to Germany with an academic post.

Some outcomes of contacts between the Nuremberg prosecutors and 'Paperclip' were decidedly negative. In August 1947 Wolfgang Lutz (the physiologist researching the effects of cold on large mammals in Munich, whom Alexander had interrogated) was given Robert Benford's address by the prosecutor McHaney. Lutz requested a position in America to pursue research on revival from low temperatures and rescue from stratospheric conditions – the very topics of the criminal Dachau research. The Americans declined to employ him.[95]

Schäfer, acquitted unreservedly at Nuremberg, was hired first to represent the absent Strughold on a temporary basis at Heidelberg. One of his first acts in September 1947 was to provide a denazification testimonial for Strughold.[96] He returned to pharmacological research, working in Hamburg. In 1949 he moved to the USA under 'Paperclip', where he worked for the USAF conducting seawater drinking experiments and radiobiological research. In 1951 he was evaluated as 'a most ineffective research worker' and as having 'very little real scientific acumen'. The USAF favoured his return to Germany.[97] The one unequivocally not guilty defendant was judged to be scientifically incompetent.

Blome at first practised as a specialist for skin diseases and urology in Dortmund, while having links to 'Matchbox', a British scheme for recruiting German scientists. But he opted for a US Army Chemical Corps project on 21 August 1951. This was cancelled when the US consul in Frankfurt denied immigration clearance, because he had been a blatant Nazi activist. He was given the position of camp doctor at the European Command Intelligence Center at Oberursel.[98] Goudsmit ranked him 'definitely as a third rate scientist' and reminded the State Department that 'Blome was a primary target of the Alsos Mission ... The man was our private prisoner for a while and I had a rather bad impression of him.' These links between the Trial and US strategic research show a military medical establishment willing to use erstwhile Luftwaffe and Wehrmacht researchers, but that the transfer could sometimes turn out to be decidedly unproductive.

Rein declined the chair of physiology at Heidelberg in April 1946, because of the build-up of Göttingen as a scientific centre, and suggested Strughold as an alternative. The American commander of the Aero Medical Center strongly supported the scheme for Strughold to have a joint appointment. The University overcame its hesitations and Strughold combined work for the Americans with university teaching. At a time when German wartime aviation medicine was subject to legal scrutiny at the Medical and Milch Trials, Strughold's chair was opportune.[99]

Once the Trial was over, Strughold was enticed to visit the United States in August 1947. His old contact Harry Armstrong launched him into space medicine. Armstrong did everything to prevent Strughold returning to Heidelberg. The take-off of Strughold's research for the US military was helped when, in September 1948, the Joint Intelligence Objectives Agency cleared him of being a Nazi or a war criminal. Alexander also took the view that he was innocent. There followed a tug-of-war between the University of Heidelberg and the Air University at Randolph Field, Texas to retain Strughold. Heidelberg was bereft of its physiologist, but benefited from its 'Strugi' advising Washington about the dire nutritional situation in Germany, sending over food supplies for the malnourished students, warning that the US press was critical of the University as reappointing former Nazis, and representing the University at the ceremony to appoint Eisenhower as president of Columbia University in October 1948.[100] Strughold constantly promised to return, but repeatedly extended his leave of absence. He felt that the United States was 'the only country of liberty', and – revealingly – that this was where he could carry out research in aviation physiology not possible in Germany.[101] Liberty for Strughold meant freedom to take research on human endurance to new extremes. In the end the lure of science triumphed over his patriotism: he resigned his chair at Heidelberg in August 1949.[102]

Having been won over to the cause of space medicine, Strughold toured Germany – with the support of Armstrong – from August to November 1949. His mission was to select German specialists to work at the USAF School of Aviation Medicine (allocated 34 'alien civilians' or 'paperclips'), and cold specialists for an Arctic Aeromedical Laboratory. His priority list of 22 researchers did not include any of the accused at the Medical Trial.[103] He returned to Randolph Field (where there were 30 'Paperclip' specialists in aviation medicine), and directed the high-altitude and space flight programmes. He was imaginative and innovative in his approaches to the problems of space medicine and night vision.[104] In May 1950 Ivy hosted a meeting with Strughold and other 'Paperclip' scientists on space medicine in Chicago.[105] He launched research on rapid acceleration, the effects of zero gravity and sensory deprivation.

In 1978–9 the US Department of Justice investigated allegations that Strughold had conducted human experiments in Dachau. Alexander once again advised, and consulted his 1945 diary. Strughold was again cleared.[106] Highly acclaimed in his lifetime, Strughold's reputation has taken a dive since his death in 1986, because of the indelible stain of having knowledge of the Dachau human experiments. One USAF flight surgeon recollected that Strughold made no secret of his development of high-altitude life support systems for the Luftwaffe, and that the Dachau studies were to hand on the magnitude of explosive decompressions which would crush the spinal column of an ejection seat occupant. Strughold (although incorrectly described as a 'Nazi') allegedly held Dachau data, which had not surfaced in the Medical Trial. The Nuremberg evidence showed Strughold was informed about the experiments in October 1942, and (more controversially, given the involvement of Weltz, the Luftwaffe and Himmler) that he had the authority to prevent them. Ohio State University removed

Strughold from a mural, which – banally – also included the fictitious Hippocrates.[107] Rumours continue to circulate that military medical research establishments in Britain and the USA hold concentration camp data on topics like hypothermia and poison gas. It remains unclear whether research on pressure and cold exposure, and on poison and nerve gases found to be criminal at the Medical Trial, was specifically utilised by the Americans and British. But the situation was one of intensification of military medical research using human experiments.[108]

Post-trial careers

The activities of erstwhile defendants in aviation medicine, medical research and as practitioners provoked continuing controversies. The aviation medical physician Romberg settled in Gilzum near Wolffenbüttel in the British zone in 1948. The local association of former prisoners agitated that evidence, which was not brought to light at the Trial, might exist for his further prosecution.[109] He then worked as a physician in Düsseldorf. Weltz continued his practice and academic career in Munich.[110]

Ruff gained a Habilitation in aviation medicine in 1952, and moved to Bonn as a specialist in aviation medicine. His contributions to the development of the ejector seat remained controversial, not least whether he used evidence of the Dachau pressure chamber experiments. Mitscherlich was outraged that Ruff as director of the Bundesanstalt für Luftfahrtforschung regained his former position, and he was an influential adviser to the DFG's commission for aviation research.[111] He had continuity of employment with Lufthansa, where he became the airline's head physician. Ruff held professorial rank from 1954, until forced by student protest and criticism from Schwarberg in the magazines *Bild* and *Der Spiegel* to resign his teaching position in 1966.[112] Having been a defendant at Nuremberg meant that the cloud of suspicion persisted, and the occasional storm of controversy would break.

Judge Johnson T. Crawford wrote on Rostock's behalf in November 1948 that 'His conduct during the trial was commendable and the evidence convinced the court that he was a splendid doctor and a good man'.[113] McHaney supported Rostock's appointment at a Veterans Hospital in Possenhof from 1948, and added that 'This acquittal constitutes a complete exoneration of Prof. Rostock ... the fact that he was subjected to trial in Nuremberg should in no way be held against him in his efforts to secure employment in his field of work in which he is so eminently qualified'. McHaney generously affirmed that Rostock answered at the Trial with 'frankness and honesty' – 'in sharp contrast to that displayed by substantially all the other defendants'.[114] From 1953 until his death in 1956 he was director of the Versorgungs-Hospital in Bayreuth.[115]

Rostock published his compendium of surgery, which he wrote in his Nuremberg cell, with Urban & Schwarzenberg in Munich, and he prepared a monograph on tetanus, drawing on his war experience.[116] If he had been condemned, he asked the former Luftwaffe surgeon Erwin Gohrbrand to publish the surgical compendium under his name. He planned a 'Collection of Documents on the Nuremberg Doctors' Trial' to refute the Mitscherlich-Mielke analysis. Rostock's 'objective' presentation was to be from the point of view of the defence.[117] Rostock had collaborated on a history of surgery with the historian of medicine Diepgen, to whom he confided his efforts to produce a counter-documentation to undermine the judgment, hoping that the collection would be useful for future dissertations.[118] During the Trial Rostock carefully analysed the strengths and weaknesses of each case and used this as the basis for his planned history. He enlisted the assistance of defence lawyers, and several provided collections of papers. He contacted the families of the accused to collect materials. Gisela Schmitz-Kahlmann, the former Ahnenerbe secretary, set out to vindicate the human experiments.[119] Their conspiratorial manner with codenames and smuggled letters had the ethos of a resistance network.[120]

Rostock complained to Rein about the Allied restrictions on publication, and cunningly expressed interest in the Society for Freedom of Science. This lobby against scientific social planning was gaining influence after the war. The Germans saw that it could be used to gather support from British and American researchers, who increasingly rallied to the anti-state agenda.[121] The defence realised that the Cold War evocations of freedom could be used to assert immunity from judicial scrutiny. Rostock embarked on an 'objective' account of the trial to refute Mitscherlich. He complained that Mitscherlich included only negative documentation on the defendants.[122] Rostock found that no medical journal would publish his critical verdict on the Trial in 1947.[123] As he doubted whether he could find a German publisher for his revisionist collection, he planned to deposit it in an institute for the history of medicine. One lawyer said that he had already written a book on the trials, and he too was having difficulties in finding a publisher. On 17 April 1948 Rostock explained to Karl Brandt that he remained pessimistic about finding a publisher for what he called the 'Beals-Seebrink [a pun on Sebring]-Buch' and so would deposit the whole collection with Diepgen at Mainz.[124] Rostock's collections were eventually deposited at the State Archives Nuremberg. This coincided with intense lobbying by the Adenauer government on behalf of convicted war criminals, while the American judges in the successor trials became ever more lenient.

Parole and denial

Handloser's application for parole showed that he rejected the verdict wholesale. He denied having any authority over SS doctors in the concentration camps, ignoring that the issue was his command over the Waffen-SS medical services. He was already terminally ill, and died in a Munich clinic on 3 July 1954, one day after his wife signed the parole application on his behalf.[125]

By way of contrast, Fischer, the youngest defendant, was faced with having to find an appropriate professional niche. When released on parole, his records noted that 'incarceration has made this applicant realise his guilt'. His former head of department, Ostertag, recommended him to the pharmaceutical company, Boehringer-Ingelheim, for whom he screened English-language periodicals from 1954.[126] The company gave assurances that he would not work as a physician, or be involved in the distribution of pharmaceuticals.[127] He thereafter lived quietly, taking the view that he was justly convicted but had served his sentence.

Genzken was released on parole to Baden Württemberg on 17 April 1954. Aged 68, 'his only desire is to be released to go to some small village near the Black Forest and quietly live out the remainder of his life'. He stuck to the view that he was convicted unjustly, as he was a soldier acting under orders.[128] The quietism of Fischer and Genzken stands in contrast to the high-profile efforts by Oberheuser and Rose to secure their rehabilitation, and to annul the Nuremberg verdicts.

After six years in prison Oberheuser was released in April 1952. She obtained a pension as a 'late returnee' as if from a POW camp. She returned to medical duties in the Plön Catholic hospital where she had been arrested. She then opened a medical practice at Stocksee near Kiel, continuing her connection to Plön. All would have been well for her, had not a former Ravensbrück prisoner protested against her practising. *The Daily Express* expressed outrage, and the *British Medical Journal* published protests, attacking the German government's refusal to provide pensions for survivors of the human experiments.[129] The state of Schleswig Holstein withdrew her medical qualifications, imposing a *Berufsverbot* in August 1958. Oberheuser challenged the legality of this. By 1960 the controversy had deepened: her lawsuit to restore her title was backed by influential political supporters of the child euthanasia doctor Werner Catel.[130] A letter to *The Guardian* referred not only to Oberheuser's phenol injections, but also to a professor (i.e. Ruff) involved in cold exposure continuing to be in post, as well as to others who had researched on mustard gas. The case was cited as symptomatic of 120 concentration camp doctors continuing to practise in Germany. The German state authorities and German Medical Association were at first impervious to all criticisms. The East Germans raised the Oberheuser affair as part of their campaign against Adenauer's revisionism. By June 1961 legal and public pressures, reinforced by the republication of the Mitscherlich-Mielke Trial report, forced Oberheuser to renounce her physician's status.[131]

Rose was released in June 1955, aged 64, and registered as a late returnee from the war, entitling him to a lump sum, a three-week holiday and free rail travel. He first tried to return to the RKI but found his old research team was dispersed, and his position as deputy director no longer existed. He then hoped that as a victim of American injustice he could resume his medical career. He found no opening in public health or research, and had no appetite for establishing himself as a general practitioner. He embarked on a lifelong campaign to reverse the judgment against him in the Federal German courts. He worked for a glass manufacturer, taking up the cause of glass recycling. In 1964 he won the battle to receive his state pension, and although bitter that his medical career was blighted, he used the pension to proclaim – misleadingly – that he was exonerated.[132] He was active in CDU politics and was on the governing council of Allers' *Die Stille Hilfe*, the organisation for victims of war crimes trials.[133]

Becker-Freyseng while in prison contributed to the Strughold compendium on aviation medicine. In 1951 his sentence was reduced to 10 years, and he was released in 1952. He became embroiled in an argument over whether his American sentence had any legal force affecting benefits as a late returnee in Germany. His USAF contacts continued, he travelled to the USA, and died at the age of 50 in 1961.[134]

Beiglböck's sentence was reduced to 10 years in 1949, and he was released in 1951. Ludwig Heilmeyer, who gave him laboratory facilities at Freiburg, had been an SA activist albeit critical and had an appointment at Cracow during the war. He was director of medical clinic of the department of chemical diagnostics at Freiburg from 1946. As an energetic clinician and genial laboratory researcher, Heilmeyer had enjoyed the esteem of Beiglböck's chief, Eppinger: assisting a former assistant of the deceased Eppinger was a mark of respect.[135] Heilmeyer also employed Rascher's collaborator, the mescalin and blood styptic researcher Kurt Plötner, at Freiburg, where he attained the position of associate professor. Heilmeyer studiously erased names of Jewish colleagues killed in the Holocaust from post-war textbooks.[136] It was ironic that the replacement of Nazis meant the reappointment of an anti-Semitic nationalist. During 1947 the Austrian authorities began to rehabilitate Eppinger by quashing the verdict that he was an 'illegal' Nazi, and regarded him as a 'legal' Nazi who had applied to join the NSDAP only in May 1939.[137]

Beiglböck became head of a hospital department of internal medicine in Buxtehude in 1952. The hospital was run by Allers, formerly the T-4 administrator, and then organiser of *Die Stille Hilfe*. Beiglböck continued to publish widely in medical journals, and to lobby for his rehabilitation. In the late 1950s, he began to rebuild his Austrian links. When due to lecture to the Medical Chamber in Vienna in 1962, there was public pressure on the Austrian police to prosecute him and on the Ministry of the Interior to prevent his lecture. Austrian legal authorities stressed that the charges made against him in 1946 remained open, a situation that distressed Beiglböck.[138] The Austrian authorities had to decide whether to reopen the still unresolved case in the summer of 1963.[139] In 1962 Allers was again in custody, and he was found guilty in the autumn of 1963. Beiglböck emulated his mentor Eppinger's fate by committing suicide on 22 November 1963, aged 58, and was laid to rest in his birthplace in Lower Austria. He left a bequest to *Die Stille Hilfe*.[140]

Poppendick was released on 8 February 1951. He immediately began to work as a locum in his home town of Husum. He informed the University of Münster that he was released as a prisoner of war, that he was bereft of all means and had a wife and four children to support. He applied to submit his doctorate (he was the only one of the physicians in the Trial without an MD), and asked the University to waive his fees. The ruse was that he altered the spelling of his name to Poppendiek, as he had been tried under his alias of Poppendick. He slyly presented a certificate issued by the council of Hude that 'Helmuth Karl Max Poppendiek' was free of any prior criminal convictions. His curriculum vitae mentioned his having attended the hereditary course at the KWI for Anthropology, but his period at the SS Race and Settlement Office was given as having worked in a 'military sanitary department', and he explained that he was a 'prisoner of war' from 1945 to 1951. He presented recent substitutes for certificates of medical qualifications, as having been 'lost in the war'. Verschuer was a compliant examiner, praising the quality of the analysis and the use of clinical material. The examination was completed in February 1953.[141] The topic was the Pelger gene, an area of study developed by Nachtsheim, and the dissertation dealt with a case of a two-year-old epileptic, who was a carrier of the gene. The University required 'Dr Poppendiek' to undertake to serve the cause of truth.[142] With an MD in the name of his alias Poppendiek secure by 1954, he opened a medical practice in Hude. The rehabilitation of Poppendick/Poppendiek shows that German civil, medical and academic authorities conspired to undermine and ignore the verdicts of the Medical Trial. German public opinion could not distinguish between a war criminal and a prisoner of war.

The German medical community became increasingly belligerent towards the brief period of scrutiny and Allied justice. Schaltenbrand complained about the duplicity of Alexander to the neuro-surgeon Tönnis.[143] Far fewer of the Medical Trial witnesses were prosecuted than was expected at the time, because plans for the Second Medical Trial evaporated. Others with a dark record, such as Hippke, Liebau and Reiter, were released.[144]

The French prosecuted Haagen and Bickenbach for murderous experiments at Natzweiler. They were condemned on 24 December 1952 in Metz to a life sentence of hard labour. A military court in Paris annulled the verdict, and they were retried at Lyons in 1954. Haagen insisted that none of his experimental subjects died. He was recalcitrant and claimed that he ought to be given a Nobel Prize rather than a prison sentence. Bickenbach took a markedly different defence, that he was ordered by Hirt to carry out the lethal experiments on an antidote to phosgene gas poisoning, and that he did so only after making his objections known to Karl Brandt that it was against his Christian conscience. Both argued that there were French precedents for experiments on criminals, including experiments by Pasteur, and that there was no basis in international law to condemn medical research undertaken in good faith. Both were sentenced to 20 years' hard labour.[145]

Barely a year into their sentences, Bickenbach and Haagen benefited from an amnesty in 1955. Haagen resumed his career at the Bundesforschungsanstalt für Virusforschung in Tübingen in 1956, where he remained until 1965. Bickenbach went into general practice. Those involved in human experiments and using body parts from euthanasia victims embarked on a campaign of militant denial of criminality. Schaltenbrand sought recognition – and grants – from American colleagues, while denying the criminality of medicine under National Socialism.[146] The Allies were disappointed in German disinterest in prosecuting war criminals.

The Medical Trial marked the end rather than the start of vigorous investigations of medical war crimes. German medical scientists formed ad hoc expert commissions to exonerate their colleagues. The case of Verschuer was considered twice – in 1946 and 1949 by KWG/MPG colleagues who saw no crime in research on bodies killed to order.[147] In May 1948 the German Congress for Internal Medicine set up a commission to examine scientifically and ethically whether the seawater experiments were criminal. Among the members was Heilmeyer, who supported Beiglböck on his release. The commission agreed that the selection of experimental subjects was flawed, and it was a mistake to research in a concentration camp, but that no crime occurred, because none of the experimental persons was injured. The report was sent to General Clay.[148]

By the 1950s a pattern had emerged of academic clusters of former war criminals and suspects. At the University of Freiburg, there were Beiglböck and Plötner. At Düsseldorf were the professor of hygiene Walter Kikuth and Anton Kiesselbach, who had assisted the murderous Hirt in Strasbourg, although no criminality could be proved after the war.[149] For the most part German medical faculties rejected the demands for clarification of the issue of coerced experimentation, leaving numerous unresolved cases. Racial experts found their way back into medicine, public health and welfare, as when the post-war Bavarian authorities took over the Nazi Sinti/Roma files.[150]

German medical organisations failed to condemn the wartime medical atrocities. When the former pre-1933 collaborator of Hirt and Jewish émigré, Phillip Ellinger, was half-heartedly offered renewed membership in the German Pharmacological Society, he wrote in March 1948 inquiring about its stance on the wartime human experiments.[151] While denazification meant scrutiny of political involvements, there was no systematic attempt to evaluate the involvement of pharmacologists, virologists and other medical specialists in research atrocities, let alone in clinical abuses. The German state prosecutors were loath to investigate, or as the Mengele case shows, to apprehend even notorious war criminals. How Mengele was able to leave Germany in 1948 and keep in touch with his family showed continued failings in the legal and police authorities' will to prosecute medical crimes.[152] The Auschwitz survivor Hermann Langbein found Mengele's Argentinian address on divorce papers brazenly filed in Freiburg in 1954.[153] When the Federal Republic prosecuted, the proceedings were often lax. Medical colleagues obligingly testified the defendant was medically unfit to stand trial, while diagnosing that victims of experiments were indeed suffering from a range of hereditary mental defects.[154]

17
A Fragile Legacy

Taylor roundly condemned the Nazi medical experiments as a scientific failure.[1] A number of scientific commentators endorsed this view.[2] The issue of whether the German human experiments had yielded useful results, and whether – morally – they could be used in further research, was debated throughout the Medical Trial and became part of its legacy. *The Lancet* raised this as 'A Moral Problem' just before the Trial, foreseeing the clash between scientific exploitation and duty to the patient. The entomologist Kenneth Mellanby sided with the dedicated scientist, unfairly hauled before a military tribunal.[3] *War Crimes News Digest* reported on 'Controversy in the British medical world about the preservation or destruction of the notes made by German doctors concerning experiments on prisoners'. In *The Daily Telegraph* Dr Layton, who took over the Belsen hospital in July 1945, cautioned: 'Whatever one may think about useful knowledge to Humanity coming from these experiments, it would be quite wrong to use such knowledge'.[4] Lord Horder maintained the opposite view: 'It would be a great mistake,' he said, 'to destroy them [the notes] altogether'. Hugh Clegg, editor of the *British Medical Journal*, believed 'if any good can come out of these experiments they should be published'. He endorsed Mellanby's mission to collect Nazi research findings.[5] Layton criticised Mellanby in *The Lancet* by insisting that research and sadism coincided.[6] The tendency ran in favour of evaluation by research scientists. In February 1947 *War Crimes News Digest* reported: 'The view of American scientific consultants is that the Nazi experiments had gained 'practically nothing of scientific importance'.[7]

Thompson's team in Nuremberg set about microfilming and interrogating imprisoned researchers. He compiled batches of documents for scientific evaluation, organised microfilming at Nuremberg and Dachau, and contacted key scientific witnesses. Lieutenant Clement Freud sorted the films, assisted by Latvian DP physicians, according to medical categories.[8] Alexander praised the ISC for 'collecting and cataloguing the scientific war crimes evidence'.[9] The idea was that the ISC would assemble the documents at the Pasteur Institute, and then open these for evaluation by research scientists.

Thompson urged UNESCO to send a representative to Nuremberg, but this fledgling world science organisation avoided the human experiments issue. The Americans suggested that the international scientific evaluation be made by a United Nations agency.[10] In March 1947 the Foreign Office decided that it did not want UNESCO or the UN to be involved with the ISC. Julian Huxley, director general of UNESCO, regretted that it was impossible to send a UNESCO representative to Nuremberg, although he hoped that Thompson could send reports.[11] Leslie Rowan, the private secretary to the prime minister, reported to Moran: 'We have seen a letter from Professor Huxley to Wing Commander Thompson indicating that UNESCO did not have staff to send a man to Nuremberg to report on German medical war crimes, and we therefore think it most unlikely that UNESCO would be able to take on a further commitment of this nature'.[12] Thompson became UNESCO's informal Nuremberg representative.

In the original scheme of the ISC, it was hoped that 'scientists should present the world with a full report on the atrocities committed under the guise of scientific experiments'.[13] In October 1947 co-operation of UNESCO was still envisaged. The idea was that the documents collected by Thompson for the Commission should be passed to UNESCO for preparation for publication by HMSO.[14] By November 1947 Thompson was appointed assistant to Julian Huxley at UNESCO, and requested copies of all ISC documents.[15] Even so, UNESCO took no position on the issue of human experiments in biological and medical research, although it was poised to tackle the legitimacy of the concept of race.[16] Moran insisted that the Commission's deliberations be strictly confidential, thereby torpedoing the original aim to present the evidence of human experiments 'before the scientific bodies of the world'.[17] The ISC remained a bilateral Anglo-French mission, as the US State Department declined support. Thompson became exhausted and disillusioned with the ISC, and UNESCO failed to develop the ethics of biomedical research.

When the judges promulgated the Code on human experiments in August 1947, it underlined how the Trial had displaced the ISC. Moran berated the British and French diplomats for being naive that the US would ultimately join the international committee.[18] The British embassy in Washington attributed the delay to the US elections in 1947, giving rise to the belief that approval from Congress was necessary.[19]

The UNWCC lost patience with the prevarications of the Moran Commission, and in December 1947 reported on Nazi medical war crimes. Moran divided his time between treating Churchill as opposition leader, and, as president of the Royal College of Physicians, taking the pulse of medical opposition to the National Health Service.[20] Labour politicians were circumspect in their dealings with him, especially as the government was faltering.

The terseness of the British official response can be attributed to a concern with the iniquities of state medicine. Moran deferred to the physiologist Dale, who wished to demonstrate that the state should not undermine the autonomy and freedom of the physician or scientist.[21] Dale's views reflected sentiments of the Society for Freedom in Science. This rising star in the Cold War ideological firmament doggedly denounced central state planning as totalitarian and a threat to free scientific inquiry. Faced by a Labour government that stressed central state direction of science as part of social planning, critics attacked large-scale, state-controlled research as ushering in Nazi or Soviet-style totalitarianism that alienated biologists because of the suppression of genetics. Nazi science was perceived as synonymous with regimentation. Protecting the scientist's autonomy meant shifting responsibility for Nazi human experiments onto the totalitarian state – a convenient target at a time of rising East-West tensions.[22]

Scientists like Bronk (president of the National Academies of Sciences) and Dale (president of the Royal Society) aimed to assist what Dale called 'good German research work'.[23] Dale ignored the growing mountain of evidence of medical crimes, and was impressed by Mitscherlich's concessions as regards the rectitude of his fellow pharmacologist Heubner.[24] The lawyer Nelte, defending Hörlein in the IG Farben Trial, informed Dale how the scientists Butenandt and Hahn stood solidly behind Hörlein, and that Jewish refugees were also grateful to him for assistance.[25] In February 1948 Dale drew the attention of the Scientific Committee for Germany to the exculpatory *Reichsgesundheitsblatt* rules on new remedies and clinical experiments of 1931, and naively accepted the assertion that all German scientists felt bound by the principles of the consent of the research subject.[26] In contrast to efforts to force the Germans to accept an ethical framework, Dale wanted a German Medical Research Council modelled on the British MRC. As the Medical Trial was drawing to a close, Dale visited Germany in order to persuade the Western Allies to liberalise the structures of the control of science.[27] He liaised with Bronk to elicit American support for greater autonomy in German medical research. He energetically engaged in the efforts to establish the Max Planck Gesellschaft early in 1948 as a joint US-British venture, naming it after a venerable scientist as figurehead. At the same time Dale set about undermining the ISC with the devious Moran, as the Nazi medical crimes were an impediment to scientific and medical progress.[28]

Dale wished to bury the accusations of German medical criminality, and instead supported new enabling structures for research. He fervently believed in German science as a precious resource. In July 1947, as the Medical Trial was ending, Moran initiated expert evaluation of the Trial documents by eight British experts, who had themselves been involved in military research or eugenics: Lovatt Evans had conducted chemical warfare experiments;[29] Ronald Hare had been involved in epidemic control;[30] the eugenicist C.P. Blacker evaluated the sterilisation experiments; Aubrey Lewis – a critical member of the Eugenics Society – German psychiatric research;[31] Hamilton Fairley had infected subjects with malaria and tested courses of drugs – notably chloraquine – at Cairns in Queensland, Australia, where some of the subjects were interned alien refugees from Nazi Germany. He also interrogated suspected German nerve gas researchers. Other members were W.G. Barnard; James Paterson Ross; and V.H. Ellis.[32] The panel's role was restricted to evaluating the quality of the German research rather than its criminality or ethics. (See Table 11 for expert panels.)

Just when the US judges were formulating the Nuremberg Code, the British panel was asked to assess the scientific value of the German human experiments. On 16 July 1947 Thompson distributed the evidence for scientific evaluation of 'the experimental work performed in concentration camps by the medical profession in Germany'. The experts were to consider such matters as: 'was the object of the experiment a reasonable one?', 'the contribution to the advancement of medical or scientific knowledge', 'the scientific and medical value' and 'was it imperative to use human beings as subjects in order to obtain the answer to the question with which the experiment was designed to deal?'[33] In September 1947 the ISC sent Hare the document books of the defendants Mrugowsky, Rose and Sievers, and evidence on bacterial warfare and epidemic jaundice.[34] Hare's verdict on the vaccine experiments was that they were unnecessary and that the science was defective.[35] Hare's report criticised unjustified use of human subjects, as well as the methods, competence and training of Ding, Hoven and Schiedlausky, and the brutality of Dietzsch, the convict in charge of the typhus wards. Hare considered specimens may have been contaminated by latent viruses and bacteria of other diseases, and that Ding had falsified data, and made valueless tests with toxic pharmaceuticals.[36] Hare had a higher regard for the originality of Haagen's typhus vaccine research, but found him reckless in persisting with research inducing a strong reaction, and in causing at least 50 deaths.[37] The overall conclusion was that the experiments lacked proper planning and added very little to the stock of existing knowledge.[38]

Lovatt Evans reported on physiological experiments in early September 1947. Thompson commended this as 'a wonderful job in piecing together the bits and scraps of information so as to make a fairly coherent picture'.[39] Evans concluded that the Bickenbach phosgene experiments were neither well designed nor well carried out, and that the presumed antidote was ineffective after exposure to phosgene.[40] The ISC began to lose momentum, because Moran decided that it would be best for medicine if the whole obnoxious business could be consigned to oblivion. The wider issue of the suffering resulting from medical atrocities was lost from sight by focusing on whether the only crime that the Germans committed was that of being bad scientists. The one question dealing with the victims was: 'Was it imperative to use human beings as subjects in order to obtain the answer to the questions with which the experiment was designed to deal?'[41]

Blacker's materials included Alexander's CIOS report of 1945 on sterilisation and euthanasia. He noted that a cheap method of sterilisation did not result; even though the extract from *Caladium seguinum* was a viable possibility, he accepted that the Germans were unable to grow the plant in sufficient quantities. Blacker condemned how hundreds were subjected to the castrating effects of X-rays in Auschwitz, when terrible burns and skin necroses were inflicted. He was disgusted by the injections of sclerosing substances by Clauberg in Auschwitz. Blacker concluded: 'None of the experiments have any bearing on eugenics as the subject is understood in this country.' The German research was for the most part carried out by inadequately trained and supervised underlings. Blacker regarded these experiments as far worse than the Nazi euthanasia measures; he hoped that the victims were 'mercifully killed'.[42]

Given that Moran promptly received all the experts' reports between August 1947 and March 1948, the question arises why he procrastinated.[43] Blaming Ivy for raising ethical issues in the press, Moran explained to Dale that it was simply a matter of evaluating the aims, results and originality of the German research.[44] Dale was keen to bury the war crimes issue, and Moran readily complied. Their aim was to close down and terminate all further discussions on German medical crimes, fitting in with Dale's efforts to resurrect German medical research. Dale shared the view of the MRC that, 'In Germany today there prevails a deep distrust against medical research so far it is extended on human beings and, though understandable, it endangers medical progress itself'.[45]

After a lamentable delay, in 1949 Moran produced a terse five-page report for the Foreign Office. This stood in marked contrast to the clarity of the Nuremberg Code and the 50 volumes of Trial proceedings and documents. Moran peevishly stated that a moral analysis would have required American co-operation.[46] A final meeting of the War Crimes Committee was held on 27 May 1949 to approve Moran's digest, which was in contrast to the detailed reports from the commissioned experts. The German scientists were accused of scientific incompetence, but the inference was that only a small number among their ranks were criminals.[47] The Foreign Office disapproved of the report's publication because it was so negative and open to criticism for its curt superficiality. Moran did not want to see his report buried, because of its strategic value to the medical establishment. He informed Attlee that 'It has, I think been proved up to the hilt that these experiments had no scientific value whatsoever'. On 11 September 1949 Attlee supported Moran's view that publishing the report would prevent any claim of the humanitarian use of concentration camp research.[48]

The war crimes issue played into the hands of the eugenicists. It suited Blacker's efforts to distance eugenics from Nazi racial atrocities, so that eugenics and sterilisation should remain palatable for public legislation. In December 1951 Blacker addressed the Eugenics Society to roundly condemn Nazi eugenics, and published his verdict in the *Eugenics Review*. That this issue carried a favourable review of Swedish eugenic sterilisation underlined the dichotomy between the reviled Nazi experiments and the desirability of eugenic sterilisation. This persuaded Blacker to recant his earlier approval for the German sterilisation legislation as observing the non-racist criteria.[49] Affidavits by the sterilised showed that this was carried out for racial and political reasons. He condemned the legislation for not being properly scientific.[50]

Blacker branded the Nazi human experiments as incompetent and racist, suggesting 'German scientists of weight and repute ... held aloof from the activities of the SS'. He cited the cases of the psychiatrist Oswald Bumke and the neurologist Kleist.[51] He failed to mention Rüdin, Verschuer or the many other eugenicists involved in racial adjudication for sterilisation, euthanasia and the Final Solution. In reaching his verdict a degree of self-interest was apparent. Blacker was keen on eugenic sterilisation legislation and to advance eugenic notions of population policy. He demarcated between Nazi eugenics as unscientific and eugenics itself as humane, enlightened and thoroughly scientific.[52] He co-operated closely with Julian Huxley, who also campaigned for the freedom and autonomy of science.[53] Eugenicists now sought to tackle the global 'population explosion', which was seen as destabilising as the spread of communism.

Anglo-American interest in strategic research intensified, and a certain amount of information on medical war crimes – as Mant admitted – was deemed secret.[54] The British Chemical Defence Experimental Station, Porton Down, had an active programme of human experiments. A review in 1946 stated reassuringly, 'No one was subjected to a test without being told the precise nature of the test and the possible consequences to himself. Volunteers from the volunteers were invited for tests involving the risk of injury.'[55] The British took over the German nerve gases Tabun and Sarin for experimental development. At least one British 'volunteer' lost his life, and others were not informed of the dangers. The intelligence officers interested in chemical compounds as inhibitors were anxious to exploit the Bickenbach case by questioning him, as the French had asked for extradition of an assistant Hermann Rühl who had measured the density of the gas.[56] Eugenics and biological weapons became part of the arsenal of Cold War medicine.

How the issue of human experiments flared up in the Cold War can be seen with accusations of unethical human experiments on the part of the Western powers. In January 1948 the Soviets derided the American experiments on prisoners in Illinois. Professor Planelies of the Soviet Academy of Sciences criticised the experiments as replicating the research methods of fascists. Here, the Soviets had picked up on an issue raised by the

German defence, underlining the Cold War climate of the Trial.[57] The Green Committee chaired by Ivy continued to oversee human experiments on prisoner volunteers at Stateville, involving skin grafts. In 1948 the Committee approved the feeding of salmonella cultures.[58]

American scientists ran a parallel committee of evaluation of Medical Trial records. The former US Secretary of War, Patterson, secured a set of prosecution and defence transcripts on the basis that 'the location of this material with the New York Academy of Medicine will make it available to the entire medical profession in the United States and Canada'.[59] Iago Galdston, a critic of experimental medicine, organised a commission of evaluation.[60] The Academy indexed the Trial documents to allow scientists to judge whether the Germans had produced any scientific innovations so far overlooked.[61] The scheme proposed to the ISC of a committee of leading American physiologists (Ivy, Bazett, Frank Lahey, Myerson and Surgeon General Kirk) was ignored, and the evaluation panel came from the ranks of the Academy's fellows. The hepatitis materials went to Franklin M. Hanger; the freezing experiments to Frank McGowan; typhus, yellow fever and malaria to Morton Kahn; psychiatry to Nolan D.C. Lewis and Dr Wilcox, Jr; euthanasia to Abraham Stone, sterilisation to Morton S. Brown; sociology to Saul Jarcho; public health to Thomas Dublin, and medical history to Galdston. By October 1949 the *Index of the German Medical and Scientific Documents* was published, although the experts' reports remained confidential. In contrast to the Moran Commission, they judged the pressure and cold experiments as innovative.[62]

Interest in Nuremberg and German medical atrocities was on the wane. In April 1947 the US Atomic Energy Commission approved experiments providing subjects gave consent and they could be of therapeutic benefit, and by November the Commission was speaking of 'informed consent'. In 1953 the US Secretary for Defense issued Top Secret guidelines for human experiments based on the Nuremberg Code, but their effect was limited.[63] The 1950s were a decade of medical optimism. Jay Katz has observed how the Nuremberg Code lacked substantial exegesis, and offered only fragile protection against the exploitation of patients as research subjects. The upswing of clinical research, the claims of science and society as the prime beneficiaries of experiments went with renewed laxity on patient-subject autonomy in human experiments.[64]

Militant ethics

Physicians invested clinical autonomy with a sacrosanct aura, as they fought expansion of state welfare and social security. In an astute sleight of hand, the socialisation of health services could be equated to nazification: opponents of the National Health Service took up the slogan 'Bevan or Belsen', Ivy publicised the lesson that involvement of the state in medicine leads to its brutalisation, and the WMA stressed how medical ethics were corrupted by state medicine.[65] The BMA diagnosed the evil of Nazi medical atrocities as emanating from 'science as an instrument in the hands of the state'. This suited opponents of the nascent NHS, and internationally the Cold War climate meant ethics became a rallying cry against socialisation of medical services. The WMA drew up a resolution on 'The German Betrayal and a Re-statement of the Ethics of Medicine'. It called for 'the drafting of a World Charter of Medicine' to be applied in medical education and practice.[66] The WMA's statement marked a response to the Medical Trial judgement. The WMA reviewed the range of evidence on German medical atrocities on the basis of the Medical Trial, and of evidence submitted by Jewish doctors in Palestine and three Swiss Red Cross doctors. It prescribed a revival of the Hippocratic Oath, but in a revised form.

The West German Medical Chambers replied with a resolution drawn up by the medical official Haedenkamp, who was irate at the Mitscherlich delegation. The resolution, submitted to the WMA's second assembly in Geneva in September 1948, alleged that 'Only a small portion of German physicians committed such crimes', and that – apart from the patently Nazi Blome – none of the physicians tried at Nuremberg had any involvement with German professional organisations. Thus 'the medical profession had no knowledge of the crimes and nor any means of preventing them'. Its resolution of 18 October 1947 drew attention to the committee of observers at Nuremberg, and to *Das Diktat der Menschenverachtung*, which provided external windowdressing. The Chambers undertook that: 'every doctor will receive a copy of the conclusions reached by those German doctors designated as observers to the trial'.[67] The Chambers insisted that out of a profession of 90,000 there were only 350 criminal doctors, and this was taken as exonerating the profession as a whole.[68] The Germans ignored the obligation to expel the guilty, and sought to rehabilitate the profession internationally by siding with the medical opposition to social medicine. But on the home front they aimed to suppress any discussion of German medical crimes.

The veteran medical publicist Fishbein, who now edited the *WMA Bulletin*, played up the new wave of ethical concern.[69] The Declaration of Geneva, adopted in September 1948, included the pledge: 'I will not use my medical knowledge contrary to the laws of humanity' – an oblique reference to the Nuremberg prosecutions.[70] The WMA circulated the declaration internationally. In September 1949 the Germans submitted a statement how 'certain German doctors' perpetrated 'brutal experiments on human beings without their consent' and conceding the deaths of millions. The 52nd *Deutsche Ärztetag* required every German doctor to subscribe to the declaration of Geneva. In May 1951 the WMA invited the duplicitous German Medical Chambers to affiliate.[71]

As long as the WMA made explicit reference to the Nazi era, it stressed the obligations of the physician to the individual patient and avoided references to human experiments. Alexander continued to see the issues of sterilisation and euthanasia as manifestations of totalitarianism. In 1949 he saw the atrocities in terms relevant for current medicine: 'The killing center is the *reductio ad absurdum* of all health planning based only on rational principles and economy'.[72] He warned that euthanasia and the belief in utility posed severe dangers to American medicine. Lambert Anthonie Hulst, a specialist in internal medicine from Utrecht, was WMA president in 1954. Not only had he attended the Nuremberg trials and was on friendly terms with Taylor, he was Alexander's brother-in-law, and had endured a hard war involving hiding his Jewish wife, as well as slave labour and starvation.[73] He contributed to the WMA's Principles for Research and Experimentation. These were adopted at its eighth assembly in Rome in 1954.[74] It was not only unjustified wartime experiments, but also the development of surgical techniques and active drugs, which were causing concern. Controlled clinical trials were creating new ethical problems. Consequently, the tone differed markedly from the Nuremberg Code, by stressing the need to respect scientific standards.

The revised Principles distinguished between experimentation on healthy subjects, who had to be 'fully informed', and experimentation on sick subjects. The responsibility of the research worker was paramount. This marked a shift away from protecting the subject back to the researcher, and allowed risky experiments with the approval of the person or of his/her next of kin. Above all, it was 'the doctor's conscience which will make the decision'. The key principle was to inform the person who submits to experimentation about the reasons for the experiment, and the risks involved. By covering the case of when the person was 'irresponsible', it delineated permissive conditions for experiments on children and the mentally ill. No explicit reference was made to the war crimes trials, and the principles restored the overarching authority of the medical researcher.[75]

There were moves to extend safeguards on experimental subjects to all therapy, and to make informed consent the basis of doctor-patient relations. Ivy saw the necessity of this in 1948. He observed how 'a patient is a voluntary experimental subject of the physician'.[76] The trend was to limit the requirements of the 'approval of the subject'. The bond between physician and patient allowed the physician to increase his knowledge without first informing the patient, and the public should understand that experimentation was also in their long-term interest.[77] Overall, there was a shift from protecting the subject to the responsibility of investigator.

Pope Pius XII, despite lamentable failings over the Holocaust, became forthright in his condemnation of unethical research. In 1952 he decreed that the patient had no right to compromise his physical or mental integrity in human experiments. The Pope addressed the WMA on

30 September 1954, when he called upon nations 'to avoid the horrors of atomic, bacteriological or chemical warfare'. He condemned 'ABC warfare' as a crime, particularly when its aim was the total annihilation of all human life. This led to a view that doctors were not free to experiment on living human beings or to remove organs from dead bodies. The Pope conceded that in cases of 'desperate or hopeless illness', experiments were permissible with 'the explicit or tacit consent of the patients'. He warned doctors and (a novel addition) nurses against taking undue liberties with patients for experiments, as life was endowed by the Creator. Man was the custodian but not the independent possessor of the body.[78]

The Papal *Diktat* was countered by arguments that experimentation should increase, provided that this was in line with the Declaration of Geneva's principle of dedication to the patient's health. Others argued for the benefits of more experimentation; generally, the trend was to relax the stringent Nuremberg criteria.[79] Recognising that rapid advances in research had outpaced treatment, the WMA shifted back to the moral responsibility of the physician for research and treatment. Twelve questions formulated for the physician tested his/her conscience in research or therapy.[80] Informed consent had dropped out of view. In the post-war upswing of medical research, neither medical scientists nor sponsoring organisations wanted to be reminded of the Nazi era. The ethical debate waned, as the research momentum increased. The military requirements of the Cold War intensified an atmosphere of secrecy, disregard for the individual research subject and a general lack of accountability.[81]

In December 1948 Ivy dutifully reported on Nazi war crimes to the AMA, pointing to the dangers of state medicine.[82] His later support for cancer therapy based on a natural substance so that invasive surgery could be avoided brought his sense of an obligation to advance therapy into conflict with professional authority. He stubbornly refused to concede that the experiments were being carried out with what some regarded as a quack remedy. It left Ivy marginalised, discredited and ultimately condemned by the AMA.[83] He found that by 1949 interest in medical war crimes had evaporated. *Esquire* and *Reader's Digest* were no longer interested in an article on 'Doctors without Conscience'. The editor of *The Rotarian* explained that the war crimes had 'lost their timeliness'. If presented as a critique of compulsory medical care, the argument required a balanced presentation, and again this issue no longer appeared urgent.[84]

Ivy drew parallels between the lack of clinical freedom and the Nazi state.[85] In 1949 he reflected that the Nazi medical crimes arose from a society 'ruled by a Government without ethics'. He accepted that most German physicians remained ethical, whereas the abuses arose from the Nazi Party and SS. Nazism resulted in an ethically poisoned German medicine with the lethal combination of compulsory sickness insurance and racial ideology.[86]

Ivy's contributions to the codification of research ethics became confused with the saga of his defence of Krebiozen cancer therapy from 1947. The discoverers of Krebiozen, the Durovics, kept its formulation secret, so that Ivy was accused of conspiracy in advocating a remedy whose efficacy could not be objectively evaluated. Ivy tested the remedy on a dog, himself and then a patient to prove it was benign and non-toxic. He hoped it contained traces of anti-cancer substances found in all organisms so that the drug would provide a panacea for cancer. Ivy was discredited by the 15 years of controversy, and exposed as violating scientific method.[87] His faith in the drug shows how Nuremberg left a sense that what was important was beneficent concern for the patient. Ivy can be seen as a duplicitous figure, who knowingly gave false testimony in Nuremberg, and who then endorsed a bogus drug. Or he can be understood in more sympathetic terms – as driven by practical moral concerns and a sense of duty to the patient, which he placed above absolute scientific truth.

The Nuremberg lawyers, physicians and military officials went their separate ways, absorbed by post-war society. It meant that the records of those involved in the trial were dispersed, destroyed, inaccessible or forgotten. The Cold War prioritised liberty of research, and the issue of medical war crimes had little relevance. Physicians wanted autonomy, and military medical researchers engaged on such strategically relevant topics as nuclear fall-out and nerve gases did not want to be reminded of the lessons of Nuremberg. The Cold War divide militated against pooling information on medical war crimes. The claims of science were steadily gaining ground. Alexander denounced Verschuer in an address to the New York Academy of Medicine in 1949. But it was not long after that Verschuer was elected to the American Society of Human Genetics.[88] Alexander and Thompson abandoned neuro-physiological research, shifting to psychiatry (in Thompson's case via efforts to provide psychological therapy to a recalcitrant Germany as a UNESCO official). This can be explained by their view that psychiatry was primarily dedicated to alleviating patients' illness. They lost their taste for medical research as a result of the revelations of human experiments.

History and oblivion

Attempts to document Nazi medical crimes for posterity were frustrated. Time and time again the participants referred to their role at Nuremberg as 'history making', and that the Trial was 'a unique and momentous act in history'.[89] Apart from the harassed Mitscherlich and Bayle's physiognomical *magnum opus*, none of the leading participants ever published autobiographical reflections on this crucial period in their lives, and the intention of producing a full documentation on Nazi medical crimes and of the Trial foundered. No sooner had the Trial finished than the participants found themselves in a Cold War climate, which played down Nazi atrocities and prioritised progress in experimental medicine.

Alexander had well-developed interests in the history of medicine and general history, seeking out such luminaries as the medical historian Max Neuburger and G.M. Trevelyan.[90] When the Trial began, he felt that he was a witness to history in the making. Despite 'the Snafu', he was convinced that 'we'll come out with something monumental, both historical and legal'.[91] He was overwhelmed by 'the wealth and importance of the material, from all points of view – medical, historical, psychological'.[92] He believed that the evidence collected ought to be placed in a museum like the Smithsonian: 'the stream of material which flows through our hands, which has to be grasped or it is lost forever. The documentation of all this is an overwhelming job. I am perpetually under the state of mind of a hunter who suddenly finds himself in a swarm of the most amazing birds with only one gun in his hand and a limited number of bullets to bring those trophies home.'[93]

Alexander addressed the imminent Cold War climate of the rehabilitation of Germany, and the problem of totalitarianism in 1947, but could not sustain this line of investigation.[94] Taylor hoped that Alexander would produce a comprehensive report on the medical aspects of the trials, but the hope remained unfulfilled.[95] Alexander ultimately felt tainted by entering the Nazi mentality: 'I didn't want to become famous from studying those bastards.'[96]

Beals collected documents and pamphlets for University of Washington Law Library. The librarian requested: 'I hope you will sprinkle your autograph liberally around over the material you collect for us.'[97] Taylor believed that Trial documents would assist historians and political scientists in understanding medical and racial policy, and the structure and functioning of a dictatorship.[98] The documents exposed the wasteful inefficiency and unnecessary human experiments.[99]

In 1948 the court authorities debated what to do with the Medical Trial transcripts, and prioritised a list of archive libraries. German deposits allowed the German academic community to assess the correctness of the procedures. The KWI scientists remained interested in the Nuremberg Trials. A 'Kaiser Wilhelm Institute, Dahlem' requested German-language Medical Trial documents and transcripts in December 1948.[100] Such documentation was illuminating from the point of view of legal procedure and regarding what was known about Nazi atrocities. The KWI was initially given priority over Yad Vashem, but on reflection the authorities reprioritised the latter. This vacillation reflected a deeper conflict, between justice for the victims of genocide or the demands of science.[101] There appeared a two-volume edition on the Medical and Milch Trials in the USA. Ivy and the Surgeon General praised it for its relevance to medical research. There was strong public demand for the book, but circulation was restricted and, by May 1950, copies were in short supply.[102] Although in April 1950 a German edition of trial documents was planned, this never came to fruition. In contrast to the widely available IMT proceedings, there was only circulation of the Medical Trial proceedings in highly abbreviated form and no German version, and by the time the series was completed, the convicted were being released.[103]

An American translation of *Das Diktat* vividly conveyed the essence of the case.[104] *Doctors of Infamy* reflected the discourses surrounding the Trial, as Alexander, Ivy and Taylor introduced the book. Lemkin was due to contribute a foreword. Although the book appeared without this, genocide and the link to human experiments were repeatedly stressed.[105] Alexander, Ivy and the concluding reflections by Mitscherlich and Mielke evoked the 'infamy of a few crazed psychologically twisted practitioners', who perpetrated the human experiments, and overall the volume stressed the failure of the German medical profession to halt the experiments and for any German physician to mount a vigorous protest.[106]

Bayle's monumental tome on the psychology of the Medical Trial defendants appeared in 1951. Through good relations with the US prosecutor McHaney, he gained privileged access to the defendants, and unique documents. He related physique and character, and focused on the flawed psychology of individual deviants.[107] Crimes and ethics passed into medical literature, as the defendants were dismissed as pathological cases.

Alexander cited his Nuremberg observations as regards 'social situational stress and motivation'. The defendants' ethical and social framework became their superego. Alexander incorporated observations on Nazi sadism into his textbook of 1953 on *The Treatment of Mental Disorders*. He commented on how Nazism provided a refuge for neurotics:

> In the case of one scholarly rather passive man, a physician who suffered from a peptic ulcer and a domineering wife, the Nazi philosophy relieved him of all self-doubts and convinced him that he was indeed a superior racial being. He was able to throw off his passivity, his ulcer, and his wife, and take up new patterns of aggressive masculinity and outright sadism, which he easily justified Nazi-wise. After capture by our Army and during his trial in Nuremberg first his meekness, then his wife, and shortly thereafter his gastric symptoms returned.[108]

Alexander reduced Nazism to a 'group psychosis'. He associated Nazi Germany and Communist Russia with 'sadistic-aggressive patterns of control' and regression to anal-sadistic patterns. Along with the mainstream of his profession, he looked forward to a new era of invasive therapies, as he endorsed the value of frontal lobotomy and electroshock (these were in favour in the early 1950s), and reflected on the hitherto unrealised potential of therapeutic drugs for psychiatry.[109] Overall, the cases at Nuremberg were used to argue for autonomy of science and to mount a critique of socialised medicine. 'Science was pure – it was politics that was corrupting.'[110]

The Mitscherlich case: the observer accused

The German medical profession and universities had the opportunity to draw up an ethical declaration on the general issues surrounding the Medical Trial. They might have emulated a declaration of guilt by liberal Protestant theologians, and issued a statement on the record of medicine under National Socialism, confronting the collective failure to oppose the nazification of medical research. German medical faculties were at first expected to deliver either joint or individual declarations at the close of the Trial.[111] The medical faculties saw a choice between a public disclaimer – that the Trial did not relate to the profession in general – or strategic silence.

Mitscherlich provided mainly documentation – albeit pointed – leaving the reader to draw conclusions, and to respond to the evidence and verdicts. He convinced the Heidelberg anatomist Hermann Hoepke that the American and British authorities would welcome a public statement on the Trial's implications for German medicine. Hoepke persuaded Rein to support a declaration by the German medical faculties, and a draft response to the verdict was circulated on 12 August 1947. The plan was to issue it in the name of the representatives of the faculties (represented by Rein for the British zone, Hoepke for the American zone, Janssen of Freiburg for the French zone, and Heubner for the Russian zone). The declaration amounted to a disclaimer – it condemned the doctors sentenced in Nuremberg, as having nothing to do with either science or with the educational task of German medical faculties. German doctors were no different from doctors in any civilised country. The pathologist Büchner formulated a statement, which recognised the Trial verdict as a hard lesson for German medicine; he drew the telling conclusion that German medicine must never again subordinate itself to political influences.[112] This fitted in with the prevailing mood of shifting the burden of guilt onto the invasive state.

The medical faculties from the French zone at Freiburg and Tübingen (where the dean was the pioneer of biological types Kretschmer) opposed the joint declaration. They pointed out that the Nuremberg judges had not issued a public report, and they believed that whatever goodwill might be gained internationally would be offset by adverse publicity for the German doctor. Any declaration could be misunderstood as an admission of guilt.[113] The plan for a joint statement had to be jettisoned.[114] The spokesmen of German medicine Büchner, Rein and Heubner – scientists of high professional standing – hounded Mitscherlich, as they were convinced that their exoneration would vindicate their profession and experimental medicine. The claim that the normal physician was guiltless overlooked the political mobilisation of the medical profession.

The Medical Chambers issued on 18 October 1947 a general declaration mourning the victims of medical atrocities, and recognising that 'men from our ranks' were perpetrators. German physicians would only be able to judge the situation when each received the report of the German medical delegation (but the distribution was to be subverted). In the interim, the Chambers stressed that the criminals were just a small group, and that the destructive political dictatorship carried the guilt. The Trial showed how state institutions and bureaucrats corrupted medicine and undermined the sacrosanct relationship between physician and patient. The Declaration recommended that an expert committee evaluate experiments on human subjects and self-experiments. Doctors should not accept directives or guidelines from any authority, and should act only in conformity with science and professional ethics.[115]

This declaration was a critique of Mitscherlich's position, and found expression in a new form of the Hippocratic Oath. Each newly qualified physician was to pledge total allegiance to the profession but refrain from obeying any other authority.[116] Mitscherlich's published Trial digest carried a sanitising Foreword from the West German Chambers, dated March 1949 and bearing the stamp of the Haedenkamp era. He refrained from reprinting it in later editions, merely mentioning that his report was approved by the West German Medical Chambers.

Mitscherlich was caught between the need for Germans to confront their recent history and the conservative political forces of apologetics and denial. Colleagues condemned him for fouling their profession's nest. The German medical profession tried to rehabilitate itself internationally, while retaining a belligerently conservative ethos among its ranks. The Cold War climate was opportune for German doctors, who found international support for opposing the socialisation of medicine and welfare. Mitscherlich became a casualty of the profession's increasingly vociferous politics of denial. He defended the Trial from the charge that it was 'victors' justice'. His documents showed the fairness of the legal procedures, and the rights and privileges accorded to the defence. In terms of medical and public opinion what was on trial was the Trial itself.

Mitscherlich decided to work on a second, vastly expanded edition of the book, to include the defence materials.[117] The text took the best part of a year to prepare, and problems of paper supply and the prickly attitude of the occupation authorities had to be resolved. The British did not like mention of McCance's nutritional experiments on German babies in Wuppertal.[118] Publication of *Wissenschaft ohne Menschlichkeit – Medizinische und eugenische Irrwege unter Diktatur, Bürokratie und Krieg* in 1949 was an opportunity to settle scores with his critics. The highlighting of the term *Wissenschaft* expressed Mitscherlich's critique of the epistemological basis of medicine. The edition consisted of 10,000 copies, and while still published by Lambert Schneider, it was sponsored by the West German

Physicians' Chambers. The export editions carried a note that it was based on the *Diktat der Menschenverachtung*, and that the documents were selected to show the political and psychological repercussions of the Trial.[119]

Mitscherlich felt that the book disappeared; only a few hundred copies went to booksellers, it earned few reviews and did not find its way into libraries.[120] He was the victim of a ruse to allow copies to be distributed internationally, notably to the WMA to secure the rehabilitation of German medicine, but to suppress the book in West Germany. Some physicians recollected buying it – but they may have confused it with *Das Diktat*.[121] The local Ärztekammer resented spending money on the book, and by 1949 the conservative opposition to the Nuremberg Trials had gained support. Currency reform rendered the book expensive, and restricted circulation in the Russian zone.

Haedenkamp was a veteran of the anti-socialist Hartmannbund and the DNVP, and activist in purging Jews from the medical profession. Having been ousted by the Nazis in 1939, he replaced the liberal Oelemann in the Chambers of Physicians, when in 1947 he was appointed secretary of a publicity commission.[122] Haedenkamp carefully managed the presentation of the German profession at international medical meetings. He organised the distribution of the Mitscherlich-Mielke edition, and this offers a possible explanation as to why the edition came to disappear.[123] After all, *Das Diktat* when distributed under the aegis of Oelemann, achieved a reasonable circulation, whereas *Wissenschaft ohne Menschlichkeit* vanished.

The medical historian Thomas Gerst rejects a Machiavellian plot to suppress, pulp or somehow bury the book, but points to the disinterest of regional chambers, which were unwilling to subscribe to it.[124] Haedenkamp's presence indicates that conservative medical politicians willingly connived at this stubborn refusal to endorse the legality of the Trial by a tacit refusal to distribute the book. Mitscherlich continued to maintain a balance of conflicting views in his presentation. The edition contained a resumé of the controversies with Rein and Heubner, as Mitscherlich was determined to expose the fallacies of placing medicine on the basis of inductive science. The effect of the controversies was that German courts rejected the validity of the Medical Trial proceedings, and that the German medical profession vindictively censured Mitscherlich.

Mitscherlich was circumspect in citing the opposition of the Freiburg brain pathologist Büchner to the experiments. But Büchner pounced on the new edition. The significantly altered title *Medizin ohne Menschlichkeit* in 1960 reflected Mitscherlich's sense of a changed brief from a critic of science to being an outsider from established West German medicine.[125] Its publication raised issues of diagnosis and therapy, the repression of memory and the attitudes of the post-war generation of physicians.[126] Although the West German Physicians Chambers again endorsed the book, the general feeling was one of animosity against this concise and factual account of medical failings under Nazism.[127] The conservative Catholic Büchner took on the mantle of guardian of the reputation of German medicine (and thereby of his own reputation). Mitscherlich characterised him as one of the most politically active reactionaries among German professors.[128] Büchner condemned Mitscherlich for imputing guilt to leading figures in German medicine for their conduct during the war. Büchner's stance can be compared to the conservative Catholic Adenauer, who downplayed German responsibility for war crimes. Such conduct between the 1950s and 1980s was not uncommon. Among the super-stars of German science, Max Planck and Hahn protected incriminated colleagues, and Butenandt vehemently denied any involvement of himself or his assistants in working with specimens from Auschwitz.[129]

Former defendants rallied to the anti-Mitscherlich cause. Rose took the view that the Trial was quintessentially *'Unrechts-Justiz'*. When summoned to give evidence in an attempt by Rose to have the Nuremberg verdicts quashed, Mitscherlich found that the judge aggressively defended the legitimacy of justice under National Socialism and that experiments on prisoners were justified in the war. Mitscherlich expressed consternation that after 15 years the West German authorities no longer accepted documented evidence given at Nuremberg.[130] He felt that he had become the defendant, harassed by a conspiracy of former defendants, medical scientists and judicial authorities.[131] Rarely, did German medical faculties respond to the issue of medical criminals still being on the loose. In 1961 Halle annulled the degree of the Auschwitz doctor Horst Schumann for crimes against humanity as documented by Mitscherlich,[132] and the Universities of Frankfurt and Munich withdrew the doctorates of Mengele.[133]

Mitscherlich found the situation frustrating in that he was surrounded by legal apathy. He approached the medical division of the Rockefeller Foundation for support for emigration to the United States, and spent six months in the US in 1951. The Foundation's support for Mitscherlich demonstrated its shock at the atrocities of German medical research. His attempt to reconstruct the epistemological basis of medicine by exposing its inhumane roots had failed, and he felt isolated and persecuted.[134] Alice von Platen abandoned Germany for Britain to train as a psychoanalyst, specialising in group analysis.

A small number of concerned individuals soldiered on in the fight to eradicate the reinstatement of Nazi doctors. The former Hamburg paediatrician Rudolf Degkwitz – forced to emigrate to New York – pursued the issue of the perpetrators of children's euthanasia in Hamburg.[135] But the republication of Mitscherlich's book occurred within a hostile era of German medicine concerned to recover its reputation as a world leader in medical research, while crushing all efforts to reform the organisation and ethos of medical practice.

Mitscherlich never realised his ambition of attaining a chair in a German medical faculty, and through this to exercise a wider influence on medicine. The critical sociologists Horkheimer and Theodor Adorno brought him to Frankfurt, thereby linking critical sociology with critical psychoanalytical medicine. The state of Hessen and the DFG financed the Sigmund Freud Institute for Psychoanalysis and Psychosomatics in Frankfurt, where it opened in April 1960. State Minister President Georg August Zinn was supportive of Mitscherlich as well as of the state prosecutor Fritz Bauer and the Frankfurt Auschwitz Trial.[136] It was also in April 1960 that *Medizin ohne Menschlichkeit* underlined the critical distance from the German medical establishment. Only in 1966 did he attain a chair of psychology in the Frankfurt faculty of social science, but despite the acceptance of the discipline of psychosomatics, Mitscherlich remained an outcast from the German medical establishment.[137]

Compensation

Nuremberg left an unresolved legacy in terms of the 'permanent injury and mutilation', 'disabling injury' and enduring pain of the victims of the wartime experiments.[138] Survivors' organisations felt that the perpetrators were judged, but that nothing was done to help the victims. The Bonn government's support of the rehabilitation of war criminals, and its providing them with pensions was in sharp contrast to its denial of any responsibility for the victims of experiments. In 1950 John Fried, Legal Consultant to the [Nuremberg] Tribunals, visited UNESCO, when his attention was drawn to the case of Jadwiga Kaminska. She was one of the Ravensbrück Rabbits, who suffered gravely from the effects of the German injections and wounding. Alexander hoped that Kaminska could be brought to Boston for treatment. She approached the Commission on the Status of Women. On 4 July 1950 the UN Human Rights Division passed a resolution on the plight of victims of the so-called scientific experiments. Egon Schwelb reported on 'Crimes perpetrated by the German Medical Profession' in July 1946 for the UNWCC. Fried and Schwelb joined forces to work through the UN to gain compensation.[139]

The question whether claims were best dealt with by international organisations was debated at the Social Committee of the United Nations in March 1951. Czechoslovakia, Poland, the USSR considered that support and care for the victims should come through the governments of the states where they lived. Delegates from Belgium, France, the United Kingdom and United States lobbied for the United Nations to intervene. The division of opinion shows a clear East-West divide, not least because of the poor relations between the Bonn government and the Soviet bloc. The case for action through the UN won the day; the UN in turn suggested that the International Refugee Organisation (IRO) administer the compensation fund, and that the WHO advise on medical aspects of the problem.[140] Whereas the trials had been devolved to Allied military and state authorities, the compensation issue became internationalised: however contradictory, the consistent aim was to embarrass the Soviet Union and satellite states.

The UN liaised with the IRO, WHO, WMA and the International Tracing Service, and collected details of 237 survivors of human experiments, but numbers rapidly rose.[141] The ITS took responsibility for non-Jewish Austrian and German victims, and the National Catholic Welfare Conference acted for Polish and non-German and non-Austrian victims of non-Jewish religion, and dealt with some 100 victims in 13 countries. Fried energetically pursued the issue with the UN Human Rights Commission, and Taylor and Alexander remained in touch with the Rabbits. With Fried's backing, Alexander arranged for surgical treatment for some of the Rabbits in Boston in 1951.[142] Jadwiga Kaminska was brought from Brussels, Helen Piasecka from Cleveland, Ohio and Janina Iwanska from Paris. They suffered from a distressing range of painful physical and psychological complaints, and Alexander arranged surgical, gynaecological and neurological treatment, as well as psychiatric consultation. In one case, a neuroma – a painful whirl tumour in the scarred area – was removed. Alexander supported treatment of four more Rabbits in 1952 and again in 1959.[143] He took the view that the victims' 'moral and mental future are dependent on the security a pension would bring them'. He contacted survivors of experiments to make sure that their cases were included in the German survey of the victims of experiments. He was heartened that Father Michalowski, who survived the pressure experiments in Dachau and was living in Chicago, received compensation.[144]

The Federal Republic of Germany was lax in pursuit of perpetrators, and despite pressure from the United States for compensation, engaged in a protracted process of evaluating victims. Pensions and compensation were largely denied to medical victims, just as it was to Roma and homosexuals.[145] Racial hygienists such as Nachtsheim or experimenters as Ernst Schenk, himself a medical criminal, advised against compensation for sterilisation victims.[146] The victims received no specific compensation, but were eligible for a small amount of compensation for having suffered at least one year in a camp with ensuing disability affecting earnings. The German authorities showed a decided lack of sympathy for cases of psychological damage.[147]

The IRO found that most victims suffered permanent injury to health, and their immigration requests were barred on account of the injuries.[148] Under international pressure, the Federal Republic conceded that it was prepared to offer humanitarian assistance to justified cases. This was less than generous. The government denied the right to a pension and limited compensation payments to a meagre level. It offered in 1951 *ex gratia* compensation to victims of the experiments, provided they had been persecuted on grounds of race, religion, opinions or political convictions. The amount averaged 5,860 DM (ca. £500), although a maximum of 25,000 DM was set. Of 1,537 claims (by 1958), 423 grants were made, 403 were rejected and 707 claims were still outstanding.[149] The Belgians and British pointed out that one could be a victim of medical experiments without falling into the German government categories.[150]

Under regulations of 1953 and 1956 the Bonn government denied compensation on the grounds that the experiments were not harmful, or that the victim was not currently in need. At first sterilisation victims and all former Resistance combatants were automatically excluded, but then given the lowest rate of compensation. While 87 sterilisation victims received 2,000 DM, only one received compensation for sulphonamide experiments, albeit at a far higher rate. Of the 106 cases compensated, malaria accounted for nine, and typhus (one of the highest causes of experiments) for just two, injections two, hormone transplantation one, low temperature one, serum one, X-rays two, sulphonamide one and phlegmon for two compensated victims.[151] Compensation for time spent in slave labour was also denied. Eventually, only five of the Rabbits received compensation from the Bonn government under the 1951 regulations. Of the seven who had affidavits Iwanska, Helena Piasecka and Sokulska were among these, but not the four whose wounds were so effectively demonstrated at Nuremberg.[152]

Schwelb as deputy director of the UN Division of Human Rights Commission forwarded dossiers on victims to the Federal Republic.[153] In France the ADIR, founded as an organisation of former concentration camp prisoners in 1950, was headed by the redoubtable survivors of Ravensbrück, Anise Postel-Vinay and Germaine Tillion, who drew the attention of René Cassin to the need for compensation. The ADIR in 1952 began to seek details of all deportees in France and elsewhere who had served as guinea pigs. In Britain individual physicians maintained pressure for proper care and compensation.

Caroline Ferriday in the United States organised the Friends of ADIR.[154] She energetically lobbied all members of the UN Economic and Social Council.[155] The United Nations transmitted the dossiers to Bonn. Norman Cousins, editor of *The Saturday Review* had sponsored treatment in the United States for survivors of Hiroshima. He organised the Ravensbrück Lapins Committee. After visiting Warsaw, Ferriday arranged for medical supplies for the Rabbits, and Polish committee to oversee their welfare and for reconstructive surgery.[156] The Germans blocked compensation on the pretext of the lack of diplomatic relations with Poland. In 1959 a first group of 27 Rabbits came to America for treatment, and eight more in 1960.[157] They rejected German compensation of $1,000 in 1959 as inadequate both for immediate medical costs and as not providing a pension.[158] Ferriday observed that 'the human guinea-pigs remain profoundly injured – physically, morally and socially and for the most part unable to support themselves'.[159] Instead of recognising that the Rabbits needed the security of lifetime pensions and indemnities, the German government offered a small lump sum in compensation.[160]

By the early 1960s the German government wished to declare the postwar era over and terminate compensation procedures, which still did not adequately recognise medical crimes.[161] Doctors who were former Nazis adjudicated on compensation applications. Their diagnostic categories were relics of the Nazi era.[162] Psychiatrists pointed out that by labelling a claimant a hereditary schizophrenic, the Germans were denying responsibility for the traumatic aftereffects of the experiments. At this point Thompson teamed up with the New York psychiatrists Martin Wangh, Kurt Eissler and William Niederland, who had pioneered analysis of 'survivors' syndrome', to organise the Provisional Committee for the Medical Rehabilitation of' Victims of Human Disasters in 1964. The Committee protested to the German Chancellor Erhard that 43 per cent of compensation claims were rejected by the Federal German government, which disregarded clear evidence of damage to health because of 'outmoded' medical knowledge.[163] The Committee acted as symbolic bridge between first hand observers of the atrocities and concerned social scientists and historians. In September 1964 Jay Katz asked Taylor about preparatory drafts of the Final Code.[164] The Committee invited the Yale psychologist, Lifton to address the meeting on the psychological effects on the Hiroshima and Nagasaki victims – indicating a wish to engage critically with the psychology of the victor.[165] Lifton contacted Alexander, McHaney and Taylor, as his interest was aroused by the problem of the Nazi medical psychology.[166] A meeting on Late Consequences of Masssive Traumatization, addressed by Thompson and Lifton, rekindled recognition for the victims of human experiments.[167] The Nuremberg Code at last began to achieve legal recognition, although this has been a lamentably slow process. Doublethink has persisted, as reckless human experimentation has recurred.

From 1976 Günther Schwarberg set out to identify the hitherto nameless 20 children experimented on in Neuengamme and killed at the Bullenhuser Damm school. His efforts marked the start of a new generation of critical historical studies of Nazi medicine.[168] Social historians examined such issues as the exclusion of Jews from the nazified medical establishment. A new critical engagement with medical science and health policy took shape, and accused the establishment of continuity between the Third Reich and medical and health provision in West Germany. Mitscherlich's analysis gained a new relevance, and his status as a leading psychoanalytic critic of German social norms enhanced the reputation of his analysis of the Medical Trial.

During the 1980s protests were made against German and Austrian scientific and medical institutions holding human specimens of euthanasia victims or from concentration camps. This culminated in a conference of German university ministers and rectors in 1989. In December 1990 histological specimens and brains of 33 children and youths killed in 1940 at Brandenburg-Görden and held by the Max Planck Institute for Brain

Research in Frankfurt were buried. Representatives of German academic institutions were present, rather than relatives or other Nazi victims.[169]

An attenuated ethics

As the victims were being denied compensation, there was a major upswing of experimental medical research. David Rothman observes that in its reporting of the Medical Trial, the *New York Times* moved the story from the front page in 1947 to the back page by the time of the executions in 1948.[170] Researchers rarely saw the Trial as directly relevant to their own clinical and laboratory investigations.[171] Rothman characterises the 1950s as a *laissez-faire* era in the laboratory.[172] There was a rise of defence-related and experimental research, with few cautioned as to the dangers. Chastened by his experience at the Medical Trial, McCance became more circumspect, reflecting in 1950 on the rift between the physician concerned with patient well-being and the experimental investigator's concern to solve problems. Even though aware that all experiments involve risk, he felt that observations could be responsibly conducted in a clinical context without obtaining the patient's consent. He considered that elite researchers should have the trust of colleagues and patients, and that a consent form would destroy the whole atmosphere of beneficent trust.[173] The general temper of the times was one of experimental medicine and clinical trials. In 1948 an 18-year-old serviceman was 'volunteered' to go to Porton Down as a human subject: he found that 'They never sat us down and explained anything'. Another was told he would experience only 'a bit of discomfort'.[174] One experimental subject died from the tests.[175] In Germany the situation was no better. Catel conducted experiments on tuberculous children in 1947–8 in the Ferdinand-Sauerbruch Hospital in Wuppertal-Elberfeld to test the new TB therapy of Domagk. Two patient deaths resulted.[176] The lack of condemnation signified a return to reckless experiments.

Ivy feared the evils of bureaucratised and unethical Nazi science could recur. He fired off a missive to President Lyndon B. Johnson in 1965, indicating that US government agencies were guilty of crimes against humanity by suppressing the patent (and highly dubious) cancer drug Krebiozen, as he considered that clinical trials demonstrated its therapeutic benefits. He warned, 'Must we wait until the image of the United States resembles that of Nazi Germany before we act?'[177] The lesson Ivy drew from Nuremberg was that it was necessary to sustain clinical freedom for the medical researcher. Here, Ivy put public interest over that of the AMA.[178]

Not until 1959 did the Harvard professor of anaesthesia research, Henry Beecher, voice alarm at the epidemic of distressing, risky and at times fatal experiments, and at the spurious rationales for these.[179] Martin Gross pronounced the Nuremberg Code a failure.[180] In 1967 a British clinician, Maurice

Pappworth, published *Human Guinea Pigs. Experimentation on Man*, which exposed research malpractice. He pointed out that hospital doctors overlooked that the subjects of their research were sick people hoping to be cured; there was no sense of their rights. The *BMJ, Lancet* and *World Medical Journal* all failed to recognise the extent of dangerous experimental treatments.

The Royal College of Physicians denied Pappworth a fellowship until shortly before he died. He alluded to Ivy's post-war concerns about 'scientific curiosity submerging morals', and stressed that the German medical criminals 'included many professors of medicine and others in official positions of power in the medical hierarchy of the Third Reich.' By this time the Nazi experiments were a spectre raised by advocates of human experiments to indicate precisely what they were not doing. Pappworth responded that no moral line could be drawn between the malpractices between 1939 and 1945, and abuses occurring between 1945 and 1965. The growth of an abusive research culture meant that he doubted whether the Nuremberg Code could ever be accepted as a legal precedent.[181]

It was also in 1959 that the medical ethicist and lawyer Jay Katz encountered the Nuremberg Medical Trial judgment, which prompted his wide-ranging reconsideration of the problem of human experimentation in society. Katz's achievement was to generalise the problem of human experimentation and examine the American context of the 1960s. Although the Nazi atrocities were his starting point, he agreed with Beecher that their cruelty was extreme, and the circumstances exceptional. This separated Nazi medical crimes from the bulk of unethical developments in medicine and law.[182] The effect was to brand Nazi medical crimes as an issue of pseudo-science and fanaticism, rather than to draw lessons from how research agendas could become inhumane. Katz proposed a National Human Investigation Board in 1973 to regulate research involving human subjects.[183] It represented a symbolic response to the demand of the Auschwitz prisoner doctors for an investigative committee.

The smouldering injustice of civil rights in the USA fired a new concern with race inequalities, injustice and persecution in medical research. An American court in Kentucky first referred to the Nuremberg judgment in 1969 in deciding a claim concerning a kidney transplant, and in 1973 a court in Detroit made use of the Nuremberg Code in a case concerning psychosurgery.[184] Even so, the Tuskegee study on the non-treatment of syphilis between 1932 and 1972 (an experiment in reverse by the US Public Health Service) showed that illegitimate and unethical research continued, until it was pointed out in the late 1960s that the study could be compared with German medical experiments at Dachau.[185]

Science as an increasingly powerful political form claimed immunity from legal and public scrutiny. Overall, we see a recasting of medical research and ethics in the chill of the deepening Cold War. Medicine shifted away from social planning to conformity to the ideology of a free world. The effort to establish principles of freedom for scientific and medical researchers required a hard exercise in the judicial scrutiny of German wartime medicine, and this in turn meant that the Allied record on research ethics was questioned. The new stress on the autonomy of the research subject and patient arose. But set against these was intensification of the experimental basis of medicine in such areas as pharmacology and the application of biology to medical research. Scientists had increasing appetites for state funding, while adopting notions of the freedom of science to guarantee autonomy and immunity from public scrutiny. The experimental approach to medical science went with schemes to harness science for the defence of the free world. The upswing in human experiments ignored the safeguards enshrined in the Nuremberg Code. National security came to override the autonomy, freedoms and rights of the individual, and the quest for justice for victims of National Socialism.

The build-up of concern with patient rights and medical abuses in research and therapy prompted the disinterring of the buried legacy of documentation and debate surrounding the Medical Trial. Critical analysis of medical malpractice called for a revised view of medicine under National Socialism. We still do not know how many victims of medical atrocities there were. In 1999 a historical commission finally established that 270,000 individuals were killed by the Japanese Unit 731. By way of contrast, the victims of German human experiments have remained in the shadows of oblivion. Although Taylor highlighted 'the nameless dead' and that there were tens of thousands of victims of medical atrocities, there has been neither adequate compensation, commemoration nor historical reconstruction. Despite the self-organisation of the Ravensbrück Rabbits, the efforts of Schwarberg to identify hitherto nameless child victims, evocative historical snapshots by Ernst Klee of perpetrator-victim confrontations and sporadic newspaper revelations of undocumented experiments, the tendency has been to marginalise and minimise the German human experiments. That German physicians financed the German- and English-language edition of Medical Trial documents opens the way to resolving the issue of the actual dimensions of the medical atrocities, recompense for the dwindling numbers of survivors, and a worthy commemoration of victims.[186] The exemplary US President's Commission on the Radiation Experiments reviewed all instances of coercive experiments in public and military institutions.[187] We need to move away from placing the onus of complaint and proof on the already damaged victim, and investigate fully the causes and circumstances, and above all care for and compensate victims.

The final belated effort to compensate under the auspices of the *Stiftung Erinnerung und Zukunft* has encountered the problem of the inadequate research on the extent and types of human experiments. Victims have again protested against the negligible lump sums offered against a life coping with disabilities, childlessness and emotional after-effects.[188] The underlying agreement is flawed, because the problem of human experiments was misunderstood at the time of the slave labour agitation. Fixation on experiments screens out other categories of research-driven atrocities such as sterilisation, euthanasia and the harvesting of body fluids and body parts. As long as the life histories of victims remain sparsely documented, even the well meaning may inadvertently underestimate the extent and significance of Nazi atrocities in medical research. The Nuremberg Medical Trial and Code, and the ISC, marked the start of efforts to understand the conduct, motives, circumstances and extent of human experiments. The next stage should have been international guarantees of compensation and care for the victims, and binding agreements on the humane conduct of medical research. Far-sighted and humane efforts have been repeatedly undermined, and set against concerns that publicising the atrocities would destabilise – rather than strengthen – the future of clinical research. Informed consent remains a fragile asset. The effects of an unrestrained impulse to obtain medical knowledge resonate disturbingly.

Tables

1. The Accused

	Born	Age at Trial	NSDAP	NSÄB	SA	SS	Sentence (commuted)
Genzken	1885	62	1926	1933	30–31	1936	Life (20 yrs)
Handloser	1885	62	–	-	–	–	Life (20 yrs)
Weltz	1889	58	1937	1937	–	–	Acquitted
Schröder	1891	55	–	–	–	–	Life (15 yrs)
Rostock	1892	55	1938	1937	–	–	Acquitted
Blome	1894	53	1922/31	1931	1931	–	Acquitted
Pokorny	1895	52	–	–	–	–	Acquitted
Rose	1896	51	1922/30	1931	1922	–	Life (15 yrs)
Gebhardt	1897	49	1933	1933	–	1935	Death
Poppendick	1902	44	1932	1933?	–	1932	10 yrs
Hoven	1903	43	1937	–	–	1934	Death
Brack	1904	42	1929	–	23–27	1929	Death
Brandt, Karl	1904	42	1932	1932	1933	1934	Death
Mrugowsky	1905	41	1930	–	1930	1931	Death
Sievers	1905	41	1928/9	–	–	1935	Death
Beiglböck	1905	41	1933	1938	1934	–	15 (10 yrs)
Ruff	1907	40	1938	–	–	–	Acquitted
Brandt, Rudolf	1909	38	1932	–	–	1933	Death
Becker-Freyseng	1910	37	1933	1933	SA	–	20 (10 yrs)
Oberheuser	1911	36	1937	1937	–	–	20 (10 yrs)
Romberg	1911	36	1933	–	33–36	–	Acquitted
Schäfer	1912	36	–	–	–	–	Acquitted
Fischer	1912	35	1939	–	–	1934	Life (15 yrs)

Key: NSÄB = National Socialist Physicians League

2. Social Profile of Defendants at the Nuremberg Medical Trial

Name	Father's Occupation	Religion	Specialism	Occupation
Rose	Post Official	Ev to 1936 Gotterkenntnis/ Atheist 1946	Bacteriology	Tropical Medicine
Genzken	Pastor	Ev to 1936/ theosophy		Waffen-SS Sanitary Office
Hoven	Farmer	Ev. To 1925		Camp Doctor Buchenwald
Mrugowsky	Physician	Ev	Bacteriology	Waffen-SS Hygiene Institute
Blome	Manufacturer	Ev to 1934	Dermatology	Deputy Reich Medical Führer
Poppendick	Rail Official	Ev to 1935/Ggl		SS Race Office
K. Brandt	Police Official	Ev to 1936	Surgery	Reich Health Commissioner
Gebhardt	Physician	RC	Surgery	Dir. Hohenlychen
Romberg	Teacher	Ev		Inst f Aviation Medicine
Becker-Freyseng	Banker	RC		Inst f Aviation Medicine/ Luftwaffe official
Beiglböck	Physician	RC to 1938/ Ggl.	Internal Medicine	Oberarzt
Fischer	Commerce	Ev to 1941;	Surgery	Assistant Dr, Hohenlychen
Oberheuser	Engineer/ officer	Ev/ Ggl.	Dermatology	Assistant Dr, Hohenlychen
Weltz	Chemist	Ev. to 1911/ konf-los/ from 45 RC/Evang	Radiology	Aviation Medicine, Munich
Rostock	Farmer	Ev.	Surgery	Prof. Surgery Berlin
Ruff	Engineer	Ev.		Aviation Medicine, Berlin
SS Officials:				
Brack	Physician	RC to 36/ Ggl		Official, Kanzlei des Führers
R. Brandt	Werkmeister	Ev to 37/ Ggl		Secretary to Himmler
Sievers	Church Musician	Ev to 1928/33		Sec. SS-Ahnenerbe
NON-NSDAP:				
Handloser	Conductor	RC		Army Medical Officer
Pokorny	Military Officer	RC to 26/ Ev./RC 1945	Dermatology	Own practice

2. Social Profile of Defendants at the Nuremberg Medical Trial *continued*

Name	Father's Occupation	Religion	Specialism	Occupation
Schäfer	Architect	Ev.		Aviation Medicine, Berlin
Schröder	Teacher	Ev.		Luftwaffe Medical Officer

Key: Ev = evangelical; RC Roman Catholic; Ggl = Gottgläubig.

3. Academic Status

Name	University	Position/ Subject	Date of Appointment
Gebhardt	Berlin	Hon. Prof.	1935
		Ordinarius: Chirurgie	1937
Rostock	Berlin	Ordinarius: Chirurgie	1941
Brandt, K.	Berlin	Hon..Prof.	1940
Blome	Berlin	Hon. Prof	1941/2
		Krebsforschung	
Handloser	Berlin	Hon. Prof	
Mrugowsky	Berlin	Apl. Prof: Hygiene	1944
		Privatdozent	1939
Ruff	Berlin	Lehrbeauftragter:	1935
		Luftfahrtmedizin	
		habil	1937
Rose	Berlin	Lehrbeauftragter	
Schäfer	Berlin	Assistant: Inst. f	
		Luftfahrtmedizin	
Fischer	Berlin	Physician, Charite	Dec 44-Apr 45
Becker-Freyseng	Berlin	Habil.	Feb.1944
Weltz	Munich	Apl Prof: Röntgen-	1943
		Physiologie Dozent	1937
Schröder	Berlin	Hon. Prof.	1944
Gebhardt	Munich	Dozent	1932-33
Beiglböck	Vienna	Oberarzt: Innere	1939-45
		Medizin, 1. Med. Klinik	
		Apl. Prof	1944
Handloser	Vienna	Honorar-Professor.	
		Military Medicine	1938-41

4. SS/Waffen-SS Doctors Prosecuted at Nuremberg

Name	Date Joined Waffen-SS	Waffen-SS Position	Joined General SS	Final Rank
Brack	iii/iv.42	Sturmbannführer (Versorgungsoffizier)	xii.29	
K. Brandt	40	Generalleutnant der Waffen-SS	.vii.34	Obersturmbann- führer
R. Brandt	xii.40	Oberscharführer (Sergeant)	25.x.33	Standartenführer
Fischer	13xi39	Sturmbannführer (Major)	1xi33	
Gebhardt	40	Beratender Chirurg des Corpsarztes 1940-31viii43; Oberster Kliniker ix.43-45	1v33	Gruppenführer
Genzken	1iii36*	1940 Stabsführer d San-inspektion d. Waffen SS 1942 Generalleutnant der Waffen-SS	5xi33	Gruppenführer
Hoven	39	Hauptsturmführer (Captain)	34	Unterscharführer
Mrugowsky	39	Standartenführer Obersten Hygieniker, xi.40	31	
Poppendick	1.iii.42	Obersturmbannführer (Colonel); Chief physician RuSHA 41	1.vii.32	SS-Oberfuhrer (Lt. Colonel)

Not Waffen-SS: Sievers (SS Standartenführer)
*From summer 1940 incorporated into Waffen-SS

5. Arrest

Name	Date of capture	By whom	Where
Becker-Freyseng	1. 2.16.ix.46	1.GB 2.US	1. 2.Heidelberg Aero Medical Center
Beiglböck	22.iii.45	GB	Lienz/ Drau, Austria
Blome	16/17.v.45	US	Munich
Brack	20.v.45 8.vi.45/6	US	Nr Stuttgart as Hermann Ober Nr Traunstein
Brandt, Karl	23v45	GB	Schloss Glücksburg, Flensburg (NB Conti arrested 19v45 at Flensburg)
Brandt, Rudolf	20.v.45	GB	Bremervörde Fallingbostel
Fischer	4.viii.45	GB	Rosche, Kreis Uelzen
Gebhardt	20.v.45 16/17.v.45	GB	Bremervörde Marine-Lazarett Murwik, Flensburg
Genzken	15v45	GB	Schloss Gravenstein, Denmark
Handloser	8v45	GB	Eutin
Hoven	11/21.iv.45/5.vi.45	US	Nr Buchenwald
Mrugowsky	13v45	GB	Feldlabor. Waffen SS, Winnert b. Husum
Oberheuser	20. vii.45	GB	Stocksee, Schleswig-Holstein
Pokorny	5.ix.46	US	Munich
Poppendick		GB	CIC 1 Neuminster
Romberg	8.x.45	GB	Gilzum nr Braunschweig
Rose	8.v.45	US	Lazarett Sonnenberg, Kitzbühl/ Tirol
Rostock	5.v.45	US	Garmisch Eisstadion
Ruff	16.ix.46	US	Heidelberg Aero Medical Center
Schäfer	1. iv.45 2. 16.ix.46	US	Bad Pyrmont Heidelberg Aero Medical Center
Schröder	1. 8.v.45 2. 16.ix.46	US	Kitzbühl Heidelberg Aero Medical Center
Sievers	1.v.45	US	Waischenfeld/ Oberfranken
Weltz	9/11.ix.45	US	Munich-Freising

6. Interrogations for the Nuremberg Medical Trial

Name	USSBS/ Alsos/ FIAT/ IMT/ Police Interrogations Before NMT	Number For NMT	First For NMT 1946	Last NMT Interrogation 1947
Mrugowsky	3	13	19 July	16 May
Sievers	5	24	16 August	24 September
R. Brandt	2	18	20 August	4 September
Handloser	2	10	24 August	16 August
Brack	1	13	4 September	7 August
Pokorny	0	6	12 September	28 July
Genzken	0	12	17 September	5 August
Schröder	2	15	19 September	16 August
Becker-Freyseng	0	10	24 September	7 August
Ruff	0	12	28 September	6 August
Blome	4	11	2 October	17 April
Schäfer	0	6	4 October	17 July
K. Brandt	12	15	9 October	3 September
Weltz	0	8	9 October	11 January
Poppendick	0	7	11 October	6 August
Rostock	1	4	14 October	7 August
Hoven	3	10	16 October	17 October in Landsberg
Gebhardt	4	15	17 October	15 August
Romberg	0	6	29 October	7 August
Rose	2	7	30 October	2 October in Landsberg
Beiglböck	3	7	30 October	28 February
Oberheuser	1	6	31 October	7 December 1946
Fischer	10	6	31 October	7 December 1946
Total	54	241		

7.   Defence Lawyers

Defence Lawyer	IMT	NMT	Counsel at later War Crimes Trials	NSDAP
Bergold, Friedrich	Bormann	—	Milch	–
Boehm, Georg	SA	Poppendick		–
Flemming, Fritz [Assisted By Zwehl]	Mrugowsky			–
Fritz, Hans	Fritzsche	Rose		NSDAP
Fröschmann, Georg	Assistant for Ribbentrop	Brack	Berger/Foreign Office Mummenthey/ Pohl	NSDAP 1937–45 SA 34-45
Gawlik, Hans	SD; associate to Babel for SS	Hoven	Boberhin/ Pohl Volk/Pohl Naumann/ Einsatzgruppen Milch Biberstein/ Einsatzgruppen Klein/Pohl	NSDAP 1933–45
Hoffmann, Karl	Assistant to Babel for SS	Pokorny	Nosske/ Einsatzgruppen	NSDAP 1933-41
Kauffmann, Kurt	Kaltenbrunner	R. Brandt		NSDAP 1937–45
Marx, Hanns	Streicher	Becker-Freyseng; Schröder	Engert/ Justice	NSDAP 1933–35
Merkel, Rudolf	Gestapo	Genzken		–
Nelte, Otto	Keitel	Handloser	Hoerlein/ IG Farben	NSDAP 1933–45
Pelckmann, Horst	SS	Schäfer	Terberger/Flick	–
Pribilla, Hans	Assistant to Servatius for Leadership Corps	Rostock	Jaehne/ I.G. Farben Lautenschlaeger Schubert/Justice Tschentscher/ Pohl	NSDAP 1935–45
Sauter, Fritz	Funk; Ribbentrop; Schirach	Blome; Ruff	Lautenschlaeger Hoerlein/ IG Farben	NSDAP 1933–45 NSDAP 1937–45
Seidl, Alfred	Frank; Hess	Fischer; Gebhardt; Oberheuser	Lammers/ Foreign Office Pohl/Pohl Lammers/ Foreign Office Walther Duerrfeld/ Auschwitz Trial	

7.   Table Defence Lawyers   *continued*

Defence Lawyer	IMT	NMT	Counsel at later War Crimes Trials	NSDAP
				(later Minister in Bavaria, CSU Fraktionsvor-sitzender)
Servatius, Robert	Sauckel; NSDAP	K. Brandt	Wesse Eichmann Eirenschmalz/Pohl	–
Steinbauer, Gustav Tipp	Seyss-Inquart	Beiglböck Becker-Freyseng; Schröder	–	–
Vorwerk, Bernd		Romberg		NSDAP
Weisgerber, Josef Assistant to Merkel		Sievers	Speidel	NSDAP 1940–45; SA 31–4
Wille, Siegfried		Weltz		NSDAP candidate 1933

Source: RG 153 /86-3-1/book 3/ box 10 list of counsel; RG 238 NM 70 entry 213, box 1 defence counsel, military tribunals, Nurnberg 31 Jan. 1949

8.   Human experiments/related atrocities cited at NMT (by year)

Year	Numbers of Coerced Human Experiments/Related Atrocities
1939	3
1940	1
1941	8
1942	18
1943	25
1944	27
1945	1

9. Ivy Second World War Research

Topic	Date	Agency	Subjects	Co-Worker/s
Procedures for producing canned water	1939	Committee on Clinical Investigation/ NRC		
Explosive Decompression	1939–		Animals; Humans at Wright Field	J.J. Smith
Free Fall from 32500 feet; effects of fall from different attitudes from 40000 feet with oxygen	1940–41	Decompression Subcttee		
Exploratory Project on 'Drinkability' of Diluted Seawater. Project X-100	Nov 1942	Naval Medical Research Institute	10 including ACI[vy]	Self- experiment with lab co-workers
Purification of Sea Water by Chemical Process. Project X-100	April 1943	Naval Medical Research Institute		
A Comparison of the Research Now Available for Securing Drinking Water in a Rubber Raft	April 1943	Naval Medical Research Institute	Ivy	
The Armbrust cup [on water for life rafts]	May 1943	Naval Medical Research Institute		
Potability of sea water after desalination. Project X-100	June 1943	Naval Medical Research Institute		Consolazio. Futcher Pace
The 'Tablet Life Ration' for the Shipwrecked. Project X-100	June 1943	Naval Medical Research Institute		
Demineralization of Sea Water by Permutit Project X-100	June 1943	Naval Medical Research Institute		Consolazio
Improved Method for the Chemical Demineralization of Sea Water. Project X-100		Naval Medical Research Institute		

9. Ivy Second World War Research *continued*

Topic	Date	Agency	Subjects	Co-Worker/s
Pressure Breathing. Research Project X-116	Feb. 1943	Naval Medical Research Institute	5 including AC Ivy	Adler, Burkhardt, Grodins Atkinson
Re drug B2B Effect of B2B and Preoxygenation Plus B2B on the Incidence of 'Bends' and 'Incapacitating Bends and Chokes' at 40,000 ft for one hour	1942		205 males (2 control gps of 50, av age 23 yrs	Atkinson, Adler, Burkhardt
Effect of dextro-amphetemine (Dexedrine) and of Preoxygenation plus dextro-amphetemine on the incidence of 'bends' and of 'incapacitating bends'.	1942			Ferris
Benzedrine and fatigue	1942–3			
Carbondioxide and decompression sickness	1942			
Pressure breathing on incidence of bends at 38,000 ft for 2 hrs w exercise Incidence of bends at 47,500ft w pressure breathing	Rept 9/3/44	OSRD	7 subjects 50 subjects	
Effect of pressure Breathing on bends at 38,000 ft and 47,500 ft	Cancelled 7/8/44	OSRD Div Aviation Med		
Relation of diet and nutritional factors on altitude tolerance such as aeroembolism and decompression sickness	Nov 30, 1942–Aug 1944	OSRD		A.J. Atkinson H.F. Adler W.L. Burckhardt L. Thometz

9. Ivy Second World War Research   *continued*

Topic	Date	Agency	Subjects	Co-Worker/s
Drugs and Fatigue: Effect of Benzedrine, Pervitin and Caffeine on staying awake and on performance after march or flights	Feb-Oct, 1942	OSRD		R.H. Seashore
Effect of Pecuniary Incentives on the influence of 'intolerable' cases of bends and chokes	Rept 23/3/45	Cttee on Med Res OSRD		Burkhardt, Thometnz
Adaptation to pressure. Seven subjects	18 months; NMT 2/9280	Decompression Sub-Committee	Conscientious objectors	
Vitamin B complex deficiency			Conscientious objectors	
Testing of Stratolator	15iv-15vi45	OSRD Cttee on Med Research		

Sources: RACRU Bronk 303-U Ivy Contract Box 26 file 26
Ivy papers
Fulton Papers: Fulton Ivy Correspondence, Box 91
NMT 2/9199, 9440
NMT 2/9298, 9328
NARA RG 52 Records of the Bureau of Medicine and Surgery Research Division Box 5 Entry no WN352934
National Academies of Science

10.   International Scientific Commission: Meetings 1946-47

Date of Meeting	British	French	US
31 July – 1 Aug 46	K.Mant – War Crimes Investigation Unit S.Smith – Edinburgh A.G.Somerhough – War Crimes Section Thompson – FIAT (British)	R. Legroux– Pasteur Institute P. Tchernia – Navy Dept P. Lépine – Pasteur Institute H. Piedelièvre – Alfort College	A.C.Ivy – physiologist J.M.McHaney- lawyer, Nuremberg J.C.Duvall – US Army War Crimes Group, Wiesbaden
16 Oct 46	Mant Somerhough	F. Bayle – Navy Dept surgeon H. Simmonet – Alfort College Legroux Tchernia Lépine Piedelièvre A.Touffait – War Crimes Research	A.G. Hardy – lawyer, Nuremberg,
15 Jan 47	W.G.Barnard – Royal College of Physicians Mant Moran – Royal College of Physicians Smith, Somerhough Thompson	Bayle Legroux Tchernia Lépine Piedelièvre Touffait	T. Taylor–US Chief of Counsel, Nuremberg L. Alexander – neurologist, Expert for the Prosecution, Nuremberg H.S. Leger–Paris office, US Chief of Counsel

11.   National Expert Committees: Names/Topics

British	French	US (proposed)	New York Academy of Medicine, Dec 47-49
Lord Moran	Francois Bayle	Andrew Ivy	Franklin M. Hanger/
Sir Henry Dale	(naval surgeon),	Leo Alexander	Hepatitis
(President of the	H. Simonnet	George Minot	Frank McGowan/
Royal Society)	(Professor of	(Harvard Medical	Freezing
W.G. Barnard	Physiology,	School)	Experiments
(Royal College of	Alfort College),	Bazett (University	Morton Kahn/
Physicians)	René Legroux	of Pennsylvania)	Typhus, Yellow
Sydney Smith	President of	Frank Lahey	Fever, Malaria
(University of	the Scientific	(Lahey Clinic	Nolan D.C. Lewis/
Edinburgh) Sweeney	Council	Boston) Abraham	Psychiatry
(St Thomas'	Pasteur Institute),	Myerson	Wilcox ,Jr/
Hospital)	Paul Tchernia	(State Research	Psychiatry
....	(Laision Recherche	Committee Boston)	Abraham Stone/
Advisers:	Ministère de	Surgeon General	Euthanasia
C.P. Blacker/	la Marine),	Kirk	Morton S. Brown/
Eugenics	P. Lépine		Sterilization
N. Hamilton	(Pasteur Institute),		Saul Jarcho/
Fairley/ Malaria	H. Piedelièvre		Sociology
Barnard/ Phosgene	(Professor of		Thomas Dublin/
V.H. Ellis/	Forensic		Public Health
Orthopedic Surgery	Medicine Paris),		Iago Galdston/
Aubrey Lewis/	A. Touffait		Medical History
Psychiatry	(Directeur du		
Lovatt Evans/	Service de		
Physiology	Recherche des		
Ronald Hare/	Criminels de		
Bacteriology	Guerre).		

12. The Evolution of the Nuremberg Code

Nuremberg Code 19 August 1947	Ivy I 31.7.46	AMA 12.46	Alex. I 7.12.46	Alex. II 25.1.47	Alex. III 3.4.47
1. Voluntary Consent	X	X	X	X	X
– Free power of choice			X	X	X
– Sufficient knowledge and comprehension			X	X	X
– Understanding and enlightened decision				X	
– Know nature, duration, purpose of experiment			X	X	X
– Method & means of conduct					
– Inconveniences & hazards					
– Effects on health and person	X				
Absent from Code:					
– Safeguards for mentally ill patients	-				X
				X	X
– duty for quality of consent rests upon individual who directs the experiment			X	X	X
			X	X	X
2. Fruitful results for good of society	X		X	X	X
3. Animal experimentation	X	X	X	X	X
4. Avoid unnecessary physical/ mental suffering & injury	X		X	X	X
5. No death or disabling injury except self-experiments	X		X	X	X
6. Degree of risk should not exceed humanitarian importance of problem	X		X	X	X
7. Protect experimental subject			X	X	X
8. Proper preparations to be conducted by scientifically qualified persons		X	X	X	X
9. Subject at liberty to end experiment					

12.   The Evolution of the Nuremberg Code   *continued*

Nuremberg Code 19 August 1947	Ivy I 31.7.46	AMA 12.46	Alex. I 7.12.46	Alex. II 25.1. 47	Alex. III 3.4.47
10. Scientist obliged to end experiment if injuries, disabilities, death likely					

Ivy I = Pasteur Institute July 1946
Alexander I = Ethical and Non-Ethical Experimentation on Human Beings, Nuremberg 7 Dec 1946
Alexander II = Affidavit, 25 January 1947
Alexander III = 15 April 1947
Ivy I:
1. The consent of the subject is required. Volunteers should be told of hazards and insurance should be provided.
2. Experiments should be based on animal experiments, and the results should be beneficial for 'the good of society'.
3. Experiments should avoid unnecessary physical and mental suffering and injury, should be carried out by qualified persons and should not be conducted if the outcome will be death or disability.

Archives, Interviews, Bibliography

Archives

AUSTRIA
VIENNA
Archiv der Republik (AdR) – Akten des Gauamts Wien, Bundesministerium f. Unterrricht (BfU), Kurator der wissenschaftlichen Hochschulen in Wien. Bundesministerium f. Justiz (BMfJ).
Dokumentationsarchivs des Österreichischen Widerstandes (DÖW)
University Archives (UA) Vienna: Medizinische Sitzungsprotokolle, Beiglböck, Eppinger files

CANADA
OTTAWA
Deschatelets Archives: Caron papers
National Archives Canada (NAC): RG 24 National Defence, Series C-2, FIAT microfilms; RG 25 Ministry of External Affairs
University of Ottawa Archives, File Thompson JWR
TORONTO
University of Toronto, Fisher Rare Book Library, Banting Papers
University of Toronto Archives

CZECH REPUBLIC
PRAGUE
Archives of Charles University, Prague, German University, Doktoren Matrikel
Central State Archives Prague (CSA): Series 288 Office of the Czechoslovak Delegation at the War Crimes Tribunal

FRANCE
COLMAR
Centre des Archives de l'Occupation française en allemagne et en autriche de Colmar (AOF): Services des crimes de guerre (AJ); culture (AC), Province de Palatinat
PARIS
Archives de France (AdeF): Ministère de la Justice. Direction du Service de Recherches de crimes de guerre ennemis; Procès de Nuremberg.
Archives du Centre de Documentation Juive Contemporaine(ACDJC): Interrogation records.
Institut Pasteur: Fonds Pierre Lépine.
UNESCO archives: papers relating to Julian Huxley, and the UNESCO German Programme.

GERMANY
BERLIN
Archiv der Humboldt Universität Berlin (AHUB): Personal files; Medical Faculty minutes
Bundesarchiv Berlin (BAB): Records of the former Berlin Document Center; Ahnenerbe; Reichsforschungsrat.
Bundesarchiv Berlin, Dahlwitz-Hoppegarten (BAB D-H): Personal files of the accused.

Bundesbeauftragte für die Unterlagen des Staatssicherheitsdienstes der ehemaligen Deutschen Demokratischen Republik (BfBStU).

Robert-Havemann-Archiv: Havemann papers.

Archiv und Bibliothek zur Gerschichte der Max-Planck-Gesellschaft (MPG): Personalakten Hallervorden, Spatz, Tönnis. Nachlass Adolf Butenandt, Richard Kuhn, Otto Hahn. General Administrative Papers

DÜSSELDORF

Cécile und Oskar Vogt-Archiv

FRANKFURT AM MAIN

Archivzentrum der Stadt- und Universitätsbibliothek Frankfurt-am-Main: Alexander-Mitscherlich-Archiv Mitscherlich Papers (MP)

Institut für Geschichte der Medizin: Archiv der medizinischen Fakultät Frankfurt.

FREIBURG

Bundesarchiv Militärarchiv Freiburg M20

Universitätsarchiv (UA) Freiburg: Hoven file, Medizinische Fakultät

GÖTTINGEN

Universitätsarchiv (UAG) Kur PA Lenz Fritz

HAMBURG

Staatsarchiv Hamburg Bernhart-Nocht-Institut (STAH): Rose-Nauck correspondence

HALLE

Leopoldina Archiv, Akte Strughold

Universitätsarchiv (UA) Halle: Medizinische Fakultät

HEIDELBERG

Universitätsarchiv (UA) Heidelberg: Medizinische Fakultät: Mitscherlich, Strughold files.

Lambert Schneider Archives (LS Arch)

JENA

Universitätsarchiv Jena (UAJ): Medizinische Fakultät; Skramlik, Emilian Ritter von.

KOBLENZ

Bundesarchiv (BAK) Zsg 154 Sammlung Christian Pross

MÜNSTER

Universitätsarchiv (UA) Münster: Bestand 5, Helmut Poppendiek; Nachlass Verschuer.

MUNICH

Bayerisches Hauptstaatsarchiv (BHSTA) Akten des Bayer. Staatsministeriums f. Unterricht und Kultus MK 54855 Dr. Werner Leibbrand.

Universitätsarchiv der LMU München (UAM): Leibbrand, Weltz files.

NUREMBERG

Bayerisches Staatsarchiv Nürnberg (STAN): Rep 502A KV- Verteidigung Handakten.

ROSTOCK

Universitätsarchiv

TÜBINGEN

Privatarchiv Jürgen Peiffer

ISRAEL

Central Zionist Archives Jerusalem, Zollschan papers

SWITZERLAND

GENEVA

International Committee of the Red Cross (ICRC), see Wiener Library.

World Health Organisation

UNITED KINGDOM

ABERYSTWYTH

National Library of Wales: Lord Elwyn-Jones Papers

EDINBURGH
Royal College of Physicians of Edinburgh (RCPE): Sydney Smith papers
LONDON
British Medical Association Archives: Reports on Nazi Medical Crimes.
Imperial War Museum (IWM): [Irving] Transcripts of the diaries of Erhard Milch 1933–45
Royal Society: Dale Papers (RSDP).
The National Archives: Public Record Office (TNA: PRO): Air Ministry; Cabinet Office, Foreign Office, Medical Research Council (FD1), Prime Minster's Office, War Office.
Wellcome Library for History and Understanding of Medicine (WL): Hare Papers, Lovatt Evans Papers, Moran Papers, Royal Society for Tropical Medicine.
Wiener Library: International Committee of the Red Cross microfilm records.
Mant Family Papers, Keith Mant Autobiography and notebooks.

UNITED STATES OF AMERICA
CHICAGO
American Medical Association: Ivy Report on Nazi Medical Crimes
University of Chicago, Dept of Special Collections, A.J. Carlson Papers (Misc. Arch. Col.). The Morris Fishbein Papers
DURHAM, NC
Duke University Archives (Depository for Medical Center Records), HC 36 Collected Papers and Documents. 65th General Hospital, Alexander Papers (APD)
GULFPORT, FL
Stetson University College of Law Library: Papers of Dean Harold L. Sebring
HOUSTON
Rice University: Julian Huxley Papers
LARAMIE
American Heritage Center, University of Wyoming: A.C. Ivy Papers, 1799–1984, Accession Number 8768.
NEW HAVEN
Yale University Fulton papers; Rosen papers; Fortunoff Video Archive for Holocaust Testimonies.
NEW YORK
Center for Jewish History: Collection of Raphael Lemkin (American Jewish Historical Society); John H.E. Fried papers (Leo Baeck Institute Archive)
Arthur W. Diamond Law Library, Columbia University Law School
Telford Taylor Papers (TTP-CLS)
New York Public Library (NYPL): Displaced German Scholars Papers;
Raphael Lemkin Papers; Robert Lifton Papers
New York Academy of Medicine (NYAM): administrative archives.
United Nations Archives (UNA), United Nations War Crimes Commission (UNWCC)
SEATTLE
Washington University (WUS) Beals Papers
SLEEPY HOLLOW, NY
Rockefeller Archive Center (RAC)
Rockefeller University (RU): Bronk and Cohn Papers
Rockefeller Foundation (RF) on Duke University Psychiatry and German Science
STANFORD, CA
Hoover Institution: America First Committee; Boris Pash Papers
Stanford University: John West Thompson Jnr study transcripts
WASHINGTON, DC
Library of Congress: Jackson Papers

National Academies of Science Archives: Committee on Aviation Medicine, Subcommittee on Decompression Sickness files
National Archives and Records Administration (NARA):
RG 52 Records of the Bureau of Medicine and Surgery Research Division
RG 112 Office of the Surgeon General Army. Geographical Files (Germany)
RG 153 Records of the Judge Advocate General (Army) JAG War Crimes Board – Entry 144, Persons and Places files
- Nuremberg Administrative files
RG 165 Entry 179 box 739 Case of Rascher
RG 200 ALSOS
RG 238 World War II War Crimes Records NM 70 entry 145; entry 148 Daily Bulletins OCCWC; entry 159 Berlin Branch; entry 200 '201 series'; entry 202 Berlin branch
RG 243 Entry 2 section 5 US Strategic Bombing Survey
RG 260 OMGUS, Field Information Agency
RG 319 IRR Personal; ALSOS
RG 330 Entry 1B JIOA Foreign Scientists Case File 1945–58
RG 338 707-Medical Experiments ('Heidelberg Documents')
 290-59-30/33 Parolee Case Files, Landsberg Prison
 290-59-33 Executee Files, Landsberg Prison
RG 341 190 64 24 06 Records of the Headquarters US Air Force
RG 466 Entry 54
BDC collections
M 1019 Nuernberg War Crimes Trials Interrogations, 1946–1949
M 1270 Interrogation Records Prepared for War Crimes Proceedings at Nuernberg 1945–1947
T1021/12

National Library of Medicine (NLM) MS c 135 Hugo Spatz; MS C 408 International Military Tribunal.
Niels Bohr Library, Goudsmit Papers
United States Holocaust Memorial Museum (USHMM), 1998.A.0044 Aviation Reports; Ferriday Papers

Interviews, correspondence, videotapes

Charlotte Bloch-Kennedy, March 1997
Ben Ferencz, 2000–1
Walter Freud, Oxted, UK, 1999
William C. Gibson, 1999.
Cecily Grable, Newton, Mass., May 1998
Hugh and Fred Iltis, 2004
Saul Jarcho, New York, New York, 1996
Keith Mant, Walton-on-Thames, 1996
Joseph Meier, New Jersey, 1996
Alice Ricciardi-von Platen, Cortona, 1997–2004
Wolfgang Scholz, Ladenburg, 2001
Telford Taylor, New York, 6 Nov 1996
'The 1997 Nuremberg Prosecutors Conference', Taped at the University of South Carolina School of Law, Sept 1997

Select bibliography

The Nuremberg Medical Trial 1946/47. Transcripts, Material of the Prosecution and Defense. Related Documents. English Edition, On Behalf of the Stiftung für Sozialgeschichte des 20. Jahrhunderts Edited By Klaus Dörner, Angelika Ebbinghaus and Karsten Linne, in cooperation with Karlheinz Roth and Paul Weindling. Microfiche Edition (Munich: Saur 1999). *Guide to the Microfiche Edition* (Munich: Saur, 2001).

Trials of War Criminals Before the Nuernberg Military Tribunals under Control Council Law No. 10 (Washington: US Government Printing Office, 1950), *The Medical Case*, vols 1–2.

'1946–1996 Le procès des médecins à Nuremberg', *Revue d'histoire de la Shoah*, no. 160 (1997).

Abbati, Caterina, *Ich, Carmen Mory. Das Leben einer Berner Arzttochter und Gestapo-Agentin (1906–1947)* (Zurich: Chronos Verlag, 1999).

Alexander, Leo, 'The Treatment of Shock from Prolonged Exposure to Cold, Especially in Water', Combined Intelligence Objectives Subcommittee, Target Number 24 Medical CIOS G-2 Division, Shaef rear, July 1945. Abstract in: UNWCC Summary of Information No. 41 'Medical Experiments on Human Beings ("Versuchspersonen")', September 1945. Reprinted in: *Trial of the Major War Criminals Before the International Military Tribunal in Nuernberg*, Doc. 400 PS, 25: 536–607, 1947.

Alexander, Leo, 'Medical Science under Dictatorship', *New England Journal of Medicine*, vol. 241 (1949) 39–47.

Alexander, Leo, 'War Crimes and their Motivation. The Socio-Psychological Structure of the SS and the Criminalization of a Society', *Journal of Criminal Law and Criminology of Northwestern University School of Law*, vol. 39, no. 3 (Sept–Oct 1948) 298–326.

Alexander, Leo, 'Destructive and Self-destructive Trends in Criminalised Society. A Study of Totalitarianism', *Journal of Criminal Law and Criminology of Northwestern University School of Law*, vol. 39, no. 5 (Jan–Feb 1949) 553–64.

Alexander, Leo, 'The Molding of Personality under Dictatorship. The Importance of the Destructive Drives in the Socio-Psychological Structure of Nazism', *Journal of Criminal Law and Criminology of Northwestern University School of Law*, vol. 40, no. 1 (May–June 1949) 3–27.

Alexander, Leo, 'Limitations in Experimental Research on Human Beings', *Lex et Scientia*, vol. 3 (1966) 8–24.

Annas, George J., Grodin, Michael A. *The Nazi Doctors and the Nuremberg Code* (Oxford: Oxford University Press, 1992).

Bayle, François, *Croix gamée contre caducée. Les expériences humaines en Allemagne pendant la deuxième guerre mondiale* (Neustadt: Imprimerie Nationale, 1950).

Bower, Tom, *The Pledge Betrayed. America and Britain and the Denazification of Post-War Germany* (New York: Doubleday, 1982).

British Medical Journal no. 7070 (7 December 1996) 'Nuremberg Doctors' Trial: 50 Years On'.

Ebbinghaus, Angelika, Dörner, Klaus (eds), *Vernichten und Heilen. Der Nürnberger Ärzteprozess und seine Folgen* (Berlin: Aufbau-Verlag, 2001).

Final Report of the Advisory Committee on Human Radiation Experiments (New York and Oxford: Oxford University Press, 1996).

Foreign Office, *Scientific Results of German Medical War Crimes. Report of an Enquiry by a Committee under the Chairmanship of Lord Moran MC, MD* (London: HMSO, 1949).

German Aviation Medicine. World War II. Prepared under the Auspices of the Surgeon General (US Air Force, Department of the Air Force, Washington DC: Department of the Air Force, 1950), 2 vols.

Gerst, Thomas, 'Der Auftrag der Ärztekammern an Alexander Mitscherlich zur Beobachtung und Dokumentation des Prozessverlaufs', *Deutsches Ärzteblatt*, vol. 91 (1994), B-1200–B-1210.

Gilbert, Gustave M., *Nuremberg Diary* (New York: Farrar, Straus, 1947).

Gilbert, Gustave M., *The Psychology of Dictatorship. Based on an Examination of the Leaders of Nazi Germany* (New York: The Ronald Press Co., 1950).

Goudsmit, Samuel A., *ALSOS. The Failure in German Science* (London: Sigma Books, 1947), reprinted with introduction David Cassidy as *ALSOS* (Woodbury, NY: AIP Press, 1996).

Harkness, Jon M., 'Nuremberg and the Issue of Wartime Experiments on US Prisoners. The Green Committee', *JAMA*, vol. 276, no. 20 (1996) 1672–5.

Holzlöhner, E., 'Verhütung und Behandlung der Auskühlung im Wasser', *Bericht über eine wissenschaftliche Besprechung am 26. und 27. Oktober 1942 in Nürnberg über Ärztliche Fragen bei Seenot und Winternot* (n.d., n.p.).

Huxley, Aldous, *Science, Liberty and Peace* (New York and London: Harper and Brothers, 1946). Translated as *Wissenschaft, Freiheit und Frieden* (Zurich: Steinberg Verlag, 1947).

Hunt, Linda, *Secret Agenda. The United States Government, Nazi Scientists, and Project Paperclip, 1945 to 1990* (New York: St. Martin's Press, 1991).

Ivy, Andrew C., 'The History and Ethics of the Use of Human Beings in Medical Experiments', *Science*, vol. 108 (1948) 1–5. Cf. 'Citizen Doctor', *Time*, 13 January 1947: 47.

Ivy, Andrew C., 'Nazi War Crimes of a Medical Nature. Some Conclusions', *JAMA*, 139 (1949) 131–5.

Jacob, Bruce R., 'Remembering a Great Dean: Harold L. "Tom" Sebring', *Stetson Law Review*, vol. 30 (2000) 73–173.

JAMA, 276 no. 20 (1996) 'Nuremberg Special Issue'.

Katz, Jay, *Experimentation with Human Beings. The Authority of the Investigator, Subject, Professions, and State in the Human Experimentation Process* (New York: Russell Sage Foundation, 1972).

Katz, Jay, 'The Nuremberg Code and the Nuremberg Trial. A Reappraisal', *JAMA*, 276 no. 20 (1996) 1662–6.

Klee, Ernst, *Auschwitz, die NS-Medizin und ihre Opfer* (Frankfurt am Main: S. Fischer, 1997).

Klier, Freya, *Die Kaninchen von Ravensbrück. Medizinische Versuche an Frauen in der NS-Zeit* (Munich: Knaur, 1994).

Kogon, Eugen, *Der SS-Staat. Das System der Deutschen Konzentrationslager* (Frankfurt am Main: Verlag der Frankfurter Hefte, 1946).

Leibbrand, Werner (ed.), *Um die Menschenrechte der Geisteskranken* (Nürnberg: Die Egge, 1946)

'Prison Malaria: Convicts Expose themselves to Disease so Doctors can Study it', *Life*, Overseas Edition for Armed Forces, vol. 18 no. 23, June 4, 1945, 43–4, 46.

Lifton, Robert Jay, *The Nazi Doctors. Medical Killing and the Psychology of Genocide* (New York: Basic Books, 1986).

Mant, A. Keith, 'The Medical Services in the Concentration Camp of Ravensbruck', *The Medico-legal Journal*, 18 (1949) 99–118.

Marrus, Michael, 'The Nuremberg Doctors' Trial in Historical Context', *Bulletin of the History of Medicine*, vol. 73 (1999) 106–23.

Mellanby, Kenneth, *Human Guinea Pigs* (London: Gollancz, 1945); 2nd edn (London: Merlin Press, 1973).

Mellanby, Kenneth, 'A Moral Problem', *The Lancet* (7 December 1946) 850.

Mellanby, Kenneth, 'Medical Experiments on Human Beings in Concentration Camps in Nazi Germany', *British Medical Journal* (15 January 1947) 148–50.

Mitscherlich, Alexander, 'Der Arzt und die Humanität. Erste Bemerkungen zum Nürnberger Ärzteprozeß', *Die Neue Zeitung* (20 December 1946).

Mitscherlich, Alexander 'Was ist ein Mensch Wert? – 'Medizinische' Versuche im Dritten Reich' in A. Mitscherlich, *Das Ich und die Vielen. Parteinahme eines Psychoanalytikers* (Munich: Piper, 1978), pp. 238–58.

Mitscherlich, Alexander, Mielke, Fred (eds), *Das Diktat der Menschenverachtung* (Heidelberg: Lambert Schneider, 1947).

Mitscherlich, Alexander, Mielke, Fred (eds), *Medizin ohne Menschlichkeit, Dokumente des Nürnberger Ärzteprozesses* (Frankfurt am Main: S. Fischer, 1960). 1st edn Heidelberg: Lambert Schneider, 1949 as *Wissenschaft ohne Menschlichkeit*.

Mitscherlich, Alexander, Mielke, Fred (with contributions by Ivy, Taylor, Alexander and Deutsch), *Doctors of Infamy: the Story of the Nazi Medical Crimes* (New York: Henry Schuman, 1949).

Moreno, Jonathan D., *Undue Risk. Secret State Experiments on Humans* (New York: Routledge, 2001).

Oppitz, Ulrich-Dieter, *Medizinverbrechen vor Gericht. Das Urteil im Nürnberger Ärzteprozess gegen Karl Brandt und andere sowie aus dem Prozess gegen Generalfeldmarschall Milch* (Erlangen: Palm & Enke, 1999).

Overy, Richard, *Interrogations, The Nazi Elite in Allied Hands, 1945* (London: Penguin, 2000).

Peiffer, Jürgen, *Hirnforschung im Zwielicht: Beispiele verführbarer Wissenschaft aus der Zeit des Nationalsozialismus. Julius Hallervorden – H-J. Scherer – Berthold Ostertag* (Husum: Matthiesen Verlag, 1997).

Peiffer, Jürgen, 'Assessing Neuropathological Research Carried out on Victims of the "Euthanasia Programme"', *Medizinhistorisches Journal*, 34 (1999) 339–55.

Peter, Jürgen, *Der Nürnberger Ärzteprozess im Spiegel seiner Aufarbeitung anhand der drei Dokumentensammlungen von Alexander Mitscherlich* (Münster: Lit, 1994).

Platen-Hallermund, Alice. *Die Tötung Geisteskranker in Deutschland: Aus der deutschen Ärzte-Kommission beim amerikanischen Militärgericht* (Frankfurt: Verlag der Frankfurter Hefte, 1948).

Ricciardi-von Platen, Alice, 'Geleitwort', in Stephan Kolb, Horst Seithe/IPPNW (eds), *Medizin und Gewissen. 50 Jahre nach dem Nürnberger Ärzteprozess* (Frankfurt-am-Main: Mabuse, 1998), pp. 13–14.

Roth, Karl Heinz, 'Strukturen, Paradigmen und Mentalitäten in der luftfahrtmedizinischen Forschung des "Dritten Reichs" 1933 bis 1941: Der Weg ins Konzentrationslager Dachau', vol. 15 *1999* (2000) 49–77.

Schwarberg, Günther, *The Murders at Bullenhuser Damm. The SS Doctor and the Children* (Bloomington: Indiana University Press, 1984), trans. of the German edn 1980.

Shevell, Michael, 'Neurology's Witness to History: The Combined Intelligence Operative Sub-Committee Reports of Leo Alexander', *Neurology*, vol. 47 (1996) 1096–1103.

Shevell, Michael, 'Neurology's Witness to History (Part 2): Leo Alexander's Contributions to the Nuremberg Code (1946–47)', *Neurology*, vol. 50 (1998) 274–8.

Shuster, Evelyne, 'Fifty Years Later: the Significance of the Nuremberg Code', *The New England Journal of Medicine* (1997) 1436–40.

Shuster, Evelyne, 'The Nuremberg Code: Hippocratic Ethics and Human Rights', *Lancet* 351 (1998) 974–7.

Smith, Constance C., *Guide to the Documents That Were Used As Evidence In The Trial of the Nazi Physicians. Case I of the United States Military Tribunal At Nuremberg* (New York: The New York Academy of Medicine, n.d. [1949]).

Taylor, Telford, *The Anatomy of the Nuremberg Trials. A Personal Memoir* (New York: Back Bay Books, 1992).

Tillion, Germaine, *Ravensbrück* (Cahiers du rhône, 1946); new edn (Paris: Éditions du Seuil, 1973, 1988).

The Trial of German Major War Criminals. Proceedings of the International Military Tribunal Sitting at Nuremberg Germany (London: HMSO, 1946–51), vols 1–28.

United Nations War Crimes Commission, *German Medical War Crimes. A Summary of Information* (London: World Medical Association, n.d. [1947]).

UNWCC, Summary of Information No. 55, December 1947 Notes on German Medical War Crimes.

UNWCC, *Law Reports of Trials of War Criminals. Selected and Prepared by the United Nations War Crimes Commission* (London: HMSO, 1947), vol. 1.

Weindling, Paul Julian, *Health, Race and German Politics between National Unification and Nazism* (Cambridge: Cambridge University Press, 1989).

Weindling, Paul Julian, 'Ärzte als Richter: Internationale Reaktionen auf die Medizinverbrechen des Nürnberger Ärzteprozesses in den Jahren 1946–1947', in C. Wiesemann and A. Frewer (eds), *Medizin und Ethik im Zeichen von Auschwitz. 50 Jahre Nürnberger Ärzteprozess*, (Erlangen and Jena, 1996), pp. 31–44.

Weindling, Paul Julian, 'Human Guinea Pigs and the Ethics of Experimentation: the *BMJ*'s Correspondent at the Nuremberg Medical Trial', *British Medical Journal*, vol. 313 (1996) 1467–70. Revised version in: Len Doyal and Jeffrey S. Tobias (eds), *Informed Consent in Medical Research* (London: BMJ Books, 2000), pp. 15–19. French version (trans. F. Eytan), 'Les cobayes humains et l'éthique de l'experimentation: le correspondant de BMJ au procès des médecins à Nuremberg', *JAMIF*, no. 459 (1997) 36.

Weindling, Paul Julian, *Epidemics and Genocide in Eastern Europe* (Oxford: Oxford University Press, 2000).

Weindling, Paul Julian, 'The Origins of Informed Consent: The International Commission for the Investigation of Medical War Crimes, and the Nuremberg Code', *Bulletin of the History of Medicine*, vol. 75 (2001) 37–71.

Weindling, Paul Julian, 'From International to Zonal Trials: the Origins of the Nuremberg Medical Trial', *Holocaust and Genocide Studies*, vol. 14 (2000) 367–89.

Weindling, Paul Julian, ' "Tales from Nuremberg": the Kaiser Wilhelm Institute for Anthropology and Allied Medical War Crimes Policy', in Doris Kaufmann (ed.), *Geschichte der Kaiser-Wilhelm-Gesellschaft im Nationalsozialismus. Bestandsaufnahme und Perspektiven der Forschung*, (Göttingen: Wallstein Verlag, 2000), 621–38.

Notes

Introduction

1. The National Archives, Public Record Office, Kew, London (hereafter, TNA: PRO) WO 39/470 Die Häftlingsärzte von Auschwitz an die internationale Öffentlichkeit, 4 March 1945.
2. George J. Annas, Michael A. Grodin, *The Nazi Doctors and the Nuremberg Code* (Oxford: Oxford University Press, 1992). Jay Katz, *Experimentation with Human Beings. The Authority of the Investigator, Subject, Professions, and State in the Human Experimentation Process* (New York: Russell Sage Foundation, 1972).
3. Werner Süskind, *Süddeutsche Zeitung* Nr 106 (14 December 1946).
4. Telford Taylor Papers, Arthur W. Diamond Law Library, Columbia University Law School (New York City) (hereafter TTP-CLS) 5/1/1/1, Taylor before the German American Youth Club, 22 July 1947.
5. National Archives and Records Administration, Washington DC [NARA] RG 153 box I folder 2, 85-2 United Nations, 'The Crime of Genocide', Report of the 6th Committee A/231, 10 December 1946.
6. Samantha Power, *'A Problem from Hell'. America and the Age of Genocide* (London: Flamingo, 2003).
7. Allen Menkin, 'Historical Basis of Medical Crimes in Nazi Germany', *North Carolina Medical Journal* (1968), 189–201. M.J. Lynch, *Medicine and the State* (Springfield, 1963). Comments in BAK Zsg 154 Sammlung Christian Pross box 62
8. 'Nuremberg Revisited', *JAMA*, vol. 276, no. 20 (1996), 1631. Also Barondess, *JAMA*, vol. 276, no. 20 (1996), 1661.
9. Kristie Mackrakis, *Surviving the Swastika: Scientific Research in Nazi Germany* (Oxford: Oxford University Press, 1993); Ute Deichmann, *Biologists Under Hitler* (Cambridge, Mass.: Harvard University Press, 1996); Notker Hammerstein, *Die Deutsche Forschungsgemeinschaft in der Weimarer Republik und im Dritten Reich* (Munich: Beck, 1999); Robert Proctor, *The Nazi War on Cancer* (Princeton: Princeton University Press, 1999).
10. Paul Weindling, *Health, Race and German Politics between National Unification and Nazism* (Cambridge: Cambridge University Press, 1989). Michael Kater, *Doctors under Hitler* (Chapel Hill: University of North Carolina Press, 1989).
11. Kater, *Doctors*.
12. TTP-CLS -5/1/1/1 Taylor Before the German American Youth Club, p. 4. TTP-CLS-10/0/1/13 Taylor 'Nuremberg – A Symbol of Freedom under Law', Amherst College, 15 February. 1950.
13. P.J. Weindling, *Health, Race and German Politics between National Unification and Nazism 1870–1945* (Cambridge: Cambridge University Press, 1989).
14. P.J. Weindling, *Epidemics and Genocide in Eastern Europe* (Oxford: Oxford University Press, 2000).
15. Carola Sachse (ed.), *Die Verbindung nach Auschwitz – Biowissenschaften und Menschenversuche an Kaiser-Wilhelm-Instituten* (Göttingen: Wallstein Verlag, 2004).

16. 'German Medical War Crimes: Medical Ethics and Postwar Justice', 14 March 1997, co-organised with John Senior.
17. *Le Monde Juif. Revue d'Histoire de la Shoah* (7–8 December 1997) and (May–August 1997).

1   The Rabbits Protest

1. Callum MacDonald, *The Killing of SS Obergruppenführer Reinhard Heydrich* (New York: Free Press, 1989). For photos, see Jarolslav Cvancara and Nekomu Zivot, *Nekomuu Smrt 1939–1941* (Prague: Laguna, 2002), pp. 335–9.
2. Bundesarchiv Berlin (hereafter BAB) R 31/368, Walter Dick praised for treating Heydrich.
3. Peter Witte et al., *Der Dienstkalender Heinrich Himmlers 1941/42* (Hamburg: Christians, 1999), 27 May–5 June 1942.
4. Keith Mant, 'From Nuremberg to the Old Bailey', chapter 2 'War Crimes Investigations' (unpublished autobiography). I am grateful to the Mant family for making this available. Medical Archives, Durham, NC, HC 36 Collected papers and Documents. 65th General Hospital, Leo Alexander Papers (hereafter APD) box 4, Alexander to McHaney, 'The Motives of the Sulphonamide Experiments', 5 March 1942.
5. E. Chain et al., 'Penicillin as a Chemotherapeutic Agent', *The Lancet* 239 (1940) 226–8. E.P. Abraham et al., 'Further Observations on Penicillin', *The Lancet* 2 (1941) 177–88.
6. The Habilitation thesis and examination was a higher doctorate, which gave the holder the right to lecture as 'Privatdozent'.
7. University Archive (= UA) München E–II–1413 Akten des akademischen Senats der Univ München. AHUB Personal-Akten des o. Professors Karl Gebhardt Bd 1 September 1934–43. Hans Waltrich, *Aufstieg und Niedergang der Heilanstalten Hohenlychen (1902 bis 1945)* (Blankensee: Strelitzia, 2001).
8. Alexander Papers Durham, NC (APD) Alexander to McHaney, 5 March 1947 on the motives of the sulphonamide experiments. NARA M 1019/20 Gebhardt interrogation, 17 October 1946, pp. 14–19; Gebhardt interrogation by Alexander, 3 December 1946, pp. 19–23. NMT 2/2461 Karl Brandt testimony on 4 February 1947.
9. Harry Marks, *The Progress of Experiment. Science and Therapeutic Reform in the United States, 1901–1990* (Cambridge: Cambridge University Press, 1997), pp. 100–5.
10. Witte et al., *Der Dienstkalender Heinrich Himmlers 1941/42*, 7 March, 1 and 4 September 1941.
11. Paul Weindling, 'Genetik und Menschenversuche in Deutschland 1940–1960. Hans Nachtsheim, die Kaninchen von Dahlem und die Kinder vom Bullenhuser Damm', in Hans-Walter Schmuhl (ed.), *Rassenforschung an Kaiser-Wilhelm-Instituten vor und nach 1933* (Göttingen: Wallstein, 2003), pp. 245–74.
12. Imperial War Museum (IWM) Milch diary, 26 May 1947; Gitta Sereny, *Albert Speer: his Battle with Truth* (London: Macmillan, 1995), pp. 324–5, 387, 409–19, 424–6. Philippe Aziz, *Doctors of Death. Vol. 1 Karl Brandt, The Third Reich's Man in White* (Geneva: Ferni, 1976), pp. 135–9. NARA RG 238 entry 202 box 1 file 3, Friedrich Koch interrogation, 10 March 1947.
13. NARA M 1019/20 Genzken interrogation, 20 September 1946, p. 8. AFP Alexander, 'Questions to be put to Dr. Fischer on the stand', 8 March 1947.

TNA: PRO WO 309/469 A. Martin and Carmen Mory: Report on Carl Gebhardt, 18 August 1945. = NMT 8/ 337. Peter Padfield, *Himmler. Reichs-Führer-SS* (London: Cassell, 2001), pp. 514, 539, 576.

14. NARA M 1019/17 Fischer interrogation by Alexander, 3 December 1946, p. 12.
15. Bernd Biege, *Helfer unter Hitler. Das Rote Kreuz im Dritten Reich* (Reinbek: Kindler, 2000), pp. 173–6. Ernst Gunther Schenck, *Patient Hitler. Eine medizinische Biographie* (Düsseldorf: Droste, 1989), pp. 475–6. NARA M 1019/20 Gebhardt interrogation, 3 December 1946, p. 29. Ian Kershaw, *Hitler 1936–1945: Nemesis* (Harmondsworth: Penguin, 2000), pp. 832–3.
16. Reinhard Strebel, 'Das Männerlager im K.Z. Ravensbrück 1941–1945', *Dachauer Hefte* 14 (1998) 141–74, 161. A. Keith Mant, 'The Medical Services in the Concentration camp of Ravensbrück', *The Medico-legal Journal* 18 (1949), pp. 99–118.
17. United States Holocaust Memorial Museum (USHMM) Ferriday Collection, box 2, list, 22 May 1961. Germaine Tillion, *Ravensbrück* (Paris: Editions du Seuil, 1976), p. 104.
18. Mant, 'Medical Services', p. 108. TNA WO 309/1697
19. NMT M 1019/17 Fischer interrogation, 3 December 1946, p. 6. Fischer and Gebhardt statement, 28 December 1946.
20. Mant, 'From Nuremberg', chapter 2 'War Crimes Investigations'. USHMM Ferriday Collection. W. Woelk and Karen Bayer, 'Herta Oberheuser', *Nach der Diktatur* (Essen: Klartext, 2003), pp. 253–68, 261. APD, Alexander, 'Case History of the Polish Witnesses', 17 December 1946, p. 10.
21. NARA M 1019/14 Dzido interrogation.
22. Stanislaw Sterkowicz, 'Medizinische Experimente im Konzentrationslager Ravensbruck' (= 'Eksperymenty medycne w obozie koncentrazyjnym Ravensbruck'), 22–3; NMT 10th Trial Day. 20 December 1946 2/882–3 re Oberheuser calling them Rabbits.
23. NMT 5/109.
24. Freya Klier, *Die Kaninchen von Ravensbrück. Medizinische Versuche an Frauen in der NS-Zeit* (Munich: Knaur, 1994).
25. Sterkowicz, 'Medizinische Experimente', 21. Germaine Tillion, *Ravensbrück* (Paris: Éditions du Seuil, 1973), p. 109. Dunja Martin, ' "Versuchskaninchen" – Opfer medizinischer Experimente', Claus Füllberg-Stolberg, Martina Jung, Renate Roebe and Martina Schreitenberger (eds), *Frauen in Konzentrationslagern Bergen-Belsen Ravensbrück* (Bremen: Temmen, 1994), pp. 113–22. Dunja Martin, 'Menschenversuche im Krankenrevier des KZ Ravensbrück', in ibid., 99–112. Klier, *Die Kaninchen von Ravensbrück*, pp. 218–21.
26. TNA: PRO WO 235/305, f. 50 for the BBC broadcast.
27. Germaine Tillion, *Ravensbrück* (Paris: Éditions du Seuil, 1973), p. 26.
28. Sterkowicz, 'Medizinische Experimente', 21. Klier, *Kaninchen*, p. 257. The delegate was probably Meyer.
29. Martin, 'Versuchskaninchen',120. Sterkowicz, 'Medizinische Experimente', 22.
30. AdeF BB/35/275 Sulfamides – rapport et études, Keith Mant, 'Experiments in Ravensbruck Concentration Camp carried out under the Direction of Prof. Karl Gebhardt', 17 pp. carbon copy no date; Memorandum To Mr McHaney From: Major Mant, 5 December 1946.
31. STAN Rep 502A KV-Verteidigung Handakten, Verteidigung Pokorny Nr 25 Balachowsky to R.C. Schaefer, 7 July 1947. Weindling, *Epidemics.*
32. Author's interview with Charlotte Bloch-Kennedy 1996 and 1997. Paul Weindling, 'The Scientist as Survivor: Ludwik Fleck and the Holocaust', *La Lettre de la Maison Française d'Oxford*, no. 13 (2001), 85–96.

33. Robert Proctor, *The Nazi War on Cancer* (Princeton: Princeton University Press, 1999), p. 344 n. 4.
34. NMT 4/2985–7 statement of Dr A. von Erlach, Berne, 23 February 1947 to Seidl.
35. Jean-Claude Favez, *The Red Cross and the Holocaust* (Cambridge: Cambridge University Press, 1999), p. 43.
36. Favez, *Red Cross and the Holocaust*, p. 267. NMT 4/2963 on Meyer; 4/2986 Erlach statement.
37. Sterkowicz, 'Medizinische Experimente', 25. Klier, *Kaninchen*, pp. 263–5. Raymond Palmer, 'Felix Kersten and Count Bernadotte: a Question of Rescue', *Journal of Contemporary History* 29 (1994) 39–51.
38. Sterkowicz, 'Medizinische Experimente' 23–5. AdeF BB/35/275 Memorandum To Mr McHaney from: Major Mant, 5 December 1946.
39. NARA M 1019/20 Genzken interrogation, 23 September 1946, p. 3.
40. Favez, *Red Cross and the Holocaust*, pp. 7, 38–43. NMT 29/2436 re ICRC and contacts re occupied territories. Caroline Moorehead, *Dunant's Dream: War, Switzerland and the History of the Red Cross* (London: HarperCollins, 1998), pp. 468–70.
41. 'SS-Gruppenführer, Generalleutnant Prof. Gebhardt und F. Fischer, Besondere Versuche über Sulfonamidwirkungen', in Bericht über die 3. Arbeitstagung Ost der Beratenden Fachärzte von 24–26 Mai 1943 in der Militärärztlichen Akademie Berlin, pp. 43–4.
42. Bericht über die 4. Arbeitstagung der Beratenden Ärzte vom 16. bis 18. Mai 1944, im SS-Lazarett Hohenlychen, pp. 7–9.
43. ACICR microfilm G44.01, G 44/13–11 otages, détenus politiques. Allemagne. Corréspondance générale avec Croix-Rouge 1.12.1943–30.5.1945, Resolution of the Liaison Committee of Women's international Organisations signed Elsie Zimmern, 23 October 1944; British Red Cross Society and Order of St. John, 15 November 1944. Polish Red Cross to Max Huber, 16 November 1944, J.E. Schwarzenberg reply, 28 November 1944. *An Experiment in Co-operation 1925–1945. The History of the Liaison Committee of Women's International Organisations 1925–1945* (nd, np), p. 26.
44. ACICR G 44/13–19.04 Schwarzenberg to Miss Warner, Organisation of the British Red Cross Society, 7 December 1944.
45. TNA: PRO WO 235/305 f 50.
46. ACICR G44/13–7 Corréspondance générale avec autorités allemandes, 10 January 1945–15 October 1945, frame 1395, 2 January 1945, betr. Liebesgabensendungen in Konzentrationslager.
47. ACICR G44/13–19.04 Marti, Note au Secretariat (confidentielle). 12 June 1944
48. Michael Hubenstorf, 'Anatomical Science in Vienna, 1938–1945', *The Lancet* 355 (2000), 1385.
49. ACICR G44/13–19.01 H. Rossel, Camp de Concentration de Ravensbruck, 14 October 1944, p. 7; G 44/13–19.04 H. Landolt, 3 January 1945.
50. ACICR G 44/13–14.01 Rapport du Dr Descoeudres sur le camp de Buchenwald, 17–18 April 1945. Reports of 14 August 1940, p. 4 and June 1941.
51. ACICR G 44/13–19.04 Dr H. Landolt to Suhren, 25 April 1945. Klier, *Kaninchen*, pp. 279–89.
52. ACICR G 44/13–19.03 M. Spörri, Rapport sur le camp de Concentration de Ravensbrueck, 4 May 1945, 'Les cobayes', p. 7.
53. ACICR G 44/13–19.02 Albert de Cocatrix, 16 November 1945 about a tour of inspection, 23 April to 4 May 1945. Favez, *Red Cross*, p. 270.

54. On the DRK see Basic Handbooks, *Public Health*, 256C–G. Moorehead, *Dunant's Dream*, 704–5. cf NMT 2/2390 re Grawitz – negotiated with him as DRK President in 1945 to secure safety of hospitals.

55. For varying views on Grawitz's death, see Michael Elkins, *Forged in Fury* (London: Piatkus, 1996), p. 186. Peter Voswinckel, *50 Jahre Deutsche Gesellschaft für Hämatologie und Onkologie* (Würzburg: Murken-Altrogge, 1987), pp. 39–40. Markus Wicke, 'Arzt und Mörder', *Berliner Zeitung* (16/17 January 1999).

56. NARA M 1019/20 Gebhardt interrogation, 12 November 1946, p. 3. APD Alexander to Taylor, 5 December 1945.

57. Hugh Trevor-Roper, *The Last Days of Hitler* (London: Pan Books, 1962), pp. 164–5. NARA M 1019/20 Gebhardt interrogation, 3 December 1946, p. 9. Klier, *Kaninchen*, pp. 273–4. Biege, *Helfer unter Hitler*, pp. 167–71, 265–6.

58. NARA RG 144 Gebhardt file, TNA: PRO WO 309/1766, arrest report dated, 17 May 1945. See also Gebhardt's detention report in TNA: PRO WO 309/1766.

59. ACICR G44/13–8 Correspondance avec gouvernements en allemagne, 29.8.44–16.7.45, US Dept of State Bern, 19 January 1945 on urgency of situation and need to keep civilians alive US wants frequent visits of delegates. Also 28 February 1945, State Department to ICRC.

60. NARA RG 338 USAR/JAG War Crimes Records Regarding Medical Experiments Col. Von Erlach 21 November 1946, Col. Dr Theodore Brunner à la haute Cour de Justice de Nuremberg, 24 November 1946.

61. NARA RG 238 NM 70 Entry 199 Box 3 Handloser memoranda, 16 May to Admiral Burkner, 17 May to Demobilisation Dept, 30 May, 27 June 1945 to Col. Escritt. M 1019/24 Handloser interrogation, 27 August 1946, pp. 1–3.

62. NMT 4/6364–7 statement by Luxenburger, 2 February 1947; also 46615–6 Luxenburger affidavit 24 March 1947. Zeitgeschichtlichen Forschungsstelle Ingolstadt, *Der Fall Rose. Ein Nürnberger Urteil wird widerlegt* (Asendorf: Mut Verlag, 1988), pp. 66–77. Hana Vondra, 'Malariaexperimente in Konzentrationslager und Heilanstalten während der Zeit des Nationalsozialismus', diss. Hannover, 1989, 79–90, xii–xvi for interview with Rose.

63. Gerhard Rose, *Krieg nach dem Kriege* (Dortmund, 1920).

64. NARA RG 466 Entry 54 Schröder file.

2 Allied Experiments

1. Adrienne Noe, 'Medical Principle and Aeronautical Practice: American Aviation Medicine Prior to World War II', University of Delaware PhD 1989, 239.

2. Walter Eggleston, *Scientists at War* (London: Oxford University Press, 1950), 213–18. Donald H. Avery, *The Science of War: Canadian Scientists and Allied Military Technology during the Second World War* (Toronto: University of Toronto Press, 1998).

3. University of Toronto, Banting Collection 76, box 3 correspondence, March 1940; box 20 aviation medicine research, 12 August 1940.

4. NARA M 1019/5, 28 February 1947 interrogation of Becker-Freyseng stating that one firm alone produced 40 pressure chambers, and that there were four mobile pressure chambers.

5. National Academies of Science Washington DC, Committee on Aviation Medicine (NAS CAM) J.F. Fulton to Lewis Weed, National Research Council, 1 May 1941; Weed to Fulton, 3 May 1941; Fulton to Carl Wiggers, Western Reserve University, Ohio, 1 May 1941; Sanford Larkey to L.W. Billingsley, 22

July 1942. RAC RG 8–1949 Series 717 folder 1561 Dr Hubertus Strughold, R.A. Lambert note, 1 May 1941. Carlson to Fulton, 5 May 1941. Hoover Institution America First Papers box 26 A.J. Carlson. 'Scientist's Scientist', *Time* (10 February 1941) cover.

6. Siegfried Ruff and Hubertus Strughold, *Compendium of Aviation Medicine. Reproduced under a License Granted by the Alien Property Custodian* (np, 1942).

7. NAS CAM M.C. Winternitz to Andrus, 28 March 1942.

8. NAS CAM Letters re Distribution of Compendium of Aviation Medicine, Ruff and Strughold destroyed 4 August 1960 from L.B. Flexner file. The British Central Scientific Office requested five copies.

9. NAS CAM J.F. Fulton, 'Recent Developments in Aviation Medical Research in the United States and Canada. A Confidential Report to Air-Vice Marshal Whittingham', The Air Ministry, London, 10 April 1941.

10. 'The Conscientious Guinea Pig', vol. 46, *Time* (December 1945). Also 10 July 1945. Cf. the US Army film 'Food for Fighters'. NMT 2/9490–2 for Ivy testimony.

11. NAS CAM Appendix 1, 1st meeting of Committee on Aviation Medicine 1940, Duties of the Operator of the Low-pressure Altitude Chamber, Training Routine in Toronto. This accords with the recollections of a collaborator of John Thompson, W.C. Gibson, interview with author, Oxford, June 2001.

12. Personal communication to author by W.C. Gibson.

13. Ivy, Burckhardt and Thometz research for the Committee of Medical Research of OSRD on 'Effects of Pecuniary Incentives on the Influence of "Intolerable" Cases of Bends and Chokes', Report of 23 March 1945, p. 363. NMT 2/9490–2 for Ivy testimony.

14. A.A. Skolnick, 'Discovery of 50-Year-Old Naval Logbook May Aid Follow-up Study of Radium Exposed Veterans', *JAMA*, vol. 276, no. 20 (1996) 1628–9.

15. Jonathan D. Moreno, *Undue Risk. Secret State Experiments on Humans* (New York: Routledge, 2001), p. 15.

16. Kenneth Mellanby, *Human Guinea Pigs* (London: Victor Gollancz, 1945); second edn (London: Merlin Press, 1973). Paul Weindling, 'Kenneth Mellanby', *New Dictionary of National Biography* (Oxford: Oxford University Press, in press). Paul Weindling, 'Human Guinea Pigs and the Ethics of Experimentation: the *BMJ*'s Correspondent at the Nuremberg Medical Trial', *British Medical Journal*, vol. 313 (1996) 1467–70.

17. American Heritage Center, University of Wyoming, A.C. Ivy Papers, 179–1984, Accession Number 8768 (hereafter Ivy Papers), Box 101 Andrew Conway Ivy, MD, Ph.D. An Oral History. Recorded by James David Boyle, MD, The American Gastroenterological Association, 1969.

18. Ivy Papers, Box 86 folder 24 Naval Research Institute. Folder 25 Memorandum (6 pp.), nd. 'Nazi War Crimes of a Medical Nature', 7–9. W.U. Consolazio, N. Pace and A.C. Ivy, 'Drinking Sea Water', *Smithsonian Institution Annual Report* (1945) 153–63.

19. Edward V. Rickenbacker, 'Pacific Mission', *Life Magazine* (25 January 1943) 19–26; (1 February 1943) 78–92 (esp. 83–5 re storage and catching of water over 21 days (8 February 1943) 94–106. 'Navy Uses Special Devices', *Life Magazine* (June 1943) 41–50. NARA M 1019/62 Interrogation of Schäfer by Alexander and Ivy, 22 January 1947.

20. Peter J. Kuznick, *Beyond the Laboratory. Scientists as Political Activists in 1930s America* (Chicago: Chicago University Press, 1987), pp.159–60, 245.

21. Yale University Fulton Papers, Box 29 4 June 1940 Carlson to Fulton; 6 June Fulton to Carlson; folder 1257 7 May 1942 Fulton to Capt C.S. Stephenson, Bureau of Medicine and Surgery; 9 May 1942 Fulton to Ivy.
22. Ivy papers Box 86 folder 24 JF Fulton to Ivy, 2 June 1942. Also in Fulton Papers.
23. Ivy Papers, Box 86 Naval Research Institute, Rear Admiral H.W. Smith to Ivy September 4, 1942. Ivy to Smith, 18 September 1942. Ivy resignation, 24 July 1943. Ivy report on Naval Research Institute, nd, 6pp. Box 91 folder 6 A.C. Ivy to Horace Ivy, 2 October 1974.
24. Bends refers to joint pain; chokes refers to an inability to breathe without coughing and constriction in the chest and throat.
25. NAS CAM Committee of Medical Research, OSRD Monthly Progress Report No. 1, 30 November 1942.
26. Ivy Papers, Box 86 Naval Research Institute, Ivy report on Naval Research Institute, p. 6.
27. *The Human Radiation Experiments. Final Report of the President's Advisory Committee* (Oxford: Oxford University Press, 1996), p. 54.
28. NMT 2/9489, 16 June 1947.
29. NAS CAM, Bulletin of the Subcommittee on Decompression-sickness, pp. 219, 228–41.
30. AFP Arthur Schuller, 1 May 1934, to Fred Alexander, 25 April 1934. There were ca 6,000 medical refugees in the USA from Austria and Germany. Cf. Kathleen Pearle, *Preventive Medicine: The Refugee Physician and the New York Medical Community 1933–1945* (Bremen: Universität, 1981), p. 113.
31. Josef Kastein [Julius Katzenstein], *History and Destiny of the Jews* (New York: Garden City Publishing Co. [1938]), John Thompson Library. Dedication of 6 November 1938. Milton Rosenbaum interview with author, 23 November 2001.
32. 'The first twenty years. A history of the Duke University Schools of Medicine, Nursing and Health Services, and Duke Hospital 1930 to 1950', *Bulletin of Duke University*, vol. 24, no. 7 (May 1952).

3   Criminal Research

1. Martin Gumpert, *Heil Hunger! Health under Hitler* (London: Allen and Unwin, 1940).
2. Robert A. Brady, *The Spirit and Structure of German Fascism* (London: Victor Gollancz, 1937), pp. 39–77.
3. Paul Weindling, 'Theories of the Cell State in Imperial Germany', Charles Webster (ed.), *Biology, Medicine and Society 1840–1940* (Cambridge: Cambridge University Press for Past and Present Publications, 1981), pp. 99–155, and 'Die Preussische Medizinalverwaltung und die "Rassenhygiene"', in A. Thom and H. Spaar (eds), *Medizin und Faschismus* (Berlin: Akademie für ärztliche Fortbildung, 1983), pp. 23–35. Weindling, *Health, Race and German Politics*.
4. Robert Lifton, *The Nazi Doctors* (New York: Basic Books, 1986), chapter 15 'The Experimental Impulse'.
5. Ernst Klee, *Auschwitz, die NS-Medizin und ihre Opfer* (Frankfurt am Main: S. Fischer, 1997). Paul Weindling, 'Human Experiments in Nazi Germany: Reflections on Ernst Klee', *Auschwitz. Die NS-Medizin und ihre Opfer* and *Ärzte ohne Gewissen* (1996)', *Medizinhistorisches Journal*, vol. 33 (1998) 161–78.
6. Angelika Ebbinghaus and Karl Heinz Roth, 'Vernichtungsforschung: der Nobelpreisträger Richard Kuhn, die Kaiser-Wilhelm Gesellschaft und die

Entwicklung von Nervenkampfstoffen während des "Dritten Reiches"', *1999*, vol. 17 (2001) 15–50. Ebbinghaus and Roth, 'Von der Rockefeller Foundation zur Kaiser Wilhelm/ Max-Planck-Gesellschaft: Adolf Butenandt als Biochemiker und Wissenschaftspolitiker des 20. Jahrhunderts', *Zeitschrift für Geschichtswissenschaft*, vol. 50 (2002) 389–418.

7. See the series *Geschichte der Kaiser-Wilhelm-Gesellschaft im Nationalsozialismus* with the Wallstein Verlag, Göttingen.

8. Paul Weindling, 'What Did the Allies Know about Criminal Human Experiments in the War and its Immediate Aftermath?', in Astrid Ley (ed.), *Menschenversuche* (Erlangen Museum, 2001), pp. 52–66.

9. NARA M 1019/7 Blome interrogation by Meyer, 2 December 1946, pp. 2, 7.

10. NARA M1019/68 Sievers interrogation, 16 August 1946.

11. NARA M 1019/60 Rostock interrogation, 16 October 1946. Aziz, *Brandt*, 113–14.

12. Michael Marrus, *The Nuremberg War Crimes Trial 1945–46. A Documentary History* (Boston: Bedford Books, 1997), pp. 185–7.

13. *Trials of War Criminals before Nuernberg Military Tribunals* (Washington, DC: US Government Printing Office, 1949), vol. 2, p. 631. Hippke to Obergruppenführer Wolff, 6 March 1943.

14. NARA M 1019/9 K. Brandt interrogation, 18 October 1946, p. 4.

15. Gilbert Shama and Jonathan Reinarz, 'Allied Intelligence Reports on Wartime German Penicillin Research and Production', *Historical Studies in the Physical Sciences*, vol. 32 (2002), 347–67 at 352.

16. Archives de l'Occupation française en allemagne et en autriche (hereafter AOF), Justice c. 3645 p. 287 no. 8949 Camp Dachau, Report of the Atrocities Committed at the Dachau Concentration Camp, incl. medical aspects.

17. 'Typhus Vaccines. A Crucial Experiment', *Lancet* ii (18 December 1943) 770 concerning Ding, 'Über die Schutzwirkung verschiedener Fleckfieberimpfstoffe beim Menschen und den Fleckfieberverlauf nach Schutzimpfung', *Zeitschrift für Hygiene*, vol. 124 (1943), 670–82. Weindling, *Epidemics and Genocide*, p. 356.

18. Weindling, *Epidemics and Genocide*, pp. 352–63, esp. 356. National Archives of Canada (hereafter NAC) Ottawa RG 24 National Defence D-1-b vol. 4034 file 1070–1–39 Canadian medical intelligence division circulated *The Lancet* analysis on 10 February 1944.

19. NARA RG 338 290–59–17 Medical Experiments No 127031 arrest report.

20. AdeF BB/35/260 Bayle papers 4a Assasinat de juifs à Kiev, 31 July 1946 Col. Straight re Medical Experiments. HQ to Rigney, US Chief of Counsel, re Kiev Pathological Institute.

21. George Creel, *War Criminals and Punishment* (New York: Robert M. McBride, 1944), pp. 65–71.

22. CMAC WT1/RST Royal Society for Tropical Medicine, letter, 10 June 1944 and press release, 12 May 1944.

23. Weindling, *Epidemics and Genocide*, p. 408.

24. NMT 1/75.

25. NMT 1/115.

26. Linda Hunt, *Secret Agenda. The United States Government, Nazi Scientists, and Project Paperclip, 1945 to 1990* (New York: St. Martin's Press, 1991), p. 87.

27. On Schuler's career, see Weindling, *Epidemics and Genocide*. Eugen Kogon, *'Dieses merkwürdige, wichtige Leben'. Begegnungen* (Weinheim: Quadriga, 1997), pp. 63–70.

28. NARA M 1270/19 Claus Schilling interrogation, 5 September 1945, 15–16. Hoover Institution, Stanford Ca. Pash Papers 2/7–9 Barnes, Diary, p. 24.

29. Central State Archives Prague (CSA) Fond 316 Czech Commission on War Crimes.
30. NARA RG 165 Entry 179 box 739 CI Consolidated Interrogation Report No. 10, 6 December 1945, The Case of Dr Siegmund Rascher. M 1019/68 Sievers interrogation of 28 October 1946, p. 32. Anna Maria Sigmund, *Die Frauen der Nazis II* (Munich: Heyne, 2002), pp. 267–84 for Karoline Rascher. NARA M 1019/68 Sievers interrogation, 5 September 1946, pp. 3–5. Wolfgang Benz, 'Dr med. Sigmund Rascher: eine Karriere', *Dachauer Hefte*, vol. 4, no. 4 (1988) 190–214.
31. Winfried SüSS, *Der 'Volkskörper' im Krieg* (Munich: Oldenbourg, 2003).
32. NARA RG 338 Executee Files: Landsberg Prison, Brack's prisoner's file; David irving (ed.), *Adolf Hitler, The Medical Diaries* (London: Sidgwick and Jackson, 1983), pp. 196–210. Karl Brandt, interrogation, 29 May 1945. M 1019/20 Genzken interrogation, 20 September 1946, p. 10 attributing Brandt's downfall to his objection to the Führer being prescribed arsenic tablets by Theo Morell. Aziz, *Brandt*, pp.165–73. RG Brack prisoner's file. David Irving (ed.), *Adolf Hitler, The Medical Diaries*, pp. 196–210.
33. Gitta Sereny, *Albert Speer. His Battle with Truth* (London, 1995), pp. 505, 524–5. Aziz, *Brandt*, 170–4.
34. Süss, *Der 'Volkskörper' im Krieg*.
35. 'Geneticists' Manifesto', *Journal of Heredity*, vol. 30 (1939) 371–3.
36. Gene Weltfish, 'Science and Prejudice', *Scientific Monthly* (September 1945), 210–12.
37. Aldous Huxley, *Science, Liberty and Peace* (New York and London: Harper and Brothers, 1946). Translated as *Wissenschaft, Freiheit und Frieden* (Zurich: Steinberg Verlag, 1947). A.V. Hill, *The Ethical Dilemma of Science* (London: Scientific Book Club, 1960), pp. 85–9. Cf. STAN Rep 502A KV- Verteidigung Handakten Rostock Nr 8 correspondence with Rein.
38. NARA RG 338 Parolee Case Files, Landsberg Prison, Siegfried Handloser.
39. *Basic Handbooks. The Nazi System of Medicine* (London: HMSO, 1944), p. 229.
40. Ibid., pp. 240, 243.
41. Sheldon H. Harris, *Factories of Death. Japanese Biological Warfare 1932–45, and the American Cover-Up* (London: Routledge, 1994), pp. 170–1.
42. APD 4/26, Diary for 1942–43, 21 February 1943.
43. NARA RG 153/86–3–1/ Box 11 folder 4 bk 3 H. Bowen Smith to George Rosen, 13 March 1945.
44. Yale University Rosen Papers, Subject File: War Department 1945 Order, 2 March 1945 to proceed for three months outside US; War Dept, Adjutant General's Office, 12 May 1945 Office of the Surgeon General, Special Orders for Captain George Rosen, 30 May 1945. Gaylord W. Anderson to Thomas Dublin, 30 December 1946.
45. Yale University Rosen Papers, Subject File: War Department 1945, Notes on the Kaiser Wilhelm Institut für Medizinische Forschung. Reported by Capt. George Rosen, CIOS File XX–1 [ca. April 1945]. Cf. also CIOS Consolidated Advance Field Team Assessment Report, 7 April 1945. USHMM Aero-Medical Exploitation, Interrogation of Gerhard Rose by J.B. Rice and George Rosen, 25–26 June 1945.
46. APD 4/43 address book, Alexander reported back to Rosen on 13 November 1945.
47. USHMM Aero-Medical Exploitation, Interrogation of Gerhard Rose by J.B. Rice and George Rosen, 25–26 June 1945. Rosen Papers, Yale University, Subject File: War Department 1945, interrogation of Rose by J.B. Rice and George Rosen, 7 June 1945. Eugène Ost, 'Die Malaria-Versuchsstation im Konzentrationslager Dachau', *Dachauer Hefte*, vol. 4, no. 4 (1988) 174–89.

48. NMT 4/6517 Rose Document Book III – Document 1934, Otis Schreuder affidavit, 13 March 1947.

49. USHMM Aero-medical Exploitation, Rose, Development of Hepatitis Problem during the War, 16 June 1945. Brigitte Leydecker, Burghard F. Klapp, 'Deutsche Hepatitisforschung im Zweiten Weltkrieg', *Der Wert des Menschen*, pp. 261–93.

50. UNA UNWCC File C–1, Hugo Iltis to B. Ecer. Hugo Iltis, *Der Mythos von Blut und Boden* (Vienna: Rudolf Harand, 1936). Franz Weidenreich, 'Memorandum on scientists suspected as war criminals', Weidenreich collection, American Museum of Natural History, Box 1 folder 3. My thanks to Veronika Lipphardt.

51. NMT 3/1766 and UN Archives, and UNA UNWCC Report on Sterilisation in Germany and occupied countries to be sent to the members of Committee I. Submitted by Dr B. Ecer.

52. Central Zionist Archives Jerusalem, Zollschan papers, Science Commission A 122. 4/12.

53. War Dept Pamphlet. No. 31–158, Civil Affairs Guide, Denazification of the Health Services and Medical Profession of Germany, War Dept, June 1945 [issued 26 May 1945].

54. AdeF BB/35/260 1.Questions de droit, généralités.1c, Collecte des documents et des témoignages, Fred Niebergall, Chief of Document Control Branch Evidence Division, OCC, affidavit, 3 December 1946.

55. *Denazification*, pp. 54–65.

56. 'Vorwort der Arbeitsgemeinschaft der Westdeutschen Ärztekammer', Alexander Mitscherlich, *Wissenschaft ohne Menschlichkeit* (Heidelberg: Lambert Schneider, 1949), p. v.

57. *Denazification*, pp. 64–5.

58. AOF C. 3190 P.72/a–c RP 220/2 Rhenanie-Palatinat Fiches politiques sur les médecins allemands, case of Walter Christ.

59. Paul Weindling, 'Verdacht, Kontrolle, Aussöhnung. Adolf Butenandts Platz in der Wissenschaftspolitik der Westalliierten (1945–1955)', in Wolfgang Schieder and Achim Trunk (eds), *Adolf Butenandt und die Kaiser-Wilhelm-Gesellschaft. Wissenschaft Industrie und Politik im 'Dritten Reich'* (Göttingen: Wallstein Verlag, 2004), pp. 320–46.

60. Paul Weindling, ' "Mustergau" Thüringen. Rassenhygiene zwischen Ideologie und Machtpolitik', in Uwe Hossfeld et al. (eds.), *'Kämpferische Wissenschaft'. Studien zur Universität Jena im Nationalsozialismus* (Vienna: Böhlau, 2003), pp. 1013–26. Susanne Zimmerman, *Die medizinische Fakultät der Universität Jena* (Berlin: VWB, 2000), p. 85.

61. Ernst Klee, *Was sie taten – Was sie wurden* (Frankfurt/M: S. Fischer, 1986), p. 284, also p. 318 for other euthanasia doctors Eugen Joos, 3 September 1945, Robert Müller, 2 June 1945, Erich Straub, 29 April 1945.

62. Frederick H. Kastan, 'Unethical Nazi Medicine in Annexed Alsace-Lorraine: The Strange Case of Nazi Anatomist Professor Dr. August Hirt', in George O. Kent (ed.), *Historians and Archivists. Essays in Modern German History and Archival Policy* (Fairfax: George Mason University Press, 1991), pp. 173–208, 194–6.

63. Lifton, *Nazi Doctors*, p. 311.

64. Ibid., 364–5, 408.

65. Aziz, *Brandt*, pp. 53–4. Ulrich Schlie (ed.), *'Alles, was ich weiss' aus unbekannten Geheimdienstprotokollen von Sommer 1945, Albert Speer mit einem Bericht 'Frauen um Hitler von Karl Brandt* (Munich: Herbig, 1999), pp. 10–12, 112. NARA M 1019/20 Gebhardt interrogation, 3 December 1946, p. 27. RG 242 entry 2, USSBS, Interview No. 11, Reichsminister Albert Speer (31 May 1945).

66. For Gebhardt's detention report, see TNA: PRO WO 309/1766. This document is at variance with Hugh Thomas, *The Unlikely Death of Heinrich Himmler* (London, 2001), p. 149 stating the arrest was at Bremervorde on 20 May 1945.

67. STAN KV Anklage Interrogations R 132, Rose interrogation, pp. 30–1.

68. Yveline Pendaries, *Les Procès de Rastatt (1946–1954). Le jugement des crimes de guerre en zone française d'occupation en Allemagne* (Bern: Lang, 1995).

69. NARA RG 153 record of the Judge Advocate General (Army) War Crimes Branch Entry 144 personal and subject dossiers Box 59 – 100–646 Theo Lang. RG 338 USAREUR/JAG War Crimes, Records Regarding Medical Experiments, No. 125709–17.

70. NMT 3/1767–1769 and UNWCC archives, Lang to Interallied Commission for the Investigation of War Crimes, 10 May 1945, Dr med habil Theo Lang was senior medical officer at the Kantonale Heil- und Pflegeanstalt Herisau (County Hospital) Switzerland.

71. AOF AJ 3628 p 91 d 4528 Médecins 20.4.46 French consul Bale to French ambassador in Berne, 4 February 1946.

72. William Shirer, *Berlin Diary* (New York: Knopf, 1941), p. 512 (21 September 1940), pp. 569–75 (25 November). N.D.A. Kemp, *Merciful Release* (Manchester: Manchester University Press, 2002), pp. 122–4.

73. NMT 3/1767–1769 and UNWCC archives, Lang to Interallied Commission for the Investigation of War Crimes 10 May 1945, annex B on the murder of ill and aged people in Germany, dated December 1941.

74. For Blome see NARA RG 153 Entry 144/100–665/box 59.

75. AdeF BB/35/260 Pink file 3 Procès I Biographes, affidavits, depositions, examens mentaux des accusés. Liste des médecins ayant pris part aux expériences médicales et a l'euthanasie.

76. NARA RG 319 Entry 11d Box 347 IRR Personal, Rüdin file.

77. NARA RG 319 IRR Personal Box 347 CI Arrest Report, 12 January 1946. Notice of transfer, 23 May 1946. E. Zerbin-Rüdin to author 11 July 2004.

78. NARA RG 319 IRR Personal Box Verschuer file.

79. AdeF BB/35/ 261 Folder 1a Rapports médicaux américains, OSS 'Principal Nazi Organisations Involved in the Commission of War Crimes, Nazi Racial and Health Policy'. Draft for the Use of the War Crimes Staff, 15 August 1945, p. 54.

80. P.J. Weindling, 'The Medical Publisher J.F. Lehmann and Racial Hygiene', in Sigrid Stöckel (ed.), *Die 'rechte Nation' und ihr Verleger. Politik und Popularisierung im J.F. Lehmanns Verlag 1890–1979* (Berlin: Lehmanns Media, 2002), pp. 159–70.

81. OSS, 'Principal Nazi Organisations Involved in the Commission of War Crimes, Nazi Racial and Health Policy. Draft for the Use of the War Crimes Staff', 15 August 1945, 1–54 , pp. 53–4.

82. AdeF BB/35/261 Folder 1a Rapports médicaux américains, OSS 'Principal Nazi Organisations'.

83. NARA RG 338 USAREUR/JAG War Crimes, Records Regarding Medical Experiments, Thomas R. Henry, 'Henry Visits a Camp Where Nazi Doctors Killed Humans in "Scientific Tests"', dated 23 April 1945.

84. Hoover Institution Pash Papers 2/7–9, Barnes Diary, p. 23.

85. USHMM CIOS Evaluation Report 64, 7 June 1945 by G.P. Schnabel and John E. Haavik.

86. NARA RG 153/86–3–1/ Box 11 folder 4 book 3 German Research on Malaria Inoculations, Source: British Foreign Office, 30 July 1945, logged 14 August 1945 to Division of Naval Intelligence.

87. NARA RG 238 entry 200 box 27 Schilling, letters to family August/September 1945.
88. Michael and Joachim Kaasch, 'Die Auseinandersetzung des XX. Leopoldina-Präsidenten und Schweizerbürgers Emil Abderhalden um Eigentum und Entschädigung mit der sowjetischen und der amerikanischen Besatzungsmacht (1945–1949)', *Acta Historica Leopoldina*, vol. 36 (2000) 329–84. Norman Naimark, *The Russians in Germany. A History of the Soviet Zone of Occupation* (Cambridge, Mass.: Harvard University Press, 1995), pp. 203–49.

4   Exploitation

1. TNA: PRO CAB 124/544 19 June 1945 Technical and Scientific Research in Germany After the War.
2. TNA: PRO CAB 124/544 Technical and Scientific Research in Germany after the War. Otto Oexle, *The British Roots of the Max-Planck Gesellschaft* (London: GHI 1995), p. 16.
3. Cf. House of Lords Official Report. Third Volume of Session 1944–45 (London: HMSO, 1945), pp. 246–59, 284–5, for 29–30 May 1945, on proposal of Lord Vansittart for a 'permanent Inter-Allied Committee of Scientists' 'to examine and control, and if necessary prohibit the use by Germany of, any scientific discovery or invention considered dangerous for the safety of mankind'. Aldous Huxley, *Science, Liberty and Peace* (New York and London: Harper and Brothers, 1946), p. 62. Translated as *Wissenschaft, Freiheit und Frieden* (Zurich: Steinberg Verlag, 1947).
4. Ivy Papers, Box 6 folder 12, A.C. Ivy, 'Nazi War Crimes of a Medical Nature'.
5. David MacIsaac (ed.), *The United States Strategic Bombing Survey* (New York: Garland, 1976), vols. 1–5.
6. TNA: PRO FO 945/904 Eisenhower to Combined Chiefs of Staff 15 May 1945.
7. Tom Bower, *The Pledge Betrayed. America and Britain and the Denazification of Post-War Germany* (New York: Doubleday, 1982), p. 202.
8. Ibid., pp. 76–9.
9. TNA: PRO FO 935/20 CIOS General, Medical Section of Combined Intelligence Priorities Committee, 13 July 1944, 2 August 1944.
10. TNA: PRO FO 935/20 14 Aug 1944, Combined Intelligence Priorities Committee. Field Teams for Paris: Item X Paris. 10/39 Institut Pasteur (Lwoff); 39a, 39b Fourneau who was on committee for Franco-German collaboration. Other German and French scientists believed to be working on BW, 39c Garches, Item XXIV Medical 24/18a Pasteur Institut Giroud – typhus service. Levaditi – Penicillin 24/18 b Garches – sera and anti-toxins, Union intresyndicale – pharmaceutical products.
11. Bower, *Pledge Betrayed*, pp. 7–9.
12. Sereny, *Speer*, pp. 359–61.
13. NARA RG 260 OMGUS folder 39 Lists of Homework by Dustbin detainees.
14. NARA RG 260 OMGUS Box 18 Chief Fiat daily journal, 15 May 1946; John Gimbel, *Science, Technology and Reparations. Exploitation and Plunder in Postwar Germany* (Stanford: Stanford University Press, 1990), p. 17.
15. Sereny, *Speer*, pp. 558–60.
16. John Farquharson, 'Governed or Exploited? The British Acquisition of German Technology, 1945–48', *Journal of Contemporary History*, vol. 32 (1997), 23–42, 30–1.

17. Mark Walker, *Nazi Science. Myth, Truth, and the German Atomic Bomb* (New York: Plenum, 1995), pp. 207–41.

18. Rob Evans, *Gassed. British Chemical Warfare Experiments on Humans at Porton Down* (London: House of Stratus, 2000), 115. TNA: PRO WO 195/9678. Tilley, Preliminary Report on Hildebrandt, Osterwald and Beuchelt 13 April 1946, p. 6 Sydney Smith Papers, RCPE.

19. NARA M 1087 reel 31 Medical and Other Scientific Experiments, 21 April 1945, extracts from conversations between Ernst Rodenwaldt, Hubert Luetkenhaus and Ludwig Wesch.

20. TNA: PRO FO 1031/76 FIAT Reviews on discovery of Ranke's reports. NARA RG 260 OMGUS Ranke, 'Dustbin', 11 December 1945, Full explanation and description of my work in the Military Physiological Institute.

21. AFP, letter to Phyllis Alexander, 26 May 1945.

22. Carmen Mory Lebenslauf, in Lukas Hartmann, *Die Frau im Pelz* (Zürich: Nagel und Kimche, 1999).

23. NARA RG 238 Internee Personnel Records Box 22 Entry 200 NM 70, Oberheuser.

24. Caterina Abbati, *Ich, Carmen Mory. Das Leben einer Berner Arzttochter und Gestapo-Agentin (1906–1947)* (Zurich: Chronos Verlag, 1999), pp. 146–65. TNA: PRO WO 309/469 A. Martin and Carmen Mory: Report on Carl Gebhardt, 18 August 1945. = NMT 8/337. PRO WO 235/ 306 f. 261. NARA RG 238 Entry 200 Box 8, Fritz Fischer arrest report, stamped 23 August 1945, Mory as witness. I am grateful to Hartmut Pogge von Strandmann for information about the arrest of the historian Fritz Fischer as a case of mistaken identity.

25. TNA: PRO WO 235/306, f.228–70, 8 January 1947 for evidence and cross-examination of Mory. Mant personal communication, and Mant, *From Nuremberg*, chapter 2 'War Crimes Investigations'. Abbati, *Carmen Mory*.

26. Abbati, *Carmen Mory*.

27. Mark Walker, *Nazi Science. Myth, Truth, and the German Atomic Bomb* (New York: Plenum, 1995), p. 207.

28. NARA RG 112/ 295A/Box 10 Current Intelligence, US Alsos Report on German Biological Warfare (by John Barnes, Cromartie, Carlo Henze and Hofer), introduction, pp. 12–16. Blome interrogation, 30 July 1945, pp. 3–8.

29. NARA RG 200 130–76–6 Alsos. Goudsmit papers box 6 W.F. Colby to J.F. Edell, 11 August 1945; Alsos, box 4 dcl 2, Alsos files, nd. Geissler, 755–7.

30. Leo Alexander, *The Treatment of Shock from Prolonged Exposure to Cold, Especially in Water*, Combined Intelligence Objectives Subcommittee, Target No. 24 Medical CIOS G–2 Division, Shaef rear, July 1945; reprinted in *Trial of the Major War Criminals before the International Military Tribunal in Nuernberg*, Doc. 400 PS, 25: 536–607, 1947.

31. Ibid., 26–9. Rascher to Himmler, 17 February 1943, cited in *Trial of German Major War Criminals*, Vol. 30, pp. 221–2.

32. TNA: PRO FO 1031/92 Major E. Tilley, EPES, Interrogation Reports. Preliminary Report on Prof. Dr Fritz Hildenbrandt, Dr Karlhans Osterwald and Dr Hans Beuchelt, 5 April 1946.

33. BAB NS 21/913.

34. Wolfgang Weyers, *Abuse of Man*, p. 278.

35. BAB NS 21/ 914 DFG to Sievers, 9 October 1942.

36. BAB NS 21 / 914, 13 December 1942.

37. BAB NS 21/ 914, 2 June 1942, films processed by Agfa.

38. BAB NS 21/ 914 Rascher to Siever, 11 September 1942.

39. BAB NS 21/913 Rascher to Sievers, 27 August 1942. Ibid., Nr 914 Rascher to Himmler, 3 October 1942.
40. BAB NS 21/914 SD report on Holzlöhner. File on Copenhagen.
41. BAB NS 19/1589 Rascher to Himmler, 9 August 1942; Bevollmächtigte des Reiches in Dänemark to R. Brandt, 1 June 1943.
42. BAB NS 21/914 Romberg to Sievers, 2 October 1942.
43. BAB NS 21/915 Sievers to R. Brandt, 28 January 1943; Nr 916 Rascher to Sievers, 17 May 1943.
44. Niels Bohr Library, Goudsmit Papers, Box 27 folder 26, Sievers, Wolfram, Sievers to Nini Rascher, 24 March 1942. Cf. Wolfgang Benz, 'Dr med. Sigmund Rascher: eine Karriere', *Dachauer Hefte*, no. 4 (1988).
45. BAB NS 21/917 Rascher report, 15 July 1943.
46. E. Holzlöhner, 'Verhütung und Behandlung der Auskühlung im Wasser', *Bericht über eine wissenschaftliche Besprechung am 26. und 27. Oktober 1942 in Nürnberg über Ärztliche Fragen bei Seenot und Winternot* (nd, np), pp. 42–5.
47. BAB NS 21/914 Rascher to Himmler, 3 October 1942.
48. BAB NS 21/917.
49. Roth, 'Tödliche Höhen', 132–5.
50. Ibid., 134–5.
51. Kastan, 'Hirt'. NARA M 1019/68 Sievers interrogation, 20 August 1946, 14.
52. War Crimes Commission (Research Office), 'Development of Various Scientific Projects (Medical, Entomological, Biological, etc.) in German Concentration Camps', Summary of Information No. 39, August 1945.
53. AdeF BB/35/261 File Ahnenerbe 153504 9 interrogation of Wolfram Sievers ALSOS mission Heidelberg, 1 August 1945, also with E. May. ALSOS Report: Scientific Material in the Documents of Dr W. Osenberg, Head of the Planning bureau of the Reichsforschungsrat – by Wannier and Helmkamp, in detention camp DUSTBIN of FIAT. Sievers by ALSOS mission Heidelberg, 1 August 1945. UNWCC Summary of Information No. 39, August 1945, Development of Various Scientific Projects (Medical, Entomological, Biological etc.) in German Concentration Camps and Universities (NARA RG 153 85–2 folder 2). NMT 3/ Sievers Diary.
54. Hoover Institution Pash papers, box 2 folder 3 Goudsmit to Vannevar Bush, 23.
55. Pash Papers 2–9, Goudsmit, 'Alsos' Mission. History', 7 December 1945, 31, 33. Augustine represented the French also at Nuremberg. British ALSOS experts for BW were Adam and Barnes. Papers of the British officer John M. Barnes are in Pash box 2 folder 9 Eckman to Bottomley, 11 June 1945. Barnes, 'Report', 6–17 February 1945. box 2 folder 3 Bottomley to Goudsmit, 8 June 1945.
56. NARA RG 200 ALSOS files Box 8 Goudsmit to James B. Conant, 6 February 1952. Samuel A. Goudsmit, *ALSOS. The Failure in German Science* (London: Sigma Books, 1947).
57. Hoover Institution Pash Papers 2–4 Progress Report 25 June 1945 indicates interrogation on 21 April 1945. AdeF BB/35/265 10 April 1945, Haagen file of ALSOS Mission, 31 July 1944 report on penicillin. NARA RG 238 Entry 200 Box 11 Haagen personal datasheet, Haagen gives 26 April 1945 as the date of capture. Frederick H. Kastan, 'Hirt'. Hoover Institution Pash Papers box 2 folder 2 Strasbourg Operation, 7 December 1944, 2/7–9, 27.
58. RAC RU 4501 folder 21 Rivers to Simon Flexner, 18 December 1928. APD 4/35 Alexander to McHaney, 'Adviser to the Court', 29 April 1947.
59. NMT 2/6169.

60. RAC RF RG 5/4/13/141 Haagen to W.A. Sawyer, 3 November 1930, 5 April 1932, 3 August 1933. Ralph Muckenfuss to Sawyer, 8 May 1933.
61. NARA RG 200 Alsos, box 5, Goudsmit to Roger Adams, 26 February 1946. Linda Hunt, *Secret Agenda. The United States Government, Nazi Scientists, and Project Paperclip, 1945 to 1990* (New York: St. Martin's Press, 1991), p. 11.
62. RAC RU 450.1 Scientific staff biographies folder 21 Eugen Haagen. Haagen to Sawyer, 3 August 1933. RU RG6 IHB/D Ser 4 Virus Lab, 439 Scientific Reports, Box 3.
63. RAC RF 5/4/13/141 Haagen to Russell, 12 May 1934, Haagen to Sawyer, 13 March 1933. Sawyer to Haagen, 26 March 1934, 6 August 1935. Sawyer to Gregg, 5 April 1934.
64. *Denazification of the Health Services*, pp. 51–3. Samuel A. Goudsmit, *ALSOS. The Failure in German Science* (London: Sigma Books, 1947), pp. 73–5. BAB R/28/III Nr 220 'RFR Medicine 1944–45'. Partly Filed by ALSOS. Stamped: Medical Library 23 October 1946. Medizinische Forschungsauftrage; ibid., Nr 206a.
65. Kastan, 'Hirt', 173–208, 182. NARA RG 200 Entry 82A (Alsos) Box 6 file Adams, Goudsmit to Adams, 25 October 1945. NARA M 1019/68 Sievers interrogation, 20 August 1946, 14. Irmtrud Wojak, 'Das "irrende Gewissen" der NS-Verbrecher und die deutsche Rechtsprechung. Die "jüdische Skelettsammlung" am Anatomischen Institut der "Reichsuniversität Strassburg"', Fritz Bauer Institut (ed.), *'Beseitigung des jüdischen Einflusses …'. Antisemitischen Forschung, Eliten und Karrieren im Nationalsozialismus* (Frankfurt: Campus, 1999), pp. 101–30. American Institute of Physics, Niels Bohr Library, Goudsmit Papers Box 27, folder 26: Wolfram Sievers.
66. NARA M 1019/68 Sievers interrogation, 28 August 1946, pp. 3–4. 'Deutsche Wissenschaftler vor dem Militärgericht in Lyon', *Ärztliche Mitteilungen*, no. 11 (1 June 1954) 362–4.
67. NARA RG 319 entry 82a Alsos box 4 Rose to Haagen, 2 December 1943. Haagen to Rose, 8 December 1943.
68. Ibid. Office of the United States Chief of Culture to Col. R.L. Hopkins, 11 January 1946.
69. AdeF BB/35/276 Haagen Investigation by Dr Grombacher, 28 April 1947, Strasbourg. McHaney's copy: annotated 'See Page 3! OCCWC transl of Doc NO–3846'.
70. Colmar AOF, AJ C 3623 p. 51 d. 23721 Dossier: Dr Haagen, Eugen 1946–1954. AIP Goudsmit papers August 12, 1949 Goudsmit to Henderson, 12 August 1949, Henderson to Goudsmit, 5 September 1949 regarding the location of Washington papers.
71. NARA M 1019/68 Sievers interrogation, 28 August 1946, pp. 1–4.
72. Ibid., p. 5.
73. Hoover Institution Pash Papers box 2/7 Barnes diary, 9, 11, 12 April 1945.
74. NARA RG 319 IRR Personal box. For the biography of Bickenbach, see NMT Guide, p. 76. AIP Goudsmit papers, John Barnes to Goudsmit, 11 October 1949. Goudsmit to W.J. Cromartie, 9 September 1949. M.C. Henderson to Goudsmit, 5 September 1949. Krayer to Goudsmit, 9 August 1949. Goudsmit to Otto Krayer, 12 August 1949.
75. Carl Henze and W.J. Cromartie, Medical Targets in the Strasbourg Area, CIOS Item 24 Medical, in NARA RG 112 Entry 295A Box 10 Strasbourg. *Le Camp de Concentration de Struthof. Konzentrationslager Natzweiler. Temoignages* (Essor, 1998), pp. 238–43.

76. NARA RG 243 190–62–14–02 USSBS entry 2 (section 5) box 1, Report, 4 September 1945, interrogation of Brandt by FIAT, Scientific and Technical Branch, Major Gill and Kingscote.
77. Personal communication, Keith Mant.
78. AdeF BB/35/276 Documents de l'instruction. Typhus Letter Haagen, Palace of Justice, 1 June 1947 to 'Dear Professor' [to Kempner].
79. NARA RG 238 NM70 Entry 200 OCCWC. Executive Office Nuremberg Administrative Post Internee Personnel records ('201 Files') box 11 Eugen Haagen.
80. P.J. Weindling, ' "Out of the Ghetto": The Rockefeller Foundation Confronts German Medical Sciences after the Second World War', William H. Schneider (ed.), *The Rockefeller Foundation and Biomedicine: International Achievements and Frustrations from World War I to the Cold War* (Bloomington: Indiana University Press, 2002), pp. 208–22.
81. NARA RG 238 NM70 Entry 200 OCCWC. Executive Office Nuremberg Administrative Post Internee Personnel Records ('201 Files'), Eugen Haagen personnel data sheet; Geissler, *Biologische Waffen*, p. 763; M. Hubenstorf in: *Exodus von Wissenschaften aus Berlin* (Berlin: de Gruyter, 1994), pp. 448–53.
82. AOF, Justice C 3623 p. 51 d. 23721 Dossier: Dr Haagen, Eugen 1946–1954. BAOR Zonal Office of the Legal Adviser, 17 August 1948.
83. RAC RU 450.1 Elizabeth Voigt-Haagen to Thomas Rivers and Simon Flexner, 22 February 1947.
84. E.g. NARA RG 153/86–3–1 Box 10 Fedde to W. Morse, 21 April 1949.
85. APD Interrogation of Haagen, 9 July 1947.
86. AdeF BB/35/276 Documents de l'instruction. Typhus, 21 November 1946, to McHaney from A. Hochwald. Subject: Trip to Strasbourg.
87. AdeF BB/35/265 Haagen. Documents de l'instruction, 28 August, Handloser interrogation mentions Haagen. BB/35/277 pp. 1–2 of Haagen letter to Bayle, 1 June 1947, Justiz-Palast, Nuernberg. Haagen file of ALSOS Mission, dated 10 April 1945. Mant interview on capture episode. TNA: PRO WO 309/1455 OCCWC requests for Haagen 5 December 1946, 16 and 25 January 1947. RAC RF RG 12.1 Warren diary, 2 May 1947. C 3623 p. 51 d. 23721 Dossier: Dr Haagen, Eugen 1946–54, see *Le Monde* reports by Jean-Marc Theolleyre on 'Le Procès des médecins allemands', from 16 to 20 December 1952. AIP Goudsmit papers, 12 August 1949 Goudsmit to Henderson, 12 August 1949, Henderson to Goudsmit, 5 September 1949.
88. Geissler, *Biologische Waffen*, p. 545 for an ALSOS photo of the planned installation.
89. NARA RG 153 box 59 100–621 Blome interrogation, 2 July 1945.
90. NARA RG 112/295A/Box 10 Current Intelligence; Records of the Surgeon General (Army) Biological Warfare. Specialised files, box 4 folder BW Intelligence. Geissler, *Biologische Waffen*, pp. 520–43.
91. Cf. Geissler, *Biologische Waffen*.
92. TNA: PRO FO 1031/83 Memo to Major Wilson, 8 June 1945.
93. Hoover Institution Pash Papers 2–9, History, p. 32. W.F. Colby and S.A. Goudsmit, *Reichsforschungsrat* (18 March 1945).
94. Hoover Institution Pash Papers 2–4 Discovery and Securing of the Planning Office of the *Reichsforschungsrat*, 1 June 1945.
95. TNA: PRO FO 1031/92 Maj. E. Tilley, EPES, Interrogation Reports, Preliminary Report of G–2 FIAT, EPES, SHAEF on the Osenberg Group, 14 June 1945; Second Preliminary Report of G–2 FIAT, EPES, SHAEF on the Osenberg Group.

96. Hoover Institution Pash Papers 2–9, History, 34.
97. NARA RG 200 Goudsmit folders box 1, E.W.B. Gill, German Academic Scientists and the War, 20 August 1945, pp. 7–8. BAB R 26/III/107 *Forschungsauftraege auf medizinischen Gebiet (Recherche Nr 31 im alliierten Auftrag von Professor Dr Osenberg).*
98. NARA RG 200 Entry 82A ALSOS Box 4 DCL 1.
99. MPG III. Abt. Rep. 25 Nachlass R. Kuhn Nr 57 Entnazifizierung. Amerikanischer Fragebogen 1946, note of 16 January 1946.
100. Hoover Institution Pash Papers 2/7–9 Diary, 3 April 1945. Florian Schmaltz, 'Kampfstoff-Forschung im Nationalsozialismus. Zur Kooperation von Kaiser-Wilhelm-Instituten, Militär und Industrie', PhD Bremen, 2004.
101. NARA RG 200 Entry 82A ALSOS Goudsmit folders Listing Box 1 E.W.B. Gill, German Academic Scientists and the War. Appendix C Evidence of Professor R. Kuhn, 29 June 1945. CIOS Evaluation Report, Interrogation of Professor Osenberg at Chateau de Grand Chesnay, 27 July 1945.
102. Hoover Institution Pash Papers 2/7–9, 29 May 1945, 11, 24, 27 June for interrogation. 30 June for Frau Blome.
103. Wolfram Sievers, FIAT Evaluation Report 98, 13 June 1945.
104. TNA: PRO FO 1031/72 FIAT policy. FO 1031/69 DUSTBIN Policy. FO 1031/83 re interrogation of Blome, 3 July 1945. Pash 2/7–9, 29 May 1945.
105. NARA RG B.C. Andrus Interrogation report on Karl Brandt, 29 May 1945.
106. ACDJC cxxix–20, 17–18 June 1945, Interrogation of K. Brandt by R.L Meiling = NO–332 OCCWC.
107. NARA RG 338 USAR/JAG War Crimes Records Regarding Medical Experiments No.125582–4, Interrogation of Karl Brandt, K 21 August 1945. BIOS Final Report No. 542, Interrogation of Certain German Personalities Connected with Chemical Warfare, pp. 24–5 (= NMT 8/351–2).
108. BIOS Final Report No. 542, Interrogation of Certain German Personalities Connected with Chemical Warfare, pp. 24–5 (= NMT 8/351–2). BIOS interrogated Brandt in Dachau in March 1946.
109. ACDJC Doc CDLVIII–83a Le Rapport du Major D.C. Evans et de Mr I. Haupt, CIOS Advanced Field Team (VII), Rottweil, 23 April 1945. Weindling, *Epidemics and Genocide*, p. 406. Jürgen Schäfer, *Kurt Gerstein – Zeuge des Holocaust. Ein Leben zwischen Bibelkreisen und SS* (Bielefeld: Luther-Verlag, 1999).
110. TNA: PRO FO 1031/86 T.E. Job to P.M. Wilson, subject: CW Investigations (BIOS Trip 1703). On Wirth, see also FO 1031/104. Tilley, Preliminary Report on Hildebrandt, Osterwald and Beuchelt, 5 April 1946, Smith Papers.
111. USATC files RG 112/295A/Box 10 File Strasbourg Gaylord Anderson 'German BW-Strasbourg 1 January 1945; Howard Cole to George Merck, 20 August 1945; Alsos Mission HQ 30 July 1945 re Interrogation of Blome. Carl Henze and William Cromartie, Medical Targets in the Strasbourg Area, CIOS Item 24 Medical.
112. Woelk, 'Wirth', 286–7. Tilley, 'Preliminary Report'.
113. *Der Wert des Menschen* (Berlin: Hentrich, 1989), p. 305.
114. United States Strategic Bombing Survey Report No. 65, The Effects of Bombing on Health and Medical Care in Germany.
115. NARA RG 243 US Strategic Bombing Survey Entry 2 (section 5) Box 1, Interview Siegfried Handloser No. 75 July 1945, pp. 8–9. NARA RG 243 US Strategic Bombing Survey Entry 2 (section 5) box 1 Major General Schroeder.
116. Rob Evans, *Gassed. British Chemical Warfare Experiments on Humans at Porton Down* (London: House of Stratus, 2000), pp.120–2.

117. TNA: PRO WO 195/9678 D.C. Evans, 'German CW Experiments on Human Beings', report dated 12.I.48 for Ministry of Supply, Advisory Council on Scientific Research and Technical Development, Chemical Defence Advisory Board, p. 4, copy in Wiener Library Kh (1). Report cited by Observer Films (1995). *The Secrets of Porton Down.* Granada Presentation for ITV. YVA, Gerstein file 'Le Rapport du Major D.C. Evans et de Mr I. Id. Haupt. CIOS Consolidated Advance Field Team (VII)' (Original in Archives du Centre de Documentation Juive Contemporaine, Doc CDLVIII–83a).

118. Weindling, *Epidemics and Genocide* for the origins of the gas chambers using carbon monoxide and Zyklon gas.

119. TNA: PRO FO 1031/83 E.W.B. Gill, Interrogation of Professor Heinrich Kliewe, 27 June 1945.

120. NARA RG 238 '201 Series' Box 3 Blome, Auftrage vom 8. August 1945; STAN Rep 502A KV Verteidigung Handakten Sauter 3, Blome Oberursel, January 1946.

121. L. de Benedetti and Primo Levi, 'Rapporto Sulla Organizzazione Igenienico-sanitaria del Campo di Concentramento di Auschwitz', *Minerva Medica,* vol. 2 (1946) 535–44.

122. Lucie Adelsberger, 'Medical Observations in Auschwitz Concentration Camp', *The Lancet* (9 March 1945) 317–19. Eduard Seidler (ed.), *Lucie Adelsberger. Auschwitz. Ein Tatsachenbericht* (Bonn: Bouvier, 2002).

123. Alfred Wolff-Eisner, *Über Mangelerkrankungen auf Grund von Beobachtungen im Konzentrationslager Theresienstadt* (Würzburg, 1947). See also H.G. Adler, *Theresienstadt 1941–1945. Das Antlitz einer Zwangsgemeinschaft* (Tübingen, 1955).

124. M. Dvorjetski, *Le Ghetto de Vilna. Rapport Sanitaire* (Geneva, 1946).Weindling, *Epidemics and Genocide,* pp. 396–7.

125. 'Nazi Research', *Time,* vol. 46 (15 January 1945) 68.

126. APD 1/2 Dr Ali Kuci for International Prisoners Committee, 'The Last Days of Dachau' [16–17 May 1945]; 4/20 Interrogation by Major Leo Alexander, 21 June 1945. Albert Deutsch, 'Some Wartime Influences on Health and Welfare Institutions in the United States', *Journal of the History of Medicine,* vol. 1 (1946) 318–29, 318.

127. TNA: PRO WO 39/470 Die Häftlingsärzte von Auschwitz an die internationale Öffentlichkeit, 4 March 1945.

128. BAK Pross ZSg 154 Bd 73 Jan Ochocki interrogation evidence, 3 January 1947 for Klein's role in killing the subjects of gynaecological experiments.

129. TNA: PRO FO 937/110 Documents Series No. 22, February 1946. Cf. AdeF BB/35/263 Camps de concentration. Auschwitz-Birkenau, Rapport de médecins ... Dr Samual Steinberg sur les expériences de vivants faites à Auschwitz ou il a été détenu 1942 à 1945.

130. APD 1/2 International Prisoners Committee. 4/20 Interrogation by Major Leo Alexander, 21 June 1945.

131. Alexander, 'The Treatment of Shock', pp. 160–3. Reprinted in *Trial of the Major War Criminals Before the International Military Tribunal in Nuernberg,* Doc. 400 PS, 25: 536–607, 1947.

132. Harold Marcuse, *Legacies of Dachau* (Cambridge: Cambridge University Press, 2001), p. 134.

133. TNA: PRO WO 309/374 No. 2 War Crimes Investigating Team to Deputy JAG, 29 June 1945.

134. AdeF BB/35/260 4a Procès I: instruction (correspondence) 133890 Georg Doering. Schleswig, 23 January 1947 to US Military Government.

135. Alexander, 'The Treatment of Shock' Appendix 5, pp. 160–3.
136. TNA: PRO FO 937/110 Documents Series No. 22 February 1946. Cf. CINFO report on 'SS Medical Research' NARA RG 155 Box 10 Book 1, 'SS Medical Research' 10.II.46 republished by UNWCC, Documents Series No. 44, June 1946.
137. Eugen Kogon, *'Dieses merkwürdige, wichtige Leben'*. *Begegnungen* (Weinheim: Quadriga, 1997), pp. 74–8.
138. Mant, 'From Nuremberg to the Old Bailey', ch. 2.
139. Brooks E. Kleber and Dale Birdsell, *Chemical Warfare Service: Chemicals in Combat* (Washington DC: Office of the Chief of Military History, 1966), pp. 138–40.
140. Author's interview with Keith Mant, 1997; P.J. Weindling, 'Auf der Spur von Medizinverbrechen: Keith Mant (1919–2000) und sein Debut als forensischer Pathologe', *1999. Zeitschrift f. Sozialgeschichte des 20. und 21. Jahrhunderts*, vol. 16 (2001) 129–39.
141. TNA: PRO AIR 55/169 Historical Record of Disarmament 84 Group May 1945, 6.
142. APD Leo to Phyllis Alexander 4 May 1945. Cf. the film *Nazi Concentration Camps* at the IMT on 29 November 1945, *Trial of Major War Criminals*, vol. 1, pp. 265–6. Tusa, *Nuremberg*, p. 160. Lawrence Douglas, 'Film as Witness: Screening *Nazi Concentration Camps* before the Nuremberg Tribunal', *Yale Law Journal*, vol. 105 (1995) 449–81.
143. NYPL Lifton Papers, Box M5 Letter of LA to Lifton 10 October 1978.
144. Mant, 'From Nuremberg', ch. 1. For background, see Joanne Reilly, *Belsen. The Liberation of a Concentration Camp* (London, 1998).
145. John Thompson Papers (deposited with author), letter dated 16 July 1945.
146. TNA: PRO FO 371/57641 War Criminal Auschwitz, Dr Klein. WO 309/484 note of 15 November 1945 concerning Leo Gries' evidence against Klein and Rutenol.

5 Aviation Atrocities

1. Helmuth Trischler, *Luft- und Raumfahrtsforschung in Deutschland 1900–1970. Politische Geschichte einer Wissenschaft* (Frankfurt/M: Campus, 1992), pp. 178–9.
2. APD 4/27, CIOS Target Report, University of Goettingen Physiological Institute, Assessor Oscar O. Selke, 16 April 1945.
3. USHMM Aeromedical Exploitation, H.B. Wright, memorandum, 24 April 1945.
4. Trischler, *Luft- und Raumfahrtsforschung*, 255, 267, 271.
5. Karl-Thorsten Bretschneider, 'Friedrich Hermann Rein. Wissenschaftler in Deutschland und Physiologe in Göttingen in den Jahren 1932–1952', Göttingen MD dissertation 1997, 111.
6. TNA: PRO CAB 124/544 German Science and Industry Committee. FO 1031/76 FIAT Reviews. Lt Cdr Ladbrooke to Blaisdell, 4 December 1946 re 'Probable Imprisonment and/or Trial of Prof. Rein'. WO 309/473 Allegations of illegal medical experiments, 7 November 1946.
7. Julian Bach, *Saturday Review* (August 1958). Tom Bower, *The Paperclip Conspiracy. The Battle for the Spoils and Secrets of Nazi Germany* (London: Michael Joseph, 1987). Linda Hunt, *Secret Agenda. The United States Government, Nazi Scientists, and Project Paperclip, 1945 to 1990* (New York: St. Martin's Press, 1991). R. and T. Giefer, *Die Rattenlinie. Fluchtwege der Nazis* (Frankfurt/M: Hain, 1991). Klee, *Auschwitz*, p. 253.

8. Friedrich Rein, 'Wissenschaft und Unmenschlichkeit', *Göttinger Universitätszeitung* vol. 2, no. 14 (1947), pp. 3–5; Franz Büchner, *Pläne und Fügungen. Lebenserinnerungen eines deutschen Hochschullehrers* (Munich, 1965).

9. NARA M 1019/5 Interrogation Summary n. 185 of Becker-Freyseng, 24 September 1946, 3. Interrogation, 24 September 1946, 12–16, 22–3.

10. NARA M 1019/ 22, November 1946 Schröder interrogation with Marx by von Halle, 2–6.

11. Gerhard Koch, *Humangenetik und Neuro-Psychiatrie in meiner Zeit* (Erlangen: Palm & Enke, 1993), pp. 123–8. Letter of Nachtsheim to G. Koch, 2 September 1943. H. Nachtsheim and G. Ruhenstroh-Bauer, 'Die Bedeutung des Sauerstoffmangels für die Auslösung von epileptischen Anfalls', *Klinische Wochenschrift*, vol. 23 (22 January 1944) 18–21. P. Weindling, 'Genetik und Menschenversuche in Deutschland, 1940–1950. Hans Nachtsheim, die Kaninchen von Dahlem und die Kinder von Bullenhuser Damm', in Hans-Walter Schmuhl (ed.), *Rassenforschung an Kaiser-Wilhelm-Instituten vor und nach 1933* (Göttingen: Wallstein, 2003), pp. 245–74, 247, 251.

12. NARA RG 330 Entry 18 JIOA Foreign Scientists Case File Box 164 Strughold Fragebogen, 20 March 1947; Strughold affidavit, 3 September 1947 excusing NSFK membership, 2 October 1947 denazification certificate. Hugo Spatz testimonial, 19 September 1947. MPG Abt. III Nr 84/1 Butenandt papers, Nr 1344 Butenandt to Strughold, 15 July 1946.

13. Leopoldina Archiv, Akte Strughold, Viktor Harsch, 'Leben, Werk und Zeit des Physiologischen Hubertus Strughold (1898–1986). Vortrag, gehalten am 22. Mai 1998 im Rahmen der 36. Jahrestagung der DGLRM in Berlin'.

14. MPG III Abt. Rep 25 Nachlass Richard Kuhn, Nr 55 Korrespondenz mit der KW/ MPG und Institutsangelegenheiten 1945–1949.

15. TNA: PRO FD 1/5826 War Time Scientific Investigations in Germany and German Occupied Countries, Harrington to Mellanby, 21 November 1945.

16. Michael Shevell, 'Neurology's Witness to History: The Combined Intelligence Operative Sub-Committee Reports of Leo Alexander', *Neurology*, 47 (1996) 1096–103. Michael Shevell, 'Neurology's Witness to History (Part 2): Leo Alexander's Contributions to the Nuremberg Code (1946–47)', *Neurology*, 50 (1998) 274–8.

17. APD 1/21 Alexander, 'The Socio-Psychological Structure of the SS', p. 2.

18. APD 4/42, 4/ 3 (26) study of flight personnel, 1 p.

19. Leo Alexander, 'The Treatment of Shock from Prolonged Exposure to Cold, Especially in Water', CIOS Target Number 24 Medical, July 1945, 4, 11, 67.

20. AFP Leo to Phyllis Alexander, 19 June 1944, 10 July 1944.

21. APD item 42/4 box 4 file 3 (26), book plan with Albert Ax on combat stress, 8 June 1945. AFP Leo Alexander letters to Phyllis Alexander 19 June, 10 July, 11 October, 2 November 1944.

22. AFP W.H. Everts to Leo Alexander, 18 April 1944.

23. AFP Alexander to Phyllis, 26 May 1945. Shevell, 'Neurology's Witness to History'.

24. For military neuro-surgery, see Gustav J. Fraenkel, *Hugh Cairns: First Nuffield Professor of Surgery* (Oxford: Oxford University Press, 1991), pp. 137–50.

25. AFP Helene Hulst-Alexander to Leo Alexander, 20 August 1945. APD 4/33 Alexander Logbook, 5–12 May 1947 for Vienna visit. Alexander's Diary runs from 24 June to 29 August 1945 and 'Logbook. Journey to Nuremberg as Consultant to the Secretary of War' runs from 11 November 1946 to 24 June 1947. The photocopies made by Alexander for Robert Lifton in 1978 are in the

Lifton Papers, Box 22, New York Public Library, and in the Pross Papers, Bundesarchiv Koblenz Zsg 154. The Pross copy is supplemented by an index compiled by Professor Jürgen Peiffer, Tübingen. I am grateful to Professor Lifton for facilitating access to the Diaries in 1996.

26. AFP, Leo to Phyllis Alexander, 26 May 1945.
27. MPG Abt II. Rep 20B Nr 105–1 Steinbauer to Tönnis, 2 January 1948; Tönnis to Steinbauer, 23 February 1948.
28. Alexander Mitscherlich and Fred Mielke (with contributions by Ivy, Taylor, Alexander and Deutsch), *Doctors of Infamy: the Story of the Nazi Medical Crimes* (New York: Henry Schuman, 1949), p. xxxi.
29. MPG Abt. II Rep. 20B Nr 105–1 Schriften, W. Tönnis, Die Versorgung der Hirn-, Rückenmark- und Nervenverletzten im zweiten Weltkrieg.
30. BAK ZSg 154/75 Alexander, diary, 18 May 1945, p. 18. Brandt in *Bericht über die 4. Arbeitstagung der Beratenden Ärzte vom 16. bis 18. Mai 1944 im SS-Lazarett Hohenlychen* (1944), p. 15.
31. MPG II. Abt 20B Nr 105 Tönnis to Spatz 6 April 1944; Bericht über die Organisation zur Versorgung von Hirn, Rueckenmark und Nervenverletzten, Bad Ischl 4.8.45.
32. Yale University Fulton papers F383 Hugh Cairns to John Fulton, 13 September 1944.
33. Denis Williams, Report on CIOS Trip No. 277, 25 May–4 June 1945. Neuropsychiatric Organisations in the German Airforce, CIOS Item 24 no. XXVI–81. NLM MS C 135 chart.
34. MPG IX Abt. Rep 1 Hallervorden, J., J. Peiffer, 'Gedenkrede', nd.
35. BAK ZSg 154/75 Alexander diary, 14 June 1945, pp. 174–5.
36. Klee, *Was sie taten*, pp.136–9.
37. Alexander, *Neuropathology*, pp. 42–3.
38. Matthias Weber, *Ernst Rüdin. Eine kritische Biographie* (Berlin: Springer, 1993), pp. 268–72.
39. Leo Alexander, Public Mental Health Practices in Germany. Sterilization and Execution of Patients Suffering from Nervous or Mental Disease. CIOS Item no 24. File No. XXVIII–50, pp. 1–175. August 1945, 3–7, 42–3.
40. NARA RG 319 Entry 11d IRR Personnel box 347 Ruedin file. NARA T1021/ 12, Nr 128177–8 Ruedin to Nitsche, 1 July 1942. Statement by lawyer Alfred Holl, 19 July 1946.
41. Weber, *Rüdin*, pp. 283–8.
42. APD 4/26, Alexander address book, entries in October 1942 pages. Christine Teller, 'Carl Schneider. Zur Biographie eines deutschen Wissenschaftlers', *Geschichte und Gesellschaft*, vol. 16 (1990) 464–78. Volker Roelcke, Gerrit Hohendorf and Maike Rotzoll, 'Psychiatric Research and "Euthanasia". The Case of the Psychiatric Department at the University of Heidelberg, 1941–1945', *History of Psychiatry*, vol. 5 (1994) 517–32.
43. Alexander, Public Mental Health Practices in Germany. Sterilization and Execution of Patients Suffering from Nervous or Mental Disease. CIOS Item no. 24. File No. XXVIII–50, p. 7. Alexander, 'German Military Neuropsychiatry', 77. CIOS, Miscellaneous Interviews on Medical Practice and Research in Germany, 17–19.
44. Alexander, Public Mental Health Practices in Germany. Sterilization and Execution of Patients Suffering from Nervous or Mental Disease. CIOS Item no. 24. File No. XXVIII–50, p. 28.
45. Ibid., pp. 23–4.

46. Ibid., pp. 43–5. NARA RG 165 Entry 179 box 739, Case of Rascher.
47. Alexander, Neuropathology, p. 5. G.E. Smyth, Oberfeldarzt Professor Hugo Spatz. The Department of Brain Research Kaiser Wilhelm Institute, CIOS Target 24/82a Medical.1945, pp. 3–6. MPG II Abt. Rep 20b Nr 2 Spatz (Aero Medical Centre) to Tönnis, 6 November 1946. NLM MS c 135 Hugo Spatz Reports, Der Aussenstelle f Gehirnforschung. Personalstand vom 1.5.1945.
48. MPG II Abt. Rep 20B Nr 105, Tönnis to Spatz, 9 September 1946; Spatz to Tönnis, 17 December 1946.
49. BAK Zsg 154 Sammlung Christian Pross Box 63 Spatz file, Spatz to Lagerkommandant of Lager Aibling, 28 July 1945.
50. MPG II Abt., Rep 20B Nr 105 Steinbauer to Tönnis, 2 January 1948; Tönnis to Steinbauer, 23 February 1948.
51. Alexander, 'Neuropsychiatric Organizations', 16. 'German Military Neuropsychiatry' 97–116, 128; NARA M 1019/20 Gebhardt interrogation, 3 December 1946, p. 19, concerning Eliot Cutler.
52. BAK ZSg 154/75 Alexander, 'German Military Neuropsychiatry', 83–9.
53. BAK ZSg 154/75 diary, 3 June 1945, pp. 75–8. Alexander, 'Public Mental Health Practices', B.1.
54. Alexander, 'German Military Neuropsychiatry', 118–21.
55. Ibid., 122.
56. APD 4/26 notes in address book on personnel, pages 'September 1942'.
57. APD 4/27 M.F. Wright, SHAEF Rear from Nielson, ref UKX–40341 dated 7 June 1945.
58. CIOS Consolidated Advance Field Team Assessment Report. Extract, Institut für Luftfahrtmedizin, Munich.
59. Alexander, 'German Military Neuropsychiatry', 6–8.
60. Ibid., 8.
61. Ibid., 8.
62. At Nuremberg he confronted Weltz with why he concealed the human experiments, NARA M 1019/78 Weltz interrogation, 23 November 1946, 7–8.
63. Alexander, 'German Military Psychiatry', 13.
64. BAK ZSg 154/75 Diary, 5 June 1945, p. 8. Alexander, 'Treatment of Shock', 11–12.
65. Martin Gilbert, *Atlas of the Holocaust* (Oxford: Pergamon, 1988), pp. 232–5. BAK ZSg 154/74 Alexander, diary, 31 May 1945, pp. 55–6.
66. BAK ZSg 154/75 diary, 31 May 1945, p. 58.
67. APD 4/27 Message Form, nd. Franz Büchner, *Der Eid des Hippokrates. Die Grundgesetze der ärztlichen Ethik. Öffentlicher Vortrag gehalten in der Universität Freiburg i. Br.* (Freiburg: Herder, 1947).
68. NARA M 1019/59 Romberg Interrogation, 26 November 1946. Romberg affidavit 5 December 1946 that he was in Dachau from March to May 1942.
69. APD 4/27 Target Evaluation Report Universitaets Nervenklinik Frankfurt a.M. Alexander, 'German Military Neuropathology', 67–73.
70. Ibid., 161.
71. Peiffer, *Hirnforschung*, 41–5. BAK ZSg 154/75 diary, 14 June 1945, p. 170. MPG II. Abt. Rep. 20B Nr 2 Spatz to Tönnis, 17 December 1946, 12 April 1947. Nr 105 Steinbauer to Tönnis, 2 January 1948. S.P. Harper, 'Naming of Syndromes and Unethical Activities: the Case of Hallervorden and Spatz', *The Lancet* 348 (1996) 1224–5.
72. Alexander, 'German Military Neuropathology', 20.
73. Ibid., 14.

74. Ibid., 15–16, 18–19.
75. Ibid., 17–18. Cf. RAC RG 5, Box 6 folder 55 Personal papers, diaries 12 August 1946. Heidelberg 'Stopped in to look at the Library … used by the Army in preparing material for the Nurnberg trials'.
76. NARA RG 338 290–59–17, 707–Medical Experiments.
77. Alexander, 'German Military Neuropathology', 36.
78. Ibid., 37. NARA M 1019/5 Becker-Freyseng interrogation, 24 September 1945, 11.
79. Hugh Iltis to author, 6 January 2004. Hugh Iltis to Commanding General Military Intelligence Service, 16 February 1945.
80. BAK ZSg 154/75 diary, 18 June 1945, p. 201. Hugh Iltis to parents and Wilfred Iltis, Heidelberg, 25 June 1945.
81. Alexander, 'German Military Neuropathology', 20–2. APD Alexander Secret report for the UK Base for SHAEF (Rear) for United States Medical Corps (London) for Col. Prentiss.
82. Alexander, 'German Military Neuropathology', 22–6. NARA M 1019/66, Schroeder interrogation by King, 21 October 1946, 3–4.
83. Alexander, 'German Military Neuropathology', 32, 35–6. NMT 8/1493 Milch Trial, 7 February 1947.
84. Alexander, 'German Military Neuropathology', 22, 26–7.
85. NARA M 1019/66 Schröder interrogation by Alexander, 2 December 1946, 21; also 11 January 1947 by Koch, 1.
86. Alexander, 'German Military Neuropathology', 28, 31.
87. Ibid., 32–3.
88. Ibid., 33–5.
89. APD Alexander address book note of 20 June 1945. 4/27 Target report, Institut für Luftfahrtmedizin, Munich, 23 June 1945. Also typescript notes. Diary, 20 June 1945, pp. 203–4.
90. Alexander, 'Neuropathology', pp. 38–9.
91. Ibid., p. 39.
92. BAK ZSg 154/75 diary 1/2 APD Dr Ali Kuci, 'The Last Days of Dachau', 16 May 1945, 1 page copied 17 May 1945. File 2 International prisoners committee, 16 May 1945. APD 4/27 Interrogation by Major Leo Alexander MC USA at Dachau, 21 June 1945. CIOS Target Report, dated 23 June 1945. Diary, 205–6.
93. APD 4/20 Interrogation by Major Leo Alexander MC USA at Dachau, 21 June 1945. Alexander, 'Treatment of Shock', 42–6. 4/27 Target report, Institut für Luftfahrtmedizin, Munich, 23 June 1945.
94. NARA RG 238, 201 series Box 35 Weltz file, Weltz statements 19 and 31 October 1945 on relations with Rascher.
95. BAK ZSg 154/75 diary, 22 June 1945, p. 211.
96. Bower, *Paperclip*, 235–6. APD 4/35 Alexander, Interrogation of Punzengruber, 14 February 1947.
97. Guide, 89.
98. Alexander, 'Miscellaneous Aviation Medical Matters', 14.
99. Robert L. Berger, 'Nazi Science – the Dachau Hypothermia Experiments', in John J. Michalczyk (ed.), *Medicine, Ethics, and the Third Reich: Historical and Contemporary Issues* (London: Sheed and Ward, 1999), pp. 87–100.
100. Alexander, 'Miscellaneous Aviation Medical Matters', 66–8.
101. BAK ZSg 175/75 diary, 24 June 1945, pp. 219–20.
102. AFP, Leo to Phyllis Alexander, 26 June 1945. Shevell, 'Reports', 1097.
103. RAC 303–U Bronk Air Force in Europe (Germany) 1945 U (61) 9, Bronk to Germany order 6 July, for 14 days, Civilian.

104. RAC RU 303 Bronk box 37 folder 5 Interrogations of German Personnel Interested in German Aviation, nd. Another copy in USHMM 1998 A.0044 dated 15 August 1945.
105. RAC RU 303 Bronk box 37 folder 5 Burchell to Bronk, 17 August 1945.
106. 'German Aviation Medical Research At the Dachau Concentration Camp', Technical Report No. 331–45, U.S. Naval Technical Mission in Europe, October 1945, 4.
107. USHMM Aero-Medical Research Section, US Strategic Air Forces in Europe, Interrogations of German Personnel interested in German Aviation, 15 August 1945, 10, 11, 13.
108. Ibid., 20.
109. Ibid., 22, 24.
110. D.B. Dill, 'A.C. Ivy – Reminiscences', *The Physiologist*, vol. 22, no. 5 (1979) 21–2.
111. USHMM, 1998.A.0044 Index of German Aeromedical Research Workers, Appendix 12a.
112. USHMM, Aero-Medical Research Section, US Strategic Air Forces in Europe, Preliminary Report of the Interrogation of Certain High Staff Officers of the Luftwaffe Medical Services, 16 July 1945, 4–7.
113. NARA M 1019/60 Rostock interrogation, 4 March 1947, 29.
114. NMT 2/ 6479. Cf. Schreuder affidavit.
115. RAC RU RG 303.U Bronk box U (61) 9 proposed Institute.
116. RAC RU RG 303.U Bronk box 37 folder 5 Postwar Aviation in Germany, Burchell letters to Bronk.
117. NARA RG 341 190 64 24 06 Records of the Headquarters US Air Force box 162 Otis Schreuder to Otis Benson, 25 September 1946.
118. MPG III Abt. Rep 25 Nachlass Kuhn Nr 55, confiscation order 17 March 1947 and complaint by R. Kuhn.
119. NARA RG 200 Entry 82A Box 6 file Adams, Goudsmit to Adams, 25 October 1945. MPG III Abt. Rep 25 Nachlass Kuhn Entnazifizierung 1946. Goudsmit, *Alsos*, 77, 80 re Richard Kuhn – suspected and painfully observant of Nazi formalities. I am grateful to Florian Schmaltz for information on Kuhn.
120. RAC RU RG 303. U Bronk box U (61) 9 proposed Institute.
121. A. Ebbinghaus and Karl Heinz Roth, 'Vernichtungsforschung: der Nobelpreisträger Richard Kuhn, die Kaiser-Wilhelm Gesellschaft und die Entwicklung von Nervenkampfstoffen während des "Dritten Reiches" ', *1999*, vol.17 (2001) 15–50.
122. Tom Bower, *The Paperclip Conspiracy. The Battle for the Spoils and Secrets of Nazi Germany* (London: Michael Joseph, 1987), pp. 243–5.
123. RAC RU RG 303.U Bronk box U (61) 9, HQ Germany, Office of the Surgeon, Air Surgeon Air Div. to Chief, Policy and Liaison, Air Div., 2 July 1945.
124. MPG II. Abt. Rep 0001A Reisebericht Telschow Teil 2, 1 December 1945.
125. RAC RU 303 Bronk box 37 folder 5 Burchell to Bronk, 22 September 1945.
126. RAC RU RG 303–U Bronk box 37 folder 5 Postwar Aviation in Germany, Burchell letters to Bronk.
127. Tilley, Preliminary Report on Hildebrandt, Osterwald and Beuchelt, 5.
128. TNA: PRO FO 1031/119 Re-Birth of the German Menace. Dr Gawlik, Counsel for Defence IMT Nuremberg. E. Tilley, memo 'Growing German Disregard of Allied Authority by German Political Leaders', 14 October 1946. Tilley, 'Preliminary Report', 8.
129. BAK ZSg 154/75 Pross papers, Pross comments on Alexander.

130. AFP Leo to Phyllis Alexander, 24 August 1945. The reports appeared as Reports for Combined Intelligence Objectives Subcommittee, Target Number 24 Medical CIOS G-2 Division, Shaef rear.
131. BAK ZSg 154/75 Alexander diary, 29 August 1945, 235.
132. NARA RG 153/86–3–1/ Box 11 folder 4 bk 3 Alexander to Medical Intelligence Division 20 August 1945 submitting report on Public Mental Health Practices in Germany. RG 112 Surgeon General Geographic Series 1945–46 Box 1370. RG 200 ALSOS Goudsmit files Box 6, Colby to Edell, 11 August 1945.
133. RAC RF 1.1/717/12/96 memo of Lambert, 17 December 1945. RAL diary, 5 November 1945.
134. AFP Leo to Cecy Alexander, London, 1 September 1945.
135. UNWCC Summary of Information No. 41 'Medical Experiments on Human Beings ("Versuchspersonen")'. NARA RG 153 85–2 folder 2.
136. APD 4/35. Alexander address book Surgeon General Release of cold report, 9 and 13 November 1945.
137. NARA RG 341 190 64 24 06 Records of the Headquarters US Air Force box 162 Robert Benford to Air Surgeon, 8 November 1946.
138. George Connery, 'Army Uncovers Lurid Nazi "Science" of Freezing Men', *Washington Post* (4 November 1945), section II.
139. BAK ZSg 154/75 Alexander diary, 9 November 1945.
140. J.D. Ratcliff, 'Murder for Research', *Coronet* (April 1946) 10–14. Alexander address book note for 13 November 1945, interview with Ratcliff. C. Lester Walker, 'Secrets by the Thousands', *Harper's Magazine*, vol. 193 (1945) 329–36.
141. ' "Bestien in Menschengestalt". Dachauer Häftlinge als Versuchsobjekte', *Süddeutsche Zeitung* (6 October 1945). Cf NARA M 1019/78 Weltz interrogation by Alexander, 31 December 1946, 38.
142. Reprinted in *Trial of the Major War Criminals Before the International Military Tribunal in Nuernberg*, Doc. 400 PS, 25: 536–607, 1947.
143. AdeF BB/35/272.
144. RAC RF 1.2/200A/ box 126/ folder 1114 Duke University Psychiatry, 1 November 1945, Leo Alexander to Alan Gregg.
145. RAC RF 1.2/200 A/box 126/folder/114 Boston State Hospital Alexander to R.A. Lambert, 5 March 1946; Lambert to Alexander, 22 March 1946.
146. RAC RF 2–1945/717/2103 concerning the non-response of the RF, 9 November 1945; RF 1.1/717/2/15, 6 April 1947. P.J. Weindling, ' "Out of the Ghetto": The Rockefeller Foundation Confronts German Medical Sciences after the Second World War', in William H. Schneider (ed.), *The Rockefeller Foundation and Biomedicine: International Achievements and Frustrations from World War I to the Cold War* (Bloomington: Indiana University Press, 2002), pp. 208–22. 'The Rockefeller Foundation and German Biomedical Science, 1920–40: from Educational Philanthropy to International Science Policy', in N. Rupke (ed.), *Science, Politics and the Public Good. Essays in Honour of Margaret Gowing* (Basingstoke: Macmillan, 1988), pp. 119–40.
147. Max Weinreich, *Hitler's Professors* (New York, 1946).
148. NAC C–5842, Spotlight No 1 (= Information Section Counter-Intelligence-Branch, HQ US Group CC), 3 August 1945.
149. Henry Friedlander, *The Origins of Nazi Genocide. From Euthanasia to the Final Solution* (Chapel Hill: University of North Carolina Press, 1995), p. 317.
150. TNA: PRO FO 1031/72, 12 September 1945.
151. TNA: PRO FO 935/56 FIAT reports, Minutes of a Meeting, Lansdowne House, Berkeley Square, London, 21 August 1945.

152. *Journal Officiel du Commandement en Chef français en Allemagne*, vol. 1, no. 6 (22 November 1945), Reorganisation was ordered 14 March 1946.
153. TNA: PRO FO 1031/83 E.W.B. Gill to Wilson 8 June 1945. Dustbin report no. 15, 25 August 1945.
154. NARA RG 338 USAR/JAG War Crimes Records Regarding Medical Experiments No.125576–125581. 200 copies were produced with 150 going to US agencies.
155. NARA RG 260/ OMGUS 254 Scientific Exploitation of German Specialists in Science and Technology, FIAT Scientific and Technological Branch, A.A. Kingscote, 'SS Medical Research', 4 September 1945. State Museum of Auschwitz-Birkenau, *Death Books*, 201.
156. NAC C–5842 Microfilm Ottawa BIOS Evaluation Reports Canadian Military HQ file 55/8013/3.
157. NAC Blaisdell Military Service file. University of Toronto Archives, Blaisdell file. By December 1946 he covered medical targets for the Ministry of Reconstruction, Ottawa – over 10,000 pages by December 1946 ready for microfilming. BIOS requested German Medical Documents collected by the pathologist Lt Col James Blaisdell on 9 September 1946 being items 1–593 in the Frankfurt series and 1–437 in the Kaiser Wilhelm Institute series (BIOS to Canadian Military HQ 9 September 1946 microfilm).
158. NAC RG 24 Series C–2 55/8031 Maunsell to John F. Fulton, 3 December 1945.
159. Description, 18 May 1946. Blaisdell, 'A visit to Germany to investigate wartime advances in certain areas of applied physiology'. Interrogations by Miles and Kennedy.
160. NARA RG 260 OMGUS folder 39 Lists of Homework by Dustbin detainees, 29 October 1945.
161. MPG Abt 10 B Advisory Committee on German Science.
162. NAC RG 24 National Defence, Series C–2, microfilm FIAT accession lists 55/8031/1 Blaisdell to Cuthbertson, 27 February 1946, 27 November 1945, 6 March 1946.
163. Ibid. List of Documents and Reports Relating to German Medical Science Collected and Microfilmed at Hoechst and Frankfurt, Germany by Lt Col. Blaisdell RCAMC in Collaboration with F.I.A.T. (British). Cf. NMT 8/358–60, note of January 1946 [by Thompson?].
164. NAC RG 24 Series C–2 55/8031 BIOS to Canadian Military HQ, 9 September 1946.
165. Georg Schaltenbrand, *Neurology, Teile 1–3* (= FIAT Überblick über die deutsche Wissenschaft 1939–1946) (Wiesbaden, 1948). MPG Abt II. Rep 20B Nr 105–1 Spatz to Tönnis, 6 November 1946. Werner Leibbrand, *Um die Menschenrechte der Geisteskranken* (Nurnberg, 1946).
166. TNA: PRO WO 309/ 468 and FO 1031/74 Major-General J.S. Lethbridge to Head, CI Bureau, 6 December 1945.
167. NAC RG 24 Series C–2 55/8031 Memo, 3 June 1946, Cuthbertson to Winfield.
168. TNA: PRO FO 1031/ 74 Thompson to Capt. M. Marschuk, Judge Advocate's Office, US Forces, 29 November 1945. FD 1/5826 War Time Scientific Investigations in Germany and German Occupied Countries. C.B. Mickelwait to Thompson, 7 December 1945. FO 937/165 Bradshaw to Phillimore, BWCE Nurnberg to War Office, 12 December 1945. Also in WO 309/468.
169. TNA: PRO FO 1031/74 Legal Division CCG to Legal Division OMGUS, 15 December 1945. Legal Division OMGUS 26 Dec 45. FO 937/165 cipher, 27 December 1945, CONCOMB to TROOPERS, 26 December 1945. Plans: 26 December 1945 to try Karl Brandt as part of the Hohenlychen group. FO 371/57576.

170. TNA: PRO FO 1031/74 Lethbridge to Maunsell, 28 January 1946.
171. TNA: PRO FO 937/165.
172. TNA: PRO FO 937/165 CONCOMB to TROOPERS, 26 December 1945.
173. BAK ZSg 154/71 Pross Papers, Pross interview with Telford Taylor, 8 September 1987.

6   From the International Tribunal to Zonal Trials

1. Marrus, 'Holocaust', 7.
2. Tusa, *Nuremberg*, p. 157.
3. The 1997 Nuremberg Prosecutors Conference, University of South Carolina School of Law, Sept 1997, Jackson Jnr, reel 3.
4. *Trial of German Major War Criminals* (London: HMSO, 1946), vol. 1, pp. 12–13, 22, 69.
5. *Trial of German Major War Criminals*, vol. 1, pp. 69–70 on 21 November 1945.
6. *Trial of German Major War Criminals*, vol. 6, p. 131.
7. NARA RG 200 Entry 82A Box 4 DCC 2, Robert Jackson to R.L. Hopkins, 11 January 1946 returning RFR to Haagen, 21 July 1943, Rose to Haagen, 2 December 1943, Haagen to Rose, 8 December 1943. ALSOS files RFR No 35 Correspondence with Dr Rascher concerning obtaining equipment for unspecified medical research projects.
8. *Trial of German Major War Criminals*, vol. 3, pp. 158–66, 20 December 1945.
9. NARA RG 238, 201 series Box 35 Weltz file, Franz Büchner affidavit, 28 March 1946.
10. Robert Sigel, *Im Interesse der Gerechtigkeit. Die Dachauer Kriegsverbrecherprozesse 1945–1948* (Frankfurt: Campus Verlag, 1992), pp. 71–5.
11. *Trial of German Major War Criminals*, vol. 4, pp. 204–26, 11–12 January 1946.
12. APD 4/33 Alexander, Logbook, 11 December 1946, p. 46. Mant, 'From Nuremberg', chapter 2 'War Crimes Investigations'. Blaha, 'Medizin auf schiefen Ebene', 77. NARA M 1019/7 Interrogation of Blaha by von Halle, 11 December 1946.
13. Blaha, 'Medizin auf schiefern Ebene', 200–1.
14. *Trial of German Major War Criminals*, vol. 6, pp. 131–2. Jürgen Peiffer, *Hirnforschung im Zwielicht: Beispiele verführbarer Wissenschaft aus der Zeit des Nationalsozialismus. Julius Hallervorden – H–J. Scherer – Berthold Ostertag* (Husum: Matthiesen Verlag, 1997).
15. Cf. newspaper reports in MPG Abt. IX, Hallervorden. 'Das Weltgericht in Nürnberg' [supplement of the *Lüneberger Landeszeitung*], p. 17. For exculpation, see W. Villinger and G. Schaltenbrand, 'Erklärung', *Der Nervenarzt*, vol. 24 (July 1953) 312.
16. Mitscherlich and Mielke, *Medizin ohne Menschlichkeit*, p. 196.
17. *Trial of German Major War Criminals*, vol. 5, pp. 188, 193, 249–61. Weindling, *Epidemics and Genocide*, pp. 352–7.
18. *Trial of German Major War Criminals*, vol. 21, pp. 81–2; Tusa, *Nuremberg*, p. 200.
19. *Trial of German Major War Criminals*, vol. 22, p. 88.
20. *Trial of German Major War Criminals*, vol. 22, pp. 90–1.
21. *Trial of German Major War Criminals*, vol. 22, pp. 92, 94.
22. Geissler, *Biologische Waffen*, pp. 758–9, 764, 777–91. Geissler argues against the position taken by Deichmann, *Biologists*; Moreno, *Undue Risk*, pp. 98–101.

23. Linda Hunt, *Secret Agenda. The United States Government, Nazi Scientists, and Project Paperclip, 1945 to 1990* (New York: St. Martin's Press, 1991), pp. 13, 152–5, 179–81.
24. *Trial of German Major War Criminals*, vol. 8, pp. 302–3 11 March 1946.
25. *Trial of German Major War Criminals*, vol. 5, p. 254, 29 January 1946, citing Lautenschläger.
26. Ernst Klee, *Deutsche Medizin im Dritten Reich. Karrieren vor und nach 1945* (Frankfurt: S. Fischer, 2001), p. 315.
27. NARA RG 338 Stack 290–59–33–1–3 Executee Files Karl Gebhardt, Andrus, 3 January 1946.
28. Kater, *Ahnenerbe*, pp. 34–6.
29. *Trial of German Major War Criminals*, vol. 21, p. 12.
30. *Trial of German* Major *War Criminals*, vol. 22, p. 123, 26 August 1946.
31. *Trial of German Major War Criminals*, vol. 20, p. 399; vol. 21, p. 7.
32. Tusa, *Nuremberg*, p. 433.
33. *Trial of German Major War Criminals*, vol. 21, p. 17.
34. *Trial of German Major War Criminals*, vol. 22, pp. 52–3.
35. *Trial of German Major War Criminals*, vol. 20, pp. 396–7, 8 August 1946.
36. APD 4/36 Karl Tauböck affidavit, 20 June 1945. 4/37 Report, 6 October 1945 re Jackson Prosecution staff.
37. *Trial of German Major War Criminals*, vol. 21, p. 11, 9 August 1946.
38. *Trial of German Major War Criminals*, vol. 20, pp. 1, 10, 29 July 1946.
39. *Trial of German Major War Criminals*, vol. 20, p. 25, 29 July 1946.
40. *Trial of German Major War Criminals*, vol. 20, p. 29, 29 July 1946.
41. *Trial of German Major War Criminals*, vol. 20, p. 66, 30 July 1946.
42. *Trial of German Major War Criminals*., vol. 20, 3 August 1946, pp. 257–9, 275–7.
43. TNA: PRO FO 937/110. Mugar Memorial Library Boston University No. 242 Leo Alexander Collection of Documents from the Nuremberg Medical Trial 1946–1947, (APBU) Box 55 I Karl Sperber statement on Auschwitz. TNA: PRO FO 937/110 for Sperber deposition.
44. *Medical Science Abused. German Medical Science as Practised in Concentration Camps and in the so-called Protectorate. Reported by Czechoslovak Doctors* (Prague: Orbis, 1946). Gisella Perl, *I was a Doctor in Auschwitz* (New York: International Universities Press, 1948).
45. UNA UNWCC Minutes M 81, 5 Visit to Hadamar Trial, 16 October 1945. *Law Reports of Trials of War Criminals. Selected and Prepared by the United Nations War Crimes Commission* (London: HMSO, 1947), vol. 1.
46. Earl W. Kintner, *The Hadamar Trial* (London, 1949). US Military Tribunal, Transcript of the Proceedings in Case 1, p. 4572 for Blome disclaiming responsibility.
47. Burleigh, *Death*, 272.
48. Friedlander, *Origins of Nazi Genocide*, pp. 78, 228–30, 324. STA Nürnberg Rep 502A KV- Verteidigung Handakte Fröschmann Nr 17.
49. US Military Tribunal, *Transcript of the Proceedings in Case 1*, pp. 1868–70, 1880, 1916–23.
50. BAB EVZ I 26. A2.
51. AOF AJ Caisse 3640 carton 8197 Dossier d. 8287 Neuengamme Prosecution evidence of Major Stewart.
52. NARA RG 238 NM70 – Entry 200 OCCWC. Executive Office Nuremberg Administrative Post Internee Personnel Records ('201 files') Box 5 'Translation

of last statement by Leonardo Conti'; Interrogation of Leonardo Conti taken at Nuernberg, Germany on 1 September 1945, by Major John J. Monigan (1100–1205 hours). Burton C. Andrus, *The Infamous of Nuremberg* (London: Leslie Frewin, 1989), pp. 87–8 omits the key passage on Blome but mentions other documents. NARA M 1019/7 Blome interrogation by Devries, 7 October 1946, summary, 5.

53. *Trial of German Major War Criminals*, vol. 5, p. 190 (on the twin block), pp. 194–5; vol. 6, p. 185 on selections; vol. 7, p. 98 on lethal experiments and bleeding of children; vol. 11, p. 353 on twin experiments. Cf. Gerald Astor, *The 'Last' Nazi. The Life and Times of Dr Joseph Mengele* (London: Weidenfeld & Nicolson, 1985), pp. 129–30, cf. also 141–2. Lucette Matalon Lagnado and Sheila Cohn Dekel, *Children of the Flames. Dr Josef Mengele and the Untold Story of the Twins of Auschwitz* (London: Sidgwick and Jackson, 1991), pp. 111–13.

54. *Trial of German Major War Criminals,* vol. 6, 131–3, citing evidence of Leo Alexander, Neuropathology and Neurophysiology, including Electroence-phalography, in Wartime Germany, 20 July 1945, 1–65.

55. *Trial of German Major War Criminals,* vol. 1, p. 104, 22 November 1945.

56. Telford Taylor, *The Anatomy of the Nuremberg Trials. A Personal Memoir* (New York: Back Bay Books, 1992), p. 103.

57. *Trial of German Major War Criminals*, vol. 1, p. 12.

58. R.W. Cooper, *The Nuremberg Trial* (Harmondsworth: Penguin, 1947), p. 110.

59. Collection of Raphael Lemkin, American Jewish Historical Society, Waltham, MA, and New York, NY. Center for Jewish History New York, American Jewish Historical Society, P–154 box 2 folder 12.

60. NARA RG 153 84–1 folder II, Howard Petersen to Telford Taylor, 22 May 1946.

61. TNA: PRO WO 309/476 Operation FLEACOMB: tracing of possible war criminals December 1946–February 1948.

62. Lutz Niethammer, *Entnazifizierung in Bayern; Säuberung und Rehabilitierung unter amerikanischer Besatzung* (Frankfurt am Main: S. Fischer, 1972).

63. Taylor, *Anatomy*, pp. 270–5.

64. Ibid., p. 286.

65. Ibid., pp. 279–83. Tusa, *Nuremberg*, pp. 138–40 for the Krupp issue.

66. NARA RG 153 84–1, Memo for Mr Benjamin Cohen, 18 June 1946.

67. National Library of Wales, Lord Elwyn-Jones Papers (A 1990/16), c.11 Foreign Office policy file, correspondence 1946. C. 14 Shawcross to Elwyn-Jones 18 April 1946. NARA RG 153 84–1 Box 1 JAG to US Chief of Counsel, 26 December 1945.

68. NARA RG 153 84–1 Box 1 folder 4/1 Minutes of the Chief Prosecutors 5 April 1946; 4 July 1946 on the Soviet demands. Tusa, *Nuremberg*, p. 373.

69. NARA RG 153 84–1 box 1 folder II Second International Trial Taylor to Patterson, 22 May 1946.

70. AdeF BB/35/89 Principaux points pouvant être éventuellement retenus pour la poursuite des industriels allemands. NARA RG 153 84–1 memo for Benjamin Cohen, 18 June 1946. Tom Bower, *The Pledge Betrayed. America and Britain and the Denazification of Post-War Germany* (New York: Doubleday, 1982), pp. 322–3.

71. Taylor, *Anatomy*, p. 287. Tusa, *Nuremberg*, p. 93 on the second trial.

72. NARA RG 153 Nuremberg Administrative Files 84/4–2 Memo to the President by Robert Jackson, 13 May 1946.

73. NARA RG 153/ 84–1, box 1 folder II memo of 4 July 1946.

74. Article 22 of the London agreement of 1945 NMT fiche 131. Tusa, *Nuremberg*, pp. 83–4.

75. NARA RG 153 Records of the Judge Advocate General (Army). War Crimes Branch. Nuremberg Administrative Files 84–1 Box 1 Robert H. Jackson to Robert P. Patterson, 7 February 1946; Howard Petersen to Telford Taylor, 17 June 1946.

76. LC Jackson papers, Taylor to Jackson, 25 July 1946.

77. Taylor, *Anatomy*, pp. 289–92.

78. NARA RG 153/84–1 box 1.

79. NARA RG 153 RG 153/86–3–1/Box 1 Folder 4/1 Memo to the President from Justice Jackson, 13 May 1946. Bower, *Pledge*, p. 322. Anthony Glees, 'The Making of British Policy on War Crimes: History as Politics in the UK', *Contemporary European History*, 1 (1992) 171–97.

80. NARA RG 153 Nuremberg Administrative Files 84/1 Box 1 Taylor to the Secretary of War 29 July 1946. Bower, *Pledge*, p. 323.

81. NARA RG 153/85–2 Box 3 folder 3 Telford Taylor to Charles Rugg, 27 March 1946. 85–2 folder 5 letters of applications from lawyers interested in war crimes vacancies. RG 153/ 84–1 box 1 folder 2 Taylor to Petersen, 30 September 1946 for uses of the term 'top notch'. Telford Taylor interview, New York, 6 November 1996.

82. B.B. Ferencz, 'Nurnberg Trial Procedure and the Rights of the Accused', *Journal of Criminal Law and Criminology*, 39 (1948–9) 144–51.

83. NARA RG 153 84–1 memo, 30 October 1946.

84. NARA RG 153 Box 1 F 4/1 July 1946. For the business analogy, see Tusa, *Nuremberg*, p. 84.

85. NARA RG 153 85–1 folder 1 US Chief of Council Nuremberg to Petersen, War Dept, 29 June 1946.

86. NARA RG 153 85–1 folder 1 memo to Taylor, 16 August 1946. Jackson Papers telegram for Taylor from Jackson, 16 August 1946.

87. NARA RG 153 85–1 memo to Taylor, 16 August 1946.

88. NARA RG 153 85–1 US Chief of Council Nuremberg to War Department, 17 August 1946.

89. NMT 1/156. US military decree, 24 October 1946.

90. OCCWC Press Release No. 76.

91. Klee suggests Flury used CO for euthanasia, in *Was sie taten*, pp. 279–81.

92. Lifton, *Nazi Doctors*, chapter 18.

93. NARA M 1019/24, Handloser interrogation by Rapp, 29 August 1946, p. 6. This may have been for the High Command case rather than for any medical trial.

94. LC Jackson Papers Taylor to War Dept, 3 September 1946.

95. NYPL Lifton Papers M9, 8 June 1977 telephone talk with James McHaney, 5.

96. Bower, *Pledge*, p. 323. TNA: PRO Prime Minister's Papers PREM 8/391, 10 August 1946.

97. AOF AJ 3642, p. 210 d 8456 Procès de Nuremberg vol. 3, War Crimes Trials in the French Zone by Josif Marcu Jr, special consultant, Rastatt, May 1948, Nuremberg, June 1948.

98. AOF AJ 3642 p. 215 d. 8531 USFET Instructions générales, 9 September 1946, Directeur du SRCGE to Dir Gen de la Justice, SRCGE, Baden Baden re conference to coordinate Services de liaison.

99. AOF AJ 3645 p. 238 d. 8950 Davies, British liaison officer, 12 September and 11 October 1946.

100. Genevieve de Gaulle, 'La Traverse de la Nuit', *La France en Allemagne*, no. 4 (Jan.–Feb.1947) 99–100 (also reporting on the NMT); no. 5 (March–May 1947) 125.

101. 'Le souci primordiale du Ministère public britannique est d'apporter au Tribunal la preuve irrecusable des mauvais traitements et des morts qui en sont la conséquence, mais n'a parfois l'impression de passer a côté du vrai procès et de la signification qu'il prend dans l'ensemble du plan nazi d'extermination des élites de resistance', in Rapport 2, 21 December 1946 by Aline Chalufour to Ministry of Justice, Garde des Sceaux.

102. Tribunal de première instance de Rastatt pour le Jugements des crimes de guerre, jugement no. 60/49. Melle Tillion to Rastatt Commissioner, 10 December 1947, AJ Camp de Jugenlager (annexe de Ravensbruck) 3633 p. 132 d. 6087.

103. Bower, *Pledge*, p. 111.

104. TNA: PRO WO 309/471 Somerhough to Shapcott, 6 September 1946.

105. Bower, *Pledge*, p. 117.

106. Angelika Ebbinghaus, 'Der Prozess gegen Tesch und Stabenow. Von der Schädlingsbekämpfung zum Holocaust', *1999*, vol. 13 (1998) 16–71.

107. Ibid.

108. TNA: PRO WO 309/471 Meeting of 15 May 1946, also Fritz Ter Meer interrogation.

109. TNA: PRO FO 937/165 Sir Alfred Brown, re. Medical Experiments on Human Beings, 11 January 1946.

110. TNA: PRO FD 1/5826 War Time Scientific Investigations in Germany and German Occupied Countries, Harrington to Mellanby, 21 November 1945.

111. TNA: PRO FD 1/5826 War Time Scientific Investigations in Germany and German Occupied Countries E. Mellanby to C.R. Harrington, NIMR, 3 January 1946.

112. TNA: PRO FO 1031/74 Thompson to Professor C.R. Harrington, 14 December 1945. Telegram from Mellanby to FIAT.

113. TNA: PRO FO 937/165 Bradshaw to Phillimore, BWCE Nurnberg to War Office, 12 December 1945.

114. 'Vorwort der Arbeitsgemeinschaft der Westdeutschen Ärztekammern', Mitscherlich and Mielke, *Wissenschaft ohne Menschlichkeit*, v, March 1949.

115. TNA: PRO FO 1031/ 74 Phillimore for Bradshaw BWCE Nurnberg to War Office.

116. TNA: PRO FO 937/ 165 cipher 27, December 1945, CONCOMB to TROOPERS, 26 December 1945. plans: 26 December 1945 to try Karl Brandt as part of the Hohenlychen group. FO 371/57576.

117. TNA: PRO FO 937/165 Bradshaw to Phillimore, BWCE Nurnberg to War Office, 12 December 1945.

118. TNA: PRO WO 309/468 Major General J.S. Lethbridge to Head, CI Bureau, 6 December 1945.

119. TNA: PRO FO 1031/74 eg FIAT to Allied Commission for Austria 24 March 1946. C.E. Ippen, Report on Trip to Vienna, 16–27 April 1946.

120. TNA: PRO FO 1031/74 Thompson to Capt. Nielly, 18 January 1946. Thompson to BWCE Nuremberg, nd.

121. TNA: PRO FO 1031/74 Col. L.H.F. Sanderson to R.J. Maunsell, 10 April 1946.

122. TNA: PRO FO 1031/74 E. Tilley to Thompson, 7 February 1946.

123. AOF AJ 3617 p. 15 d. 686 Karl Brandt.

124. TNA: PRO WO 309/ 471 (also FO 1031/74) J.W.R. Thompson to JAG, 14 March 1946. The British contact was Lt Col. H. Boggis-Rolfe in Berlin.

125. TNA: PRO WO 309/471 J.W.R. Thompson to JAG, 14 March 1946; Somerhough to FIAT on Medical War Crimes, 24 March 1946. Mant personal communication to the author, 8 July 1997.

126. Mant, *From Nuremberg*, chapter 3 'Personalities'.
127. Royal College of Physicians of Edinburgh (RCPE), Smith papers box. 9 folder 94, Thompson to Smith, 31 January 1946, 7 February 1946. TNA: PRO FO 1031/ 46 FIAT to BWCE Nuremberg. Thompson to Smith, 7 February 1946, 15 April 1946 on Hildebrandt, Osterwald and Beuchelt.
128. RCPE Smith papers, box 9 folder 94, Thompson to Smith, 15 April 1946.
129. TNA: PRO WO 309/471 International Scientific Commission for the Investigation of Medical War Crimes. WO 309/471 Meeting, 15 May 1946; NARA RG 280 box 18 Records of the US Occupation Headquarters, WWII, OMGUS. Records of the Executive Office, the Field Information Agency, Meeting with British and French FIAT, 15 May 1946. NARA RG 338 USAREUR/JAG War Crimes, Records Regarding Medical Experiments No. 125971–77 Experiments on Inmates in Concentration Camps.
130. TNA: PRO FO 1031/74 draft minutes of FIAT meeting, 15 May 1946.
131. TNA: PRO WO 309/ 469 In the Matter of the Alleged Killing of an American Airman, 19 January 1946. Mant, 'From Nuremberg', chapter 2 'War Crimes Investigations'. *Independent*, obituary, 30 October 2000.
132. TNA: PRO FO 1031/74, also WO 309/471 f. 13 Somerhough to FIAT, BAOR 24 March 1946; Thompson to Somerhough, 5 April 1946.
133. TNA: PRO 1031/74 draft minutes, 15 May 1946. Mant to Thompson, 26 June 1946. AdeF BB/ 35/263 Camps de Concentration, [Mant], 'Medical Experiments in Concentration Camps', testimonies of Leo Eitinger, Simon Umschweif. A. Keith Mant, 'The Medical Services in the Concentration Camp of Ravensbruck', *The Medico-legal Journal*, 18 (1949) 99–118.
134. TNA: PRO WO 235/316 Interim Report by War Crimes Investigation Unit, BAOR, nd.
135. TNA: PRO WO 309/469.
136. Ibid.
137. TNA: PRO 1031/74 draft minutes, 15 May 1946, 7.
138. TNA: PRO WO 309/471 Maunsell for FIAT to JAG, 22 May 1946.
139. TNA: PRO WO 309/ 471 Maunsell for FIAT BAOR 2 May 1946. On Straight, see Bower, *Pledge*, pp. 99–100, 104–6.
140. RCPE Smith papers Box 9 folder 94, report of 20 May 1946.
141. TNA: PRO FO 1031/74 Thompson to T. W. Schaeffer, 17 May 1946. WO 309/471 International Scientific Commission for the Investigation of Medical War Crimes. AIP Lépine Papers, Lépine to Thompson, 28 June 1946.
142. TNA: PRO WO 309/1652 Mant to Somerhough and Nightingale, 20 May 1946.
143. Giovanni Maio, 'Ärztliche Ethik als Politikum. Zur französischen Diskussion um das Humanexperiment nach 1945', *Medizinhistorisches Journal*, vol. 35 (2000), 35–80, 43.
144. 'René Legroux', *Annales de l'Institut Pasteur*, vol. 80 (1951) 332–6.
145. AIP Lépine papers. P.O. Lapie, *Notice sur la vie et les travaux de Jean Lépine* (Paris: Institut de France, 1970); RAC RF 12.1 Diaries Box 22, Gregg diary 1945, pp. 178–9 relating to Jean Lépine. Box 67 Warren diary, 3 June and Box 61 Strode diary, 18 March 1947 relating to Pierre Lépine.
146. Balachowsky, Buchenwald diary at the AIP, and Weindling, *Epidemics*, pp. 367–9, 412.
147. TNA: PRO FO 1031/74 Minutes, 15 May 1946. RAC RF 12.1 Box 67 Warren diary, 9 July 1946.
148. Ibid.

149. Taylor, *Anatomy*, p. 289. NARA RG 338 USAREUR/JAG War Crimes, Records Regarding Medical Experiments, Marcus on [UNWCC] Intelligence Report of German Medical Experiments, 20 September 1946.
150. NARA RG 153/86–3–1 book II, box 11 folder 4 Marcus to Director Civil Affairs Division, 16 May 1946.
151. NARA RG 153/86–3–1 BOX 9 OMGUS to War Department, 11 May 1946.
152. NARA RG 153/86–3–1 Box 10 Book 1, 16 May 1946. Chief, War Crimes Branch to Surgeon General, 17 May, SPMDH. *Human Radiation Experiments*, pp. 75, 93 n 3.
153. NARA RG 153/ 86–3–1 book II, box 11 folder 4 Marcus to Ivy, 3 July 1946.
154. NARA RG 153/86–3–1 book II/ box 11 folder 4 Bd 3 A.C. Ivy, A Report on the Paris Meeting of the American, British and French Governments to Consider War Crimes of a Medical Nature.
155. Telford Taylor interview with author, New York, 6 November 1996.
156. NARA RG 153/86–3–1 book II/ box 11 folder 4 Bd 3 A.C. Ivy, A Report on the Paris Meeting of the American, British and French Governments to Consider War Crimes of a Medical Nature.
157. NARA RG 338 USAREUR/JAG War Crimes, Records Regarding Medical Experiments No. 125728 Marcus memo, 22 August 1946.
158. AIP Fonds Lépine Deposition de Janina Iwanska faite par devant le Major Arthur Keith Mant, RAMC à la Commission d'investigation des crimes de guerre à Paris le 28 juin 1946; Deuxième Deposition d'Helena Piasecka, 28 juin 1946. Reproduced in NMT 8/431–458. Mant papers, notebook. NARA RG 153/ 86–3–1 book 3, box 10 letter to Taylor, 15 June 1950.
159. NMT 3/727f Helen Piasecka, 28 June 1946 = NO–864; NMT 3/733f Gustawa Winkowska, 18 September 1946 = NO–865; NMT 3/724ff Zofia Baj, 12 August 1946 = NO–871; NMT 3/709ff Kaminska to Mant, 12 August 1946 = NO–876.
160. AIP Fonds Lépine, Dr Denise Fresnel original deposition of 8 July 1946. NMT 8/448ff.
161. NMT 8/ 440.
162. TNA: PRO WO 309/470 telegram JAG, 12 July 1946. WO 309/ 470 f. 3 Military Attaché Stockholm to Judge Advocate Department, 12 July 1946.
163. TNA: PRO WO 309/ 470 f. 4. WO 309/472 Mant letter from Stockholm, 18 September 1946.
164. TNA: PRO WO 309/468 Draper to Mant, 20 August 1946.
165. Sylvia Salvesen (ed. Lord Russell of Liverpool), *Forgive – But Do Not Forget* (London: Hutchinson, 1958).
166. TNA: PRO WO 309/471 Notes on Conference re Medical Experiments carried out on human beings, held at Wiesbaden on 3 June 1946 between Col Brasser, US War Crimes and Capt Vollmar, AG War Crimes, BAOR.
167. TNA: PRO WO 309/459, 3 December 1946 Somerhough memorandum of 3 December 1946.
168. NARA RG 338 USAREUR/JAG War Crimes, Records Regarding Medical Experiments No. 126273–4 Samuel Sonnenfield, Memorandum to Colonel Bresee, Medical Experiments Performed at Dachau Concentration Camp, 17 May 1946. No. 12681 Bresee to Chief, Apprehension Section, 4 June 1946.
169. Michael Kater, *Das 'Ahnenerbe' der SS 1935–1945; ein Beitrag zur Kulturpolitik des Dritten Reiches* (Stuttgart: Deutsche Verlags-Anstalt, 1974).
170. Author's videotaped interview with Keith Mant, 13 March 1997.
171. TNA: PRO WO 235/316 Interim Report by the War Crimes Investigation Unit, BAOR on Ravensbrück Concentration Camp, pp. 1–2.

172. TNA: PRO FO 371/1946 concerning Polish request of 21 June 1946 for extradition of Ravensbrück medical staff.

173. TNA: PRO WO 235/ 307, ff. 145–206, 195–6 for denial of contact with Gebhardt, 207–18 for Lindell. Author's interview with Mant 27 April 1999 for Lindell's affair with Treite and WO 309/468.

174. TNA: PRO WO 235/316 f. 31. Salvesen, *Forgive*, pp. 110–12, 149–51, 230–3.

175. TNA: PRO WO 309/470 f. 10 telegram concerning meeting on 25 July at Hoechst.

176. AdeF Ministère de la Justice. Direction du Service de recherche des crimes de guerre ennemis BB/30/1821, Commission d'investigation pour l'etude de crimes de guerre scientifique 1946–7.

177. AdeF BB/30/1786 Folder Création et organisation du Service de recherche des crimes de guerre. TNA: PRO FO 1031/74 Paul Tchernia to Thompson, 24 June 1946. WO 309/468 German medical experiments: general correspondence.

178. Behague, neurologist and médecin légiste, formerly from Versailles. In 1944 appointed President de la Société de Neurologie de France.

179. AdeF BB/30/1786 Création et organisation du Service de recherche des crimes de guerre. *Le Monde* (25 July 1946) 4.

180. Union des médecins français, Les Atrocités 'scientifiques' allemandes Deuxième conférence, 'Les Médecins dans les Camps Nazies', organised by Henri Dessille and Gilbert-Dreyfus.

181. AOF AJ c. 3645 p. 287 no. 8949 Rastatt warrant against Plötner, 23 July 1946, escaped from Rastatt on 19 July 1946. US Naval report on Dachau.

182. TNA: PRO WO 309/471 Arrêté du 19 juin 1946. Paul Weindling, 'Ärzte als Richter: Internationale Reaktionen auf die Medizinverbrechen des Nürnberger Ärzteprozesses in den Jahren 1946–1947', in Claudia Wiesemann and Andreas Frewer (eds), *Medizin und Ethik im Zeichen von Auschwitz. 50 Jahre Nürnberger Ärzteprozess* (Erlanger Studien zur Ethik in der Medizin) (Erlangen and Jena: Palm und Enke, 1996), pp. 31–44.

183. AIP Fonds Lépine contains a parallel set of minutes of the International Scientific Commission.

184. Ivy, Outline of Itinerary NARA RG 153/86–3–1/ Box 11 folder 4 Bd 3. TNA: PRO FO 1031/74 Thompson to Smith, 24 June 1946.

185. NMT 2/9375, 14 June 1946.

186. Ivy, Outline of Itinerary NARA RG 153/86–3–1 Box 11 folder 4 Book 3. NARA RG 341 190 64 24 06 Records of the Headquarters US Air Force box 162 AAF Aero Medical Center Monthly Status Report No. 11, 31 August 1946. Hans-Walter Schmuhl, *Hirnforschung und Krankenmord. Das Kaiser-Wilhelm-Institut für Hirnforschung 1937–1945* (Berlin: MPG, 2000), pp. 36–7.

187. NAC microfilm C 5843.

188. National Academy of Sciences, file Aviation Medicine, Dr A.C. Ivy, Joseph Ney to Ivy, 21 August 1946.

189. Dept. of the Air Force, *German Aviation Medicine: World War II* (Washington: Dept of the Air Force, 1950).

190. NARA RG 341 190 64 24 06 Records of the Headquarters US Air Force box 162 AAF Aero Medical Center Monthly Status Report No. 11, 31 August 1946 on the progress of the Strughold monograph and list of involved personnel, nd, memorandum on Public Law No. 25 with listing of earmarked personnel. Benford to Benson, 15 October 1946.

191. NARA RG 238 entry 199 NM 70, CIC transfer of Schäfer, Ruff, Becker-Freyseng, Benzinger Schröder to IMT security, 16 September 1945. Tom Bower, *The Paperclip Conspiracy. The Battle for the Spoils and Secrets of Nazi Germany* (London: Michael Joseph, 1987), pp. 245–6.

192. NARA RG 341 190 64 24 06 Records of the Headquarters US Air Force box 162 Notes on Conversation with Capt. Sydney Titelbaum. Comments of Dr. Greg Hall on Qualifications of German Scientists available for transfer to the US, nd, and 6 November 1946.

193. MPG Archives II Abt. 20B Nr 105 R.J. Benford to Strughold, 4 February 1947.

194. TNA: PRO WO 309/471 Comments of Somerhough, 17 June 1946. The Hohenlychen Group also included Karl Friedrich Brunner, assistant medical director of Hohenlychen, and the nurse Auguste Hingst.

195. AdeF BB/35/260 4c annotations of McHaney on report to Taylor and Ervin, 6 August 1946. McHaney notes, 6 August 1946 re Medical War Crimes Meeting of 31 July.

196. TNA: PRO WO 309/471 Minutes of the Meeting to Discuss War Crimes of Medical Nature Executed in Germany Under The Nazi Regime, 31 July 1946. Interviews.

197. TNA: PRO WO 309/471 Somerhough to JAG, 6 August 1946.

198. TNA: PRO WO 309/471 Somerhough to T.P.A. Davies, Liaison Officer Baden Baden, 8 November 1946. AOF AJ 3625 p 71 d. 3452 Bayle was appointed on the French side on 31 October 1946.

199. NARA M 1019/5, Bayle affidavit, 19 June 1947.

200. NAC, Ministry of External Affairs RG 25 (A–3–6) vol. 5792, file 250 pt 1, Copy of Memo Submitted to Major Mant, 2 November 1946. Stanford University, John W. Thompson Jnr, transcript of studies.

201. TNA: PRO WO 309/471 Somerhough to JAG, 6 August 1946. AdeF BB 35 260 4c McHaney notes, 6 August 1946 on Medical War Crimes Meeting held at the Pasteur Institute, 31 July 1946.

202. AdeF BB/35/260 file 4C Mant, 18 September 1946 to McHaney, expressing hopes to complete Hohenlychen investigations by mid-October ISC meeting; reply to Mant, 30 September 1946 [also in TNA: PRO WO 309/471].

203. TNA: PRO WO 309/471 Somerhough to Judge Advocate General (JAG), 6 August 1946.

204. TNA: PRO WO 307/471 Office of JAG to DJAG BAOR, 27 August 1946; Somerhough to Shapcott, 6 September 1946.

205. APBU, Hardy to McHaney, 26 October 1946.

206. NARA RG 338 USAREUR/JAG War Crimes, Records Regarding Medical Experiments No. 12598 C.E. Straight to Adlerman, 31 July 1946; BB/35/274 Folder: Sterilisations. Interrogations et comptes-rendus d'interrogatoire Wolfson to McHaney 27, 28 June, 14 July, 5, 29 August 1946 on the plant *Caladium seguinam*. NARA RG 238 OCCWC NM 70 Entry 188 Box 2, Hardy on 'Individual Responsibility of Karl Brandt'.

207. TNA: PRO FO 1031/74 OMGUS note 26 Dec 1945. AdeF BB/35/261 M. Wolfson 22 August 1946. BB/35/264, 22 August, 'M. Wolfson to Miss Foster, subsequent – Division, Enclosed re the photostatic copies of the Dr Conti files I to VIII. Since this material is of vital importance to Mr Hardy re Dr Karl Brandt, the material is sent to you without SEA to save time. More material is forthcoming in a few days when Mr Schwenk reports to Nuremberg. M. Wolfson (US civ), Snr Research Analyst, PS Source of the Dr Conti files Ministerial Collecting Centre, Berlin-Tempelhof: Dept of the Reichs Interior (Reichsinnenministerium), Section: Health.' Also NARA RG 238 NM 70/entry 202/ box 2 papers of M. Wolfson. NARA RG 338 USAREUR/JAG War Crimes, Records Regarding Medical Experiments No. 125951–56 re WCG File No. 707 Medical Experiments 13 August 1946 re Hoven and Buchenwald, and No.

125989 War Crimes Group, Inter-office Memo, 19 August 1946. Cf. BB/35/264 4c McHaney to Taylor, 6 August 1946. Wolfson to Miss Foster, 22 August 1946, re Conti files and the Brandt case.

208. TNA: PRO FO 371/66581 note of 28 May 1947.
209. UNA PAG 3/1.0.0. 1–3, Reel 33 UNWCC M. 110, 4 31 July 1946. Breaches of Medical Ethics Doc. C.214, Schram-Nielsen to Lord Wright, 10 July 1946.
210. UNA UNWCC M.111, 2, 21 September 1946, Proposed General Survey of Medical Crimes. Doc. C.223, 30 August 1946. Letter of Taylor to Col. Springer, 21 August 1946.
211. RCPE Smith Papers, box 9 folder 94 War Crimes, H.H. Wade to Smith, 24 July and 2, 12 August 1946.
212. RCPE Smith Papers, H.H. Wade to Smith, 24 July 1946. Wade to Smith, 2 and 12 August 1946. UNWCC archives, Doc. C.223 citing letter of Smith to Wade, 9 August 1946.
213. AFP Leo to Phyllis Alexander, 5 March 1945.
214. UNWCC M 113, 5–6, 2 October 1946 Investigation and Report on Medical Crimes. For Shapcott, see Bower, *Pledge*, pp. 116, 186–7.
215. C. Lester Walker, 'Secrets by the Thousands', *Harper's Magazine*, vol. 193 (1945) 329–36. 'Citizen Doctor', *Time* (13 January 1947) 47. 'Thanatologists', *Newsweek*, vol. 28 (23 December 1946) 31. 'Herr Dr Sadist. Twenty-three Nazi Doctors Arraigned in War Crimes Trial', *Newsweek*, vol. 28 (16 December 1946) 65.
216. Julian Huxley, A.C. Haddon and A.M. Carr-Saunders, *We Europeans* (Harmondsworth: Penguin, 1939).
217. UNESCO Archive, Julian Huxley to John W. Thompson, 5 March 1947.
218. Rice University, Huxley Papers, Diary, 2 December 1946 on meeting with Hill and Dale.
219. Rice University, Huxley Papers, 17 July 1948, Huxley to 'My dear Lorenz'.
220. NYPL Lifton Papers, box M9, notes on McHaney, 29 March 1978, p. 5.

7 Pseudo-science and Psychopaths

1. Kristie Mackrakis, *Surviving the Swastika. Scientific Research in Nazi Germany* (New York: Oxford University Press, 1993), pp. 64–5.
2. NMT 8/1398 Sievers Stellungnahme.
3. AdR Vienna BmfU Kurator A2 Nr 7101, 1940–45.
4. Susanne Heim (ed.), 'Forschung für die Autarkie. Agrarwissenschaft an Kaiser-Wilhelm-Instituten im Nationalsozialismus', *Autarkie und Ostexpansion. Pflanzenzucht und Agrarforschung im Nationalsozialismus* (Göttingen: Wallstein, 2002), pp. 145–77.
5. Ulrike Kohl, *Die Präsidenten der Kaiser-Wilhelm-Gesellschaft im Nationalsozialismus* (Stuttgart, 2002).
6. Peter Alter, 'Die Kaiser-Wilhelm-Gesellschaft in den deutsch-britischen Wissenschaftsbeziehungen', in R. Vierhaus and B. vom Brocke (eds), *Forschung im Spannungsfeld von Politik und Gesellschaft. Geschichte und Struktur der Kaiser-Wilhelm-/Max-Planck-Gesellschaft* (Stuttgart: Deutsche Verlagsanstalt, 1990), p. 745 note 59. Oexle, *British Roots*, pp. 17–19.
7. Manfred Heinemann, 'Die Wiederaufbau der Kaiser-Wilhelm-Gesellschaft und die Neugründungen der Max-Planck-Gesellschaft (1945–1949)', in R. Vierhaus and B. vom Brocke (eds), *Forschung*, pp. 407–70.
8. Mackrakis, *Surviving the Swastika*, pp. 192–3.

9. Heinemann, 'Wiederaufbau', 437. Armin Hermann, 'Science under Foreign Rule. Germany 1945–1949', Fritz Kraft and Christoph Scriba (eds), *XVIIIth International Congress of the History of Science* (Stuttgart: Steiner, 1993), pp. 85–6.

10. Heinemann, 'Wiederaufbau', 430, for preliminary advice from the German Scientific Advisory Board 422, see also TNA: PRO FO 1012/424, and Royal Society Dale Papers 46.6.3, diaries 7 January 1947 Scientific Committee for Germany, also 3 February and 3 March 1947, 15 January 1948.

11. H. Dale, 'Experiment in Medicine', *Cambridge Historical Journal*, 8 (1946) 166–78, 178.

12. Walter C. Langer, *The Mind of Adolf Hitler. The Secret Wartime Report* (New York: Basic Books, 1972).

13. Tusa, *Nuremberg*, p. 161.

14. LC Jackson papers, Kelley report on Hess, 16 October 1945.

15. APD 4/35 Memorandum to Judge Sebring, 15 July 1947. George Soldan, *Der Mensch und die Schlacht der Zukunft* (Oldenburg: Gerhard Stalling, 1925).

16. AFP Leo Alexander to Phyllis Alexander, 15 September 1944.

17. Notably Tracy Putnam. AFP, Leo to Phyllis Alexander, 9 August 1944.

18. Creel, *War Criminals*, 70–7.

19. LC Jackson papers, M.C.B. Memorandum for Justice Jackson, 15 June 1945.

20. LC Jackson papers, Millet to Jackson, 11 June 1945, Jackson to Millet, 23 June 1945.

21. LC Jackson Papers Millet to Jackson, 16 August 1945, Jackson to Millet, 19 September 1945.

22. LC Jackson Papers, Outline for Procedure in Psycho-sociologic Evaluation.

23. LC Jackson Papers, Jackson to Millet, 23 June 1945.

24. Gustave M. Gilbert, *Nuremberg Diary* (New York: Farrar, Straus, 1947), pp. 3–4. Florence R. Miale and Michael Selzer, *The Nuremberg Mind. The Psychology of the Nazi Leaders* (New York: Quadrangle, 1975).

25. LC Jackson Papers, Memorandum to Paul Schroeder, 17 December 1945.

26. LC Jackson papers, Personality Studies of the Dachau physicians Fritz Hintermayer, Wilhelm Witteler, Klaus Schilling and Fridolin Puhr (22 February 1946).

27. LC Jackson Papers, Jackson to John Millet, 26 June 1946.

28. LC Jackson Papers, Millet to Jackson, 4 June 1946; Jackson to Millet, 26 June 1946.

29. LC Jackson Papers, G.M. Gilbert to Jackson, 16 October 1946. Gustave M. Gilbert, *The Psychology of Dictatorship. Based on an Examination of the Leaders of Nazi Germany* (New York: The Ronald Press Co., 1950).

30. Alexander Mitscherlich, 'Nürnberger Trichter 1945', *Gesammelte Schriften*, vol. 7 (Frankfurt: Suhrkamp, 1983) pp. 131–5 (first published in *Rhein-Neckar Zeitung*, 25 January 1946).

31. MP II2/106.4 Mitscherlich to Secretary General of the Military Tribunal, 29 January 1947.

32. AFP Leo to Phyllis Alexander, 25 August 1945.

33. *Trial of German Major War Criminals*, vol. 1, p. 300; vol. 21, p. 275. Douglas M. Kelley, *22 Cells in Nuremberg* (New York: Greenberg, 1947).

34. Oskar Vogt, 'Der Nazismus vor dem Richterstuhl der Biologie. Allen Opfern des Nazismus, den toten und den überlebenden gewidmet', unpublished paper. Vogt, 'Der Nazismus vor dem Richterstuhl der Biologie', *Schweizerische Monatsschrift*, no. 11 (November 1947). Michael Hagner, 'Im Pantheon der

Gehirne. Die Elite- und Rassengehirnforschung von Oskar und Cécile Vogt', Schmuhl, *Rassenforschung an Kaiser-Wilhelm-Instituten*, pp. 139–43.

35. Cécile und Oskar Vogt-Archiv, Düsseldorf, vol. 48 Klingelhöfer to Vogt, 23 April 1946; Kempner to Klingelhöfer, 27 April 1946; Vogt to General Secretary of the IMT, 25 May 1946; Vogt to Kempner, 5 June 1946.

8 The Nuremberg Vortex

1. James Campbell, *The Bombing of Nuremberg* (Garden City: Doubleday, 1974). Martin Middlebrook, *The Nuremberg Raid 30–31 March 1944* (London: Allen Lane, 1973).
2. AFP Leo to Phyllis Alexander, 27 November 1946.
3. Alice Ricciardi-von Platen, 'Geleitwort', Stephan Kolb, Horst Seithe/ IPPNW (eds), *Medizin und Gewissen. 50 Jahre nach dem Nürnberger Ärzteprozess* (Frankfurt-am-Main: Mabuse, 1998), pp. 13–14. Alice Ricciardi personal communication, 13 December 1998.
4. The grenade incident occurred during the later Krupp Trial. Jacob, 'Remembering', 113.
5. Mitscherlich and Mielke, in *Doctors of Infamy*, p. 151.
6. Taylor, *Anatomy*, pp. 625–32.
7. IWM, Milch diary, 26 January 1948; also agitation of Earl Carroll. Maguire, *Law and War*, pp. 150–1.
8. David Irving, *The Rise and Fall of the Luftwaffe. The Life of Luftwaffe Marshal Erhard Milch* (London: Weidenfeld and Nicolson, 1973) and *Nuremberg The Last Battle* (London: Focal Point, 1996).
9. AFP NYPL Lifton Papers M9, Lifton notes on McHaney, 29 March 1978.
10. TTP–CLS 14/8/25/539 Taylor to Harkness 14/8/25/539, 16 November 1993. Cf. 14/6/14/306 Taylor to Amon Amir, 4 April 1977. Taylor to M.A. Grodin 14/7/23/519, 6 March 1989.
11. NMT 2/9267 Hardy states, 'I am unable to read German.' Cf. NYPL Lifton Papers M9, Lifton notes on McHaney, 29 March 1978.
12. Marrus, *Nuremberg War Crimes Trial*, pp.187–8. Bloxham, *Genocide on Trial*, pp. 67–8, 88–90.
13. NARA RG 153 NM 70 Entry 188 190/19/box 2 Leo Alexander. 'Course of Life', October 1946.
14. MPG II Abt. 20 B Nr 105, Steinbauer to Tönnis, 2 January 1948.
15. UNA UNWCC M110 meeting of 31 July 1946: 'Lieut. Colonel WADE suggested that a suitable expert might be Major Leo Alexander, the author of an intelligence report on the "Treatment of Shock from Prolonged Exposure to Cold", which was a very technical piece of work.' Shortly afterwards, the UNWCC Chairman met Telford Taylor.
16. RAC RF1.2/200A/box 126/folder 1114 L. Alexander to R.A. Lambert, 5 March 1946.
17. RAC RF1.2/200A/box 126/folder 1114, 1 November 1945, Leo Alexander to Alan Gregg. RAL diary, 5 November 1945.
18. APD Leo Alexander, 'Application for Federal Employment', 9 November 1947.
19. 'Der Mord an Professor Gustav Alexander. Racheakt eines Patienten nach 27 Jahren', *Neue Freie Presse*, no. 24275 (13 April 1932). Leo Alexander, 'Why I Became a Doctor and why I Became the Sort of Doctor I am', in Noah D. Fabricant (ed.), *Why We Became Doctors* (New York: Grune, 1954). AFP Leo Alexander to Master Gusty Alexander, 31 December 1946.

20. AFP Leo to Phyllis Alexander, 13 November 1946. APD 4/33 Logbook 13 November 1946, citing 'The exploitation of foreign labour by Germany', John H. E. Fried, International Labour Office, Montreal 1945.
21. LBI MF 490 John H.E. Fried papers reels 1, 3–6.
22. Archives of Charles University, Prague, German University, Doktoren Matrikel Bd 5, S. 8. CSA, Series 288 Office of the Czechoslovak Delegation at the War Crimes Commission Box 39 London Embassy to Ecer, 26 April 1946; Ecer, 22 June 1946. AFPB, Alexander to Phyllis Alexander, 27 November 1946. Zeck, 'Nuremberg', 352.
23. Cf. Hedy Epstein website. Personal communications.
24. USHMM Ben Ferencz papers, Autobiographical notes.
25. Ben Ferencz to author, 11 January 2001.
26. Transcript of radio interview with Alexander, 10 January 1947; Tusa, *Nuremberg*, pp. 219–20.
27. NARA RG 238 entry 146 box 1, Thomas K. Hodges, affidavit, 3 November 1947.
28. Yale University, Fortunoff Video Collection, OO: 42.
29. USHMM video 06/22/92, and interview. Ed. Bruce M. Stave and Michelle Palmer, 'Witnesses to Nuremberg' (Twayne Publishers). Interview at Thomas J. Dodd Research Center, University of Connecticut.
30. LC Jackson papers, Taylor to Jackson, 2 September 1946. Taylor, *Anatomy*, pp. 418–20. Jacob, 'Remembering', 111.
31. NMT 1/158, OMGUS order, 25 October 1946. W. Paul Burman (ed.), *The First German War Crimes Trial. Chief Judge Walter Beals' Desk Notebook of the Doctors' Trial* (Chapel Hill: Documentary Publications, 1985).
32. WUS Beals Papers, Carnegie Endowment for International Peace to Beals, 10 March 1947.
33. Maguire, *Law and War*, pp. 172–6.
34. Victor Clarence Swearingen, 'Nurnberg War Crimes Trials', *Kentucky State Bar Journal*, vol 12 (1947) 11–20.
35. LC Jackson papers, Taylor to Jackson, 21 May 1947. University of Washington, Walter Beals Papers Acc. 126, Box 1 folder 16, Eugen Flegenheimer to Beals, 17 November 1946.
36. Bruce R. Jacob, 'Remembering a Great Dean: Harold L. "Tom" Sebring', *Stetson Law Review*, vol. 30 (2000) 73–173, pp. 165–6. Reflections and Memories: Justice H.L. Tom Sebring and Justice E. Harris Drew, 2 November 1990, Stetson University College of Law, *Stetson Law Review*, Honors, Dean Sebring, 5 October 2000. My thanks to Pamela Burdett, Stetson Law Library for arranging access to these sources.
37. WUS Beals papers, box 1 folder 26 Beals to Gunn, 4 November 1946. Executive Session, 14 December 1946, p. 12. Peiser was an alternative. NYPL Displaced German Scholars, Box 19 Kempner, born 1899, arrived USA September 1939. Robert M.W. Kempner, *Ankläger einer Epoche: Lebenserinnerungen* (Frankfurt/M: Ullstein, 1983). Dick Pöppmann, 'Robert Kempner und Ernst von Weizsäcker im Wilhelmstrassenprozess', *Jahrbuch zur Geschichte und Wirkung des Holocaust* (2003), 163–97.
38. WUS Beals papers box 1 folder 14 Rockwell to Beals, 20 November 1946. Box 1 folder 38 Beals to Taylor, 19 May 1947. Box 1 folder 26 Beals to Gunn, 4 and 21 November 1946. NMT 2/3ff.
39. UNA UNWCC, 37.
40. Tusa, *Nuremberg*, 3. TTP–CLS Nuremberg Photograph album, OMTPJ – P–4.

41. NARA M 1019/60 interrogation of Rostock, 4 March 1947, p. 13.
42. Douglas Kelley, *22 Cells in Nuremberg* (New York: Greenberg, 1947), pp. 8–10.
43. Cf. Burton Andrus, *I Was the Nuremberg Jailer* (New York: Coward-McCann, 1969); Tusa, *Nuremberg*, pp.126–8 re daily routines and conditions at IMT. Alexander to Hart, 11 March 1947 re food rations.
44. STAN Rep 502A KV- Verteidigung Handakte Hoffmann Nr 26, Tagebuchaufzeichnungen Pokornys während des Prozesses, 5 and 8 January 1947. Eugen Kogon, *Der SS-Staat. Das System der Deutschen Konzentrationslager* (Frankfurt am Main: Verlag der Frankfurter Hefte, 1946).
45. Oberheuser was absent on 28, 29, 30 January, 5, 6, 7, 10, 20, 21 February, 25, 26 March, 8 April, 21 June (afternoons), 11, 12, 13, 24, 25, 26, 27, 28 February, 3, 4, 5, 6, 7, 10, 11, 12, 13, 14, 17, 18, 19, 20, 21 March, 9, 10 April, 1, 2, 5, 15, May, 16, 27, 28, 30 June, 1, 2 July 1947. Rudolf Brandt was ill on 16, 19 February 1947. Gebhardt on 15, 18 May, 21 June. The prison doctors were Charles J. Roska and Roy A. Martin, see NMT 6/1048.
46. STAN Rep 502A KV-Verteidigung Handakten Hoffmann Verteidigung von Rudolf Brandt Nr 7 Verschiedenes. Manuscript von Brandt.
47. Irmgard Müller (former legal assistant at Nuremberg), personal communication.
48. Swearingen, 'Nuernberg War Crimes Trials', 15.
49. NARA RG 338 USAREUR/JAG War Crimes, Records Regarding Medical Experiments, No.125898–95 Hardy to Straight and Straight, Medical Experiments Cases, 7 September 1946, pp. 3, 5.
50. TNA: PRO WO 309/468.
51. NMT 6/14 for objections by Servatius to the term 'personal physician'.
52. Ernst Klee (ed.), *Dokumente zur 'Euthanasie'* (Frankfurt am Main: S. Fischer, 1985). Klee, *Was sie taten*. NARA RG 238 World War II War Crimes Records NM 70 Entry 200 '201 series' Viktor Brack interrogation, 15 July 1946, Fragebogen.
53. NARA RG 153 NM Entry 188 190/19/box 2 Wolfson to Schwenk, 1 October 1953.
54. NARA RG 338 USAR/JAG War Crimes Records Regarding Medical Experiments No.125707 McHaney to Straight, 6 September 1946. M1019/60 interrogation of Otto Rossmann, 4 September 1946.
55. NARA RG 153/84–1/box 1 folder 2 Taylor to Petersen, 30 September 1946.
56. Marrus, *Trial*, pp. 224–6.
57. TNA: PRO WO 309/468. AdeF BB 35/260, 4c McHaney to Ivy, 30 September 1946.
58. NARA RG 238 Office of the Chief of Council for War Crimes, Berlin Branch, entry 202 box 2 NM 70 OMGUS to Berlin Documents Center, 9 October 1946.
59. IWM Milch diary entry, 27 May 1947, comment of Speer to Milch.
60. Helmuth Trischler, 'Aeronautical Research under National Socialism', in Margit Szöllösi-Janze (ed.), *Science in the Third Reich* (Oxford: Berg, 2001), pp. 79–110.
61. NMT 8/2821 Servatius, 'Probleme der Monster-Prozesse', 8.
62. TNA: PRO WO 309/468 Taylor to McHaney, tentative selection of defendants for Medical Atrocities Case, 9 September 1946. NARA RG 238 Office of the Chief of Council for War Crimes, Berlin Branch, entry 202 box 2 NM 70 A.G. Hardy to: All Research Analysts. Subject: Tentative selection of defendants for medical atrocities case, 10 September 1946.
63. TNA: PRO WO 235/305.
64. TNA: PRO WO 235/307 ff. 291–4 for resentment of Rosenthal and Schiedlausky of Oberheuser.
65. NARA M 1019/68 Sievers interrogation 28 August 1946, pp. 11–12.

66. AdR Akten des Gauamts Wiens Nr 148565 Wilhelm Beiglböck.
67. AdR BMfJ 36116/46 Foreign Office Legal Division to Federal Ministry of Justice, 30 July 1946. Alexander secured a permit to visit Lienz and Vienna in May 1947 to interrogate witnesses for the seawater case. OMGUS permit, 3 May 1947, and was in Lienz, 15 May 1947.
68. APD 4/35 Nuremberg War Crimes Papers Alexander to Hardy, 'Austrian Police Records and Witnesses in Regard to the Seawater Case'.
69. *Der Fall Rose*, 7–8.
70. AdR 02/BMfU 1HR G2 Nr 61758. Decree of 6 June 1945.
71. Eduard Pernkopf, 'Nationalsozialismus und Wissenschaft', *Wiener klinische Wochenschrift* (1938) 545–8. M. Hubenstorf, 'Anatomical Science in Vienna, 1938–45', *The Lancet* 155 (22 April 2000) 1385–6. AdR Ministerium f Unterrricht 2111–III 8/46 Eppinger, Berufung gegen Entlassung und politische Beurteilung.
72. AdR Akten des Gauamts Wien Nr 6806.
73. AdR Ministerium f Unterricht K 10/19 Ministerium f Unterricht Personalakten Hans Eppinger.
74. AdR Ministerium f Unterricht 43384–1946 Privatdozent Dr. Wilhelm Beiglböck politische Belastung und Strafsache. DÖW Nr 19321/2 Berka, February 1946.
75. APD 4/35 Alexander memorandum on Witness Albert Gerl, 19 June 1947.
76. Rabofsky, in Erika Weinzierl and Karl R. Stadler (eds), *Justiz und Zeitgeschichte* (Vienna, 1982), pp. 515–18. AdR Hochschulwesen 19328–1946 Bundesministerium f Inneres to Bundesministerium f Unterricht, 4 August 1946; statement by Eppinger, 9 March 1946; memo by Alfred Indra, 10 September 1945 NARA M 1019/24 Handloser interrogation, 5 March 1946 p. 16. NYPL Lifton papers, cutting from *New York Times* on Eppinger.
77. University of Vienna, personal file Hans Eppinger. TNA: PRO FO 1020/466B f. 33B Federal Ministry for Social Administration, 11 June 1946; Howard M. Spiro, 'Eppinger of Vienna: Scientist and Villain?', *Journal of Clinical Gastroenterology*, vol. 6 (1984) 493–7. M. Michael Thaler, 'Looking Back at Looking Aside: The Life and Posthumous Career of the Famous Professor Hans Eppinger', unpublished paper. Beiglböck, 'Hans Eppinger zum Gedächtnis', *Acta hepatologica*, vol. 6 (1957–8) 177–80, C.L, 'What's in a Name? The Eppinger Prize and Nazi Experiments', *The Hastings Center Report*, vol. 14, no. 4 (1984) 3–4.
78. TNA: PRO FO 1020/466A War Criminals: Individual Cases f. 33A Criminal Proceedings re. Beiglböck to be passed, 3 July 1946 to Legal Division. AdR BMfJ 36.755/46 Note der Britischen Legal Division.
79. TNA: PRO FO 1020/466A f. 33A–G.
80. Klee, *Auschwitz*, p. 9.
81. APD 4/35 Alexander to Hardy, Austrian Police records and Witnesses in Regard to the Seawater Case, 11 April 1947. Alexander Logbook, '14 May 1947 to Klagenfurt. Reported to Major R.R. Cooke, M.C. JAG Branch, War Crimes Section, BTA Klagenfurt (BTA 198)'.
82. Christine Teller, 'Carl Schneider. Zur Biographie eines deutschen Wissenschaftlers', *Geschichte und Gesellschaft*, vol. 16 (1990) 464–78. Volker Roelcke, Gerrit Hohendorf and Maike Rotzoll, 'Psychiatric Research and "Euthanasia". The Case of the Psychiatric Department at the University of Heidelberg, 1941–1945', *History of Psychiatry*, vol. 5 (1994), 517–32.
83. *Badische Zeitung* (3 January 1947).
84. BAK ZSg 154/75 Alexander diary for 1945, pp. 118, 136–7.
85. AdeF BB 35/260 Fritz Flemming to Demnitz, Behringwerke, 9 and 16 December 1946; 4 and 11 January 1947.

86. NARA M 1019/7 Bieling interrogation by von Halle, 3 April 1947.
87. On Bieling's postwar career, see Weindling, *Epidemics and Genocide*, pp. 409–11.

9   Internationalism and Interrogations

1. NARA M 1019/59 Romberg interrogation, 11th personality interrogation by Alexander on 27 February 1947, p. 29.
2. NARA M 1019/20 Gebhardt interrogation, 23 July 1947, p. 27.
3. NARA M 1019/20 Gebhardt interrogation, 23 July 1947, pp. 28, 31.
4. TNA: PRO WO 309/762, September 1946 authority to transfer to US authorities, AJM Harris.
5. TNA: PRO WO 307/471 Somerhough to H. Shapcott, JAG Office, 6 September 1946.
6. NMT 8/ 2819 Brunner was a staff member of Gebhardt.
7. TNA: PRO WO 309/471 Comments of Somerhough, 17 June 1946. Tentative List of Defendants for Medical Atrocities Case. McHaney to Mant, 30 September 1946. Hingst was part of the 'Hohenlychen Group', identified by Somerhough in February 1946; cf. WO 309/468 and NMT 8/427. IMT interrogation Auguste Hingst, 18 September 1945. Rolf Rosenthal had been sentenced by the SS to penal servitude for carrying out abortions in 1943. APBU deposition of Kowalewska and Ramdohr on Rosenthal's extreme sadism, 13 November 1946. Deposition of Dr. Rolf Rosenthal to Keith Mant, 10 October 1946.
8. AdeF BB/35/260 4c and TNA:PRO WO 309/471 McHaney to Mant, 30 September 1946. McHaney did not appreciate that Gebhardt and Fischer worked at Hohenlychen.
9. AdeF BB/35/275 Sulfonamides – rapport et études. Mant sent on 13 October 1946 depositions of Sokulska, Winkowska, Piasecka, Iwanska, Maczka (2), Baj, Nedvedova-Nejedla, Kaminka, Report on Ravensbruck Concentration Camp.
10. NARA RG 238 NM 70 entry 144 NMT Executive Session No. 5, 20 November 1946, pp. 11–12.
11. NARA RG 238 Office of the Chief of Counsel for War Crimes, Berlin Branch, entry 202 box 2 NM 70 Schwenk to Taylor, 'Report on My Trip to London' (25 October 1946–3 November 1946). OCCWC to E.H. Schwenk, 18 November 1946 Documents Brought from London to Berlin.
12. TNA: PRO WO 309/470 f. 9, Mant to Brigadier Shapcott, 14 December 1946.
13. BAK ZSg 154/73 Mant report 'Experiments in Ravensbrück Concentration Camp Carried Out under the Direction of Professor Karl Gebhardt'. Mant to McHaney, 5 December 1946. Deposition of Gebhardt sworn before Major Mant, 7 December 1946. TNA: PRO WO 309/471 Thompson to Somerhough, 25 February 1947.
14. TNA: PRO WO 309/468. Lord Russell of Liverpool, *Scourge of the Swastika* (London: Cassell, 1954).
15. AdeF BB/35/260 4c McHaney to Hardy, 25 October 1946, item 134749.
16. AOF Justice d. 8948bis Caisse 3645 pacquet 237 Buchenwald, 4 novembre 1946, SRCHE to Dir Gen Justice, Baden Baden.
17. AdeF BB/35/276 Documents de l'instruction. Typhus, 21 November 1946 'To McHaney, from A. Hochwald. Subject: Trip to Strasbourg'.
18. AdeF BB/35/276 Documents de l'instruction. Typhus Letter Haagen, Palace of Justice, 1 June 1947 to 'Dear Professor' [to Kempner].

19. UNA UNWCC file 37. War Crimes News re 'THE SECOND NUREMBERG TRIALS'. Includes 'The Trial of 23 Doctors' Re US Press Release of 21 November 1946.
20. NMT 6/38–9. Major Theodore Donner and Lt Col. Joseph Chalupsky of Czechoslovakia, Charles Dubost and Constant Quatre of France, Col. Arnaud, Baron Van Tuyll van Sarooskerken and Captain J.J.M. Witlox, and Dr Stanislawa Piotrowski of Poland. 6/170 for Lord Wright on 9 May 1947.
21. NARA RG 238 entry 145 190/12/05 Box 4 OCC Daily Bulletin, 6 March 1947.
22. Cf. APD 1/3 *Bruxelles-médical*. Charles Sillevaerts, 'Le Second Procès de Nuremberg', *Bruxelles-médical*, vol. 26 (29 December 1946) 1597; vol. 27 (5 January 1947) 23; (12 January 1947) 65; (19 January 1947) 118; (26 January 1947) 169; (2 February 1947) 223; (9 February 1947) 285; (16 February 1947) 339; (2 March 1947) 424; (9 March 1947) 528; (16 March 1947) 584; (23 March 1947) 655; (30 March 1947) 729; (6 April 1947) 778; (13 April 1947) 832; (20 April 1947) 888; (4 May 1947) 991; (18 May 1947) 1111; (25 May 1947) 1178; (1 June 1947) 1231; (8 June 1947) 1289; (15 June 1947) 1351; (22 June 1947) 1405; (29 June 1947) 1451; (6 July 1947) 1541; (13 July 1947) 1596; (20 July 1947) 1647; (24 August 1947) 1847; (31 August 1947) 1892; (7 September 1947) 1951; (14 September 1947) 2006; (21 September 1947) 2057; (28 September 1947) 2122; (5 October 1947) 2159; (12 October 1947) 2225. Extracts in NMT 8/2110–2140.
23. A. Ravina, 'La fin du procès des médecins allemands criminels de guerre', *La Presse médicale* (18 October 1947), Nr 61, pp. 705–6.
24. CSA fond 316 Czechoslovak Commission on the Investigation of War Criminals, Nor c 1 Annex 192 Box 45 folder 3 Nuremberg Court Radiogram, 31 October 1946.
25. WUS Beals papers, box 1 folder 3, Ecer to Beals, 12 December 1946.
26. APD 4/33 Alexander, Logbook, p. 201.
27. Bower, *Pledge*, pp. 202–3.
28. Cf. Hunt, *Secret Agenda*, p. 152.
29. NARA RG 238 Entry 159 Box 3 interrogations procedures.
30. NARA RG 338 NM 70 entry 144 Executive Session No. 5, 20 November 1946, pp. 3–4.
31. NARA M 1019/9 R. Brandt interrogation by Rapp, 19 October 1946, pp. 3–4.
32. Fortunoff Collection Yale, Herbert Meyer interview.
33. NARA M 1019/7 Blome interrogation by Devries and Hardy, 24 October 1946.
34. NARA RG 338 USAREUR/JAG War Crimes, Records Regarding Medical Experiments No. 125995 Suggestions Relative to Interrogation and the Secural of Information of Scientific Interest.
35. NARA M/1019/62 Interrogation of Schäfer by Alexander and Ivy, 22 January 1947.
36. NARA M 1019/50 Oberheuser interrogation, 28 December 1946, p. 17.
37. NARA M 1019/60 Rose interrogation by Alexander, 1 February 1947, p. 30.
38. NARA M 1019/60 Rose interrogation by Alexander, 27 December 1946.
39. RAC RF 1.2/200A/ box 126/ folder 1114 Duke University Psychiatry, 1 November 1945 Leo Alexander to Alan Gregg.
40. AFP Leo to Phyllis Alexander, 31 December 1946.
41. NARA M 1019/7 Blome interrogation, 26 February 1946, pp. 47–9, 51.
42. APD 4/40 Radio interview, 27 January 1947.
43. NARA M 1019/60 Rose interrogation, 27 December 1946, pp. 44–7.
44. NARA M 1019/61 Ruff interrogation, 23 January 1947, pp. 13–17.

45. NARA M 1019/5 Becker-Freyseng interrogation, 27 November 1946, pp. 17–19.
46. BAK ZSg 154 Sammlung Pross, Bd 73 Schröder memo concerning interrogation of 2 December 1946.
47. NARA M 1019/20 Gebhardt interrogation, 23 July 1947, pp. 29–31. The psychologist Wanda von Baeyer elaborated the concept to Alexander.
48. APD Memorandum to McHaney, 'Suggestions for Inclusion in the opening statement of the Pohl Case (WVHA) "The Concentration Camps as Training Grounds for Criminal Behavior and as Proving Grounds for Readiness to Criminal Behavior on the Part of SS Personnel"', 26 March 1947, p. 23. Kershaw, *Hitler. Nemesis*, p. 772.
49. On 28 April 1947 Alexander spoke at the Nurnberg Service Club on 'The Social Psychology of the Nazis' (= Daily OCC Bulletin no. 63, 28 April 1947 in: NARA RG 238 entry 145 Box 4).
50. APD 4/34 Interrogation of Rudolf Brandt, 23 November 1946.
51. Alexander, 'Psychology', 3, 128.
52. APD 4/33 Alexander Logbook 186–7, early February 1947.
53. TNA: PRO WO 309/471 Somerhough to British Liaison Office Baden-Baden, 8 November 1946. Mant fondly remembered Bayle as a supplier of cigars and copies of trial proceedings, personal communication.
54. STAN Rep 502A KV Verteidigung Handakten Fritz 24, Rose, Schriftprobe, 27 December 1946. Executive Session No. 5, 20 November 1946, p. 11.
55. Bettina Ewerbeck, *Gasbrand* (Ulm, 1955). *Gasbrand* means gangrene.
56. APD 4/33 Alexander, Logbook, 26 December 1946, 9 January 1947.
57. Bayle, *Psychologie*, 318. These interrogations are listed in STAN KV Rep 502 VI gen 10 Verzeichnis der Interrogations Aug-Dez 1946 chronologisch; gen 11 January-April 1947; 12 May-August 1947;13 September-December 1947; 14 January-May 1948. Transcripts of the later interrogations by Bayle and Thompson have not been located.
58. François Bayle, *Croix gamée contre caducée. Les expériences humaines en Allemagne pendant la deuxième guerre mondiale* (Neustadt: Imprimerie Nationale, 1950). François Bayle, *Psychologie et Ethique du National-socialisme. Etude Anthropologique des Dirigeants SS* (Paris: Presses Universitaires de France, 1953).
59. MP II2/112.20 Mitscherlich to Oelemann, 4 December 1946; II2/106.2 Mitscherlich to Military Tribunal No. 1, 9 December 1946.
60. Platen, 'Ärzteprozess', 29.
61. MP II2/106.4 Mitscherlich to Tribunal 29 January 1947. MP II.2/112.25 Mitscherlich to Oelemann, 21 February 1947.
62. NYPL Lemkin papers Reel 4.
63. NARA RG 238 Entry 155 Box 1 Taylor 'An Outline of the Research and Publication Possibilities of the War Crimes Trials', November 1948.
64. NAC RG 25 External Affairs Series A–3–b, Volume 5792 File 250(s) Request from Wing Cdr John Thompson to Remain in Europe on Special Work.
65. NARA M 1019/68 Sievers interrogation by Thompson, David Walker, Meyer, Ramler 5 February 1947; M 1019/7 Blome interrogation by Thompson, 8 February 1947.
66. NARA M 1019/7 Blome interrogation by Thompson with Flight Officer David Walker; Sauter and Ramler present, 8 February 1947, p. 12.
67. NARA M 1019/68 Sievers interrogation by Thompson, David Walker, Meyer, Ramler, 5 February 1947; M 1019/7 Blome interrogation by Thompson with Flight Officer David Walker; Sauter and Ramler present, 8 February 1947.

68. Cf. Ernst Wagemann, *Narrenspiegel der Statistik: die Umrisse eines statistischenWeltbildes* (Hamburg: Hanseatische Verlagsanstalt, 1942) pp. 103–5, 173–9. Doz. med habil Carl Hermann Lasch, 'Krebskrankenstatistik. Beginn und Aussicht', Habilitation, Rostock, 1940.

69. NARA M 1019/7 Blome interrogation by Thompson with Flight Officer David Walker; Sauter and Ramler present, 8 February 1947, pp. 1–9.

70. NARA M 1019/7 Blome interrogation by Thompson with Flight Officer David Walker; Sauter and Ramler present, 8 February 1947, pp. 15–23.

71. NARA M 1019/68 Sievers interrogation, 5 February 1947, p. 1. Hielscher affidavit, 10 March 1946. Sievers gave Ernst Ebert as a second witness in a questionnaire of 5 March 1946.

72. NARA M 1019/68 Sievers interrogation, 20 August 1946, p. 7.

73. NARA M1019/8 Brack, 12 September 1946, p. 7; on Globocnik, 13 September, pp. 12–14.

74. NARA M 1019/20 Gebhardt interrogation, 5 November 1946, pp. 21–5.

75. NARA M 1019/20 Gebhardt interrogation, 25 September 1945, pp. 12–13. NARA RG 243 US Strategic Bombing Survey Entry 2 (section 5) box 1 Interview Major General Schroeder.

76. NMT 2/2400 3 February 1947.

77. NARA M 1019/20 Gebhardt interrogation, 18 October 1945 am, 7–8.

78. NARA M 1019/68 Sievers interrogation 29 August 1946 pm.

79. NARA M 1019/9 R. Brandt interrogation, 4 September 1946, pp. 9, 15.

80. NARA M 1019/20 Genzken and Mrugowsky interrogation, 23 September 1946.

81. NARA M 1019/68 Sievers interrogation, 20 August 1946, p. 8.

82. NARA M 1019/9 R. Brandt interrogations.

83. Mant recollects that Alexander viewed Rascher as schizophrenic, Mant, *From Nuremberg*, chapter 2 'War Crimes Investigations'.

84. NARA M1019/60 Rostock interrogation by Alexander, 4 March 1947, p. 30.

85. APD 4/34 Neff interrogation summary 13 December; 1/3 Alexander, Summary of Interrogation of Neff, 14 December 1946.

86. NARA M 1019/68 Sievers interrogation, 20 August 1946, pp. 9–11.

87. NARA M 1019/60 Rostock interrogation, English text, pp. 7, 14, 19, 31–33. 1019/20 Genzken and Gebhardt interrogations. 1019/20 Gebhardt interrogation.

88. NARA M 1019/24 Interrogations of Handloser by Rapp, e.g. 24 September 1946.

89. Sueskind in *Suddeutsche Zeitung* Nr 106, 14 December 1946, commenting on the Trial as 'ein duesteres Kapitel moderner Kulturgeschichte', and that it was officially not an 'Ärzteprozess' but 'ein Prozess gegen SS Ärzte und die deutsche Wissenschaftler' with 3 SS administrators, and only ten SS members.

90. NARA M 1019/20 Gebhardt interrogation, 18 October 1945, pp. 4–8.

91. NARA M 1019/78 Weltz interrogation 11 January 1947, pp. 11–12.

92. Cf. Karin Orth, *Das System der nationalsozialistischen Konzentrationslager* (Hamburg: Hamburger Edition, 1999).

93. NARA M 1019/9 R. Brandt interrogation, 4 September 1946, p. 14.

94. NARA M1019/60 Rostock interrogation by Alexander, 4 March 1947, p. 10.

95. Aziz, *Karl Brandt*,p. 13. Karl Brandt, Angeborener Verschluss der Gallenausfuhrgänge, med Diss. Freiburg 1929, 51 Lebenslauf.

96. Isabel Heinemann, *'Rasse, Siedlung, Deutsches Blut'. Die Rasse- und Siedlungshauptamt der SS und die rassenpolitische Neuordnung Europas* (Göttingen: Wallstein Verlag, 2003), pp. 16–17.

97. NARA M 1019/8 interrogation of Brack by Rodell, 13 September 1946.
98. NARA M 1019/8 interrogation of Brack by Rodell, 13 September 1946, p. 14.
99. Cf. Dieter Pohl, *Von der Judenpolitik zum Judenmord. Der Distrikt Lublin des Generalgouvernement 1939–1944* (Frankfurt/M., 1996). *Täter und Gehilfen des Endlösungswahns. Hamburger Verfahren wegen NS-Gewaltverbrechen 1946–1996.* Ed. By Helge Grabitz (Hamburg: Ergebnisse-Verlag, 1999).
100. NARA M 1019/7 Blome interrogation by Rapp, 22 October 1946, pp. 1–3.
101. Kurt Blome, *Arzt im Kampf. Erlebnisse und Gedanken* (Leipzig: Ambrosius Barth, 1942).
102. NARA RG 238 NM 70 Entry 200 OCCWC. Executive Office Nuremberg Administrative Post Internee Personnel Records ('201 Files') box 3 'A short description of the official functions of the Reichsärztekammer, Dustbin/Kransberg 17.8.45'.
103. NARA M 1019/7 Blome interrogation by Devries, 7 October 1946, summary pp. 3–4.
104. NARA M 1019/ 7 Blome interrogation by Rapp, 22 October 1946, p. 1.
105. APD 4/33 Alexander logbook 5 December 1946, 'Biography of Defendant Kurt Blome'.
106. Lagnado and Dekel, *Children of the Flames*, pp. 120–9.
107. NYPL LP 8 June 1977 telephone talk with James McHaney, p. 9.
108. B. Skinner Mandelaub, 'Court Archives History', 15 November 1949, NARA RG 238/ entry 145/ box 2.
109. TNA: PRO WO 235/83.
110. NARA RG 153 box 59, 100–621 Olga Lang, 5 November 1946.
111. AdeF BB/35/260 letters to trial authorities.
112. NARA M 1019/60 Rostock interrogation, 4 March 1947, p. 26; M 1019/7 Blome interrogation, 26 February 1946, p. 35.
113. *War Crimes News*, 'THE SECOND NUREMBERG TRIALS', re 'The Trial of 23 Doctors', US Press Release of 21 November 1946.
114. NARA M 1019/ 9 R. Brandt interrogation by Rapp, 24 October 1946, pp. 4–5.
115. TTP-CLS Box 6, Iwan Devries to Kempner, 21 February 1949.
116. LC Jackson Papers, Gilbert to Jackson, 16 October 1946.
117. BAK Sammlung Pross ZSg 154/73 'Bericht des Badener Rundfunks vom 21. Febr. 1947. 19.40 Uhr'.
118. Joachim Mrugowsky, MS History of typhus. Transcribed by Dr Gertrud Rudat. I am grateful to Professor Hartmut Mrugowsky for making this available.
119. Hunt, *Secret Agenda*, 92.
120. STAN Rep 502A KV- Verteidigung Handakten Hoffmann Nr. 7 Rudolf Brandt, Erste Gedankenzusammenstellung über Erziehungsfragen, 21 July 1946.
121. Defence lawyers were present for Becker-Freyseng on 27 November 1946.
122. Platen, 'Aus dem Rechtsleben. I Anklage', p. 29.
123. STAN Rep 502A KV- Verteidigung Handakten Sauter Nr 3, Bettina Blome to Kurt Blome, 14 March 1947.
124. AFP Leo to Phyllis Alexander, 27 November 1946.
125. NARA RG 238 NM 70 Entry 144 Executive Session, 13 November 1946.
126. *Badische Zeitung*, 3 January 1947.
127. Tusa, *Nuremberg*, p. 124.
128. NARA RG 238 NM 70 Entry 144 Executive Session, 12 November 1946.
129. STAN Rep 502A KV- Verteidigung Handakten Hoffmann Nr 6, Rudolf Brandt to Hoffmann, 17 March 1948: 'since 21 March 1947 you have not answered any of my 10 letters or acknowledged receipt'.

130. STAN Rep 502A KV- Verteidigung Handakten Rostock Nr 3 Plädoyer-Entwurf, letter from Pribilla on 17 March 1947.
131. STAN Rep 502A KV- Verteidigung Handakten Rostock Nr 3 Plädoyer-Entwurf Pribilla to Rostock, 17 March 1947.
132. STAN Rep 502A KV- Verteidigung Handakten Fröschmann Nr 23 Revision und Gnadengesuch, 27 August 1947. Fröschmann to Dir M Schlegelmilch.
133. NARA RG 338 Stack 290, Row 59 Compartment 33, shelf 1–3 Executee Files: Landsberg Prison: Karl Brandt.
134. Personal information.
135. NARA RG 338 USAR/JAG War Crimes Records Regarding Medical Experiments No.127367–8. Re letter to Franz Gross of CIBA on behalf of Romberg.
136. NARA RG 338 290–59–17 Medical Experiments '707 files' intercepted letter regarding Pokorny's divorce, No. 127370, 3 January 1947.
137. TNA: PRO WO 309/762 Dr Karl Brandt, 'Prepared Defense Statement for Nazi Criminal on Trial Dictated from Nurnberg Lawyer's Office to German in Hamburg', 12 February 1947.
138. WUS Beals papers Box 1 folder 3 Clay to Beals, 9 May 1947.
139. WUS Beals papers Box 1, Fröschmann to Beals, 7 August 1947.
140. STAN Rep 502A KV Verteidigung Pokorny Nr 23 Verfahrenfragen. Sauter 1, 3 March 1947 to Wirtschaftsamt; Hunt, *Secret Agenda*, 87. Complaint of 13 January 1947.
141. NARA RG 238 NM 70 Entry 144 Executive Session, 7 December 1946, pp. 3–4.
142. STAN Rep 502A KV- Verteidigung Handakten Fröschmann Nr 23 Besprechungs-Notiz, 25 February 1948.
143. NARA M 1019/9 interrogation by Alexander, 26 April 1947, p. 19.
144. NARA RG 238 NM70 – Entry 200 OCCWC '201 files' Box 5 Elfriede Conti letter of 17 January 1946; reply of 13 May 1946.
145. BAK Sammlung Pross ZSg 154/73, note of 16 November 1946.
146. Lilly Pokorná-Weilová was a radiologist who qualified in 1922. Cf. *Zdravotnická Ročenka Československá* (1938), 625.
147. NARA RG 466 Entry 54 Box 13 Hella von Schultz to Bickel ICRC delegate to Germany, 10 April 1946. Rose statement, April 1946.
148. NARA RG 466 Entry 54 Box 13 Suggestion for Sonderauftrag for Rose, 27 June 1946; Hella von Schultz to Military Government Wurttemberg-Baden, 13 September 1946. von Schultz to Rose, 24 June 1946.
149. NARA RG 466 Entry 54 Schröder file.
150. NMT 8/1054–6 Bettina Blome to Sauter, 30 November 1946.
151. STAN Rep 502A KV- Verteidigung Handakten Sauter Nr 3, Bettina Blome to Kurt Blome, 14 March 1947, also nd.
152. e.g. R. Brandt, NMT 4/2628.
153. NARA M 1019/60 Interrogation of Ruff by H. Meyer, 25 February 1947.
154. STAN Rep 502A KV- Verteidigung Handakten Sauter 3, Bettina Blome to Kurt Blome, 14 March 1947.
155. STAN Rep 502A KV- Verteidigung Handakten Rostock 10 Rostock to von Weizsäcker, 16 July 1947.
156. Staatsarchiv Hamburg 352–8/9 Bernhard-Nocht-Institut Schriftwechsel in die Angelegenheit Dr med Gerhart Rose Band 1 Dr Hella von Schultz to Nauck, 28 January 1946. Nauck to Generalarzt d. R. Prof Dr Rose, 1 March 1946, 10 March 1946.

157. Weindling, *Epidemics*, p. 408. STAH 352–8/9 Bernhard Nocht-Institut, Schriftwechsel in der Angelegenheit Dr med Gerhart Rose, Nauck to Rose, 30 May 1946.
158. STAN Rep 502A KV- Verteidigung Handakten Fritz Nr 24, Rose draft for Bayle's Stammbuch, 27 December 1946.
159. TTP-CLS-14/8/25/539 Taylor to Harkness, 16 November 1993.
160. NMT 6/263, 19 July 1947; 264 on 19 August 1947.
161. AdeF BB/35/271 guerre bactériologique – mémoire et étude, Hardy to Taylor, 9 April 1947.
162. AdeF BB/35/271 Taylor to US ambassador, 9 April 1947 for transmission to the Soviet Political Advisor (Ministry of Justice, deleted).
163. NARA RG 153/86–3–1/Box 11/ folder 4/book 3 Andrew Ivy, Outline of Itinerary.
164. NARA RG 153/86–3–1/box 10/book 1 Taylor to War Department, 1 November 1946.
165. STAN Rep 502A KV- Verteidigung Handakten, Fritz Nr 35.
166. NARA RG 238 box 1 entry 199 defence council, Walter Rapp to Major Teich, 27 November 1946.
167. NARA RG 338 Parolee case files Landsberg Prison.
168. NARA RG 153/86–3–1/box 10/book 1. RG 153/86–2–2/box 9 book 2, D. Marcus to Ivy 7 November 1946.
169. TNA: PRO WO/470 f. 9 Mant to Shapcott, 14 December 1946.
170. APD 4/34 Alexander Memorandum to Telford Taylor, Suggestions for a Discussion of the Thanatology Genocide Angle, 5 December 1946. NARA RG 238 Entry 155 Box 1 Alexander, The Perversion of the Attitude Towards Death (thanatolatry).
171. APD 4/40 Wolfe Frank radio interview with Alexander, 10 January 1947.
172. NMT 2/50–123 Telford Taylor, 'Opening Statement of the Prosecution', 9 December 1946.

10 Science in Behemoth: The Human Experiments

1. NARA RG 153 86–2–2 box 9 folder 2 'Herr Dr. Sadist. Twenty-three Nazi Doctors Arraigned in War Crimes Trial', *Newsweek*, vol. 28 (2 December 1946) 65. Press Release, 19 November 1946, re experiments on human subjects.
2. Franz Neumann, *Behemoth, The Structure and Practice of National Socialism, 1933–1944* (New York, 1966), 1st edn 1942, pp. xii, xv, 98, 111–12. Harry Marks, *The Progress of Experiment* (Cambridge: Cambridge University Press, 1997), p. 101. RAC RU Cohn Papers 450 C661-U Paige Box 1 (18) correspondence with Neumann.
3. Claire Hulme and Michael Salter, 'The Nazi's Persecution of Religion as a War Crime: the OSS's Response within the Nuremberg Trials Process', cited as internet publication, www-camlaw.Rutgers.ed/publications/law-religion/v31.htm, paras. 32–7.
4. Taylor, *Anatomy*, pp.49, 90. Hulme and Salter, 'Religion as a War Crime', notes 63 and 64 on Carl Schorske's recollections.
5. Taylor, 'The Meaning of the Nuremberg Trials', 9.
6. NMT 2/11185 on Genzken as the fifth in the hierarchy.
7. Astrid Ley (ed.), *Menschenversuche* (Erlangen Museum, 2001). D'Addario photos are reproduced in *Medizin und Gewissen*, pp. 459–67.

8. NARA M 1019/9 Rudolf Brandt interrogation, 19 August 1946, English text, p. 6.
9. Alice Ricciardi-von Platen, author's interview, tape 2 side 2 for shocked impressions of Oberheuser.
10. NARA M 1019/17 Fikentscher interrogation, 6 September 1946.
11. NARA M 1019/20 Gebhardt interrogation, 5 November 1946, pp. 14, 16.
12. Michael Kater, *Doctors under Hitler* (Chapel Hill, 1989), pp. 54–61.
13. NMT 2/ 11549 Oberheuser, concluding plea.
14. NARA M 1019/50 Oberheuser interrogation, 28 December 1946, p.18.
15. Herta Oberheuser, 'Zuckungs- und Wulstschwelle des musculus rectus abdominalis des Frosches und ihre Beeinflussung durch die Narkose', Med Diss Bonn 22 März 1937, 'Aus dem physiologischen Institut der Univ Bonn.'
16. NARA M 1270/13 Oberheuser interrogation, 27 September 1945, p. 6.
17. NARA M 1019/5 Beiglböck Interrogation, 28 February 1947.
18. Ludmila Hlavackova and Petr Svobodny, *Biographisches Lexikon der deutschen Medizinischen Fakultät in Prag 1883–1945* (Prague: Charles University, 1998), p. 163.
19. NMT 2/2380, 3 February 1947.
20. For Rose's career see Weindling, *Epidemics and Genocide*, pp. 139, 242–4, 334, 360f, 379–81.
21. The lawyer Weisgerber was born in Metz in 1893, and his assistant Bergler in Mühlhausen in 1895; Karl Brandt was also born there in 1904 and Schäfer in 1911.
22. Gerhard Rose, *Krieg nach dem Kriege* (Dortmund, 1920).
23. BAB Dahlwitz-Hoppegarten ZB2/ 1875 Rose Lebenslauf.
24. APD 1/9 Alexander to Hardy, 'Questions to be put to Dr. Blome on the stand', 12 March 1947.
25. NARA RG 153/86–3–1/Box 11/ folder 4/book 3 Ivy, 'A Report', 21.
26. NARA M 1019/ 68 Sievers interrogations on involvement of RFR in Dachau research, August 1946.
27. AHUB Akten der med Fak Nr 77, 1943–45.
28. AHUB Dekanat Charité 0100/21a Protokolle der Fakultätssitzungen 1946–54, 16 April 1947. Rektorat Nr 76 Parteimitglieder on dismissal of Gebhardt and Rostock. Personalakten 325 Wolfgang Heubner Bd 1. Johanna Kneer, 'Wolfgang Heubner (1877–1957) Leben und Werk', med. Diss. Tübingen, 1989.
29. AHUB Akten der medizinischen Fakultät betr Kurt Blome. Nr 254.
30. NMT 8/999 Faculty meeting, 3 September 1947.
31. *Süddeutsche Zeitung* (5 April 1947), no. 32 'Hippokrates als Warner' by W.E. Süskind. STAN KV-Verteidigung Rep 502A Handakten Rostock Nr 3 Rostock to Pribilla, 14 April 1947.
32. AHUB Akten der med. Fak. betr Stumpfegger Nr 128. B.6 vol 12 Josef Koestler SS-Lazarett Hohenlychen. Treite, Percival/ Percyval. Dozent Dr file closed, 1 June 1943. Personal-Akten med Fak Nr 91.
33. NARA M 1019/20 Gebhardt interrogation, 3 December 1946, pp. 29–31.
34. NARA M 1270 Fischer interrogation, 21 September 1945, 4 January 1947, 26.
35. BAB NS 21/ 916 Sievers, 22 May 1943.
36. APD 4/34 Wiskott to Alexander, 26 November 1946; Alexander to Taylor, 30 November 1946. Wiskott advised Alexander on his investigations of aviation physiology in June 1945, see address book note for 20 June 1945.
37. NARA M 1019/20 Gebhardt interrogation 5 November 1946, p. 13.
38. Kater, *Doctors*, p. 319.
39. AdeF BB/35/260 Rudolph Nissen, 116 East 58th Street, NY, 5 April 1947 to Telford Taylor. Rudolf Nissen, *Helle Blätter – Dunkle Blätter. Erinnerungen eines*

Chirurgen (Stuttgart: Deutsche Verlags-Anstalt, 1969), pp. 84, 163–7. Dorothea Liebermann-Meffert and Hubert Stein, *Rudolf Nissen* (Heidelberg: Barth, 1992).

40. Ferdinand Sauerbruch, *Das war mein Leben* (Bad Wörishofen: Kindler und Schiermeyer, 1951), pp. 556–8.

41. STAN KV Rep 502A Verteidigung Handakten Rostock Nr 3, Rostock to Pribilla, 14 March 1947.

42. AHUB UK Pers. R 225 Bd 2 Bl. 121 Rostock to Dieterici, 31 January 1947. AHUB Charité 0100/21a Protokolle der Fakultätssitzungen 1946–54, 14 June 1950.

43. MPG Abt. III Rep 84/1 Nr 1344 Ebbecke to Dekan Tübingen, 13 June 1946.

44. Kater, *Doctors*, pp. 130–1.

45. Niels Bohr Library, Goudsmit Papers box 27 folder 26, Sievers note, 4 February 1942.

46. NARA M1019/78 Weltz interrogation, 6 November 1946, p. 6.

47. BAK NS 21/ 913 Rascher to Himmler, 9 August 1942. Ibid., Nr 917 Sievers to Dachau Kommandant – to allow a visit from Blome and Strassburger on 9 August 1943. BAK ZSg 154/73 Pross Bd 73 Alexander, Interrogation of Punzengruber, 14 February 1947.

48. NARA M 1019/59 Romberg 26 Nov 1946, 7; M 1019/7 Blome interrogation, 2 October 1946, pp. 29–35, 53. NMT 3/391 Sievers to Rudolf Brandt, 27 September 1943. Gerhard Aumüller (ed), *Marburger Medizinische Fakultät im Dritten Reich* (Munich: K. G. Saur, 2001), pp. 558–62, 638–9 excludes the subsequent Polygal episode. *The Nuremberg Medical Trial 1946/7. Guide to the Microfiche-Edition*, p. 26. Cf. Kater, *Ahnenerbe*, pp. 101f, 231–43. NMT 3/396–7 Sievers to Rudolf Brandt, 21 March 1944, draft letter to Strassburg medical faculty. NMT 3/410 Sievers diary, 22 February 1944, 3/411–419 Polygal meetings, 22 March–8 December 1944.

49. NMT 2/9375–6, 14 June 1946, when Ruff points out the life-saving implications of reducing time spent at 12,000 metres.

50. NMT 2/10920, McHaney closing statement, 14 July 1947.

51. NARA RG 153/86–3–1/Box 11/ folder 4/book 3 Ivy, Questions Relative to the Strategy of the Trials, p. 22.

52. NYPL Lifton Papers M9 notes on McHaney, 29 March 1978, p. 5.

53. Author's interview with Keith Mant, 20 March 1996.

54. AFP Leo to Phyllis Alexander, 18 December 1946.

55. 'Bares Himmler Ire at Beating by U.S. Science', *Chicago Tribune* (30 January 1947).

56. NARA M 1019/20 Genzken interrogation, 20 September 1946, p. 9.

57. NARA M 1019/36 Kogon interrogation by von Halle, 28 November 1946, p. 7.

58. *Trial of German Major War Criminals*, vol. 1, p. 4. Bloxham, *Genocide*, pp. 86–7.

59. NARA M 1019/68 Sievers interrogation, 20 August 1946, p. 6.

60. Ibid., p. 4.

61. NARA M 1019/9 Rudolf Brandt interrogation by Rapp, 19 August 1946, p.18.

62. NMT 4/2577–2579 Karl Brandt, Final Plea, 19 July 1947.

63. Fritz Redlich, *Hitler. Diagnosis of a Destructive Prophet* (Oxford: Oxford University Press, 1999), pp. 130–2, 209, 252.

64. NMT 2/2438 Brandt's loss of post as escort physician.

65. ACDJC CXXIX–20 Karl Brandt interrogation, 17–18 June 1945.

66. BDC Karl Brandt to Wolff, 26 January 1943. Pohl to Karl Brandt, 20 March 1943. Rudolf Brandt to Grawitz, 7 October 1943; Grothmann to Gluecks, 8 February 1944.

67. BDC Karl Brandt to Himmler, 9 June 1944.

68. NARA RG 153/87–0/Box 13 Frank C. Waldrop in '23 Nazis Tried for Scientific Mass Murder', *New York Herald Tribune*, 10 December 1946, quoted in *Daily Trial Report*, No. 83.

69. NARA RG 319 entry 82A (Alsos) medicine files, correspondence of Rostock with Kliewe concerning type cultures; Heilgas experiments in Theresienstadt ghetto; position of medical research in January 1945.

70. STAN Rep 502A KV- Verteidigung Handakten Rostock Nr 2, Draft of final plea, 30 March 1947.

71. NARA RG 238 NM 70 entry 159, Executive session, 13 November 1946.

72. STAN Rep 502A KV Verteidigung Handakten Bergold Nr 4 Karl Wolff on experiments.

73. IWM Milch diary, 31 January 1947.

74. Oppitz, *Medizinverbrechen*, pp. 281–347 for the text.

75. IWM Milch diary extracts, 7 August 1945 on the atom bomb produced at great cost by Jews; 26 January 1948 about the 'Jewish prosecution'.

76. Bower, *Pledge Betrayed*, pp. 246–7; Hunt, *Secret Agenda*, p. 87; NARA M 1019/60 Ruff interrogation, 5 December 1946, pp. 19–22.

77. BAK ZSg 154/73 Pross Ivy, 'Memorandum on Ruff'. APD 4/34 Alexander, 'Pain induced during the pressure experiments by Rascher, Romberg and Ruff'.

78. NMT 8/1344 Sauter to Rein, 17 January 1945.

79. NARA M 1019/78 Weltz interrogation, 9 October 1946, p. 21.

80. NARA M 1019/78 Weltz interrogation by Alexander, 11 January 1947, p. 10.

81. NARA M 1019/59 Romberg interrogation, 26 November 1946, pp. 7–10.

82. APD 4/34 re Hornung, 20 December 1946.

83. Friedlander, *Origins*, pp. 68–9.

84. NMT 4/1428f, 3 February 1947.

85. NARA M1019/9 Rudolf Brandt interrogation by Alexander, 26 April 1947.

86. Marrus, 'Holocaust', 26.

87. NMT 5/36–38.

88. NARA M 1019/ 68 interrogation of Sievers, 16 August 1946, 16. NARA M 1019/68 Sievers interrogation, 3 September 1946, pp. 4–6, 13; 12 September 1946, pp. 3–4.

89. NMT 8/1383–6 Sievers, 'Das Ahnenerbe'.

90. NMT 4/2204.

91. NARA M 1019/54. Pohl interrogation, 10 July 1946.

92. NMT 2/2438 Brandt's chart on duration of experiments, 3 February 1947.

93. Blaha, 'Medizin auf schiefe Ebene', pp. 71, 192 of his autobiographical testimony: 'Wir waren zwar Zeugen und Opfer dieser Geschehnisse, aber wir konnten nicht das ganze Ausmass begreifen'.

94. APD 4/33 Alexander logbook, 13 March 1947.

95. Günther Schwarberg, *The Murders at Bullenhuser Damm. The SS Doctor and the Children* (Bloomington: Indiana University Press, 1984). Schwarberg, *Meine zwanzig Kinder* (Göttingen: Steidl, 1996).

96. Robert Proctor, *The Nazi War on Cancer* (Princeton: Princeton University Press, 1999), 344 n. 4.

97. Weindling, *Epidemics and Genocide*, pp. 355–9.

98. Cf. K. Brandt's table NMT 4/8586 confirming this.

99. Jürgen Peiffer, 'Assessing Neuropathological Research Carried out on Victims of the "Euthanasia Programme"', *History of Psychiatry*, 34 (1999) 339–55. Heinz Faulstich, 'Die Zahl der "Euthanasie"-Opfer', in Andreas Fewer and Clemens Eickhoff (eds), *'Euthanasie' und die aktuelle Sterbehilfe-Debatte* (Frankfurt: Campus, 2000), pp. 218–34.

100. NARA RG 338 USAREUR/JAG War Crimes, Records Regarding Medical Experiments No. 125978–87, AAF Aero Medical Center, Monthly Status Report No. 9, 30 June 1946.
101. Bower, *Paperclip*, pp. 245–6. Hunt, *Secret Agenda*, p. 84. NARA RG 238 190–12–39 Arrest Report Schroeder 16 Sept. 1946. For a photograph of Aero Medical Center scientists see Rolf Winau, *Medizin in Berlin* (Berlin: de Gruyter, 1987), 350. NLM MS C 408 International Military Tribunal, Robert Benford covering letter, 6 August 1977, and List of Personnel Involved in Medical Research and Mercy Killings, nd [1946].
102. NARA M 1019/59 Romberg interrogation, 29 October 1946, p. 32.
103. NARA M 1019/59 Romberg interrogation, 26 November 1946, p. 4.
104. NARA M 1019/60 Interrogation Rostock, 4 March 1947, pp. 28–9.
105. NARA RG 338 290–59–30/33 Parolee Case Files, Landsberg Prison Box 35 Weltz, letter of Büchner, 28 March 1946.
106. Ibid. J.S. Alderman, War crimes group Inter-office memo, subject Dr A.G. Weltz, 19 August 1945.
107. NARA M 1019/78 Weltz interrogation, 9 October 1946, 6 November 1946.
108. NARA M1019/78 Weltz interrogation, 6 November 1946, pp. 10–11. Padfield, *Himmler*, pp. 374–6.
109. NARA M 1019/59 Romberg interrogation by Alexander, 26 November 1946, pp. 13–14.
110. NARA M 1019/59 Romberg Interrogation, 26 November 1946. Romberg affidavit 5 December 1946: in Dachau from March to May 1942.
111. Yale University, MS 1236 Fulton Papers Series III box 271 folder 149, *A Bibliography of Aviation Medicine*, 1946.
112. NARA RG 330 Entry 18 JIOA Foreign Scientists Case File Box 11, Benzinger files.
113. MPG Archiv Abt III Rep 84 Nachlass Butenandt, Theodor and Ilse Benzinger. NARA RG 330 Entry 18 JIOA Foreign Scientists Case File Box 11, Benzinger files. Benzinger attended the Nuremberg meeting on cold in October 1943. Cf. Achim Trunk, *Zweihundert Blutproben aus Auschwitz* (Berlin: MPG, 2003).
114. STAN Rep 502 KV Anklage Interrogations B.56 Benzinger, Theodor 19 and 20 September 1946.
115. Yale University MS 1236 Fulton Papers, *Bibliography of Aviation Medicine*, vol. II, 1946.
116. NARA RG 341 190 64 24 06 Records of the Headquarters US Air Force box 162 Nazi 'Doctors' to be Tried Next, *Stars and Stripes* (12 October 1946).
117. NARA RG 330 Entry 18 JIOA Foreign Scientists Case File Box 11, Benzinger files.
118. BAK ZSg 154 Sammlung Pross Bd 73 Hardy to Alexander, nd, arising from Ivy memo on Ruff.
119. NARA M 1019/61 Ruff interrogation by Alexander, 22 January 1947.
120. NMT 2/9267, 13 June 1947.
121. *Badische neueste Nachrichten* (17 June 1947), 'Der Ärzteprozess. Angeklagter verhört Sachverständigen'.
122. Alexander, 'Treatment of Shock', 21–2.
123. Ibid., 21–9.
124. AFP Leo to Phyllis Alexander, 27 November 1946. Vienna University archives Nationalen WS 1925, forms of Alexander and Beiglböck confirm this. Beiglböck did not recognise Alexander when first interrogated by him – see M 1019/5 interrogation, 27 November 1946, p. 6.

125. Vienna university personal file, recommendation by Eppinger, 1 February 1943; Dean of Medical Faculty to Minister, 19 March 1943; Rust to Rector University of Vienna, 23 June 1944.
126. NMT 2/9476, 16 June 1947 concerning the Vienna experiments. DÖW Nr. 19321/2.
127. AHUB Universitätsarchiv Personalakten 325 Bd 1. Eppinger today is commemorated by a street close to Heubner's Berlin institute.
128. NARA M 1019/62 Schäfer interrogation by Alexander, 17 July 1947.
129. NARA M 1019/5 Interrogation Becker-Freyseng by von Halle, 20 November 1946; by Alexander, 27 November 1946, pp. 1–14. TNA: PRO FO 1020/466A f. 33B Federal Ministry for Social Administration, 11 June 1946. APD 'List of Names and Other Identifying Data of Seawater Experiment Victims'.
130. NMT 2/9112, 11 June 1947.
131. NARA M 1019/7 Interrogation of Blaha by von Halle, 11 December 1946, p. 8.
132. NARA M 1019/ 123 Alexander interrogation Joseph Vorlicek.
133. APD 4/35 Alexander memorandum on Witness Albert Gerl, 19 June 1947.
134. DÖW Nr 2573. Erika Thurnher, *Nationalsozialismus und Zigeuner in Österreich* (Salzburg: Geyer Edition, 1983), pp. 196–201.
135. NMT 2/9056–9151, 11 June 1947. See, for example, 9113–14 for deletions of curves.
136. NMT 2/9058; 6/ 211–12 for 10 June 1947; 213–14 for court scrutiny of the charts on 11 June; 6/223–4 for 17 June; 233 for 23 June the prosecution requested that the charts be sent for expert examination.
137. NMT 2/2958–60, 12 June 1947.
138. NARA RG 52 Records of the Bureau of Medicine and Surgery Research Division Box 5 Entry no. WN352934 Ivy reports on potability of seawater after desalination.
139. NMT 2/9214–16, Ivy, 12 June 1947.
140. NMT 2/9215–17, 12 June 1947. NARA M 1019/5 Ivy interrogation of Beiglböck. W.U. Consolazio, N. Pace and A.C. Ivy, 'Drinking Sea Water', *Smithsonian Institution Annual Report* (1945), 153–63. Cf. Ivy on desalination in *US Naval Institute Bulletin* and in *Proceedings Chicago Institute of Medicine*.
141. NMT 8/629, Mant to McHaney, 5 December 1946.
142. NARA M 1019/ 20 Genzken and Mrugowsky interrogation, 23 September 1946, p. 5.
143. APD 4/35 Alexander, 'Additional Questions to be put to Dr. Fischer on the Stand', 11 March 1947.
144. NARA M 1019/ 20 Genzken interrogation, 30 November 1946, p. 2.
145. NARA M 1019/20 Genzken interrogation, 23 September 1946, pp. 19–20. NARA M 1019/ 20 Genzken and Mrugowsky interrogation, 23 September 1946, pp. 6–7, 13–14.
146. NARA M 1019/ 20 Genzken and Mrugowsky interrogation, 23 September 1946.
147. Author's interview with Wolfgang Scholz, 2002.
148. APD 4/35 Alexander to Hardy, Subject: Document Book (Rose), 11 April 1947. Weindling, *Epidemics and Genocide*, pp. 379–81.
149. NMT 4/6556ff Rose, Document Book III, 46.
150. NMT 4/4603ff Rose, Document Book I, Document 12, 65–8.
151. NMT 4/4603ff Rose, Document Book I, 69, 73.
152. Ibid., 74–5.
153. APD 4/35 Alexander to McHaney, 'Adviser to the Court', 29 April 1947.
154. Waldemar Hoven, 'Versuche zur Behandlung der Lungentuberkulose durch Inhalation von Kohlekolloid', MD, Freiburg, 1943.

155. UA Freiburg B53/774 Waldemar Hoven.
156. NARA M 1019/29 Hoven interrogation by Alexander, 29 November 1946.
157. NARA M 1019/29 Hoven interrogation by Hardy, 22 October 1946, pp. 17–18, 201 Files Hoven, Alexander, Hardy, 22 October 1946.
158. NARA M 1019/20 Sievers interrogation, 16 August 1946.
159. NARA M 1019/68 Sievers interrogation, 17 August 1946, p. 2. Niels Bohr Library, Box 27, folder 26: Wolfram Sievers, Philipp von Luetzelburg to Rascher, 17 February 1943.
160. NARA M 1019/68 Sievers interrogation, 29 August 1946 pm, pp. 5–6.
161. NARA M 1019/68 Sievers interrogation, 11 October 1946, pp. 3–4, 6–12.
162. NARA M 1019/68 Sievers interrogation, 20 August 1946, p. 13.
163. NARA M 1019/20 Gebhardt interrogation, 17 October 1946, p. 6. M 1019/20 Gebhardt interrogation, 18 October 1945 am and pm, pp. 1–4.
164. NARA M 1019/68 Sievers interrogation by Meyer, 27 February 1947, pp. 1–4. On Feix, see *Frankfurter Neue Presse* (20 December 1946). APD 4/35 Feix interrogation by Alexander, 31 January 1947.
165. NARA M 1019/68 Sievers interrogation, 29 August 1946 pm, pp. 1–2. 3 September 1946, p. 15. M 1019/7 Blome interrogation, 26 February 1946, p. 4. Klee, *Deutsche Medizin*, pp. 323–4. *Münchener Medizinischer Wochenschrift* (28 January 1944), p. 46.
166. NMT 4/3660. Wolfgang Scholz interview with author.
167. NMT 4/3659–62 affidavit Mrugowsky, 16 January 1947; 4/3663–4 affidavit Reiter 24 January 1947.
168. STAN Rep 502A KV-Verteidigung Rostock Nr 3 Pribilla to Rostock, 21 March 1947.
169. STAN Rep 502A KV-Verteidigung Rostock Nr 3 Rostock to Pribilla, 18 March 1947. AHUB UK Pers Bd 2 Bl. 122–3 Rostock to Dieterici, 27 February 1947.
170. STAN Rep 502A KV-Verteidigung Handakten Rostock Nr 10 Rostock to von Weizsaecker, 16 July 1947.
171. STAN Rep 502A KV-Verteidigung Handakten Hoffmann Nr 25 Servatius to other defence lawyers, 19 December 1946.
172. APD 4/33 Alexander Logbook, 25 November 1946: 'Telephone conversation with Wiskott and Rein, who are not too eager to cooperate.'
173. STAN Rep 502A KV-Verteidigung Handakten Fritz Nr 22 Sachliche Korrespondenz, H von Schulz file, to E. Payn, Walton-on-Thames, 30 November 1946 re. Miss Hislap and Dr Sturton, and Society for Tropical Medicine. Brumpt to Fritz, 19 November 1946.
174. AdeF BB/35/260 Fritz Flemming to Behringwerke, 9 December 1946, Demnitz to Flemming, 16 December 1946, Flemming to Demnitz, 4 January 1947, Demnitz to Flemming, 11 January 1947. Weindling, *Epidemics and Genocide*, p. 401.
175. Tusa, *Nuremberg*, p. 417.
176. NARA M 1019/20 Gebhardt interrogation, 5 November 1946, p. 24.
177. 'Prison Malaria: Convicts Expose themselves to Disease so Doctors can Study it', *Life*, vol. 18, no. 23 (4 June 1945), 43–4, 46.
178. NMT 4/1796 Karl Brandt documents, *Life*; 4/1900ff and 4/2152ff for *Reader's Digest*. Becker-Freyseng documents from *Time*. 4/282f. Weltz used four extracts from *Reader's Digest*: 4/8222, 4/8223ff, 4/8226, 4/8227.
179. NMT 4/2098.
180. Tusa, *Nuremberg*, pp. 251, 259.
181. NMT 2/9382–6.

182. A.C. Ivy, 'Ethics Governing the Service of Prisoners as Subjects in Medical Experiments', *JAMA*, cited as reprint in the Ivy Papers. Ivy Papers, Ivy, 'Some Ethical Implications of Science'.

183. Moreno, *Undue Risk*, pp. 32–4, citing Nathan Leopold, *Life Plus 99 Years* (1958). NMT 2/9384 for discussion of Leopold.

184. Hornblum, *Acres of Skin*, pp. 80–2.

185. TNA: PRO FO 371/65099 G. Jenkins to General Robertson, 19 May 1947. R.A. McCance and R.F.A. Dean, 'Response of New-born Children to Hypertonic Solutions of Sodium Chloride and of Urea', *Nature*, vol. 160 (1947) 904. R.A. McCance, 'Blood-urea in the First Nine Days of Life', *Lancet,* vol. 252 (1947) 787–8. This was admitted as a defence document for Karl Brandt as document 110 and exhibit no. 44.

186. Nicholas Ramscar, 'Human Experiments', FHS Dissertation in Physiological Sciences, Oxford University, 2003. TNA: PRO FO 1013/1961.

187. TNA: PRO FO 1013/1961. NMT 2/6191, 10345–6, 4/2038. 6/1588 refusal of request to summon McCance on 24 April 1947.

188. Margaret Ashwell (ed.), *McCance and Widdowson*: *A Scientific Partnership of 60 Years 1933 to 1993* (London: British Nutrition Foundation, 1993), pp. 164–6, 168–9 (Dorothy Rosenbaum).

189. NARA M 1019/ 20 Genzken interrogation by Alexander, 30 November 1946, p. 8.

190. TNA: PRO FD 1/5826 War Time Scientific Investigations in Germany and German Occupied Countries E. Mellanby, 3 and 24 November 1947.

191. STAN Rep 502A KV- Verteidigung Handakten Fritz Nr 28, Einwand der 'Freiwilligkeit' der Versuchspersonen bei Menschenversuchen im Ausland.

192. STAN Rep 502A KV- Verteidigung Handakten Fritz Nr 35, Selbstversuche am eigenen Körper und berufliche Infektionen während der Laufbahn als Hygieniker.

193. STAN Rep 502A KV- Verteidigung Handakten Fritz Nr 28 Material zu Menschenversuchen. Beziehungen zu Fleckfieberversuchen in Buchenwald.

194. STAN Rep 502A KV- Verteidigung Handakten Fritz Nr 35 Verschiedenes.

195. NARA RG 153/86–3–1 Box 10 Rose to Haagen, 13 December 1943.

196. STAN Rep. 502A KV-Verteidigung Handakten Fritz Nr 22 Richard Haas, Bad Wiessee to H. von Schulz, 25 February 1947. Weindling, *Epidemics and Genocide*, pp. 345–52.

197. STAN Rep 502A KV- Verteidigung Handakten Fritz Nr 23 K. Mühlens to Fritz, 16 February 1947 deposition that with Rose in Africa in 1941–2, also in Salonika. H. Gins to Fritz, 8 February 1947.

198. STAH 352–8/9 Bernhard-Nocht-Institut, Schriftwechsel in die Angelegenheit Dr med Gerhart Rose, Band 1 Letter of Dr Hella von Schulz to Nauck, 28 January 1946, Nauck to Rose, 30 May 1946 Nauck wishes to help as far as possible.

199. STAN Rep 502A KV- Verteidigung Handakten Fritz Nr 26 'Fragen an Herr Professor Rose'. Personal communication, Wolfgang Eckart regarding Rose and Anopheles. NMT 2/6475–81, also Blaurock and Otis B. Schreuder affidavits.

200. AdeF BB/35/260 Alexander to McHaney, 'Countering the Defense that the Germans were Experimenting on Prisoners Condemned to Death', 23 November 1946. 'Ethique médicale: Expériences sur des condamnés', 7 January 1947, A Horlik-Hochwald to McHaney: defence of experimentation on criminals and persons condemned to death.

201. NMT 4/2558. Cf. W.P. Havens et al., 'Experimental Production of Hepatitis by Feeding Icterogenic Materials', *Proceedings of the Society of Experimental Biology and Medicine* (November 1944) 206. Josef Neefe and John Stokes, 'An Epidemic of Infectious Hepatitis Apparently due to a Water Borne Agent', *JAMA*, vol. 128 (1945) 1063–75. Josef Neefe et al., 'Hepatitis Due to Injection of Homologous Blood products in Human Volunteers', *Journal of Clinical Investigation*, vol. 23 (1944) 836–55.
202. NMT 4/2558.
203. NMT 4/1382, 1483–90 Defence Brack; NMT 4/1560–69 = Albert Q. Maisel, 'Bedlam 1946. Most U.S. Mental Hospitals are a Shame and a Disgrace', *Life* (6 May 1946), 36–44.
204. NMT 4/1523–4, 18 February 1947. Ulrich Herbert, *Best* (Bonn: Dietz, 1996), pp. 413–19.
205. NMT 4/2590 Brandt final plea.
206. APD 4/33 Alexander logbook, 5 February 1947, p. 187.
207. AOF Colmar AJ 3617 P. 15 d. 686 Karl Brandt re Bickenbach AJ 3623 P. 51 d. 23721 Dossier: Dr Haagen, Eugen 1946–1954 for reports on the Metz trial. Dr G. Boggaerts, Brussels, 6 October 1945, 'Rapport sur le Camp de Concentration Natzweiler (Struthof)'; Kastan, 'Unethical Nazi Medicine', 191. I. Simon, 'Les médecins criminals nazis Bickenbach, Haagen et Hirt du camp de Struthof et le procès de Metz', *Revue d'histoire de la médecine hébraïque*, vol. 6, no. 3 (1952–3), 133–42.
208. NMT 2/11512–17 concluding plea of Gebhardt.
209. NMT 4/ 2849 Gebhardt's lawyer argued that it was not a matter of medical ethics but of law.
210. NARA M 1019/20 Gebhardt interrogation, 17 October 1946, p. 18.
211. NMT 4/3025–49 on 6 May 1947.
212. NMT 8/629 Mant to McHaney, 5 December 1946. Gebhardt response to Mant, NMT 4/3052.
213. NARA M 1019/20 Interrogation, 30 November 1946, p. 16.
214. NARA RG 243 US Strategic Bombing Survey Entry 2 (section 5) Box 1 Handloser Interview no. 75, July 1945.
215. NMT 4/3523, 3526–7.
216. NMT 4/3642 affidavit Gutzeit.
217. NMT 4/3650 Sievers, 8 January 1947.
218. APD 4/33 Alexander logbook, pp. 186–7.
219. Paul Weindling, 'Human Guinea Pigs and the Ethics of Experimentation: the *BMJ*'s Correspondent at the Nuremberg Medical Trial', *British Medical Journal*, vol. 313 (1996) 1467–70.
220. 'Lord Horder says retain Nazi data', *The Daily Telegraph* (14 december 1946), p. 5.
221. STAN Rep 502A KV- Verteidigung Handakten Fritz Nr 34 Mellanby to Fritz, 9 July 1947.
222. NMT 4/103f Becker-Freyseng Document Book no 20. APD 1/20 OCCWC, Special Release No. 99, 11 January 1947.
223. APD 1/20 OCCWC Public Relations Office, Special Release No. 99. NARA RG 153/ 87–0/ box 13 Daily Press Review No. 83, 8 January 1947.
224. Ibid. K. Mellanby. 'A Moral Problem', *The Lancet* vol. 251 (1946) 850. Leo Michalowski on 21 December 1946, NMT 2/931–946. *The Lancet* letters on 'A Moral Problem' the Dachau survivor Leo Michalowski's evidence.
225. W.S.S. Ladell, 'Effects on Man of Restricted Seawater Supply', *The Lancet*, 2 (1943) 441–4, NMT 4/801ff.

226. APD 4/33 Alexander Logbook, 89, 7 February 1947 conversation with Steinbauer, re Ladell's letter.
227. APD 4/33 Ladell to Steinbauer, 25 January 1947.
228. APD 4/35 Alexander to McHaney, Suggestions for Questions of Dr W.C.S. Ladell, 25 February 1947.
229. APD 4/33. Ladell to Steinbauer, 25 January 1947. Alexander memo, 25 February 1947: 'Suggestions for Questions of W.C.S. Ladell'.
230. NARA RG 153/ 86–3–1/box 10, book 3 'Versuchspersonen KZ-Dachau gesucht'. Raimund and Georg Papai affidavits, 28 July 1947, Xaver Reinhart, 28 August 1947, Jean Senes, 8 July 1948.
231. NARA M 1019/20 Gebhardt interrogation, 3 December 1946, p. 16.
232. NARA M 1019/9 Rudolf Brandt interrogation, 24 March 1947.
233. NMT 4/2686–2689 Closing plea.
234. NARA M 1019/20 Gebhardt interrogation, 5 November 1946, pp. 13, 18–22. APD 4/35, Alexander, 'Questions to be Put to Dr Fischer on the Stand', 8 March 1947.
235. NARA M 1019/20 Genzken interrogation, 20 September 1946, p. 3.
236. APD 4/35, Alexander, 'Cross-examination of Dr. Ruff', 28 April 1947.
237. STAN Rep 502A KV-Verteidigung Fritz 35 Selbstversuche am eigenen Körper; Verschiedenes. Information von Rose über Person und Stellung.
238. NMT 4/2460.
239. BAK ZSg 154 Pross Bd 73 Alexander notes on Mrugowsky and Ding. Weindling, *Epidemics and Genocide*. Klee, *Auschwitz*, pp. 321–40.
240. NMT 4/5973.
241. NYPL Lifton Papers Box 5, Alexander file, 12 July 1947, 'An Unnumbered Document'.

11 The Medical Delegation

1. Mitscherlich Papers (MP) II 2/7.1 Mitscherlich to Anspacher, 11 December 1946.
2. NARA RG 238 World War II War Crimes Records NM 70 entry 148 Executive Session no. 5, 20 November 1946, 13. *Badische Zeitung* (3 January 1947). This also struck Alice Ricciardi-von Platen.
3. Alexander Mitscherlich and Fred Mielke (eds), *Das Diktat der Menschenverachtung* (Heidelberg: Lambert Schneider, 1947). Mitscherlich and Mielke (with contributions by Ivy, Taylor, Alexander and Deutsch), *Doctors of Infamy*.
4. UAH PA 5032 Alexander Mitscherlich, 3 April 1946. Alexander Mitscherlich, *Ein Leben für die Psychoanalyse* (Frankfurt, 1980), pp. 134–6. Mitscherlich, *Wissenschaft*, pp. ix, 299. Eberhard Damm, 'Alfred Webers "Freier Sozialismus"', Jürgen Hess, Hartmut Lehmann and Volker Sellin (eds), *Heidelberg 1945* (Stuttgart: Franz Steiner, 1996), pp. 329–47. Alfred Weber, 'Freier Sozialismus. Ein Aktionsprogramm', in Alexander Mitscherlich and Alfred Weber (eds), *Freier Sozialismus* (Heidelberg, 1946), pp. 37–94.
5. Mitscherlich, *Ein Leben*, pp. 81–5, 95–8. Rebecca Schwoch, *Ärztliche Standespolitik im Nationalsozialismus. Julius Hadrich und Karl Haedenkamp als Beispiele* (Husum: Matthiesen, 2001).
6. Alexander Mitscherlich, *Freiheit und Unfreiheit in der Krankheit. Das Bild des Menschen in der Psychotherapie* (Hamburg: Claussen & Goverts, 1946), pp. 78–9.

7. Viktor von Weizsäcker, 'Über medizinische Anthropologie', *Arzt und Kranker* (Leipzig: Koehler & Amelang, 1941), pp. 35–61.

8. Michael Hubenstorf, *'Tote und Lebendige Wissenschaft', Von der Zwangssterilisierung zur Ermordung II* (Vienna: Böhlau, 2002), pp. 237–420, n. 405.

9. Walter Wuttke-Groneberg, 'Von Heidelberg nach Dachau', *Medizin im Nationalsozialismus* (Berlin West: Verlagsgesellschaft Gesundheit, 1980), pp.113–38; also Heinrich Huebschmann, pp. 141–4. Karl-Heinz Roth, 'Psychomatische Medizin und "Euthanasie": Der Fall Viktor von Weizsäcker', *1999, Zeitschrift für Sozialgeschichte des 20. und 21. Jahrhunderts*, vol. 1 (1986), 65–99. Roth 'Replik', *1999*, vol. 5, no. 2 (1990) 163–5.

10. Peter Maguire, *Law and War: An American Story* (New York: Columbia, University Press, 2000), pp. 195–8.

11. STAN Rep 502A KV- Verteidigung – Handakten Nr 10 Rostock to von Weizsäcker, 16 July 1947.

12. Viktor von Weizsäcker, *'Euthanasie' und Menschenversuche* (Heidelberg: Lambert Schneider, 1947).

13. Thomas Gerst, 'Der Auftrag der Ärztekammern an Alexander Mitscherlich zur Beobachtung und Dokumentation des Prozessverlaufs', *Deutsches Ärzteblatt*, vol. 91 (1994), B–1200–B–1210, B–1201.

14. MP II2/112.22 Koch to Oelemann, 9 January 1947. II2/112.23 Oelemann to Mitscherlich, 31 January 1947. Platen audiotape 1. Platen to author 28 July 2002.

15. NMT 8/630–2 De Bary to Oelemann, 25 October 1946.

16. NMT 8/634 draft resolution for 2 November 1946.

17. MP II2/112.1 Mitscherlich to Karl Oelemann, 7 November 1946.

18. MP II2/138.1a Mitscherlich to Medical Faculty University Heidelberg, n.d. [1946].

19. NMT 8/985 Berlin Faculty, 27 November 1946.

20. MP II 2/112.11 Marburg, 18 November 1946; II2/112.12 Giessen, 17 November 1946.

21. MP II2/112.15 D. Ackermann to Oelemann, 20 November 1946; II2/112.17 Klin. Direktion Mainz to Oelemann, 27 November 1946. Karl-Thorsten Bretschneider, 'Friedrich Hermann Rein. Wissenschaftler in Deutschland und Physiologe in Göttingen in den Jahren 1932–1952', Göttingen MD dissertation 1997, 96 citing Faculty response of 19 November 1946.

22. MP II2/112.16 Dekan to Oelemann, 22 November 1946.

23. MP II2/7.1 Mitscherlich to Anspacher, 11 December 1946.

24. MP II2/106.1 Oelemann, 27 November 1946. 106.2 Mitscherlich to Military Tribunal No. 1, 9 December 1946. Cf. Dekanat Charité 100/21 Protokolle der Fakultätssitzungen 1946 0100/21a Protokolle der Fakultätssitzungen 1946–54. The Faculty minutes for this period are missing at Rostock.

25. MP II2/112.30 Oelemann to Mitscherlich, 26 May 1947.

26. UA Halle Rep 29 Nr 293 Bd 2 Halle in name of President of Provinz Sachsen to Rector of Halle, 2 December 1946.

27. Ibid. Budde to Rector, 14 January 1947.

28. NMT 8/997 Arbeitsgemeinschaft der med und jur Fak. Zur Beobachtung und Stellungnahme zum Ärzteprozess to Internationalen Gerichtshof Nuremberg, 7 July 1947. Mitscherlich reply, 5 August 1947.

29. UA Halle Rep 29 Nr 304 Faculty meeting, 10 September 1947.

30. UA Halle Rep. 6 Nr 2676 Tätigkeitsbericht des 1. Studentenrates der Martin Luther Universität über seine Amtsperiode, 17 November 1947, p. 2.

31. UA Heidelberg PA 2866 Alexander Mitscherlich.

32. Jeffrey Herf, *Divided Memory: The Nazi Past in the Two Germanys* (Cambridge, Mass., 1997), pp.74–9.
33. Robert Havemann, 'Menschen als Versuchstiere: Zum Nürnberger Ärzteprozess', *Der Kurier* (28 January 1947).
34. MP II2/142.1 Von Skramlik to Mitscherlich, 14 April 1947. UAJena Skramlik, Emilian Ritter von Bestand D Nr 1087 Bd 1.
35. Information from Sigrid Oehler-Klein. UA Rostock Med Fak 1027. Fakultaetssitzungen, with missing minutes from 30 April 1946 to 22 October 1947.
36. MP II/24.1 Brugsch to Mitscherlich, 2 April 1947. Mitscherlich to Brugsch, 25 April 1947. II2/124.1 Mitscherlich to Rompe, 9 May 1947; cf. Dekan Med Fak Jena to Mitscherlich, 14 April 1947. Theodor Brugsch, *Arzt seit fünf Jahrzehnten* (Berlin: Ruetten & Loening, 1957), pp. 346–54. Theodor Brugsch, 'Und wieder ist Krieg – Befreiung zu neuem Beginn', Günter Albrecht and Wolfgang Hartwig (eds), *Ärzte. Erinnerungen, Erlebnisse, Bekenntnisse* (Berlin: Der Morgen, 1973), pp. 179–206.
37. NMT 8/991 Berlin Faculty Minutes, 16 April 1947.
38. NMT 8/703 Entnazifizierungskommission Abt. Ärzte beim Magistrat der Stadt Berlin 5 November 1947. NMT 8/704–5 Deutsche Ärztekommission to Entnazifizierungskommission, 22 November 1947.
39. UA Heidelberg PA 5032 Alexander Mitscherlich, Jaspers to Mitscherlich, 9 May 1947. PA 2866 Alexander Mitscherlich, Jaspers support for a Clinic for Biographical Medicine, 14 July 1946.
40. MP II/97.1 Mitscherlich to Bürgermeister Levie, 17 December 1946. Personal communications Alice Ricciardi. Mitscherlich, *Ein Leben*, pp. 147–8.
41. These were Benz, Jensen, Koch and Spamer. MP II2/112.21 Mitscherlich to Oelemann, 16 December 1946.
42. MP II/105.1 Mielke to Mitscherlich, 16 February 1947; II2/105.2 Mielke to Mitscherlich, 6 April 1947; II2/105.3 Mielke to Mitscherlich, 9 April 1947.
43. Alexander Mitscherlich, 'Der Arzt und die Humanität. Erste Bemerkungen zum Nürnberger Ärzteprozeß', *Die Neue Zeitung* (20 December 1946). MP II2/115.1 Platen to Mitscherlich, 10 January 1947.
44. MP II2/112.34 von Platen to Mitscherlich, 28 December 1946.
45. MP II 2/12.1 Charlotte Becker-Freyseng to Mitscherlich, 9 October 1947.
46. BAK ZSg 154/75 Alexander diary, p. 122.
47. MP II2/112.34 von Platen to Mitscherlich, 28 December 1946. NMT 8/662 Platen to Mitscherlich, 10 January 1947.
48. Alice Ricciardi-von Platen, audio tape 2.
49. MP II2/35.1 Mitscherlich to *DMW*, 3 February 1947. II2/41.1 Mitscherlich to Dreyer, Public Relations Nuremberg, 3 February 1947; II2/87.3 Mitscherlich to *DMW* editor Heinz Köbke, 17 February 1947. II2/87.6 Köbke to Mitscherlich, 18 March 1947.
50. MP II2/87.1 Editor to W. Spamer, 17 January 1947.
51. MP II2/87.4 Heinz Köbke to Mitscherlich, 24 February 1947.
52. Gerst, 'Der Auftrag', 1202.
53. MP II2/87.7 Mitscherlich to Köbke, 21 March 1947.
54. Gerst, 'Der Auftrag', B–1202 gives 25,000 for the edition. NMT 8/691 for Mitscherlich stating that 30,000 were issued.
55. J. Peiffer, 'Diktat der Menschenverachtung', *Ende und Anfang*, vol. 2, no. 4 (15 May 1947).

56. The original German is: 'Aus der Deutschen Ärztecommission ... Eine Dokumentationsbericht gegen 23 SS-Ärzte und Deutschen Wissenschaftler'.

57. MP II2/35.1 Mitscherlich, 3 February 1947 to *DMW*.

58. NMT 8/ 654 Mitscherlich to Oelemann, 1 May 1947.

59. MP II2/105.2 Mielke to Mitscherlich, 6 April 1947.

60. Gerst, 'Der Auftrag', B–1204. MP II 2/63.2 Jupp Harmacher to Mitscherlich, 30 May 1947.

61. NMT 8/ 685–93. Protokoll der Arbeitstagung der Ärztekammern, 14 und 15. Juni 1947.

62. 'We were Nestbeschmutzer' – personal communication Alice Ricciardi. Mitscherlich, *Ein Leben*, p. 147.

63. MP II2/138.2 Kurt Schneider to Mitscherlich, 30 May 1947.

64. Mitscherlich, *Ein Leben*, pp. 144–7.

65. Karl-Heinz Leven, 'Der Freiburger Pathologe Franz Büchner 1941–Widerstand mit und ohne Hippokrates', in Bernd Grün, Hans-Georg Hofer and Karl-Heinz Leven (eds), *Medizin und Nationalsozialismus. Die Freiburger Medizinische Fakultät* (Frankfurt: Peter Lang, 2002), pp. 362–95. The 1941 lecture was republished in modified form in: F. Büchner, 'Der Eid des Hippokrates. Die Grundgesetze der ärztlichen Ethik', in *Das christliche Deutschland 1933–1945* (Freiburg, 1945), pp. 9–31. Platen, *Tötung*, p. 126.

66. UA Heidelberg PA 5032 Alexander Mitscherlich , Büchner affidavit, 28 March 1946.

67. UA Heidelberg PA 5032 Jaspers to Alexander Mitscherlich, 3 June 1947.

68. MP II2/78.1 and 2 Mitscherlich to Jaspers, 8 and 12 May 1947. Neumann, 'Freiburger Ordinarien im Zweiten Weltkrieg', in Grün et al., *Medizin*, pp. 408–11. Steven P. Remy, *The Heidelberg Myth The Nazification and Denazification of a German University* (Cambridge, Mass.: Harvard, 2003), p. 227.

69. MP II2/26.1 W. Bappert to Landgericht Freiburg, 24 April 1947. Text of Weltz affidavit. II2/8.1a, Büchner Stellungnahme zu dem Buche 'Medizin ohne Menschlichkeit', 17 November 1960. II2/26.7 Mitscherlich to Landgericht Freiburg, 5 May 1947.

70. MP II2/15.2 Mitscherlich to G. Biermann, 24 May 1947. Lambert Schneider Archive (LS Arch) Geiler et al. to Lambert Schneider and Mitscherlich, 16 May 1947. Geiler to Plum, 16 May 1947.

71. *Das Diktat*, slip inserted between pp. 70 and 71. UA Freiburg B53/33 Büchner to Enkelkring, 10 May 1947. Büchner to Janssen, 12 May 1947.

72. MP II2/71.1 Heubner to Mitscherlich, 17 April 1947. II2/71.9 Heubner, 'Richtigstellung der Broschüre von Mitscherlich und Mielke'.

73. Johanna Kneer, 'Wolfgang Heubner (1877–1957) Leben und Werk', med. Diss. Tübingen, 1989, p. 84.

74. Ibid., 80–5.

75. MP II2105.4 Mitscherlich to Mielke, 6 May 1947.

76. 'Protest von Heubner und Sauerbruch', *Göttinger Universitäts-Zeitung*, no. 3 (1948), 6–7. MP II2/149.2 Mitscherlich letter, 29 November 1947. II2/71.3 Heubner deposition, 13 June 1947. MP II2/71.4 Sauerbruch deposition, 14 June 1947. II2/71.8 Mitscherlich to Heubner, 20 June 1947.

77. MP II2/71.14 comparison of Achelis' demands and the court decision. II2/71.16 Schneider to Achelis, 20 August 1947. II2/71.17 Achelis to Mitscherlich, 26 August 1947. II2/71.18 Mitscherlich to Achelis, 1 September 1947; II2/71.19 Achelis to Mitscherlich, 28 October 1947; II2/71.20 Mitscherlich to Achelis, 25

November 1947; Achelis to Mitscherlich, 23 February 1948; II2/71.23 Achelis to Landgericht Berlin, 6 April 1948.

78. UA Freiburg B53 Nr 227 f. 164 Sitzung, 29 April 1947; B53/33 Dekan Freiburg to Hochschulreferent Kilching, 13 May 1947; Dekan to Rektor Freiburg, 13 May 1947; statement on Mitscherlich's citation of Büchner, n.d.

79. MP II2/30.2 Campenhausen to Mitscherlich, 7 June 47.II/30.2a Campenhausen to Weber, Strughold, Hess, 7 June 1947. 2b to Witte 12 June 1947.

80. F.H. Rein, 'Wissenschaft und Unmenschlichkeit', *Göttinger Universitäts-Zeitung*, no. 14 (20 June 1947) 6–8. Karl-Thorsten Bretschneider, 'Friedrich Hermann Rein. Wissenschaftler in Deutschland und Physiologe in Göttingen in den Jahren 1932–1952', Göttingen MD dissertation 1997, 97.

81. MPG Abt. III/14A/4305. Cf. Court case – Hahn papers, Studnitz to Hahn.

82. MP II2/164.2 Weizsäcker to Rein, 5 September 1947.

83. A. Mitscherlich, 'Unmenschliche Wissenschaft', *Göttinger Universitäts-Zeitung*, no. 17/18 (15 August 1947) 6–7. Rein, 'Vorbeigeredet', ibid., 7–8. Mitscherlich, 'Absicht und Erfolg', *Göttinger Universitäts-Zeitung*, no. 3 (1948) 4–5. MP II2/60.1 Mitscherlich to *Göttinger Universitäts-Zeitung*, 14 July 1947. II2/60.6 Mitscherlich to Bollnow, 29 November 1947.

84. 'Protest von Heubner und Sauerbruch', *Göttinger Universitäts-Zeitung*, no. 3 (1948), 6–7. Mitscherlich reply in *Göttinger Universitäts-Zeitung*, no. 10 (23 April 1948).

85. Platen, audiotape 3.

86. MP II2/112.21 Mitscherlich to Oelemann, 16 December 1946.

87. MP II2/112.24 Mitscherlich to Oelemann, 6 February 1947. MP II2/105.5 Mielke to Mitscherlich, 6 May 1947.

88. Alice von Platen, 'Der Nürnberger Ärzteprozess II Verteidigung', *Hippocrates* (1947), 202.

89. Klee, *Was sie taten*, pp. 193–5.

90. MP II2/115.2 Platen to Mitscherlich, 10 August 1947.

91. Alice Ricciardi-von Platen, tape 1 side 2.

92. Alice Ricciardi-von Platen, tape 2 side 2.

93. BAK ZSg 153/73 Conversation with Countess Platen, 26 February 1947.

94. Alice Ricciardi-von Platen, tape 2 side 1. Eugen Kogon, 'Der Kampf um Gerechtigkeit', *Frankfurter Hefte*, vol. 3 (1948), 373–83.

95. Eugen Kogon, 'Gericht und Gewissen', *Frankfurter Hefte*, vol. 1 (April 1946), 25–37.

96. NMT 8/698 Platen to Mitscherlich, 10 August [1947]. MP II2/115.4 Platen to Mitscherlich, 27 October 1947. II2/115.5 Platen to Mitscherlich, 3 November 1947.

97. Alice Ricciardi-von Platen, tape 2 side 2. *Frankfurter Hefte*, in Kogon, *Begegnungen*, pp. 92–3.

98. Alice Ricciardi-von Platen, 'Geleitwort', *Medizin und Gewissen. 50 Jahre nach dem Nürnberger Ärzteprozess* (Frankfurt am Main: Mabuse-Verlag, 1998), pp. 13–14.

99. Alice Platen-Hallermund, *Die Tötung Geisteskranker in Deutschland: Aus der deutschen Ärzte-Kommission beim amerikanischen Militärgericht* (Frankfurt: Verlag der Frankfurter Hefte, 1948), p. 84.

100. Ibid., pp. 85–7.

101. Ibid., pp. 86–7.

102. Ibid., pp. 89–90.

103. Ibid., p. 106.

104. Ibid., pp. 8–9, 11–12.
105. Ibid., pp. 91–2.
106. Ibid., p. 20.
107. Ibid., p. 37.
108. UA Heidelberg PA 2866 Alexander Mitscherlich.
109. Stephen Spender, *European Witness* (London: Hamish Hamilton, 1946), p. 218.
110. Bloxham, *Genocide on Trial*, pp. 156, 179.
111. APD 4/40 Alexander and Wolfe Frank interview for Radio Newsreel London, 10 January 1947, and German interview, 21 January 1947.
112. STAN Rep 502A KV-Verteidigung – Handakten Sauter Nr 4 Ewersbeck-Blome letters, n.d.
113. Georg Hohmann, *Ein Arzt erlebt seine Zeit* (Munich: J.F. Bergmann, 1954), pp. 170–4.
114. Fred Mielke, 'Der Nürnberger Prozesse und der deutsche Arzt', *Bayerisches Ärzteblatt*, vol. 2 (17 December 1947) 1–2.
115. STAN Rep 502A KV- Verteidigung – Handakten Rostock Nr 2 Prozess-Material Opinion of W.E. Süskind re. Baden Radio commentary, 25 February 1947.
116. W.E. Süskind, 'Hippokrates als Warner', *Süddeutsche Zeitung*, no. 32 (5 April 1947). STAN KV- Verteidigung Rep 502A Handakten Rostock Nr 2 Rostock to Pribrilla, 14 April 1947.
117. STAN Rep 502A KV- Verteidigung – Handakten Rostock Nr 8, Rostock recommended Dr Koch in the January 1947 issue of the *Südwestdeutschen Ärzteblatt*, Weizsäcker in *Psyche* on euthanasia and human experiments, *Diogenes, Freie Studentische Zeitschrift* (4. Heft 1947), the commentary on *Das Diktat der Menschenverachtung* by W. Kahle, and works by Jaspers and Radbruch. Cf. Rostock to Rein, 12 August 1947.

12   A Eugenics Trial?

1. NYPL Lemkin papers, reel 2 Autobiography, p. 4.
2. Michael Marrus, 'The Holocaust at Nuremberg', *Yad Vashem Studies*, vol. 26, cited as online publication.
3. William Seltzer, 'Population Statistics, the Holocaust, and the Nuremberg Trials', *Population and Development Review*, vol. 24 (1998), 511–52. Cf. *New York Times* (8 January 1945).
4. Marrus, *Nuremberg*, pp. 154–64. Marrus, 'Holocaust', 16.
5. Taylor, 'The Meaning of the Nuremberg Trials', 14.
6. *Trial of German Major War Criminals*, vol. 3, pp. 176–7. On Hofmann, see Isabel Heinemann, *'Rasse, Siedlung, Deutsches Blut'. Die Rasse- und Siedlungshauptamt der SS und die rassenpolitische Neuordnung Europas* (Göttingen: Wallstein Verlag, 2003), pp. 75, 561ff.
7. Marrus, 'Holocaust', 4–6.
8. John Fried, 'Nuremberg and the Holocaust', *Toward a Right to Peace* (Northampton, Mass: Aletheia Press, 1994), pp. 15–32.
9. Marrus, *Nuremberg*, pp. 191–3. Bloxham, *Genocide on Trial*, p. 64.
10. Michael Marrus, 'The Nuremberg Doctors' Trial in Historical Context', *Bulletin of the History of Medicine*, vol.73 (1999) 106–23.
11. NYPL Lemkin papers reel 2, Autobiography, 2. Powers, *Problem*, p. 51.
12. Raphael Lemkin, *Axis Rule in Occupied Europe* (New York, 1944), p. x.
13. Ibid., pp. xi–xii.

14. AdeF BB/35/263 Gunn forwards memo of Dr Lemkin of 21 January 1947 'Planning of a Special Trial on Abduction of Women into Prostitution', memo submitted to McHaney, 29 January 1947.
15. NYPL Lemkin papers, reel 2 Autobiography, p. 4.
16. NYPL Lemkin papers, reel 1 Lemkin to Gypsy Lore Society, 2 August 1949.
17. NYPL Lemkin papers, reel 4 Lemkin to Telford Taylor, 28 September 1945.
18. NYPL Lemkin papers, Lemkin memo, 11 January 1947.
19. NYPL Lemkin papers, reel 2 Autobiography, pp. 4–5. Steven L. Jacobs (ed.), *Not Guilty. Raphael Lemkin's Thoughts on Nazi Genocide* (Lewiston: Mellen, 1992).
20. Raphael Lemkin, 'Genocide – a Modern Crime', *Free World*, no. 9 (April 1945), 39–43. 'Le crime de génocide', *Revue de droit international*, vol. 24 (October-December 1946) 213–23. 'Genocide as a Crime under International Law', *American Journal of International Law*, vol. 41 (January 1947) 145–51.
21. NARA RG 153/ 86–2–2/ Box 9 book 2 Walter Rapp, Memorandum for Colonel Taylor, 26 April 1946. Powers, *Problem*, pp. 51–2 cites the less favourable recollections of Ferencz.
22. NARA RG 153/ 86–2–2/ Box 9 book 2 Lemkin to Marcus, 13 January 1947.
23. AdeF BB/30/1786 Meeting of Touffait, Lemkin and Gunn, 28 May 1946. UNA UNWCC M 108, Minutes of 108th meeting held on 19 June 1946, Present Dr Lemkin with Lt Kintner – USA. RCPE Smith Papers, box 9 folder 94 War Crimes, H.H. Wade to Smith, 24 July and 2, 12 August 1946.
24. NARA RG 153 Nuremberg Administrative files, 85–2, book 1 'The Crimes of Genocide', 10 December 1946. UNA UNWCC, M 129, 18 December 1946.
25. UN Economic and Social Council. 'Prevention and Punishment of Genocide. Historical Summary 2 Nov 1946–20 Jan 1947', E/621 (26 January 1948). Lawrence J. LeBlanc, *The United States and the Genocide Convention* (Durham NC, 1991), pp. 19–23. The Convention was adopted by the United Nations General Assembly in Paris on 9 December 1948. The United States ratified the Convention in 1988. *The United Nations and Human Rights 1945–1995* (New York: United Nations, 1995), pp. 18–22.
26. NMT 3/697, 701 statement of Schiedlausky, 7 August 1945. APD 4/33 Alexander Logbook, 175–8, 28 January 1947. APD 4/35 Katzenellenbogen interrogation by Alexander, 6 January 1947. TTP-CLS 14/7/22/475 Telford Taylor to *New York Times Book Review*, 8 October 1986.
27. NMT 4/1902.
28. NMT 4/1905–6 Géza von Hoffmann. 4/1911ff.
29. The Committee of the American Neurological Association for the Investigation of Eugenical Sterilization: Abraham Myerson, James B. Ayer, Tracy J. Putnam, Clyde E. Keeler and L. Alexander, *Eugenical Sterilization. A Reorientation of the Problem* (New York: The Macmillan Company, 1936).
30. APD 4/40 Alexander radio interview transcript, 21 January 1947. 4/33 Alexander Logbook, 11 November 1946.
31. Lemkin was at Duke University from 1941 to 1942.
32. NYPL Lemkin papers, reel 2, Summary of the activities of Raphael Lemkin, 6.
33. APD 4/33, Alexander Logbook, 13 November 1946.
34. Cf. *New York Times* (21 November 1946).
35. NYPL Lemkin papers reel 3 History of the Genocide Convention, 3–5.
36. Raphael Lemkin, 'Genocide as a Crime under International Law', *The American Journal of International Law*, vol. 41, no. 1 (January 1947) 146–51, footnote 6.
37. NARA RG 153 Box 16 folder 4 book 1 89–2 Telford Taylor, The Meaning of the Nuremberg Trials. An address at the Palais de Justice, Paris, 25 April 1947.

38. WUS Beals Papers Box 1 folder 32 Beals to Gunn, 27 February 1947; Box 1 folder 6 Gunn to Beals, 7 March 1946; folder 16 Carnegie Endowment for International Peace to Beals, 10 March 1947.

39. NARA RG 153 84–1 Lemkin memo to Taylor, 30 October 1946; WJC to Secretary of the Army, Kenneth Royall, 21 November 1947.

40. NARA RG 153 86–3–1, Box 10 book 1 Lemkin, The Importance of the War Crimes Concept for the Doctors Case. Memorandum for Col. David Marcus, 10 January 1947. NMT 8/1802–4.

41. NARA RG 153/86–3–1/Box 10/ Book 1.

42. WUS Beals Papers Box 1 folder 6 Gunn to Beals, 3 January 1947 [actually dated 3 December 1946 but in reply to a letter of 21 December 1946].

43. Leo Alexander, 'The Socio-Psychological Structure of the SS', *Folia Psychiatrica*, vols. 1–2 (1948), 2–14. APD 4/37 Alexander, 'The Socio-Psychological Structure of the SS' (read 12 June 1945, Amsterdam), 45, 47.

44. Alexander, 'Socio-Psychological Structure', 49, 56.

45. Ibid., 69.

46. NARA M 1019/68 Sievers interrogation, 20 August 1946, p. 7.

47. NMT 3/410–442 Sievers diary. Cf. Kater, *'Ahnenerbe'*, pp. 261–4.

48. NARA M 1019/50 Jan Ochocki interrogation, 13 January 1947.

49. NMT 2/110; Brack interrogation citing Globocnik 2/587.

50. Robert Kempner, *Ankläger einer Epoche. Lebenserinnerungen* (Frankfurt/M: Ullstein, 1983).

51. J.M. Inbona, 'Le procès des médecins allemands. Leur responsabilité dans la technique du génocide', *La Presse Médicale*, no. 21 (12 April 1947) 251–2.

52. AdeF BB/35/260 4a Procès I: instruction (corréspondance) 133890 Geo Doering. Schleswig, 23 January 1947 to US Military Government.

53. APD 4/34 Hornung interrogation by Meyer, 20 December 1946. Alexander memorandum to McHaney, 20 December 1946.

54. STAN Rep 502A KV- Verteidigung Handakten Hoffmann Nr 6, Kersten, 29 October 1947.

55. NMT 5/2604, Clemency plea, 7 April 1948.

56. STAN Rep 502A KV- Verteidigung Handakten Sauter Nr 7, Blome note, 13 January 1947.

57. AFP Leo Alexander to Phyllis Alexander, 10 December 1946.

58. NARA RG 153 NM 70 Entry 188 190/19/box 2 Alexander to McHaney, 'General nature of the evidence', 18 December 1946. Alexander on Dzido, see 18 December 1946. For clinical examination, see clinical notes, 17 December 1946. Alexander, memo on Jadwiga Dzido to McHaney and Hardy; also Memo on Kusmierczuk to McHaney and Hardy, n.d., Clinical Notes 17 December, RG 238 Entry 188/190/191 Box 2. Broel-Platen, clinical examination by Alexander RG 238 Entry 188/190/191 Box 2.

59. Sebring, 'The Medical Trial', 9–10.

60. APD 4/35 Alexander to McHaney, 'Suggestions for Questions of Dr WCS Ladell', 25 February 1947.

61. NMT 2/719, Neff on 18 December 1946; 2/1128 Holl on 3 January 1947.

62. NMT 2/9017, 10 June 1947.

63. APD 4/33 Alexander to McHaney, 'Neuro-psychiatric examination of the witness, Karl Hoellenreiner', 28 June 1947. NMT 6/1612 court order of 28 June 1947; 6/1622–3 Court order of 21 July 1947.

64. AFP Leo to Phyllis Alexander, 1 July 1947. APD 4/35 Alexander to McHaney, 'Two Important Witnesses against Dr. Haagen', 28 May 1947.

65. AFP Leo to Phyllis Alexander, 1 July 1947
66. NMT 2/1183–7, 1224 on Ding's Acridin experiments. NMT 2/593.
67. Weindling, *Epidemics and Genocide*, pp. 363–5, 427.
68. NYPL Lifton Papers M9, notes on McHaney, 29 March 1978, 7.
69. NMT 4/6384ff, Eugen Gildemeister April 1941 to December 1942.
70. NMT 4/6403ff RKI report, 1943, 73–5.
71. Mechthild Rössler, 'Konrad Meyer und der "Generalplan Ost" in der Beurteilung der Nürnberger Prozesse', in Mechthild Rössler and Sabine Schleiermacher (eds), *Der 'Generalplan Ost'. Hauptlinien der nationalsozialistischen Planungs- und Vernichtungspolitik* (Berlin: Akademie Verlag, 1993), pp. 356–65; cf. acquittal of Konrad Meyer, 10 March 1948. Heinemann, *'Rasse, Siedlung, Deutsches Blut'*, p. 10.
72. Lemkin, 'Genocide as a Crime under International Law', 146–51, n. 6.
73. BfBStUZAST K 153 70/47 Bd 1, 3 re sterilisation trial in Schwerin.
74. Hans Harmsen, 'The German Sterilization Act of 1933', *Eugenics Review*, vol. 46 (1955) 227–32.
75. Gisela Bock, *Zwangssterilisation im Nationalsozialismus* (Opladen: Westdeutscher Verlag, 1986), pp. 244–5.
76. *Das Reichsgesundheitsamt im Nationalsozialismus. Menschenversuche und 'Zigeunerforschung' zwischen 1933 und 1945* (Berlin, 1998).
77. AdeF BB/35/260 Concerning Prof. Rudin Ernst (numbered 140084), Col. Amen, Field Int Section, 7 September 1945.
78. NMT 3/1767–1769 and UNA UNWCC, Lang to Interallied Commission for the Investigation of War Crimes, 10 May 1945.
79. Matthias Weber, *Ernst Rüdin. Eine kritische Biographie* (Berlin: Springer, 1993), pp. 261–7. Volker Roelcke, 'Programm und Praxis der psychiatrischen Genetik an der Deutschen Forschungsanstalt für Psychiatrie unter Ernst Rüdin', in Hans-Walter Schmuhl (ed.), *Rassenforschung an Kaiser-Wilhelm Instituten vor und nach 1933* (Göttingen: Wallstein Verlag, 2003), pp. 38–67.
80. NARA RG 338 USAR/JAG War Crimes Records Regarding Medical Experiments No. 125571–4 Straight to McHaney and Hardy, 5 November 1946.
81. NARA RG 338/ JAG War Crimes Records Regarding Medical Experiments No. 127376–9.
82. NARA RG 338 290–59–17 Medical Experiments No. 127376, 28 Letters from people who were sterilised, 28 February 1947.
83. Inbona, 'Le procès des médecins allemands. Leur responsabilité dans la technique du génocide', 251–2. AOF AJ c. 3645 p. 287 no. 8949 'Pouvez-vous aider la justice française?', *Bulletin de liaison de l'amicale de Dachau, Comite de l'Île de France'*, (Fédération nationale des deportés et internés resistants et patriotes), no. 6 (August 1946).
84. NARA M 1019/68 Schwind, Herbert Alfred interrogation by Meyer and Alexander, 12 December 1946, p. 4.
85. Fortunoff Video Collection, Yale University, Rodell interview.
86. AdeF BB/35/274 Sterilisation. Rapports et documents Leo Alexander, 22 November 1946. NMT 2/600, 3/543ff.
87. Born 20 May 1894 Dettweiler (bas-Rhin).
88. NMT 2/605–8, 4/5957–8.
89. AdeF BB/35/274 Declaration at School of Medicine by Christian Champy, Professor of Medicine, Paris [= DOC NO–521]. Hermann Stieve, 'Paracyclische Ovulationen', *Zentralblatt für Gynäkologie*, no. 7 (1944) 260. Klee, *Auschwitz*, pp. 97–111.

90. Weindling, *Health, Race and German Politics*, pp. 552–70. Hans-Peter Kröner, *Von der Rassenhygiene zur Humangenetik. Das Kaiser-Wilhelm-Institut für Anthropologie, menschliche Erblehre und Eugenik nach dem Kriege* (Stuttgart: Gustav Fischer, 1998).

91. NARA RG 153 Records of the Judge Advocate General (Army) War Crimes Branch, Box 93 Persons and Places 1944–49 File 100 – 106 JAG War Crimes Office; Astor, *Last Nazi*, p. 127; Kröner, *Rassenhygiene*, pp. 92–4.

92. NARA RG 338/ 290–59–17–5 USAR EUR/JAG War Crimes. Records Regarding Medical Experiments, 125576–125580 Major A.A. Kingscote for L. Gill, 'SS. Medical Research', 4 September 1945. 125925–125933 Concentration camp custodial and medical personnel, n.d. [1946].

93. Weindling, *Health, Race and German Politics*, pp. 563–4. Benno Müller-Hill and Ute Deichmann, 'The Fraud of Abderhalden's Enzymes', *Nature*, vol. 393 (1998) 109–11. Benno Müller-Hill, 'The Blood from Auschwitz and the Silence of the Scholars', *History and Philosophy of the Life Sciences*, vol. 21 (1999) 331–65.

94. NMT 8/2816.

95. Paul Weindling, 'Ärzte als Richter: Internationale Reaktionen auf die Medizinverbrechen des Nürnberger Ärzteprozesses in den Jahren 1946–1947', in C. Wiesemann and A. Frewer (eds), *Medizin und Ethik im Zeichen von Auschwitz. 50 Jahre Nürnberger Ärzteprozess* (Erlanger Studien zur Ethik in der Medizin) (Erlangen and Jena: Palm und Enke, 1996), pp. 31–44.

96. United Nations War Crimes Commission (Research Office), *Bulletin No. 18*, 26 November 1945.

97. NARA RG 153/ 100–621/ Box 59, JAG War Crimes Branch, Persons and Places File, OSS List of German Doctors – War Criminals, Freiherr von Verschür, Otmar, 30 May 1946.

98. NARA RG 319 Entry 11d IRR Personal Box 401 Otmar von Verschuer Memorandum, 25 July 1946. 13 August 1946 release order on authority of G–2.

99. Paul Weindling, 'Genetik und Menschenversuche in Deutschland, 1940–1950. Hans Nachtsheim, die Kaninchen von Dahlem und die Kinder von Bullenhuser Damm', in Hans-Walter Schmuhl (ed.), *Rassenforschung an Kaiser-Wilhelm Instituten vor und nach 1933* (Göttingen: Wallstein Verlag, 2003), pp. 245–74.

100. Weindling, *Health, Race and German Politics*, p. 560 for Verschuer's Reich Research Council grants. TNA: PRO FO 1012/428 'Liquidation of German War Potential Committee. Kaiser Wilhelm Institutes in the American Zone', 14 October 1946.

101. Weindling, *Health, Race and German Politics*, pp. 553–8.

102. Ibid., pp. 568–9.

103. NARA RG 319 Entry 11d IRR Personal Box 401 Otmar von Verschuer, CIC letter, 12 July 1946.

104. NMT 4/1736 Ludwig Schmidt to Clay, 28 August 1947.

105. NMT 4/2604 Brack Clemency plea, 7 April 1948.

106. NMT 4/1618 Thea Brack, 7 June 1947.

107. Mitscherlich, Mielke, *Das Diktat*, p. 162.

108. 'Vertriebene Wissenschaft', *Neue Zeitung* (3 May 1946). Simone Hannemann, *Robert Havemann und die Widerstandsgruppe 'Europäische Union'* (Berlin: Robert-Havemann-Gesellschaft, 2001), pp. 111–15.

109. NL Havemann, Bd 10a Havemann to Ludwig Wörl, 19 September 1946. Havemann to Nachtsheim, 13 July 1946. Paul Weindling, '"Tales from

Nuremberg": the Kaiser Wilhelm Institute for Anthropology and Allied Medical War Crimes Policy', in Doris Kaufmann (ed.), *Geschichte der Kaiser-Wilhelm-Gesellschaft im Nationalsozialismus. Bestandsaufnahme und Perspektiven der Forschung*, 2 Bde. (Göttingen: Wallstein Verlag, 2000), pp. 621–38. AdeF BB/35/260 KWG Havemann, 15 August 1946 to Betreuungsstelle für Sonderfälle, Frankfurt/M. Attached to this: letter to Dr Freimann, Frankfurt a. M. from Max Weinreb. Copy of letter to Freimann, 28 October 1946 concerning Otto Wolken as a witness against Mengele.

110. NL Havemann, Bd 13 Rostock to Havemann, 18 May 1947. Havemann to Wirth, 8 June 1947. Havemann to Rostock, 2 June 1947.
111. NARA RG 338 USAREUR/JAG War Crimes, Records Regarding Medical Experiments, No. 125898–95 Hardy to Straight and Straight, Medical Experiments Cases, 7 September 1946, 3, 5.
112. APD 4/35 M. Wolfson to Leo Alexander, 7 January 1947.
113. NARA M 1019/ 50 Nikolas Nyiszli deposition by von Halle, 8 October 1947.
114. R. Pommerin, *Sterilisierung der Rheinlandbastarde* (Cologne, 1979).
115. STAN Rep 502A KV- Verteidigung Handakten Sauter 3 Material zu Menschenversuchen, Blome statement, 23 June 1946.
116. BStU ZM 1639 Akte 2.
117. BDC Helmut Poppendick file, 19 December 1944.
118. NMT 4/6097 Poppendick Closing plea, 63.
119. NMT 2/5743.
120. NMT 2/5701, 5743.
121. NMT 4/6004.
122. NMT 4/5973.
123. Weindling, *Health, Race and German Politics*, pp. 476–7.
124. Weindling, *Health, Race and German Politics*, p. 502. F. Lenz, 'Gedanken zur Rassenhygiene (Eugenik)', *Archiv für Rassen- und Gesellschaftsbiologie*, vol. 37 (1943) 84–109. Lenz Papers, Lenz to Verschuer, 12 and 17 June 1940. NMT 4/1982; Aly and Roth, 'Gesetz'.
125. UAG Kur PA Lenz Fritz, 18 January 1946, Rector to military control officer, Göttingen. Lene Koch, *Racehygiejne I Danmark 1922–1956* (Copenhagen, 1996) on Kemp.
126. MPG Abt III Rep 20B Nachtsheim papers Lenz to Nachtsheim, 28 December 1945, 11 August 1946, 2 September 1946; Nachtsheim to Lenz, 21 August 1946.
127. NMT 2/5657and 4/5957–8 as Poppendick Exhibit 1. NMT 4/5957 Defence Book Poppendick Affidavit Lenz, 20 January 1947. It was ironic that details of Lenz's appointment as a university professor were given as 'name and location of university unknown' on 5 November 1946.
128. *International Military Tribunal* vol. 5, 121, vol. 20, 276; vol. 21, 5–6, vol. 23, 193.
129. NARA M 1019/54 Poppendick interrogations, 30 November 1946, 20 and 23 July 1947, p. 3.
130. NARA M 1019/54 Poppendick interrogation, 30 November 1946, p. 14.
131. NARA M 1019/54 Poppendick interrogation, 20 June 1947, pp. 10–13.
132. NARA M 1019/7 Blome interrogation, 26 February 1946, pp. 47–9.
133. NMT 4/6009–10 testimony of Fritz Schwalm, 3 May 1947.
134. Heinemann, '*Rasse, Siedlung, Deutsches Blut*', pp. 566–7.
135. NMT 4/6090.
136. Cf. NMT 4/4603 Poppendick Defence Book, extract from Jores, *Clinical Endochrinology*.

137.   NMT 4/6012 Kogon evidence, 22 April 1947. Website 'Gay Holocaust' for comprehensive documentation.http://users.cybercity.dk/~dko12530/hunt_for_Danish _kz_htm. Wolfgang Roll, *Homosexuelle Häftlinge im Konzentrationslager Buchenwald* (Buchenwald, 1991).

138.   Website 'Gay Holocaust', pp. 7–8.

139.   APD 4/37 CIC report on Medical Sterilisation, 6 July 1945. APD 4/36 Taubröck statement at Nuremberg, 20 June 1945.

140.   G. Madaus and F.E. Koch, 'Tierexperimentelle Studien zur Frage der medikamentlosen Sterilisierung (durch Caladium seguinum)', *Zeitschrift für die gesamte experimentelle Medizin*, 109 (1941) 68–87. Michael G. Kenny, 'A Darker Shade of Green: Medical Botany, Homeopathy, and Cultural Politics in Interwar Germany', *Social History of Medicine*, vol. 15 (2002) 481–50.

141.   APD 4/35 Alexander to Hardy, 10 March 1947. Koch, in *Zeitschrift f d experimentelle Medizin* (1941).

142.   APD 4/36 Karl Taubock affidavit, 20 June 1945. 4/37 Report, 6 October 1945, re Jackson Prosecution staff.

143.   NARA M 1019/54 Pokorny interrogation, 12 September 1946. Selma Steinmetz, *Österreichs Zigeuner im NS-Staat* (Vienna: Europa, 1966).

144.   AFP Leo to Phyllis Alexander, 31 December 1946.

145.   APD 4/33 Alexander Logbook, ca 21 March 1947 for twin and racial research attributed to Wirth; Alexander saw the ballet *Coppelia* in Paris on 16 January 1947 after a meeting of the ISC for Medical War Crimes.

146.   L. Alexander, 'Science under Dictatorship', *March of Medicine*, vol. 14 (1949) 51–106.

147.   Kröner, *Von der Rassenhygiene zur Humangenetik*, pp. 99, 124, 130–1. Ernst Klee, *Auschwitz. Die NS-Medizin und ihre Opfer* (Frankfurt a.M.: S. Fischer, 1997), p. 486. B. Müller-Hill, *Tödliche Wissenschaft* (Reinbek: Rowohlt, 1984), pp. 9, 73, 159, 164 on Magnussen.

148.   Weindling, *Health, Race and German Politics*, pp. 586–7. Gerald Astor, *The 'Last' Nazi. The Life and Times of Dr Joseph Mengele* (London: Sphere Books, 1998), pp. 128–9; Lucette Matalon Lagnado and Sheila Cohn Dekel, *Children of the Flames. Dr Josef Mengele and the Untold Story of the Twins of Auschwitz* (London: Sidgwick and Jackson, 1991), pp. 119–20. Kröner, *Rassenhygiene*, pp. 99, 121–5, 130–1 based on NARA RG 238 Collection of WW2 War Crimes Records, IMT Office of US Chief of Council. RG 338/290–59–17–5 USAR EUR/JAG War Crimes. Records Regarding Medical Experiments, Supplementary list of persons who may be involved in Medical Experiment Case, Document No. 125478.

149.   NARA RG 238, Office of the Chief of Counsel for War Crimes, Berlin Branch, General Records, Entry 22, box 2, NM 70 M. Wolfson, memo to E.S. Schwenk, 3 October 1946.

150.   NARA RG 238, Office of the Chief of Counsel for War Crimes, Berlin Branch, General Records, Entry 22, box 2, NM 70 Wachenheimer's undated requests, one also for Kurt Riegel and the other for Liebau, are located between similar requests, dated 4 and 8 November 1946. Hedy Epstein, website: www.hedyepstein.com.

151.   Quoted in Kröner, *Rassenhygiene*, p. 124.

152.   Ibid. AdeF BB/35/269 documentation relative aux expériences médicales, memo of Wolfson, 7 November 1946, Wolfson to Alexander, 12 December 1946, Alexander to Wolfson, 13 December 1946.

153.   Lagnado and Dekel, *Children*, pp. 126–9 record the pre-trial report has disappeared.

154.   APD 4/35 M. Wolfson to L. Alexander, 7 January 1947.

155. APBU M. Wolfson list, 12 February 1947.
156. Kröner, *Rassenhygiene*, p. 129.
157. Achim Trunk, *Zweihundert Blutproben aus Auschwitz* (Berlin: MPG, 2003).
158. Lagnado and Dekel, *Children*, pp. 120, 129, 288–9.
159. APBU Sperber deposition.
160. Cf. Astor, *Last Nazi*, pp. 144–5; Lagnado and Dekel, *Children*, 103–7, 110–13, 115–17, 131. Gisela Perl, *I was a Doctor in Auschwitz* (New York: International Universities Press, 1948).
161. NARA RG 153 Records of the Office of the Judge Advocate General (Army) War Crimes Board Persons and Places File, (Dossier File) 1944–1949 Box 93, file No. 100 – 1184, Perl – forwarded letters to War Crimes Group US Chief of Counsel Nuremberg; Perl to War Department, 11 January 1947, Gunn to Perl, 27 January 1947 that her letter forwarded to the US Chief of Counsel Nuremberg; 7 October 1947 Perl to Gunn 'I have learned that the trial of the greatest "mass murderer" Dr Mengerle will be held very soon in Nurnberg'; E.H. Young to Perl that the Auschwitz trial is in the hands of the Polish Government, 8 December 1947; Taylor to Perl, 19 January 1948 re your letter of 8 December 1947 'we wish to advise our records show Dr Mengerle is dead as of October 1946'; also Young to Perl, 12 February 1948.
162. APD 4/33, Alexander Logbook, March entries.
163. NARA RG 153/87–0–1/box 13 to Deputy Military Governor OMGUS, 14 March 1947.
164. APD 4/33, Alexander Logbook, 179–86, interrogation of 29 January 1947.
165. MPG Abt. III Rep 84/1 Nr 627 Hanns Marx to Butenandt, 21 April 1947; Butenandt to Hanns Marx, 4 July 1947.
166. Trunk, *Zweihundert Blutproben*.
167. Heinemann, '*Rasse, Siedlung, Deutsches Blut*', pp. 533–5.
168. NMT 8/1386. Heinemann, '*Rasse, Siedlung, Deutsches Blut*', pp. 537–9.
169. MPG Abt III Rep.20B Nachtsheim papers, Nachtsheim to Lenz, 16 October 1948; Weindling, *Health, Race and German Politics*, pp. 568–70.
170. Nikolai Krementsov, *Stalinist Science* (Princeton: Princeton University Press, 1997), pp. 149–90.

13 Euthanasia

1. NMT 8/670 Mielke to Mitscherlich, 16 February 1947. Faulstich, 'Zahl'.
2. NMT 2/2393–5. Winfried Süss, *Der 'Volkskörper' im Krieg* (Munich: Oldenbourg, 2003), pp. 283–5, 409.
3. STAN Rep 502A KV- Verteidigung Handakten Sauter Nr 3 Blome deposition, Darmstadt, 23 June 1946.
4. Friedlander, *Origins*, pp. 68–70, 321.
5. His figure appears to be credible as historians uncover more and more about decentralised forms of medical killing. Cf. Faulstich, 'Zahl'.
6. Alexis Carrel, *Man, The Unknown* (London: Hamish Hamilton, 1938), p. 296.
7. Udo Benzenhöfer, 'Der Fall "Kind Knauer"', *Deutsches Ärzteblatt*, vol. 95 (1998) B–954–5. Benzenhöfer, 'Genese und Struktur der NS-Kinder und Jugendlicheneuthanasie', *Monatschrift Kinderheilkunde*, vol. 10 (2003) 1012–19. Cf. Friedlander, *Origins*, p. 39.
8. Friedlander, *Origins*, p. 44.
9. STAN Rep 502A KV- Verteidigung Handakten Fröschmann Verteidigung Brack Nr 18 'Bedlam 1946', Material über die Zustände in US-amerikanischen

Nervenheilanstalten, Most US Mental Hospitals are a Shame and a Disgrace, Albert Q. Maisel.

10. US Military Tribunal, NMT 2/2501, 4 February 1947, Transcript of the Proceedings in Case 1, p. 2424.

11. STAN Rep 502A KV- Verteidigung Handakten Fröschmann Verteidigung Brack Nr 22 Strafsache gegen Dr Werner Heyde, testimony that Brack claimed to view matters as a doctor, 1 April 1947.

12. APD 4/33, 4 February 1947.

13. NMT 2/7688, 14 May 1947.

14. Burleigh, *Death*, p. 273.

15. Klee, *Was sie taten*, p. 310.

16. NARA T1021/ 12 Schneider to Brack, 10 December 1941.

17. STAN Rep 502A KV- Verteidigung Handakten Akte Fröschmann Nr 17. Rostock 10 Rostock to von Weizsäcker, 16 July 1947.

18. STAN Rep 502A KV- Verteidigung Handakten, Akte Sauter Nr 6 Bettina Blome to Sauter, 30 November 1946.

19. NMT 2/ 4605, 14 March 1947.

20. NMT 2/2317, 30 January 1947.

21. Ibid.

22. Götz Aly, *Endlösung* (Frankfurt, 1995), pp. 314–15.

23. NARA RG 153 NM 70 Entry 188 190/19/box 2 Hardy, 'Individual responsibility of Prof. Dr. Karl Brandt'.

24. STAN Rep 502A KV- Verteidigung Handakten Sauter Nr 6, concerning *Arzt im Kampf*, p. 222. APD 4/35 Alexander memo, 4 December 1946, Biography of the Defendant Kurt Blome. Questions to be put to Dr Blome, 12 March 1947.

25. APD 4/35, 12 March 1947, Blome questions.

26. Aziz, *Brandt*, p. 17. US Military Tribunal, Transcript of the Proceedings in Case 1, NMT 2/2380.

27. NMT 2/2430.

28. STAN Rep 502A KV- Verteidigung Handakten Fröschmann Verteidigung Brack Nr 17 Beweisnahme Brack.

29. STAN Rep 502A KV- Verteidigung Handakten Fröschmann Verteidigung Brack Nr 19 Plädoyer für Brack, 'Euthanasie. Für und Wider'; Tandler, *Gefahren der Minderwertigkeit*. Revision und Gnadengesuch, 27 August 1947, Fröschmann to M. Schlegelmilch.

30. Ernst Klee, *Dokumente zur 'Euthanasie'* (Frankfurt: S. Fischer, 1985), pp. 304–5.

31. NMT 2/1597–9.

32. NARA M 1019/7 Boehm interrogation by H. Meyer, 5 November 1946, p. 13.

33. NARA M 1019/7 Boehm on 5 November 1946, 4 and 28 February 1947.

34. NARA M 1019/7 Boehm interrogation by H. Meyer, 5 November 1946, 28 February 1947, deposition 28 February 1947.

35. Burleigh, *Death*, p. 274. Weindling, *Epidemics and Genocide*, pp. 23, 241, 293.

36. STAN Rep 502A KV- Verteidigung Handakten Fröschmann Verteidigung Brack 20 'Das Euthanasie Problem' Bd 1 including Carrel, 12, 317–18.

37. NMT 6/400 denied, 4 December 1946, and 401, 402, 403.

38. STAN Rep 502A KV- Verteidigung Handakten Sauter Nr 5 Blome to Sauter, 16 July 1947.

39. APD 4/35 Memorandum to Sebring, 15 July 1947. Notes on Sievers interrogation 25 November 1946. Eleventh personality interrogation by Alexander, 27 February 1947.

40. NMT 2/7680, 14 May 1947.

14   Experiments and Ethics

1. NMT 5/26, 19 August 1947.
2. NMT 5/52. Jay Katz, 'The Nuremberg Code and the Nuremberg Trial. A Reappraisal', *JAMA*, vol. 276, no. 20 (1996) 1662–6.
3. NMT 5/23. Michael Marrus, 'The Nuremberg Doctors' Trial in Historical Context', *Bulletin of the History of Medicine*, vol. 73 (1999) 106–23, 110. Alexander Mitscherlich and Fred Mielke, *Medizin ohne Menschlichkeit* (Frankfurt am Main: S. Fischer, 1960). Alice Platen-Hallermund, *Die Tötung Geisteskranker in Deutschland: Aus der deutschen Ärzte-Kommission beim amerikanischen Militärgericht* (Frankfurt: Verlag der Frankfurter Hefte, 1948). François Bayle, *Croix gamée contre caducée. Les expériences humaines en Allemagne pendant la deuxième guerre mondiale* (Neuwied: Imprimerie Nationale, 1950). More recent literature includes Jürgen Peter, *Der Nürnberger Ärzteprozess* (Frankfurt am Main: Lit Verlag, 1994). Ulrich-Dieter Opitz, *Medizinverbrechen vor Gericht* (Erlangen: Palm & Enke, 1999).
4. APD 4/35 Taylor to Fenger, 16 April 1947.
5. Evelyne Shuster, 'Fifty Years Later: the Significance of the Nuremberg Code', *The New England Journal of Medicine*, vol. 337 (1997) 1436–40. J.M. Harkness, 'The Significance of the Nuremberg Code', *The New England Journal of Medicine*, vol. 338 (1998) 995–6. NMT 5/23.
6. 'Biomedical Ethics and the Shadow of Nazism: a Conference on the Proper Use of the Nazi Analogy in Ethical Debate', *Hastings Center Reports*, vol. 6, no. 4 (1976) Supplement, 1–20, 6. Katz, 'Reappraisal', 1662–6.
7. Sebring Papers, 'The Medical Case. Nazi War Crimes Trials, Nurnberg, Germany', pp. 10–11. The paper derives from Sebring's period as Dean of Stetson University College of Law.
8. TTP–CLS 14/8/25/539 Jon Harkness to Taylor, 16 November 1993, Taylor to Harkness, 29 November 1993, Harkness to Taylor, 29 December 1993. *Final Report of the Advisory Committee on Human Radiation Experiments* (New York: Oxford University Press, 1996), pp. 75–7, 93–4. Jon M. Harkness, 'Nuremberg and the Issue of Wartime Experiments on US Prisoners. The Green Committee', *JAMA*, vol. 276, no. 20 (1996) 1672–5.
9. Paul Weindling, 'From International to Zonal Trials. The Origins of the Nuremberg Medical Trial', *Holocaust and Genocide Studies*, vol. 14 (2000) 367–89.
10. NARA RG 338 290–59–17, 707–Medical Experiments No. 125991–3, 'Are the "Experiments" Performed Crimes and Barbarities?'
11. Norman Howard-Jones, 'Human Experimentation in Historical and Ethical Perspectives', in Zbigniew Bankowski and Howard-Jones (eds), *Human Experimentation and Medical Ethics* (Geneva: XVth CIOMS Council for International Organization of Medical Sciences, 1982), pp. 453–95.
12. As translated in Michael A. Grodin, 'Historical Origins of the Nuremberg Code', *The Nazi Doctors and the Nuremberg Code*, pp. 121–44, 130–1. Christian Bonah, 'Le drame de Lübeck', in Christian Bonah, E. Lepicard and V. Roelcke (eds), *La médicine expérimentale au tribunal* (Paris: CPI, 2003), pp. 65–94.
13. *Volkswohlfahrt* (11 June 1931) col. 607.
14. Grodin, 'Historical Origins', 128–32. NMT 2/9141–9145, 9170–9171.
15. MP II2/45.1 Eichholtz to Mitscherlich, 22 November 1946.
16. BAB R 26 III/197 RFR Correspondence re: drugs, paints and dehydrated foods, Dr Rascher, September 1943–June 1944, 5 June 1944 Sievers to Plötner, Eichholtz memo on Polygal and the need for a stronger pectin product.

17. 'That this has been the Magna Carta of German physicians since 1931, and has remained continuously in force since then.' Royal Society, London, Dale Papers 20.2.16 Rost to Dale, 2 December 1946. Weindling, ' "Tales from Nuremberg"'. TNA: PRO CAB 124/545 Office of the Minister of Reconstruction. 'Royal Institution. Correspondence on the Control of German Scientific Research and Development', 25 October 1945, Dale to Martin Flett, 14 November 1945. Kneer, 'Heubner', 51–2. .

18. RS Dale Papers 20.2.17 Dale to Rost, 2 January 1947.

19. RS Dale papers 20.2.18 Dale to Moran, 6 February 1948.

20. TNA: PRO FO 1031/92 Maj. E. Tilley, EPES, Interrogation Reports Preliminary Report on Prof. Dr Fritz Hildenbrandt, Dr Karlhans Osterwald and Dr Hans Beuchelt, 5 April 1946, 5–6, 41.

21. TNA: PRO WO 309/ 471. AIP Fonds Lépine for annotated minutes.

22. Royal College of Physicians of Edinburgh, Sydney Smith papers, box 9, folder 94. Smith memo of 20 May 1946.

23. TNA: PRO WO 309/471 Comments of Somerhough, 17 June 1946.

24. TNA: PRO WO 235/307 re Treite claiming consent.

25. Ivy Papers, Box 86 folder 25 Memorandum (6 pp), n.d. [ca. 1945].

26. APD 1/9 Alexander, 'Ethical and Non-Ethical Experimentation on Human Beings', 15 April 1947. Canadian National Archives, John W. Thompson, military service record.

27. Yale University MS 1236 Fulton papers series I box 91 folder 1257 Ivy file, Fulton to C.S. Stephenson, 7 May 1942.

28. Ivy papers, Bronk correspondence with Ivy, 2 May 1945.

29. Fulton papers folder 1258, 1943–51 Ivy to Fulton, 16 January 1946.

30. Ivy Papers box 82 folder 23, *The Bulletin for Medical Research, Special Memorial Issue*, Ivy annotations on Carlson's sense of responsibility to mankind. Carlson, 'The Necessity for Animal Experimentation', p.17, reprinted from *Proceedings of the Annual Congress on Medical Education and Licensure, Chicago, February 11 and 12 1946*. University of Chicago, Dept of Special Collections, A.J. Carlson Papers (Misc. Arch. Col.) A.C. Ivy to Carlson, 20 January 1950. Susan E. Lederer, *Subjected to Science* (Baltimore: Johns Hopkins University Press, 1995), pp. 117–19, 185.

31. Ivy Papers, box 6 folder 12, A.C. Ivy, 'Some Ethical Implications of Science', Presented to the Central Association of Science and Mathematics Teachers, 28 November 1947. Ivy, 'The History and Ethics of the Use of Human Subjects in Medical Experiments', *Science*, vol. 108 (2 July 1948) 1–5.

32. NMT 2/9202 Ivy on 12 June 1947.

33. RAC RF 12.1 Diaries box 22, Gregg diary 1945, 178–9, 21 October 1945, visit to Lyons.

34. AdeF BB/30/1821 21 June 1946, R Legroux letter suggests meeting at Institut Pasteur at 18.00 on 2 July in the laboratory of Lépine.

35. Weindling, *Epidemics and Genocide*, pp. 323–8.

36. Ivy Papers, Box 88 folder 2, A.C. Ivy, 'The Message of Pasteur', 30 September 1954, pp. 3–4 for the move from animal to human experiments.

37. AdeF BB/30/1821 Commission d'investigation pour l'étude des crimes de guerre scientifique, 1946–7. TNA: PRO WO 309/471 Arrêté du 19 juin 1946 portant institution auprès de la Présidence du Gouvernement (Etat-Major général de la Défense Nationale) d'une commission d'investigation pour l'étude des crimes de guerre scientifique. Minutes of Meeting to Discuss War Crimes of Medical Nature Executed in Germany under the Nazi Regime, 31 July 1946.

38. NARA RG 153/86–3–1/Box 11 folder 4 Book 3 Ivy, Report, Outline of Itinerary, 29 July 1946.

39. University of Ottawa Archives, File Thompson J.W.R., 5 September 1946, Danis to Thompson on 'Outline of the Principles and Rules of Experimentation'. For further material on medical ethics, see Deschatelets Archives, Ottawa, Caron papers.

40. NAC RG 25 External Affairs Series A–3–b, volume 5792 file 250(s) Request from Wing Commander John Thompson, J.W. Holmes to Norman Robertson, 13 August 1946.

41. University of Ottawa Archives, J.W.R. Thompson file, 23 December 1946 and 3 June 1947, Thompson to Danis.

42. TNA: PRO WO 309/471 Minutes of Meeting to Discuss War Crimes of Medical Nature Executed in Germany under the Nazi Regime, 31 July 1946.

43. TNA: PRO WO 309/471 Minutes of Meeting to Discuss War Crimes of Medical Nature Executed in Germany under the Nazi Regime, 31 July 1946. The file also contains a set of edited draft minutes, indicating that Ivy's caution was excluded from the final minutes. The minutes of 1 August 1946 note that the previous day's minutes were agreed 'after the acceptance of certain alterations'. Weindling, 'Ärzte als Richter', 31–44.

44. TNA: PRO WO 309/ 471 Minutes of Meeting to Discuss War Crimes of a Medical Nature Executed in Germany under the Nazi Regime, Appendix B. For a variant text, see the Outline in NARA RG 338 290–59–17 Medical Experiments No. 125990.

45. TNA: PRO WO 309/471 Second Meeting of the International Scientific Commission for the Investigation of Medical War Crimes, minutes dated 26 October 1946.

46. TNA: PRO WO 309/468, 30 September 1946, letter to Mant concerning Taylor's agreement to 'try the principal doctors who engaged in medical experimentation on involuntary human beings'. AdeF BB/35/260, 4c McHaney to Ivy, 30 September 1946.

47. NARA RG 153/86–3–1/Box 11/ folder 4/book 3 Andrew Ivy, Outline of Itinerary.

48. A.C. Ivy, 'Nazi War Crimes of a Medical Nature', *Phi Lamba Kappa Quarterly*, vol. 22 (1948) 5–12.

49. NARA RG 338 290–59–17, 707–Medical Experiments Nos 125990–125995. The authorship of the elaborated Rules for Ethical Experimentation is unclear.

50. NARA RG 153/86–3–1/Box 11/ folder 4/book 3 Andrew Ivy, Outline of Itinerary.

51. Ibid. Ivy, Outline of Itinerary, 20.

52. NARA RG 153/86–2–2/box 9 book 2, Special Release No. 63, 19 November [1946] OCCW, Public Relations.

53. NARA RG 153/ 86–2–2/ Box 9 book 2 David Marcus to Ivy, 7 November 1946.

54. NARA M 1019/29 Hoven interrogation by McHaney.

55. NARA M 1019/66 Schröder interrogation by Meyer, 2 October 1946, pp. 29–30.

56. 'Biomedical Ethics', *Hastings Center*, Taylor on p. 6.

57. NARA RG 153/86–3–1/Book 1/Box 10 Taylor to War Department, 1 November 1946.

58. NARA RG 153/86–2–2/ Box 9 book 2 Alexander to C.T. Perkins, 13 November 1946; Alexander to Walter Cramer, George Washington University, 14 November 1946; Alexander to Winfred Overholzer, St Elizabeth's Hospital, Washington DC, 14 November 1946; Alexander to William Talisferro, Rusk Medical School, Chicago, 14 November 1946; Alexander to William Rappleye, Columbia University, 14 November 1946; Alexander to Alan Chesney, Johns

Hopkins Medical School 14 November 1946; Alexander to Board of Scientific Directors, Rockefeller Institute of Medical Research, 15 November 1946.

59. Ivy, Report on War Crimes of a Medical Nature, AMA Archives, 10, 13, 14. For the AMA response see NMT 2/9310.

60. *JAMA*, vol. 133 (1946), 35. *Human Radiation Experiments*, p. 77.

61. NMT 2/9310, Ivy on 13 June 1947.

62. TNA: PRO WO 309/471 Somerhough, 6 September 1946.

63. TNA: PRO WO 309/471 A.G. Somerhough to Judge Advocate General, 'Proposed International Scientific Commission for the Investigation of War Crimes of a Scientific Nature', 6 August 1946.

64. The delegation consisted of Henry Dale, Sydney Smith (Edinburgh), Sweeney (St Thomas' Hospital), and W.G. Barnard (Royal College of Physicians, London). For background, see *Scientific Results of German Medical Warcrimes. Report of an Enquiry under the Chairmanship of Lord Moran, MC, MD* (London: HMSO, 1949).

65. NARA RG 153/86–2–2/box 9/book 2 Patterson to Secretary of State. 17 December 1946. RG 153/86–3–1/box 11 folder 4 book 3 Secretary of War to Secretary of State, 23 February 1947. RG 153 86–3–1/ box10 folder 2 book 2. Taylor to Fenger, 16 April 1947, 4/45.

66. Mant to author, 10 October 1994.

67. MP II2/106.1 Oelemann 27 November 1946. II2/106.2 Mitscherlich to Military Tribunal No. 1, 9 December 1946.

68. NARA RG 153 86–3–1 Book 1 Box 10 War Dept to USCC Nuremberg, 16 November 1946. 'The Brutalities of Nazi Physicians', *JAMA*, vol. 133 (23 November 1946) 714–15.

69. NARA RG 153 86–3–1 Book 1 Box 10, Coded message of 16 November 1946. RG 153/86–3–1/box 10/book 1. RG 153/86–2–2/box 9 book 2, D. Marcus to Ivy, 7 November 1946.

70. AdeF BB/35/260, folder 3d Gunn to Taylor, 4 December 1946. 'The Brutalities of Nazi Physicians', *JAMA*, vol. 132 (23 November 1946) 714.

71. Ivy Papers, Box 4 folder 4 The Report of the Governor's Committee on the Ethics Governing the Service of Prisoners as Subjects in Medical Experiments, 8 December 1947.

72. Alexander Mitscherlich, 'Der Arzt und die Humanität. Erste Bemerkungen zum Nürnberger Ärzteprozeß', *Die Neue Zeitung* (20 December 1946).

73. MP II2/35.1 Mitscherlich to *DMW* Redaktion, 3 February 1947.

15   Formulating the Code

1. NMT 2/9300, Hardy, 13 June 1947.

2. 'Biomedical Ethics', *Hastings Center Reports*, p. 6.

3. NMT 2/2458, 4 February 1947.

4. NMT 2/2463–4, 4 February 1947.

5. NMT 2/2464–5, 4 February 1947.

6. NMT 2/2453, 4 February 1947.

7. NMT 2/2455–8, 4 February 1947.

8. NARA M 1019/20 Impact of atomic bomb. Cf. Gebhardt interrogation by Alexander, 3 December 1946, pp. 24–5. Final plea for Brandt by Servatius, in Katz, *Experimentation*, pp. 302–3.

9. IWM, Milch diary extracts, 7 August 1945.

10. NMT 2/10360 and 4/1952 citing papers by Elisabeth Pfeil on national euthanasia in More's *Utopia*.
11. NMT 2/11490 Becker-Freyseng plea.
12. NMT 2/2670 Karl Brandt on 6 February 1947 on whether he could have persuaded Hitler to stop experiments.
13. NYPL Lifton Papers, M9 Auschwitz research diary – long luncheon with James McHaney, 29 March 1978, p. 10. Telephone talk with James McHaney, 8 June 1977.
14. STAN KV Rep 502A KV- Verteidigung Handakten, Fritz Nr 28 Rose Einwand der Freiwilligkeit der Versuchspersonen bei Menschenversuchen in Ausland.
15. NARA M 1019/60, 23 January 1947, p. 8.
16. NMT 2/9352, 13 June 1947.
17. NMT 4/35 Alexander to McHaney, Questions of Schröder, 28 February 1947; NARA M 1019/66 Schröder interrogation by Alexander, 21 January 1947, pp. 12–13.
18. TTP–CLS 14/4/2/35 Alexander Hardy to Ivy, 4 May 1950.
19. APD 4/41 Johannes Maria Hoecht, 'Remarks to the Trial against the Physicians', *Die Begegnung*, vol. 2.
20. STAN Rep 502A KV- Verteidigung Handakten Rostock Nr 3 Plädoyer-Entwurf, Pribilla to Rostock, 17 March 1947; Rostock to Pribilla, 16 July 1947 and reply.
21. APD 4/40 Alexander radio interview, 27 January 1947.
22. APD 4/35 Alexander to Harold Kurtz, Translation Branch, 4 March 1947. Werner Bergengruen, 'Die Letzte Epiphenie', from *Dies Irae*.
23. STAN Rep 502A KV- Verteidigung Handakten Sauter Nr 3 Material zu Menschenversuchen, 'Grundsaetzliches zur Frage der Menschenversuche' (alt).
24. STAN Rep 502A KV- Verteidigung Handakten Sauter Nr 3, Bettina Blome to Kurt Blome, n.d.
25. STAN Rep 502A KV- Verteidigung Handakten Sauter 3, Blome, 'Menschenversuche', 11 March 1947.
26. STAN Rep 502A KV- Verteidigung Handakten Sauter 3 Material zu Menschenversuchen, 'Grundsaetzliches zur Frage der Menschenversuche' (alt).
27. STAN Rep 502A KV- Verteidigung Handakten Sauter Nr 3, Material zu Menschenversuchen, 'Grundsaetzliches zur Frage der Menschenversuche' (alt); 'Menschenversuche' 11 March 1947.
28. Der Arzt sei Seelsorger des Volkes. Angelika Ebbinghaus, 'Strategien der Verteidigung', in Ebbinghaus and Klaus Dörner (eds), *Vernichten und Heilen* (Berlin: Aufbau, 2002), p. 432.
29. STAN Rep 502A KV- Verteidigung Handakten Hoffmann Nr 23, Tagebuchaufzeichnungen Pokornys während des Prozesses from 2 January 1947, 28.
30. *Trials of War Criminals Before the Nuernberg Military Tribunals under Control Council Law No. 10* (Washington: US Government Printing Office, 1950), *The Medical Case*, vol. 2, 629–30, Himmler to Milch, 13 November 1942.
31. Sebring papers, 'Nurnberg, Germany, War Crimes Trials', n.d. Sebring, 'Germany Today, Yesterday and Tomorrow'.
32. NMT 4/ 3691 and 3720, affidavit of H. Siegmund, 29 January 1947. NARA M 1019/59 Romberg interrogation, 26 November 1946, p. 3.
33. STAN Rep 502A KV- Verteidigung Handakten Hoffmann Nr 27 Notizen Pokornys über Persönlichkeit und Lebenslauf; NARA M 1019/8 interrogation of Brack, 4 December 1946, p. 6.
34. NARA M 1019/20 Gebhardt interrogation, 23 July 1947, p. 18.

35. NARA M 1019/29 Hoven interrogation by Alexander, 31 January 1947, p. 6.
36. NARA M 1019/17 Fischer interrogation, 4 January 1947, p. 26.
37. NARA M 1019/20 Genzken interrogation by Alexander, 23 January 1947, p. 8.
38. NARA RG 338/290–59–33/ Executees Landsberg Prison, Wolfram Sievers to Hella Sievers, 19 August 1945, to Nona Illing, 2 September 1945.
39. NARA M 1019/50 Oberheuser interrogation by Alexander, 28 December 1946.
40. NARA M 1019/17 Fischer interrogation, 11 January 1947, p. 2.
41. Cf. *Böhme-Brevier* (Leipzig: Insel, 1939).
42. NMT 2/10860, 4/4052–7 Affidavit Meyer-Abich, author of Ideen und Idealen der biologischen Erkenntnis, on 1 March 1947; 4/5059–61 democratic credentials of Meyer-Abich; definition of holism 4/5062–5070.
43. Andreas Frewer, 'Die Euthanasie-Debatte in der Zeitschrift Ethik 1922–1938', Andreas Frewer and C. Eickhoff (eds), *'Euthanasie' und die aktuelle Sterbehilfe Debatte* (Frankfurt, 2000), p. 102.
44. STAN Rep 502A KV- Verteidigung Handakten, Hoffmann Verteidigung Rudolf Brandt Bd 6, Mrugowsky to Hoffmann, 21 September 1947.
45. STAN Rep 502A KV- Verteidigung Handakten Sauter Nr 4 Blome to Sauter re summing up as an attack on the SS, 16 July 1947.
46. STAN Rep 502A KV- Verteidigung Handakten Sauter Nr 5 Korrespondenz.
47. NMT 24 pages on 6 May 1947, to 4/3049.
48. APD Ivy, memo for Hardy, 9 June 1947.
49. NARA M 1270/4 Fischer interrogation, 5 November 1945, p. 4.
50. Ibid., p. 8.
51. NMT 2/9040, 10 June 1947.
52. NMT 2/9009–10, 10 June 1947.
53. APD 4/34. Interrogation of Wolfgang Lutz on 5 December 1946.
54. NMT 4/287–312 Luxenburger and Halbach, 14 April 1947. It was originally called Beispiele von Menschenversuchen in der englisch-amerikanischen medizinischen Forschung, see STAN Rep 502A KV Verteidigung, 6 Material zu Blome.
55. NMT 2/10668.
56. NMT 2/8126.
57. J.R. Neefe and J. Stokes, 'An Epidemic of Infectious Hepatitis Apparently due to a Water-borne Agent. Epidemiologic Observations and Experiments in Human Volunteers', *JAMA*, vol. 128 (1945) 1063–75.
58. NMT 4/298.
59. NMT 4/287–312.
60. NMT 4/312.
61. Victor Heiser, *An American Doctor's Odyssey* (New York: Norton, 1936), p. 149.
62. Sauter asked for cross-examination of Ivy by defendants, 13 June 1947, NMT 2/9256–62. 13 June 1947. Ruff used Marshland, 'Collapse at High Altitude', Air Surgeon, November 1944, p. 3. Cf. NMT 2/9349.
63. NMT 2/10667–10668 30 June 1947.
64. NYPL Lifton Papers M9, McHaney, 29 March 1978, p. 9.
65. Katz, 'Nuremberg Code', 1664.
66. Ivy Papers, Box 89 folder 5, Ivy to Irving Ladimer, Roger W. Newman, W.J. Curran, 23 March 1964.
67. NMT 2/9277–8, 13 June 1947.
68. NMT 2/9312, 13 June 1947.
69. NMT 2/9311–2. *The Medical Case*, vol. 2, pp. 82–3. University of Chicago, Department of Special Collections, The Morris Fishbein Papers, Box 63 folder 6, Louis A. Buie to Fishbein, 16 February 1934.

70. NMT 2/9281, 13 June 1946, objection by Sauter.
71. Jon M. Harkness, 'Nuremberg and the Issue of Wartime Experiments on US Prisoners. The Green Committee', *JAMA*, vol. 276 (1996) 1672–5. NMT 2/9284–8, 13 June 1947.
72. 'Prison Malaria: Convicts Expose Themselves to Disease so Doctors can Study it', *Life*, Overseas Edition for Armed Forces, vol. 18, no. 23 (4 June 1945), 43–4, 46. NMT 4/315. *Life Magazine* (4 June 1945) [and 22 July 1946].
73. NMT 4/10668, 30 June 1947.
74. Paul de Kruif, *Microbe Hunters* (London: Cape, 1930), 1st edn 1927, p. 282.
75. STAN Rep.502A KV Sauter 3. Menschenversuche, Blome, 'Grundsätzliches zur Frage der Menschenversuche' (old draft).
76. APD 4/35 Alexander to Robert Toms, 17 February 1947, concerning a *Reader's Digest* article on the prophecies of de Kruif. Louella Parsons was a gossip columnist.
77. NMT 2/9207–8, 9302–6, 13 June 1947.
78. Fishbein Papers box 98 folder 2, Ivy to Fishbein, 21 April 1947; Harkness dates Ivy's letter incorrectly as 27 April. Sauter at NMT 2/9281, 13 June 1947. Ivy 2/9336–7 and 9389 links the Green Committee to December 1946.
79. NMT 2/9284, 13 June 1947.
80. Fishbein papers box 98 folder 2 'Prisoners' and Ivy to Fishbein, 24 June 1947.
81. NMT 2/9293–8, 13 June 1947.
82. NMT 2/9298, 13 June 1947.
83. NMT 2/9429, 9432, 16 June 1947.
84. NMT 2/9434–5, 16 June 1947.
85. NMT 2/9437, 16 June 1947.
86. Fishbein papers, Ivy to Fishbein, 12 November 1947.
87. NMT 2/9430, 16 June 1947. Rose in reality also used tropica.
88. Harkness, 'Nuremberg'.
89. Fishbein papers, revised draft, 17 November 1947, of Report of Governor's Committee on the Ethics Governing the Service of Prisoners as Subjects in Medical Experiments.
90. 'Ethics Governing the Service of Prisoners as Subjects in Medical Experiments', *JAMA*, vol. 136 (1948) 447–58. Reprinted in Irving Ladimar, *Clinical Investigation* (Boston, 1963), 461–5. Ibid., 466 for statement of disapproval.
91. Bayerisches Hauptstaatsarchiv, Akten des Bayer. Staatsministeriums f. Unterricht und Kultus MK 54855 Dr. Werner Leibbrand. UA Munich Prof. Dr Werner Leibbrand E.II.2240.
92. Fridolf Kudlien, 'Werner Leibbrand als Zeitzeuge. Ein ärztlicher Gegner des Nationalsozialismus im Dritten Reich', *Medizinhistorisches Journal*, vol. 16 (1986) 332–52.
93. Werner Leibbrand, *Der göttliche Stab des Äskulap* (Salzburg: Otto Müller Verlag, 1939), p. 31. NMT 2/2451 general issue of human experiments and their criminality, and references to Leibbrand's evidence.
94. NMT 8/622. Unpublished autobiography of Leibbrand, p. 345.
95. STAN Rep 502A KV-Verteidigung Handakten 'Prof Leibrandt über den SS-Ärzte-Prozess', reported 9 December 1946.
96. APD 4/35 'Memorandum for Mr McHaney. Subject: Questions for interrogation of Dr. Leibbrand. From: Dr Alexander', 14 January 1947. Werner Leibbrand (ed.), *Um die Menschenrechte der Geisteskranken* (Nürnberg: Die Egge, 1946), pp. 3–4.

97. NMT 2/9409–12, 16 June 2003. Evelyne Shuster, 'Fifty Years Later: the Significance of the Nuremberg Code', *The New England Journal of Medicine*, vol. 337 (1997) 1436–40, 1438.

98. Ludwig Edelstein, 'The Hippocratic Oath: Text, Translation and Interpretation', *Supplements to the Bulletin of the History of Medicine*, no. 1 (1943), reprinted in *Ancient Medicine* (Baltimore: Johns Hopkins University Press, 1967), pp. 3–65.

99. A. Haug, *Neue Deutsche Heilkunde (1935–36)* (Husum: Matthiesen Verlag, 1985). D. Bothe, *Neue Deutsche Heilkunde 1933–1945. Dargestellt anhand der Zeitschrift Hippokrates und der Entwicklung der volksheilkundigen Laienbewegung* (Husum: Matthiesen Verlag, 1991). Carsten Timmermann, 'Rationalizing "Folk Medicine" in Interwar Germany: Faith, Business and Science at Dr. Madaus & Co.', *Social History of Medicine*, vol.14 (2001) 459–82. Timmermann, 'A Model for the New Physician: Hippocrates in Inter-War Germany', in D. Cantor (ed.), *Reinventing Hippocrates* (Aldershot: Ashgate, 2001), pp. 307–29.

100. Case 2 Milch Trial, 1 February 1947.

101. Milch Trial, 11 February 1947.

102. NMT 4/290.

103. NARA M 1019/7 Blome interrogation, 26 February 1946, p. 39.

104. NARA M 1019/ 59 Romberg 11th personality interrogation by Alexander, 27 February 1947, p. 32.

105. NARA M 1019/7 Boehm interrogation by H. Meyer, 4 February 1947, p. 14.

106. Franz Büchner, *Der Eid des Hippokrates. Die Grundgesetze der ärztlichen Ethik. Öffentlicher Vortrag gehalten in der Universität Freiburg i. Br.* (Freiburg: Herder, 1947). Cf. Eugen Kogon, 'Ärzte als Knechte des Todes', *Frankfurter Hefte* (1947); reprinted Kogon, *Ideologie und Praxis der Unmenschlichkeit* (Weinheim: Quadriga, 1995), pp. 167–70.

107. NARA M 1019/5 Becker-Freyseng interrogation by Alexander, 1 February 1947, pp. 14–20.

108. Ivy Papers, Box 6 folder 8, A.C. Ivy, 'Basic Principles. The Significance of the Moral Philosophy of Medicine', Commencement Address, University of Nebraska College of Medicine, 22 March 1947.

109. NMT 2/9314–5, 13 June 1947.

110. NMT 2/9316, 13 June 1947.

111. Ivy Papers, Box 6 folder 12, A.C. Ivy, 'Some Ethical Implications of Science', presented to the Central Association of Science and Mathematics Teachers, 28 November 1947. Mitscherlich and Mielke, *Doctors of Infamy*.

112. Ivy Papers, Box 7 'Does Science have an Ethics?' Address to Roosevelt College, December 1951.

113. NARA RG 153/86–3–1/box 11 folder 4 book 3 International Scientific Commission; Belton O. Bryan to Ivy, 25 March 1947. Note on meeting of 9 April 1947.

114. APD 4/35 Memorandum to: Mr McHaney. Subject: Advisor to the Court, 29 April 1947.

115. Ivy Papers, Box 6 folder 12 'Nazi War Crimes of a Medical Nature', p. 10. This passage is omitted from the published version, Andrew C. Ivy, 'Nazi War Crimes of a Medical Nature. Some Conclusions', *JAMA*, 139 (1949) 131–5. Also Ivy, 'Nazi War Crimes of a Medical Nature', *Phi Lambda Kappa Quarterly*, vol. 22, no. 4 (1948) 5–12.

116. Ivy Papers, Box 6 folder 12, Ivy, 'Nazi War Crimes of a Medical Nature', 17.

117. For Ivy as witness on 12–16 June, NMT 2/9196–9495. *Trials of War Criminals Before the Nuernberg Military Tribunals under Control Council Law No. 10*

(Washington: US Government Printing Office, 1950), *The Medical Case* vol. 1, pp. 994–1004; vol. 2, 82–6. Jay Katz, *Experimentation with Human Beings* (New York: Russell Sage Foundation, 1972), pp. 292–306.

118. APD 4/34.
119. APD 4/34 Alexander memorandum to Telford Taylor, Suggestions for a Discussion of the Thanatology Genocide Angle, 5 December 1946. NARA RG 238 Entry 155 Box 1 Alexander, The Perversion of the Attitude Towards Death (thanatolatry).
120. AFP Leo to Phyllis Alexander, 27 November 1946.
121. APD 4/40 Wolfe Frank radio interview.
122. Mant papers, War Crimes Investigation Unit notebook, chart of experiments.
123. APD 4/34 memorandum to Taylor, McHaney and Hardy, 'The Fundamental Purpose and Meaning of the Experiments in Human Beings of which the Accused in Military Tribunal No. 1, Case No. 1 have been Indicted: Thanatology as a Scientific Technique of Genocide'.
124. WUS Beals Papers Box 1 folder 6 Gunn to Beals, 3 December 1946 but in reply to a letter of 21 December 1946.
125. Lemkin, 'Genocide as a Crime under International Law', 146–51.
126. APD 1/12 'One Aim of the German Vivisectionists: Ktenology as a Scientific Technique of Genocide' Released for Publication on 31 December 1946, and read before the International Scientific Commission (War Crimes) in Paris, on 15 January 1947.
127. 'Thanatologists', *Newsweek* (23 December 1946) 31. See also *Newsweek* (2 December 1946).
128. APD, 4/ 33 Logbook, p. 79.
129. Ibid., p. 105.
130. For Reed see Lederer, *Subjected to Science*, pp. 19–23, 132–4.
131. Leo Alexander, 'Ethics of Human Experimentation', *Psychiatric Journal of the University of Ottawa*, vol. 1 (1976) 40–6.
132. AIP Fonds Lépine, International Scientific Commission (War Crimes), 15 January 1947. APBU Legroux to Alexander, 9 January 1947.
133. Ibid. Cf. *Harpers Magazine* (1946) 329–33.
134. APD 4/33 Alexander Logbook, 21–22 January 1947.
135. AdeF BB/ 335/ 260 folder 4c Special Release No. 104, 23 January 1947, OCCW.
136. APD 4/35 Logbook, p. 178. Archivzentrum der Stadt- und Universitätsbibliothek Frankfurt-am-Main, Alexander-Mitscherlich-Archiv, II 2/106.7 Leo Alexander, Affidavit, 25 January 1947.
137. Leo Alexander, 'Ethical and Non-ethical Experimentation on Human Beings. General Ethical, Medico-Legal and Scientific Considerations in Connection with the Vivisectionists' Trials Before the Military Tribunal in Germany', pp. 37. Alexander's text was stamped approved for publication by Civil Affairs Division, War Crimes Branch, The Pentagon, 14 March 1947, and 'no objection to publication on grounds of military security, War Department Public Relations Division 17 Mar 1947', NARA RG 153 Records of the Judge Advocate General (Army) 86–3–1/box 10/book 3. Also Damon M. Gunn to Lorwin note of 14 March 1947, and Gunn to Alexander, 21 March 1947.
138. APD 1/9, 1–29 with a corrected title 'Ethical and Non-ethical Experimentation in [*sic*] Human Beings. General Ethical, Medico-Legal and Scientific Considerations in Connection with the Vivisectionists' Trial Before the Military Tribunal in Germany'.
139. APD 4/33.

140. Leo Alexander, 'Limitations of Experimentation on Human Beings with Special Reference to Psychiatric Patients', *Diseases of the Nervous System*, vol. 27, pt 7 suppl. (1966) 61–5, 62. The memorandum was published in French translation in Bayle, *Croix gamée contre caducée*, pp. 1430–6. Although the date of the document is given as April, the text also coincides with the document approved on 14 March 1947.

141. Grodin, 'Historical Origins of the Nuremberg Code', 121–44.

142. Ivy Papers, Box 89 folder 5 Ivy to Ladimer, 23 March 1964.

143. APD 1/9, 15 April 1947.

144. APD diary 1945, p. 61, Alexander met Luxenburger, 2 June 1945.

145. 'Biomedical Ethics and the Shadow of Nazism', *Hastings Center Report*, 6 (4, Supplement) (1976) 1–20, esp. p. 6. TTP–CLS 14/6/12/257 Taylor to P.R. Breggin, 23 July 1974; Breggin to Taylor, 27 July 1974; Taylor to Breggin, 2 August 1974.

146. WUS Beals Papers, Fishbein to Beals, 5 April 1947, 24 April 1947, 20 May 1947. Box 1 folder 37 Beals to Fishbein, 1 May 1947.

147. APD 4/33 Alexander Logbook, 223, 24 April 1947. Cf. letter by Alexander to McHaney, 29 April 1947.

148. NMT 2/9236–9244, 9248–50 Sebring questioning Ivy.

149. NMT 2/9329–30, 13 June 1947.

150. NMT 2/9120–1, 11 June 1947.

151. NMT 2/9498, 17 June 1947.

152. NMT 2/9243, 12 June 1947.

153. NMT 2/9244, 12 June 1947. Fishbein Papers, box 1 folder 39 Beals to Fishbein, 29 May 1947. Box 1 folder 41 Beals to Fishbein, 30 June 1947.

154. Shuster, 'Fifty Years Later', 1437.

155. APD 4/35 Memorandum to Judge Sebring, 15 July 1947.

156. Irving Ladimar and Roger W. Newman (eds), *Clinical Investigation in Medicine: Legal, Ethical and Moral Aspects* (Boston University: Law-Medicine Research Institute, 1963). Ivy Papers, Box 89 folder 5, Ivy to Irving Ladimer, R.W. Newman and W.J. Curran, 23 March 1964.

157. Ibid.

158. NARA RG 153/ 86–3–1/ box 11 folder 4 book 3 International Scientific Commission meeting with Ivy, 9 April 1947. Robert M. Springer to Secretary of State, 31 January 1947 on Appointment of American Medical Committee.

159. APD 4/35 Taylor to Fenger, 16 April 1947.

160. NAC RG 25 box 51 92 file 250 (S) pt 1, J.W. Thompson to Lester Pearson, Under Secretary of State for External Affairs, 16 November 1946.

161. RCPE Smith Papers re visit to Germany.

162. APD 4/35 V.A. Fenger to T. Taylor, 28 March 1947; Taylor to Fenger, 16 April 1947; Alexander to Fenger, 15 April 1947.

163. A.C. Ivy, 'Nazi War Crimes of a Medical Nature', *Phi Lambda Kappa Quarterly*, vol. 22, no. 4 (1948) 5–12.

164. 'Doctors on Trial', *BMJ* (25 January 1947) 143.

165. WUS Beals Papers, Box 1 folder 16 Fishbein to Beals, 20 May 1947.

166. UNESCO Archives 'Human Rights', 256–7, Paris, March 1947; 258–72 UNESCO Committee on the theoretical bases of Human Rights, July 1947.

167. Peter Bartrip, *Themselves Writ Large. The British Medical Association 1832–1966* (London: BMJ Publishing, 1996), pp. 294–7. Alfred Cox, *Among the Doctors* (London, n.d. [1949]), pp. 189–90.

168. *BMJ*, ii (5 October 1946) 496, 503, 506.

169. BMA Archives, 'War Crimes and Medicine', June 1947.
170. *War Crimes and Medicine. Statement by the Council of the Association for Submission to the World Medical Association* (London: BMA, June 1947).
171. UNA UNWCC reel 36 Notes on German Medical War Crimes. UNWCC, Summary of Information No 55, Dec 1947 Notes on German Medical War Crimes. UNWCC, German Medical War Crimes. A Summary of Information, London: BMA, n.d. [1948].
172. WHO archives, Louis Bauer to W.P. Forrest, special assistant to the Director General of the WHO 31 May 1949. My thanks to James Gillespie for this reference.

16 Cold War Medicine

1. IWM Milch diary, 18 August 1947.
2. STAN Rep 502A KV- Verteidigung Handakten Rostock Nr 4 Entwürfe zum Schlusswort, letters in the event of condemnation; Nr 10 Voraussage des Prozess-Ausgangs, Rostock to Jutta Rach, Augsburg, August 1947. AHUB Rostock Pers Akte Bd 2, Kommentar des Badischen Rundfunkes, 11 April 1947.
3. Harry M. Marks, *The Progress of Experiment* (Cambridge: Cambridge University Press, 1997), pp. 136–63. J. Rosser Mathews, *Quantification and the Quest for Medical Certainty* (Princeton: Princeton University Press, 1995), p. 142.
4. Peter Kremer, 'Mr. Sebring Of Stetson', *St. Petersburg Times* (28 November 1965), 11.
5. APD 4/35 Alexander, 'Closing Statements', 15, 12 July 1947.
6. MP II2/105.8 Mielke to Mitscherlich, 22 August 1947.
7. Hunt, *Secret Agenda*, pp. 245–8.
8. NMT 5/27–8 Sentence.
9. NMT 6/1620 court ruling on defence motion, 14 July 1947.
10. NMT 5/116, 5/197.
11. NMT 5/115–16.
12. NMT 5/47–9.
13. NMT 5/64–5. Scholtz interview with author.
14. NMT 5/66.
15. NMT 5/72–4.
16. APD 1/3 Sillevaerts, 'The Judgement', *Bruxelles médical* (31 August 1947).
17. NMT 5/126.
18. NMT 5/127–8.
19. NMT 5/128.
20. NMT 5/201.
21. NMT 5/201–5.
22. NMT 5/206.
23. NMT 5/24 Bullenhusen Damm atrocity
24. Cf. Maguire, *Law and War*, pp. 211–12.
25. 'Nuremberg War Trials Subject of Discourse by Dr. Alexander at Custer', *The Long Island Mattituck Tribune*, vol. 76, no 49 (21 August 1947).
26. University Archives Freiburg, B53/ 227, Medical Faculty Minutes 29 April 1947, f. 165 Fall Hoven. B54/3350 Karl Brandt personal file.
27. Waldemar Hoven, 'Versuche zur Behandlung der Lungentuberkulose. Durch Intulation von Kohlekolloid', Med. Diss Freiburg i. B. 1943. Klee, *Auschwitz*, pp. 40–1.

28. UAM Akten des Akademischen Senats der Universität München, UAM E–II–N Dr med habil Georg-August Weltz.
29. 'Prof. Dr. med. Wilhelm Beiglböck verstorben', *Buxtehuder Tageblatt* (25 November 1963). STAN Rep 502A KV- Verteidigung Handakten Rostock Nr 9/10 Charlotte Becker-Freyseng to Rostock, 2 May 1948.
30. APD 4/33 Alexander Logbook, 18 April 1947 Rose on the stand. General Clay's visit.
31. STAN Rep 502A KV- Verteindigung Handakten Rostock Nr 5 Gisela Schmitz-Kahlmann to Rostock, 23 April 1948 as secretary to Servatius for the 'Sammelstelle zur Verteidigung des Ärzteprozesses', and notes on the lack of an appellate authority. Frank M. Buscher, *The U.S. War Crimes Trial Program in Germany* (New York: Greenwood Press, 1989), pp. 33, 52–4.
32. NMT 5/288, OMGUS Berlin 22 November 1947. 5/289 Clay order. See clemency appeal of 2 September 1947. Jean E. Smith (ed.), *The Papers of General Lucius D. Clay: Germany 1945–1949* (Bloomington, 1974), pp. 658–9.
33. STAN Rep 502A KV- Verteidigung Handakten Rostock Nr 9 Korrespondenz zum Freispruch von Rostock, telegrams and letters of congratulation.
34. STAN Rep 502A KV- Verteidigung Handakten Rostock Nr 5, Rostock to Karl Brandt, 7 October 1947.
35. 'Eine Zusammenstellung von Dokumenten über Menschenversuchen in anderen Laendern gemacht und der Presse zugeleitet'.
36. STAN Rep 502A KV- Verteidigung Handakten Rostock Nr 5. Rostock to Servatius, 20 April 1948.
37. STAN Rep 502A KV- Verteidigung Handakten Rostock Nr 5. Charlotte Becker-Freyseng to Rostock, 2 May 1948; reply, 20 May 1948.
38. STAN Rep 502A KV- Verteidigung Handakten Rostock 21 May1948 Brandt to 'Lieber Rostock'. Rostock to Servatius, 24 May 1948. Rein recommends reading the interrogation of Ivy over several days.
39. NARA RG 238 Entry 200 NM 70 201 files, Pokorny release to Landsberg, 20 August 1947. STAN Rep 502A KV- Verteidigung Handakten Rostock Nr 10 Anton Oberheuser to Rostock, 25 May 1948. Charlotte Becker-Freyseng to Rostock, 2 May 1948.
40. STAN Rep 502A KV- Verteidigung Handakten Rostock Nr 5 'Ungerin' to Rostock, 11 March 1948, 'KB-chens Laune ist naturlich wieder besser u. er ist zeitweise direkt wieder frech. Das erste Zeichen, dass er auf dem Weg der Besserung ist'.
41. STAN Rep 502A KV- Verteidigung Handakten Froeschmann Nr 23, R. Servatius, 'Besprechungs-Notiz', 25 February 1948.
42. STAN Rep 502A KV-Verteidigung Handakte Rostock Nr 5, Karl Brandt Schlusswort, 19 July 1947, copy dedicated to Rostock: 'Ich bin Arzt. Und vor meinem Gewissen steht diese Verantwortlichket für Mensch und Leben'.
43. Sauerbruch, *Das war mein Leben,* p. 558.
44. STAN Rep 502A KV- Verteidigung Handakten Rostock Nr 5 Material zu Karl Brandt, u.a. psychiatrisches Gutachten, telegram re Brandt's clemency petition on 6 October 1947. Rostock to Anni Brandt, 7 October 1947.
45. STAN Rep 502A KV- Verteidigung Handakten Nr 6 Private Gnadengesuche für Karl Brandt: physiologists, pathologists and surgeons. Achelis (the dismissed professor of physiology, Heidelberg); Buerkle de la Camp, Bochum Krankenhaus; H. Coenen Munster; Domagk, IG Farben; E.K. Frey, Munich surgeon; Dr med F. Jaeger, Ludwigshafen; Joetten, Münster; Nonnenbruch, Klais, Oberbayern; Prof. Dr Ing K. Quasebart, Berlin; Prof. Dr med V. Redwitz, Bonn; Prof. Dr med Reichmann, Bergmannsheil, Bochum; W. Schulemann,

Braunschweig; Herbert Siegmund, Munster Pathological Institute; A. Stoermer, Munich; Uhlenhuth, Freiburg; Werner Wachsmuth, Berchtesgarten; Werner Zabel; L. Zukschwerdt, Goeppingen, Zentralkliniken.

46. STAN Rep 502A KV- Verteidigung Handakten Rostock Nr 5, Clay to Irmgard Theisen, Munich, 10 May 1948.

47. NARA RG 153/86–2–2/Box 9 book 2, EFL memorandum. NMT 8/2808 Servatius, 'Probleme der Monster-Prozesse in Nuremberg'.

48. NARA RG 153/ 86–3–1/box 10, book 3 E.J. Carroll to Harry S Truman, 18 May 1948.

49. NARA RG 153/86 box 7, Discussed by Howard W. Ambruster.

50. STAN Rep 502A KV- Verteidigung Handakten Rostock Nr 5, Servatius to Rostock 5 March 1948; Rostock to Servatius, 15 March 1948.

51. NMT 4/2728–9.

52. K. Mellanby, 'A Moral Problem', *The Lancet,* vol. 251 (7 December 1946) 850.

53. NARA RG 153/ 86–2–2/ box 9 book 2 Emmy Hoven to President Truman, 20 February 1948.

54. Kogon, *Begegnungen*, pp. 57, 64, 99. Kogon, 'Recht und Gnade', *Ideologie und Praxis der Unmenschlichkeit*, pp. 290–2.

55. NYPL Lifton papers Box M9 Hoven to McHaney, 1 October 1947.

56. NARA RG 153/ 86–2–2/ Box 9 book 2 Emmy Hoven to President Truman, 20 February 1948. Norbert Frei, *Vergangenheitspolitik* (Munich: dtv, 1999), pp. 137–8, 154.

57. NMT 5/318 order for Karl Gebhardt.

58. AOF AJ 3617 P. 15 d. 686 Karl Brandt.

59. NARA RG 338 290–58–33 Executee files Viktor Brack, last words.

60. 'Selbstbildnis eines "Kriegsverbrechers"', *Deutsche Hochschullehrer-Zeitung*, vol. 1, no. 1 (1962) 5–9. NARA RG 338 290–59–33, Executee Files: Landsberg Prison: Karl Brandt, last words.

61. '7 Germans Hanged for Medical Tests', *New York Times* (3 June 1948). Annas and Grodin, *Nuremberg*, p. 106. Aziz, *Brandt*, p. 252 for text of the intended speech.

62. *Stars and Stripes* (Thursday, 3 June 1948).

63. Ibid.

64. AOF AJ 3624 p 62 d 3009, 29 December 1949.

65. Stephan and Norbert Lebert, *My Father's Keeper. The Children of Nazi Leaders – an Intimate History of Damage and Denial* (London: Abacus, 2001).

66. Maguire, *Law and War*, p. 184.

67. Frei, *Vergangenheitspolitik*, pp. 163–4.

68. Ibid., pp. 188–91.

69. Mechthild Rössler, 'Konrad Meyer und der "Generalplan Ost" in der Beurteilung der Nürnberger Prozesse', *Generalplan Ost*, pp. 356–64.

70. TTP–CLS 14/4/2/35 Ivy to Hardy Box 52, 29 April 1950. Hardy to Ivy, 4 May 1950.

71. Ulrich Brochhagen, *Nach Nürnberg* (Hamburg: Junius, 1994), pp. 57, 91.

72. Thomas Alan Schwartz, 'Die Begnadigung deutscher Kriegsverbrecher, John J. McCloy und die Häftlinge von Landsberg', *Vierteljahreshefte für Zeitgeschichte*, vol. 38 (1990) 375–414. Schwartz, *America's Germany* (Cambridge, Mass., 1991), pp. 160–74.

73. NARA RG 338 Entry 250 Parolee Case files, Schroeder, James B. Conant, order of parole, 30 March 1954.

74. NARA 338 Entry 250 Executee files, Gebhardt, Lloyd A. Wilson Prison Director Landsberg, 24 February 1948.

75. TTP–CLS 14/4/2/35 Drexel Sprecher to Taylor, 5 October 1950.
76. MPG II Abt. Rep 20B Nr 105 Hirnforschung (Tönnis), Steinbauer to Tönnis, 20 January 1948, Tönnis to Steinbauer, 23 February 1948.
77. TTP–CLS–14/4/3/51 Taylor to Francis Biddle, 9 April 1951. TTP–CLS–14/4/3/53 Taylor to Mrs Eleanor Roosevelt, 19 June 1951. Cf. Taylor , 'The Nazis Go Free: Justice and Mercy or Misguided Expediency?', *The Nation* (24 February 1951) 170–2.
78. Carola Sachse, ' "Persilscheinkultur". Zum Umgang mit der NS-Vergangenheit in der Kaiser-Wilhelm/Max-Planck-Gesellschaft', Bernd Weisbrod (ed.), *Akademische Vergangenheitspolitik* (Göttingen: Wallstein, 2002), pp. 217–46.
79. Günther Wieland, 'Verfolgung von NS-Verbrechen und kalter Krieg', in *'Keine Abrechnung'*, pp. 197, 201. Brochhagen, *Nach Nurnberg*, p. 165.
80. Zimmermann, *Universität Jena*, pp. 159–74. Tony Paterson, 'Stasi files "hid proof that showed doctor was Nazi murderer" ', *The Independent* (14 February 2004), 30.
81. Bower, *Pledge Betrayed*, chapter 10.
82. Bloxham, *Genocide*, chapter 4, 'The Failure of the Trial Medium', pp. 129–81.
83. Nissen did so in 1953. Cf. Nissen, *Helle Blätter – Dunkle Blätter. Erinnerungen eines Chirurgen*, pp. 166–7, 270–1.
84. Sheldon Harris, *Factories of Death*, pp. 208–23. Till Barninghausen, *Medizinische Humanexperimente der japanischen Truppen für biologische Kriegsführung in China 1932–1945* (Frankfurt: Peter Lang, 2002).
85. *Life* (9 December 1945) 49–52.
86. *German Aviation Medicine. World War II. Prepared under the Auspices of the Surgeon General*, US Air Force, Department of the Air Force (Washington DC: Department of the Air Force, 1950), 2 vols.
87. Hermann Rein, 'Physical Methods of Gas Analysis', *German Aviation Medicine. World War II*, pp. 107–28.
88. Heubner's contribution is omitted from the bibliography in Kneer, 'Heubner', 150.
89. Benzinger, 'High Altitude Flight Breathing', *German Aviation Medicine. World War II*, pp. 429–44; 'Explosive Decompression', pp. 394–425; 'Physiological Effects of Blast', pp. 1225–59.
90. Siegfried Ruff, 'Brief Acceleration: Less Than One "Second" ', *German Aviation Medicine. World War II*, pp. 584–97.
91. Hermann Becker-Freyseng, 'Aeromedical Instruction of Flying Personnel', *German Aviation Medicine. World War II*, pp. 1069–79.
92. Oskar Schroeder, 'Air Evacuation of the Wounded', *German Aviation Medicine. World War II*, pp. 1133–8.
93. NARA RG 341 190 64 24 06 Records of the Headquarters US Air Force Benford to Benson, 15 October 1946.
94. NARA RG 330 Entry 18 JIOA Foreign Scientists Case File 1945–58, Box 11 Theodor Benzinger, statement on NSDAP, 10 June 1947.
95. NARA RG 341 Box 129 Lutz to Benford, 19 August 1947, 3 October 1947.
96. NARA RG 330 Entry 18 Box 164 Strughold, testimonial of Dr. med. habil. K.E. Schaefer, Dozent, dated 23 September 1947. UA Heidelberg PA 235 Strughold 1946–48, 23 September 1947.
97. NARA RG 319 11D Dossiers box 194 Konrad Schaefer, letter dated 27 March 1951. RG 319 Security Intelligence and Investigation Dossiers box 19, Project Paperclip file XE 065587 Konrad Schaefer, Special Contract for Employment of German Nationals, May 1949.

98. NMT 8/2876 re Paperclip, 25 June 1951, NMT 8/2885–6, 27 November 1951. Cf. Geissler, *Biologische Waffen*, pp. 757–8. Friedrich Hansen, *Biologische Kriegsführung im Dritten Reich* (Frankfurt: Campus, 1993).

99. UA Heidelberg H–III–584/1 Die ordentliche Professur f Physiologie, Rein to Rektor KH Bauer, 8 April 1946.

100. UA Heidelberg PA 1206 Medizinische Fakultät Personalakten Strughold, Hubertus 1946–1951. PA 6024 Hubertus Strughold 1946–58.

101. NARA RG 330 Entry 18 Box 164 Strughold, Security Certificate, Director JIOA, 29 September 1948. AFP Alexander to Maryann Jessup MacConochie, 20 September 1979.

102. UA Heidelberg PA 1206 Harry Armstrong, Commandant, to Hoepke, 30 December 1947, That Strughold will not return for the January examinations.

103. NARA RG 341 Office of the Surgeon General. Historical Branch, Box 129 Request for German specialists, 24 June 1949, Arctic Aeromedical laboratory O/R CSG–13; interrogation of Strughold, 8 December 1949. RG 330 Entry 18 Box 164 Strughold. Access to an interrogation of 8 December 1949 was withheld.

104. Clarence G. Lasby, *Operation Paperclip. German Scientists and the Cold War* (New York: Atheneum, 1971), p. 258. RAC RG 2 – 1953 200 24 150 Gregg to Strughold, 20 April 1953: 'Thanks … reprints on Space Medicine and the Day-Night Cycle. I found both of them interesting and they reminded me pleasantly of our supper in Germany two or three yrs ago when the conversation turned upon the same subjects.' Brooks E. Kleber and Dale Birdsell, *Chemical Warfare Service: Chemicals in Combat* (Washington DC: Office of the Chief of Military History, 1966), pp. 257–8.

105. Robert Benford, *Doctors in the Sky* (Springfield, Ill.: G.C. Thomas, 1955), p. 229.

106. AFP Jeffrey Mausner to Alexander, 7 and 13 August 1979; Alexander to Jeffrey Mausner,16 and 31 August 1979. Statement of Jefim Moschinski.

107. 'Portrait of Nazi Prompts Protest. Jewish Group Asks University to Remove Doctor's Image', *New York Times* (26 October 1993) p. A16. Protest by Anti-Defamation League, September 1995, see www.adl.org/ presrele/hola_52/ 2533_522.asp.

108. Rob Evans, *Gassed. British Chemical Warfare Experiments on Humans at Porton Down* (London: House of Stratus, 2000), chapter 4 'Cleaning up after the Nazis'.

109. BA Dahlwitz Hoppegarten VgM 10077 Bd 17 Briefe 24 July 48, Dezernat f politische Häftlinge an das Buchenwald Komittee Berlin; Police Interrogation Romberg, 25 August 1947.

110. MP II2/31.1 Elisabeth Castonier to Mitscherlich, 28 October 1960. II2/31.3 Castonier to Mitscherlich, 22 November 1960.

111. MP II/31.2 Mitscherlich to E. Castonier, 15 November 1960. Roth, 'Flying Bodies – Enforcing States: German Aviation Medical Research 1920–1970 and the DFG', paper for DFG conference, Heidelberg, 2003.

112. MP II 2/140.1 Schwarberg to Mitscherlich, 13 July 1960, reply 29 July 1960. 'Ruff unter Druck', *Der Spiegel*, vol. 14, no. 42 (1960) 53–4.

113. STAN Rep 502A KV- Verteidigung Handakten Rostock Nr 12, Johnson T. Crawford to D. von Greiser, 10 November 1948.

114. STAN Rep 502A KV- Verteidigung Handakten Rostock Nr 5, Rostock, 26 February 1948 McHaney to Dr Koessler; Rostock to Karl Brandt, 18 March 1948.

115. Rostock died on 17 June 1956 in Bad Tölz.

116. Paul Rostock, *Kompendium der Chirurgie* (Berlin: Urban & Schwarzenberg, 1948). Rostock, *Tetanus* (Berlin: de Gruyter, 1950). STAN Rep 502A KV-Verteidigung Handakten Rostock Nr 5, Rostock to Karl Brandt, 18 March 1948.

117. STAN Rep 502A KV- Verteidigung Handakten, Rostock Nr 5, Rostock to Karl Brandt, 17 April 1948.
118. AHUB Dekanat Charité 0100/21a Protokolle der Fakultätssitzungen 1946–54, 16 July 1947. NMT 8/2777–8 Rostock to Diepgen, 29 August 1947.
119. STAN Rep 502A KV- Verteidigung Handakten Rostock Nr 9, Rostock to Servatius August 1947. Rostock to G. Schmitz-Kahlman, 30 April 1948.
120. NMT 8/2771–2789 18 July 1947 Dokumentenzusammenstellung. Nürnberger Ärzteprozess.
121. William McGucken, *Scientists, Society and State: the Social Relations of Science Movement in Great Britain* (Columbus: Ohio State University Press, 1984).
122. STAN Rep 502A KV- Verteidigung Handakten Rostock Nr 8, Rostock to Rein, 9 July 1947.
123. STAN Rep 502A KV- Verteidigung Handakten Rostock Nr 10, Hildegard Remensperger to Rostock, 17 March 1948, reply 10 April 1949.
124. Ibid., Rostock to Karl Brandt, 17 April 1948.
125. NARA RG 466 Entry 54 Handloser, Gnadenerweis, 2 July 1954.
126. Archiv Peiffer Ostertag to Fischer, 5 February 1951, Fischer reply, 14 March 1951. Ostertag to Herbert Hoff, Augsburg, US Army, 539th General Hospital, 20 March 1951.
127. NARA RG 338 Parolee Files Fischer parole file, Boehringer to US Parole Officer, 13 March 1954.
128. NARA RG 466 Entry 54 box 3 Genzken file.
129. USHMM Ferriday Collection, *Daily Express* (4 March 1958). Eric Townsend, 'Medical War Crimes', *The Times* (8 August 1958). I.H. Milner, 'Medical War Crimes', *BMJ* (10 May 1958) 1121, (24 May 1958) 1127; (14 June 1958) 1420; (5 July 1958) 51; (12 July 1958) 108; (26 July 1958) 246–8; (9 August 1958) 390; (16 August 1958) 460; (20 September 1958) 744. 'Herta Oberhauser', *BMJ* (26 November 1960) 1612. *Daily Telegraph* (5 December 1960).
130. USHMM Ferriday Collection, *Der Spiegel* (9 November 1960).
131. USHMM Ferriday Collection Benjamin Portnoy 'Blots on German Medical Register' in *Manchester Guardian* (13 June 1959), 6. Oppitz, *Medizinverbrechen*, p. 55. Annette Weinke, *Die Verfolgung von NS-Tätern im geteilten Deutschland* (Paderborn: Schonigh, 2002), p. 81.
132. *Der Fall Rose*, pp. 92–9. BAK Zsg 154 Sammlung Christian Pross, Bd 63 'Autobiographische Notiz des Prof. Dr. med. Gerhard Rose'.
133. *Der Fall Rose*, p. 101. 'Prof. Gerhard Rose wird 80 Jahre alt', *Schaumburger Zeitung* (30 November 1976).
134. NMT 8/ 2859, 2862 Becker-Freyseng to Zentrale Rechtschutzstelle Bonn, 12 May 1954.
135. NMT 4/824f for affidavit by Heilmeyer, 18 February 1947.
136. Peter Voswinckel, *50 Jahre Deutsche Gesellschaft für Hämatologie und Onkologie* (Würzburg: Murken-Altrogge, 1987), pp. 29–30, 38, 62. Ludwig Heilmeyer, *Lebenserinnerungen* (Stuttgart: Schattauer, 1971). Zimmermann, *Jena*, p. 87. Guido Jakobs and Karen Bayer, 'Vertriebene jüdische Hochschullehrer – Rückkehr erwünscht?', *Nach der Diktatur*, p. 123. Werner Creutzfeld information 11 July 2004.
137. AdR 02/BmfU PA Hans Eppinger
138. 'Intervention des Innenministeriums: Vortrag des SS-Arztes Beigelböck ist abgesagt', *Neues Österreich* (18 September 1962). DÖW 6324 Oberstaatsanwaltschaft Graz to Bundesministerium f. Justiz, 12 October 1962.
139. AdR 04/BMfJ Nr 148565 Wilhelm Beiglböck to BMfU re Antrag Georgine Witte, 14 November 1947.

140. Vienna University Archives, Personal file on Wilhelm Beiglböck, *Buxtehude Tageblatt* (25 November 1963 and 6 December 1963). Oppitz, *Medizinverbrechen*, p. 54.

141. UA Münster Bestand 55 Nr 1548, Lebenslauf, 26 November 1952.

142. Ibid., undertaking of 6 February 1954

143. MPG II. Abt. Rep. 20B re Schaltenbrand to Tönnis, 14 February 1948.

144. MP II2/88.3 Kogon to Mitscherlich, 9 December 1960.

145. 'Deutsche Wissenschaftler vor dem Militärgericht in Lyon', *Ärztliche Mitteilungen*, no. 11 (1 June 1954) 362–4.

146. W. Villinger and G. Schaltenbrand, 'Erklärung', *Der Nervenarzt*, vol. 24 (July 1953) 312. Weindling, 'Out of the Ghetto', p. 209.

147. Weindling, *Health, Race and German Politics*, pp. 568–9. Hans-Peter Kröner, *Von der Rassenhygiene zur Humangenetik. Das Kaiser-Wilhelm-Institut für Anthropologie, menschliche Erblehre und Eugenik nach dem Kriege* (Stuttgart: Gustav Fischer, 1998).

148. Alexander Mitscherlich, 'Was ist ein Mensch Wert? – "Medizinische" Versuche im Dritten Reich', in Mitscherlich, *Das Ich und die Vielen. Parteinahme eines Psychoanalytikers* (Munich: Piper, 1978), pp. 238–58. 285. STAN Rep 502A KV-Verteidigung Handakten, Rostock Nr 9/10, Charlotte Becker-Freyseng to Rostock, 2 May 1948; Hildegard Remensperger, 17 March 1948.

149. Bayer and Woelck, 'Der Anatom Anton Kiesselbach', *Nach der Diktatur*, pp. 289–302. Silke Stelbrink, 'Walter Kikuth und das Hygiene-Institut', *Nach der Diktatur*, pp. 303–23.

150. *Das Reichsgesundheitsamt im Nationalsozialismus. Menschenversuche und 'Zigeunerforschung' zwischen 1933 und 1945* (Berlin, 1998).

151. Jakobs and Beyer, 'Vertriebene', 134–5.

152. Lagnado and Dekel, *Children*, pp. 138–53.

153. Ibid., pp. 176–7.

154. Klee, *Was sie taten*.

17   A Fragile Legacy

1. NARA RG 153 box 16 folder 4 bk 1 89–2 Telford Taylor, 'The Meaning of the Nuremberg Trials. An Address Delivered (in French) at the Palais de Justice Paris, 25 April, 1947', 6.

2. Robert Havemann, 'Menschen als Versuchstiere: Zum Nürnberger Ärzteprozess', *Der Kurier* (28 January 1947).

3. 'A Moral Problem', *The Lancet*, 251 (30 November 1946) 798. K. Mellanby, 'A Moral Problem', *The Lancet*, 251 (7 December 1946) 850. Cf. 'Josephine Bell', *Murder in Hospital* (London: Penguin, 1941). Also Denis Herbert, 'A Moral Problem', *The Lancet* 252 (1947) 85.

4. T.B. Layton, 'A Moral Problem', *The Lancet* 251 (14 December 1946) 882. 'Doctors Split on Nazi Test', *Daily Telegraph* (13 December 1946). 'Lord Horder Says Retain Nazi Data', *Daily Telegraph* (14 December 1946), p. 5. 'Doctors to Keep Nazi Secrets. Freezing Tests on Human Beings', *Daily Telegraph* (16 December 1946), p. 5. 'BMA Expert Going to Nuremberg', *Daily Telegraph* (17 December 1946), pp. 1, 6.

5. *War Crimes News Digest*, No. 22 (31 December 1946).

6. *The Lancet*, 251 (1946) 882.

7. *War Crimes News Digest* (6 February 1947). Cf. Sidney Hilton, 'A Moral Problem', *The Lancet*, 252 (4 January 1947) 43.

8. Clement Freud, *Freud Ego* (London: BBC, 2001), pp. 62–8. International Refugee Organisation, *Professional Medical Register* (Geneva, n.d. [ca. 1947]).

9. NARA RG 153/86–3–1/box 10/folder II book 2 Alexander to Gunn, 15 February 1947.

10. NARA RG 153/ 86–3–1/box 11 folder 4 book 3 International Scientific Commission, 9 April 1947.

11. UNESCO Archives, Paris, Julian Huxley to J.W.R. Thompson, 5 March 1947.

12. Wellcome Library PP/CMW/Eb, 26 March 1947 Leslie Rowan to Moran, quoting from D.C. McAlpine to Rowan.

13. TNA: PRO WO 309/468, Lt Col. Kenneth E. Savill to Lt Col. H.H. Wade, 6 September 1946.

14. TNA: PRO WO 309/468. Draper memo after visit by Thompson, 4 October 1947.

15. TNA: PRO WO 309/470 notes on forwarding ISC copies.

16. UNESCO, *Statement on Race* (issued 18 July 1950) (Paris, 1950).

17. TNA: PRO 309/1652 Mant to Somerhough, 20 May 1946.

18. TNA: PRO FO 317/66582 British Embassy Paris to War Crimes Section 2, August 1947. Helsby, 20 August 1947. FO 371/66581.

19. TNA: PRO FO 371/ 66581 note of 26 March 1947.

20. Lord Moran, *Winston Churchill. The Struggle for Survival 1940–1965* (London: Constable, 1966), pp. 333–6.

21. Charles Webster, 'The Metamorphosis of Lord Dawson of Penn', in D. and R. Porter (eds), *Doctors, Politics and Society: Historical Essays* (Amsterdam: Rodopi, 1993), pp. 212–28.

22. J.R. Baker, *The Scientific Life* (London: George Allen and Unwin, 1942). J.R. Baker, *Science and the Planned State* (London: George Allen & Unwin, 1945), p. 9.

23. Royal Society Dale Papers (RSDP) 93 HO 36.3.4 Dale to Eichholtz, 30 March 1949.

24. RSDP 93 HO 38.25.6 Dale to Bülbring, 2 February 1948.

25. RSDP 93 HO 46.6.14 Nelte to Dale, 3 August 1947.

26. RSDP 93 HO 20.2.18 Dale to Moran, 6 February 1948; 93 HO 20.2.20 Worsfold to Dale, 23 February 1948.

27. TNA: PRO CAB 124/1928 Scientific Committee for Germany.

28. RSDP 93 HO 20.2.21, Moran to Dale, 5 January 1949.

29. WL PP/CLE Charles Lovatt Evans, Box 1 A.11 Service at Porton.

30. WL Moran Papers PP/HAR/B.7 Moran to Hare, 14 July 1947.

31. Aubrey Lewis, 'German Eugenic Legislation: An Examination of Fact and Theory', *Eugenics Review*, 26 (1934–5) 183–91.

32. B. Loff and S. Cordner, 'World War II Malaria Trials Revisited', *The Lancet*, vol. 353 (1999) 1597. TNA: PRO WO 309/470 list of experts.

33. TNA: PRO WO 309/471 J.W.R. Thompson, International Scientific Commission (W.C.), circular 16 July 1947.

34. WL PP/HAR Hare Papers, Hare to Thompson, 26 September 1947.

35. WL PP/HAR Hare to Moran, 31 October 1947.

36. WL PP/HAR/B7 Report on the Scientific Value of Experiments of a Bacteriological Nature Carried out in Concentration Camps in Germany during the War. Cf. Hare to Moran 2 March 1948 indicating an approximate date of the report.

37. Ibid., 45–6, 51.

38. Moran, *Scientific Results*, p. 4.

39. WL PP/HAR Hare papers Thompson to Hare, 5 September 1947. Hare to Thompson, 26 September 1947.
40. TNA: PRO FO 1060/570 Ruehl's Case, 29 March 1950.
41. Foreign Office, *Scientific Results*, p. 3. Tilley, 'Preliminary Report', 8.
42. C.P. Blacker, '"Eugenic" Experiments Conducted by the Nazis on Human Subjects', *The Eugenics Review*, vol. 44 (April 1952) 9–19, 17.
43. RSDP 93 HO 20.2.27.
44. RSDP 93 HD 20.2.14 text of Runderlass in *Reichs-Gesundheitsblatt*, 11 June 1931. 93 HO 20.2.23 Dale to Moran, 12 January 1949; Moran to Dale, 14 January 1949. Moran to Dale, 2 February 1949. Janet E. Brown to Dale, 8 February 1949. Reports returned to Moran.
45. TNA: PRO FD 1/5826 War Time Scientific Investigations in Germany and German Occupied Countries.
46. Moran, *Scientific Results*, pp. 4–5. Draft text in RSDP 93 HD 20.2.22.
47. Foreign Office, *Scientific Results of German Medical Warcrimes. Report of an Enquiry under the Chairmanship of Lord Moran MC, MD* (London: HMSO, 1949).
48. TNA: PRO PREM 8/1111.
49. C.P. Blacker, 'Eugenics in Germany', *Eugenics Review*, vol. 25 (1933–4) 157.
50. C.P. Blacker, ' "Eugenic" Experiments Conducted by the Nazis on Human Subjects', *The Eugenics Review*, vol. 44 (April 1952) 9–19 (cf. note in vol. 44 (1952) 125–6).
51. Blacker, ' "Eugenic" Experiments', 18.
52. Ibid.. C.P. Blacker, *Eugenics: Galton and After* (London: Duckworth, 1952), pp. 267–8.
53. Pauline Mazumdar, *Eugenics, Human Genetics and Human Failings* (London: Routledge, 1992), pp. 247–8.
54. A.K. Mant, 'Medical Experiments in Nazi Camps'. Paper to the Chelsea Clinical Society, 9 March 1948. Mant, 'Medical War Crimes in Nazi Germany', *St Mary's Hospital Gazette* (1961) 216–21.
55. H. Cullumbine, 'Chemical Warfare Experiment Using Human Subjects', *BMJ*, no. 576 (19 October 1946).
56. TNA: PRO FO 1032/941 Note of 18 August 1949.
57. CU Fishbein Papers, clipping 'Sowjetgelehrter protestiert', *Tägliche Rundschau* (18 January 1948). OMGUS clipping dated '19. Jan 1947'. Citing *JAMA*, vol. 132/7, 162. Lt Col. Rapalski, OMGUS to Fishbein, 30 January 1948.
58. CU Fishbein Papers, Ivy to committee, 29 March 1948. Proposed Studies on the Pathogenicity of Selected Salmonella. Fishbein to Ivy, 2 April 1948. For later skin experiments, see Hornblum, *Acres of Skin*.
59. NARA RG 238 entry 159 NM 70 Box 2 Robert P. Patterson to Office of Chief of Counsel Nuremberg, 19 November 1948. Ibid. Box 7, Patterson to Taylor. Taylor, 19 November 1948. War Crimes Branch to Patterson, 13 December 1948. Minutes of Survey Committee on Disposal of Records of Nurnberg Trials, 14 December 1948.
60. New York Academy of Medicine (hereafter NYAM), AR. Minutes 1946–1949, 15 December 1947, 26 January 1948 on German Scientific and Medical Documents. AR. NYAM Committee on Medical Information. Minutes and Reports to the Council 1948–1952, 15 February 1948 for Committee members, 26 October 1949 for the final report on Documents Used in Evidence in Trial of Nazi Physicians. This report has not been located. Constance C. Smith, *Guide to the Documents That Were Used As Evidence In The Trial of the Nazi Physicians. Case I of the United States Military Tribunal at Nuremberg* (New York: The New

York Academy of Medicine, n.d. [1949]). My thanks to Dr Saul Jarcho for his recollections of US evaluations of captured German medical documents.

61. NYAM AR. N.Y.A.M. Committee on Medical Information. Minutes and Reports to the Council 1946–1949, 15 December 1947 and 26 January 1948, German Medical and Scientific Documents.

62. NYAM, AR. Minutes 1946–1949, 15 December 1947, 26 January 1948, 25 April and 26 October 1949 on German Scientific and Medical Documents. Smith, *Guide* (1949), 'Foreword'. *Guide* (1956) 'Preface', p. ii.

63. *Human Radiation Experiments*, pp. 45–51.

64. Katz, 'Statement', *Human Radiation Experiments*, p. 545.

65. Ivy, 'Nazi War Crimes of a Medical Nature. Some Conclusions', *JAMA*, vol. 139 (1949) 131–5. Webster, 'Dawson'. For background see Greta Jones, *Science, Politics and the Cold War* (London: Croom Helm, 1988).

66. BMA Archives, British Medical Association, *War Crimes and Medicine. Statement by the Council of the Association for Submission to the World Medical Association June 1947* (London: BMA, 1947). 'War Crimes and Medicine', *WMA Bulletin*, vol. 1, no. 1 (April 1949) 4–13.

67. BMA Archives, GA 7, World Medical Association, War Crimes and Medicine. The German Betrayal and a Re-statement of the Ethics of Medicine.

68. 'Vorwort der Arbeitsgemeinschaft der Westdeutschen Ärztekammern', Mitscherlich and Mielke, *Wissenschaft ohne Menschlichkeit*, v, March 1949. Cf. *Medizin ohne Menschlichkeit,* p. 13.

69. Cf Morris Fishbein, 'Medical Ethics and Medical Care', *Journal of the Tennessee State Medical Association*, vol. 23 (1930) 362–8; *WMA Bulletin* (1949–54), Editor's comments.

70. 'Report on Medical Ethics', *WMA Bulletin*, vol. 1, no. 3 (October 1949), 109. Perley, 'The Nuremberg Code', Annas and Grodin, *Nazi Doctors*, pp. 149–71.

71. 'Membership in The World Medical Association for the Japan and Western German Medical Associations', *WMA Bulletin*, vol. 3 (July 1951), 204, 210.

72. Alexander, 'Science under Dictatorship', 100.

73. AFP Helene Hulst-Alexander to Leo Alexander, 20 August 1945. TTP–CLS 8/1/2/55 L.A. Hulst to Telford Taylor, 18 November 1955, 'Leo is very well and very active, writing books'.

74. WMA document 17.6/54 quoted by Perley et al. 'Nuremberg Code', 154–5.

75. *World Medical Journal*, vol. 2, no. 1 (January 1955) 14–15. 'Human Experimentation', ibid., 21. The committee consisted of G. Campailla (Italy), G. Dekker and Heringa (Netherlands), Otto Rasmussen (Denmark), P. Lifrange (Belgium), S.C. Sen (India), F. Blasingame (USA), H. Lodhia (Burma). There is no reference to 'fully informed, free consent'.

76. A.C. Ivy, 'The History and Ethics of the Use of Human Subjects in Medical Experiments', *Science*, vol. 108 (2 July 1948) 1–5.

77. 'Human Experimentation', *World Medical Journal*, vol. 4, no. 5 (September 1957) 299–300.

78. 'Address by His Holiness, Pope Pius XII', *World Medical Journal*, vol. 2, no. 2 (March 1955) 111–12.

79. William A.R. Thomson, 'Editorial Responsibility in Relation to Human Experimentation', *World Medical Journal*, vol. 2, no. 3 (May 1955) 153–4.

80. 'Human Experimentation', *World Medical Journal*, vol. 2, no. 3 (May 1955) 167–8.

81. Rothman, *Strangers*, p. 62. Hornblum, *Acres of Skin*, p. xvi.

82. Ivy, 'Nazi War Crimes of a Medical Nature. Some Conclusions', *JAMA*, vol. 139 (1949) 131–5.

83. P.S. Ward, '"Who will bell the cat?" Andrew C. Ivy and Krebiozen', *Bulletin of the History of Medicine*, vol. 58 (1984) 28–52.

84. Fishbein Papers, Box 16 folder 11, Ivy to Fishbein, 15 February 1949 re T. Arthur Turner; Turner to Ivy, 11 February 1949; *The Rotarian*, 20 January 1949; Fishbein to Ivy, 14 and 26 March 1949.

85. Ivy Papers, Box 6, folder 12, Albert H. Stroupe, 'The Nuremberg Code – An Ivy Contribution'. Evelyne Shuster, 'The Nuremberg Code: Hippocratic Ethics and Human Rights', *Lancet,* 351 (1998) 974–7.

86. Ivy papers Box 6 folder 10, p. 1 'The Meaning of Medical Ethics Learned and Emphasized at The Nurnberg Trials', 18 August 1949.

87. Ward, 'Who will bell the cat?' Elinor Langer, 'The Krebiozen Case. What Happened in Chicago?', *Science*, vol. 151 (1966) 1061–4.

88. Leo Alexander, 'Science under Dictatorship', *March of Medicine*, vol. 14 (1949) 51–106.

89. WUS Beals papers box 1 folder 16 Dehull Travis to Beals, 5 December 1946. AFP Leo to Phyllis Alexander, 8 January 1947.

90. AFP Leo to Phyllis Alexander, 27 January 1945.

91. AFP Leo to Phyllis Alexander, 16 December 1946.

92. AFP Leo to Phyllis Alexander, 21 December 1946.

93. Ibid.

94. APD 1/21 'The Socio-Psychological Structure of the SS'. Lecture to Amsterdam, 12 June 1947, and Boston Society of Psychiatry and' Neurology, 16 October 1947.

95. APD 4/35 Taylor to Fenger, 16 April 1947.

96. NYPL Lifton papers, box M22 'August 29, 1978 – Talk with Leo Alexander'.

97. WUS Beals papers box 1 folder 5, M.G. Gallagher to Beals, 26 December 1946, 21 March 1947, 12 May 1947, 19 July 1947.

98. T. Taylor, 'The Nuremberg War Crimes Trials: an Appraisal', 7 April 1949, cited as reprint, pp. 27–8. NARA RG 238 Entry 155 Box 1 Taylor, 'An Outline of the Research and Publication Possibilities of the War Crimes Trials', November 1948.

99. Ibid., 30–1.

100. NARA RG 153/ 100–621/entry 159 Box 7 Office for the Chief of Counsel for War Crimes, Request for Record of Trials, 6; Minutes of a Meeting on Disposal of Records of Nurnberg Trials, 14 December 1948.

101. WUS Beals papers box 1 folder 16 Dehull Travis to Beals, 5 December 1946.

102. TTP–CLS–14/4/2/35 Drexel Sprecher to Ivy, 10 May 1950.

103. LC Jackson papers box 112 Alderman, in *Columbia Law Review*, vol. 51 (1951) 407ff. NARA RG 238 Entry 155 box 1 Taylor MS An Outline of the Research and Publication Possibilities of the War Crimes Trials, November 1948. 20 April 1950, 89–2 book 3 box 17 folder 1.

104. Mitscherlich and Mielke, *Doctors of Infamy*.

105. Center for Jewish History, New York. Collection of Raphael Lemkin, P–154, Box 1 Folder 19, 31 August 1948 Henry Schuman to Lemkin for agreeing to provide a foreword on genocide.

106. A.C. Ivy, 'Statement', *Doctors of Infamy*, pp. x–xi.

107. Bayle, *Croix gamée contre caducée.*

108. L. Alexander, *Treatment of Mental Disorder* (Philadelphia: Saunders, 1953), p. 36.

109. Alexander, *Treatment*, cf. pp. v, 38, 196, 287, 489.

110. Ibid., 63. NYPL LP M5 letter of Alexander to Lifton in Munich, 10 October 1978.

111. UA Halle Rep 29 Nr 293 Bd 2 Budde to Rector, 14 January 1947.

112. AUF Rein to Janssen, 14 September 1947; AHUB Dekanat Medizinische Fakultät, 12 August 1947. UA Freiburg Rein B53/33 Büchner Entwurf; also draft declaration; Voit (Dekan Mainz) to Janssen, 26 September 1947; Hoepke to Dekan

Freiburg, 30 October 1947; Hoepke to Dean Freiburg, 28 August 1947. H–III–201/2 Einladungen zum Fak. Sitzungen 1946–53, 5 September 1947, Heidelberg Medical Faculty to discuss Nürnberger Urteil gegen die Ärzte.

113. UA Freiburg B53/33 Kretschmer (Dekan Tübingen) to Janssen (Dekan Freiburg) 16 and 27 September 1947; Janssen to Hoepke, 8 September 1947; Janssen to Kretschmer, 3 September 1947.

114. UA Halle Rep Nr 29 293 Bd 11 Hoepke to Dean Medical Faculty Halle, 30 October 1947.

115. Ibid., pp. vi–vii. Rolf Schlögell, 'Entschliessung der Westdeutschen Ärztekammern zum Nürnberger Ärzteprozeß', *Hippokrates* (1947) 49.

116. Mitscherlich, *Wissenschaft*, p. vi.

117. MP II2/105.9 Mitscherlich to Mielke, 30 August 1947.

118. MP II2/105.16 Mitscherlich to Mielke 5 June 1948; II2/105.19 Mielke to Mitscherlich, n.d. [January 1949].

119. MP II2/105.34 and 35a Mielke to Mitscherlich, 30 August 1948.

120. MP II 2/23.2 Mitscherlich to Julius Bredenbeck, 21 October 1960. II 2/145.1 and 2 Mitscherlich to K.H. Stauder, 9 and 25 May 1960.

121. MP II2/65.5 Hartner to Mitscherlich, 24 March 1961.

122. Kater, *Doctors*, pp. 21, 183, 272. Rebecca Schwoch, *Ärztliche Standespolitik im Nationalsozialismus. Julius Hadrich und Karl Haedenkamp als Beispiele* (Husum: Matthiesen, 2001). NMT 8/706 Mielke to Haedenkamp, 12 December 1947. Thomas Gerst, 'Der Auftrag der Ärztekammern an Alexander Mitscherlich zur Beobachtung und Dokumentation des Prozessverlaufs', *Deutsches Ärzteblatt*, vol. 91 (1994) N–1200–B1210.

123. MP II2/105.29 Mielke to Mitscherlich March/ April 1949. Peter, *Nürnberger Ärzteprozess*, pp. 65–8.

124. Gerst, 'Der Auftrag', B–1209.

125. MP II2/10.1a Mitscherlich to Fritz Bauer with Rektor Thieme Freiburg, Entwurf, 6 February 1961.

126. MP II2/14.1 Wolfgang Bethke to Mitscherlich, 30 April 1960.

127. MP II2/28.3 Bundesärztekammer to Mitscherlich, 3 February 1960.

128. MP II/8.1 Mitscherlich to von Baeyer (Dean of Heidelberg), 15 February 1961. II2/8.1b Mitscherlich to Hartner, 22 November 1960. Rector Freiburg to Hartner, 22 November 1960. II2/10.1 Mitscherlich to Fritz Bauer, 4 February 1961. II2/50.2 Mitscherlich to Rudolf Hirsch of S. Fischer, 23 January 1961. II2/50.7 Mitscherlich to Hirsch, 21 December 1961.

129. Cf. Weindling, *Health, Race*, p. 563. Trunk, *Zwei Hundert Blutproben*. Weindling, 'Verdacht, Kontrolle, Aussöhnung. Adolf Butenandts Platz in der Wissenschaftspolitik der Westalliierten (1945–1955)'.

130. MP II2/70.3 Einvernahme durch den Vorsitzenden des Bundes-Disziplinar-Kammer 7, 21 October 1960. II2/70.6 Mitscherlich to Staatssek. Rosenthal-Pelldram 29 October 1960; II2/88.2 Mitscherlich to Kogon, 29 October 1960.

131. MP II2/88.3 Kogon to Mitscherlich, 9 December 1960; II2/88.4 Mitscherlich to Kogon, 10 December 1960.

132. UA Halle Rep 29 Nr. 564.

133. Institute of the History of Medicine, Frankfurt/M, Medical papers on doctoral degrees.

134. RAC RF RG 12.1 Officers' diaries, Gregg diary, 17 September 1949. Von Weizsäcker and Alexander, 19 September 1949.

135. MP II2/33.3 Degkwitz 12 July 1961.

136. Fritz Bauer Institut (ed.), *'Gerichtstag halten über uns selbst ...' Geschichte und Wirkungsgeschichte des ersten Frankfrter Auschwitz-Prozesses* (Frankfurt/M: Campus, 2001).

137. Falk Berger, 'Zur Biographie einer Institution. Alexander Mitscherlich gründet das Sigmund-Freud-Institut', in Tomas Plänkers et al. (eds), *Psychoanalyse in Frankfurt am Main* (Tübingen: diskord, 1999), pp. 349–72. Hermann Argeander, 'Zur Geschichte des Sigmund-Freud-Instituts', ibid., pp. 373–84.
138. AFP NMT 5/27.
139. APD J.E. Fried to Alexander, 25 September 1950, 19 October 1950. Alexander to Fried, 9 October 1950. Schwelb joined the UN Human Rights Commission from UNWCC in July 1947.
140. UN, Economic and Social Questions, in *Yearbook of the United Nations* (1951), p. 505. *The Plight of Survivors of Nazi Concentration Camps (Fifth TNA: Progress Report by the Secretary-General)*, United Nations Economic and Social Council (12 February 1958).
141. By 1953 there were 468 claimants, *Yearbook of the United Nations* (1953), p. 417), and by 1958 over 1,500.
142. AFP John Fried to Alexander, 19 October 1950.
143. AFP Jadwiga Kaminska-Procka to Alexander, 28 January 1951. Case record, 11 November 1952; Alexander, 'The Physician and the State', 1982. 'Nazis' Guinea Pigs at Beth Israel. Tortured Women Repaired', *Boston Sunday Advertiser* (11 January 1959).
144. AFP Alexander to Jadwiga Kaminska-Procka, 3 February 1954. USHMM Ferriday Papers Box 3 File 3 1952–1957 Friends of ADIR and UN Victims, memo on ADIR, 20 December 1957. Alexander to Ferriday, 10 December 1957.
145. Pross, *Wiedergutmachung*, p. 103.
146. Ibid., pp. 104–5.
147. Ibid., pp. 122–3, 150–1.
148. Maguire in *BMJ* (14 June 1958) 1420.
149. 'Plight of Survivors of So-Called Scientific Experiments in Nazi Concentration Camps', Economic and Social Questions, *Yearbook of the United Nations* (1953), p. 417. USHMM Ferriday Papers, Box 2 M.H. Armstrong Davison, 'Medical War Crimes', *British Medical Journal* (10 May 1958) 1121. Eric Townsend, 'Medical War Crimes', *The Times* (8 August 1958).
150. *Yearbook of the United Nations* (1951), p. 506.
151. USHMM Ferriday Papers Box 3 File 3 1952–1957 Friends of ADIR and UN Victims, memo on ADIR, 20 December 1957. *Yearbook of the United Nations* (1952), p. 437.
152. USHMM Ferriday Papers List of surviving Ravensbrück Lapins.
153. USHMM Ferriday Papers Box 2 FC Schwelb to Ferriday, 3 May 1957, 31 December 1957.
154. USHMM Ferriday Papers Box 3 File 3 1952–1957 Friends of ADIR and UN Victims, memo on ADIR, 20 December 1957.
155. USHMM Ferriday Papers Box 2 Ferriday to Iadwiga [Dzido-]Hassa, 15 April 1958.
156. USHMM Ferriday Papers Box 2 Ferriday to Iadwiga Dzido-Hassa, 3 June, 5 July 1958.
157. 'The Lapins in America', *Saturday Review* (1959), pp. 20, 38–40 cited as cutting in Ferriday Papers. Erica Anderson, 'These Women were Nazi Guinea Pigs', *Look* (17 March 1959). AFP cutting 'Nazis' Guinea Pigs at Beth Israel. Tortured Women Repaired', *Boston Sunday Advertiser* (11 January 1959).
158. USHMM Ferriday Papers Box 2 'Lapins Reject Bonn Offer', *Washington Post* (20 May 1959). Howard Rusk, 'Science in the Nazi Era', *New York Times* (31 May 1959).
159. USHMM Ferriday Papers Box 3 File 3 1952–1957 Friends of ADIR and UN Victims, memo on ADIR, 20 December 1957.

160. Caroline Ferriday, 'To Aid Polish Victims', *The New York Times* (26 June 1959).
161. Pross, *Wiedergutmachung*, p. 110.
162. Ibid., p.142.
163. NYPL Lifton Papers A20 Provisional Committee to Chancellor Ludwig Erhard, 3 March 1965.
164. TTP–CLS–14/5/6/115 Katz to Taylor, 2 September 1964; Taylor to Katz, 11 September 1964.
165. NYPL Lifton Papers B2 Martin Wangh to Lifton, 7 February 1965.
166. TTP–CLS–14/6/16/343 Lifton to Taylor, 3 July 1979. NYPL Lifton Papers M 5 Alexander to Lifton, 10 October 1978.
167. NYPL Lifton Papers B2 Martin Wangh to Lifton, 7 February 1965.
168. Günther Schwarberg, *The Murders at Bullenhuser Damm. The SS Doctor and the Children* (Bloomington, Ind., 1984). Schwarberg, *Meine zwanzig Kinder*.
169. 'Trauerfeier für Präparate von NS-Opfern', *FAZ* (19 December 1990).
170. Rothman, *Strangers*, pp. 62–3.
171. Advisory Committee on Human Radiation Experiments Ethics Oral History Project. http://tis.eh.doe.gov/ohre/roadmap/histories/index.html.
172. Rothman, *Strangers*, pp. 51–69.
173. R.A. McCance, 'The Practice of Experimental Medicine', *Proceedings of the Royal Society of Medicine*, vol. 44 (1950–1) 189–94.
174. Rob Evans, *Gassed. British Chemical Warfare Experiments on Humans at Porton Down* (London: House of Stratus, 2000), pp. 126–7.
175. Observer Films, 'The Secrets of Porton Down', Granada Presentation for ITV, 1995. 'The Dying Moments of a Military Guinea Pig', *Guardian* (6 May 2004), p. 8 concerning Ronald Maddison's death.
176. Thomas Gerst, 'Catel und die Kinder. Versuche an Menschen – ein Fallbeispiel 1947/48', *1999*, vol. 15 (2000) 100–9.
177. Ivy papers box 90 folder 2, Ivy to President Johnson, 19 January 1965. Ivy also opposed any relaxation of abortion laws, regarding abortion as akin to Nazi atrocities. Ward, 'Who will bell the cat?'
178. BAK Zsg. 154 Bd 71 Cf. Pross observations.
179. Henry Beecher, *Experimentation in Man* (Springfield, Ill: Thomas, 1959).
180. Martin L. Gross, *The Doctors* (New York: Random House, 1966), chapter 7.
181. Maurice Pappworth, *Human Guinea Pigs* (London: Routledge, 1967), pp. 148, 185.
182. Jay Katz, *Experimentation with Human Beings. The Authority of the Investigator, Subject, Professions, and State in the Human Experimentation Process* (New York: Russell Sage Foundation, 1972).
183. *The Human Radiation Experiments. Final Report of the President's Advisory Committee*, p. 547.
184. George J. Annas, 'The Nuremberg Code in U.S. Courts: Ethics versus Expediency', *Nazi Doctors and the Nuremberg Code*, pp. 201–21.
185. 'Testimony by Peter Buxton from the United States Senate Hearings on Human Experimentation, 1973', in Susan M. Reverby (ed.), *Tuskegee's Truths. Rethinking the Tuskegee Syphilis Study* (Chapel Hill and London, 2000), p. 175.
186. 'Spendenaufruf: Dokumente der Nürnberger Ärzteprozess sollen übersetzt werden', *1999*, no. 4 (1996) 6.
187. *The Human Radiation Experiments. Final Report of the President's Advisory Committee* (Oxford: Oxford University Press, 1996).
188. 'Capping the Cost of Atrocity: Victim of Nazi Experiments says $8,000 isn't Enough', *New York Times* (19 November 2003), B1, B8.

Index